Springer Proceedings in Mathematics & Statistics

Volume 355

This book series features volumes composed of selected contributions from workshops and conferences in all areas of current research in mathematics and statistics, including operation research and optimization. In addition to an overall evaluation of the interest, scientific quality, and timeliness of each proposal at the hands of the publisher, individual contributions are all refereed to the high quality standards of leading journals in the field. Thus, this series provides the research community with well-edited, authoritative reports on developments in the most exciting areas of mathematical and statistical research today.

More information about this series at http://www.springer.com/series/10533

Vivek Laha · Pierre Maréchal · S. K. Mishra
Editors

Optimization, Variational Analysis and Applications

IFSOVAA-2020, Varanasi, India, February 2–4

Springer

Editors
Vivek Laha
Institute of Science
Banaras Hindu University
Varanasi, Uttar Pradesh, India

S. K. Mishra
Institute of Science
Banaras Hindu University
Varanasi, Uttar Pradesh, India

Pierre Maréchal
Institut de Mathématiques de Toulouse
Paul Sabatier University
Toulouse, France

ISSN 2194-1009 ISSN 2194-1017 (electronic)
Springer Proceedings in Mathematics & Statistics
ISBN 978-981-16-1818-5 ISBN 978-981-16-1819-2 (eBook)
https://doi.org/10.1007/978-981-16-1819-2

Mathematics Subject Classification: 49-XX, 49M29, 49M37, 90-XX, 49J53

This Springer imprint is published by the registered company Springer Nature Singapore Pte Ltd.
The registered company address is: 152 Beach Road, #21-01/04 Gateway East, Singapore 189721, Singapore

To my mother, Mrs. Malovika Laha,
to my wife, Dr. Reena Baral Laha
and to my daughter, Shrinika.

—Vivek Laha

To Adrien, Vincent, Elise,
Raphaël and Marianne.

—Pierre Maréchal

To my grandson, Viaan.

—S. K. Mishra

Foreword

It is my pleasure to contribute a short foreword to this proceedings volume. This work is a result of collaboration between India and France in applied mathematics, at the interface with physics, computer science and engineering. Both countries held during February 02–04, 2020, the **Indo-French Seminar on Optimization, Variational Analysis and Applications (IFSOVAA2020)**. This seminar was organized by the Department of Mathematics of the Institute of Science, Banaras Hindu University at Varanasi, India, in collaboration with the Institut de Mathématiques de Toulouse de l'Université Paul Sabatier, under the auspices of the **Indo-French Centre for the Promotion of Advanced Research (IFCPAR/ CEFIPRA)** supported by the Department of Science and Technology, Government of India, and the Ministry for Europe and Foreign Affairs of France. The goal was to promote collaborative research between India and France in the area of Variational Analysis and Optimization.

This volume consists of 19 articles on various aspects of optimization theory written by Indian and French researchers and is based on the invited talks. It is my hope and expectation that it will provide an effective learning experience and reference resource for young students and researchers in India and France. If these contributions lead to discovering new directions of research in optimization, they will have served their purpose.

Finally, I would like to congratulate my colleagues Vivek Laha, S. K. Mishra and Pierre Maréchal for this excellent initiative.

October 2020

Michel Théra
University of Limoges
Limoges, France

Federation University Australia
Ballarat, Australia

Preface

Variational analysis is a powerful tool which uses the combination and extension of methods from convex optimization and the classical calculus of variations to a more general theory for applications. The theory of variational analysis is a trending area of research nowadays due to its increasing applications in a number of important optimization problems such as vector optimization, copositive optimization, topology optimization, set optimization, portfolio optimization, particle swarm optimization, fuzzy optimization, semi-infinite optimization, minimax programming, unconstrained optimization, variational inequalities, mathematical programs with equilibrium and vanishing constraints, robust optimization, interval-valued programming, convex and nonconvex optimization, nonsmooth analysis and the related topics. The tools of variational analysis and generalized differentiation allow us to derive necessary and sufficient conditions of optimality of a feasible solution for some difficult optimization problems which in general are not possible by classical methods of analysis. Moreover, it provides a suitable infrastructure to construct useful algorithms to detect optimal or approximate solutions from a point of view for applications.

This book contains chapters based on the invited talks during Indo-French Seminar on Optimization, Variational Analysis and Applications (IFSOVAA2020) organized by Department of Mathematics, Institute of Science, Banaras Hindu University, Varanasi, India, in collaboration with Institut De Mathématiques De Toulouse, Universite Paul Sabatier, France, during February 02–04, 2020. The leading experts both from France and India in the areas of optimization and variational analysis have contributed extraordinary chapters leading to the most recent developments in the field of the study. The subjects covered in this book include set optimization, multiobjective optimization, mathematical programs with complementary, equilibrium, vanishing and swiching constraints, copositive optimization, interval-valued optimization, sequential quadratic programming, bound-constrained optimization, variational inequalities, etc. These problems of real-life origin have a wide variety of applications in different branches of applied mathematics, engineering, economics, finance, medical sciences, robot motion planning, morphological image analysis, computer-aided design and manufacturing, consumer demand, medical image registration, uncertain optimization, coherent risk measures in financial mathematics, optimal control, global analysis, career development theories, probability and statistics, computational geometry, data fitting, inverse problems, food processing, retail chain management and so on. The construction of each chapter is in such a way that it not only provides a detailed survey of the topic but also builds systematic theories and suitable algorithms to deduce the most recent findings of the literature. Needless to say, this book will serve as a useful reference for the scholars, professors and researchers both from academia and industry working in the area as a significant contribution to knowledge.

We are extremely grateful to our main sponsor Indo-French Centre for the Promotion of Advanced Research (IFCPAR/CEFIPRA) supported by the Department of Science and Technology, Government of India, and the Ministry for Europe and Foreign Affairs, Government of France, without whose financial support this collaborative research seminar was not possible. We are also thankful to the Institute of Science, Banaras Hindu University, Varanasi, India, for its financial assistance to support the seminar. We wish to thank the local organizing committee, the chairs, the speakers and the participants of the seminar whose collective efforts made the seminar a huge success. We are cordially thankful to all the contributors and reviewers for their hard work and dedication to construct this high-quality proceeding. At long last, we warmly express gratitude toward Springer for their assistance in publishing this book.

Varanasi, India
Toulouse, France
Varanasi, India
August 2020

Vivek Laha
Pierre Maréchal
S. K. Mishra

Contents

Contributors

S. Aggarwal LBSIM, New Delhi, Delhi, India

Qamrul Hasan Ansari Department of Mathematics, Aligarh Muslim University, Aligarh, India

Jaya Bisht Department of Mathematics, Institute of Science, Banaras Hindu University, Varanasi, India

Nantu Kumar Bisui Department of Mathematics, Indian Institute of Technology Kharagpur, Kharagpur, India

Jyoti Dhingra Darbari Department of Mathematics, Lady Shri Ram College for Women, University of Delhi, Delhi, India

S. P. Dash NIC Office, Bhubaneswar, India

Amit Kumar Debnath Department of Mathematical Sciences, Indian Institute of Technology (BHU), Varanasi, India

J.-P. Dussault Département d'Informatique, faculté des Sciences, Université de Sherbrooke, Sherbrooke, Canada

Debdas Ghosh Department of Mathematical Sciences, Indian Institute of Technology (BHU), Varanasi, India

Anshu Gupta School of Business, Public Policy and Social Entrepreneurship, Dr. B. R. Ambedkar University Delhi, Delhi, India

M. Haddou INSA Rennes, CNRS, IRMAR - UMR 6625, Univ Rennes, Rennes, France

P. C. Jha Department of Operational Research, University of Delhi, Delhi, India

Vishal Kapoor Indian Institute of Technology Kanpur, Kanpur, India

A. Kaul ASMSOC, NMIMS University, Mumbai, Vile-Parle (West), Mumbai, India

Khushboo Department of Mathematics, University of Delhi, Delhi, India

Abhishek Kumar Department of Electrical Engineering, Indian Institute of Technology (BHU), Varanasi, India

Manish Kumar Department of Mathematics, Birla Institute of Technology and Science-Pilani, Hyderabad, India

Rahul Kumar Department of Mathematics, Government Chandravijay College, Dindori, India

Vivek Laha Department of Mathematics, Institute of Science, Banaras Hindu University, Varanasi, India

C. S. Lalitha Department of Mathematics, University of Delhi South Campus, New Delhi, India

Pierre Maréchal Institut de Mathématiques de Toulouse, Université Paul Sabatier, Toulouse, France

Samit Mazumder Department of Mathematics, Indian Institute of Technology Kharagpur, Kharagpur, India

Vatsalkumar N. Mer Indian Statistical Institute, New Delhi, India

T. Migot Département d'Informatique, faculté des Sciences, Université de Sherbrooke, Sherbrooke, Canada

Priyanka Mishra Department of Mathematics, Indian Institute of Technology Patna, Patna, India

S. K. Mishra Department of Mathematics, Institute of Science, Banaras Hindu University, Varanasi, India

Rakesh Kumar Misra Department of Electrical Engineering, Indian Institute of Technology (BHU), Varanasi, India

C. Nahak Department of Mathematics, Indian Institute of Technology Kharagpur, Kharagpur, India

Devaki Nandan Indian Institute of Technology Kanpur, Kanpur, India

S. K. Neogy Indian Statistical Institute, New Delhi, India

Nomita Pachar Department of Operational Research, University of Delhi, Delhi, India

Geetanjali Panda Department of Mathematics, Indian Institute of Technology Kharagpur, Kharagpur, India

Yogendra Pandey Department of Mathematics, Satish Chandra College, Ballia, India

Mark Christhian Barrueta Pinto School of Business, Universidad Peruana de Ciencias Aplicadas (UPC), Lima, Peru

R. Ramanathan Business and Management Research Institute (BMRI), University of Bedfordshire, University Square, Luton, Bedfordshire, UK

R. Sadhu Department of Mathematics, Indian Institute of Technology Kharagpur, Kharagpur, India

Nidhi Sharma Department of Mathematics, Institute of Science, Banaras Hindu University, Varanasi, India

Pradeep Kumar Sharma Department of Mathematics, University of Delhi South Campus, New Delhi, India

Shiwani Sharma Department of Operational Research, Faculty of Mathematical Sciences, University of Delhi, Delhi, India

Walter Cedric Simo Tao Lee Institut de Mathématiques de Toulouse, Université Paul Sabatier, Toulouse, France

Devender Singh Department of Electrical Engineering, Indian Institute of Technology (BHU), Varanasi, India

Harsh Narayan Singh Department of Mathematics, Institute of Science, Banaras Hindu University, Varanasi, India

Vinay Singh Department of Mathematics, National Institute of Technology, Aizawl, Mizoram, India

Balendu Bhooshan Upadhyay Department of Mathematics, Indian Institute of Technology Patna, Patna, India

Chapter 1
Linear and Pascoletti–Serafini Scalarizations in Unified Set Optimization

Khushboo and C. S. Lalitha

Abstract This chapter is devoted to the study of linear and nonlinear scalarization schemes in unified set optimization in terms of approximate minimal solutions. Characterization of a preference set relation is employed to characterize approximate minimal solutions in terms of approximate minimal solutions of linearized scalar problems. Similar characterizations are derived using a nonlinear scalarization scheme, namely the Pascoletti–Serafini scalarization scheme for set optimization, which is an extension of the scheme considered in [Pascoletti, A., Serafini, P.: Scalarizing vector optimization problems, J. Optim. Theory Appl. 42(4), 499–524, 1984] for vector optimization problems.

Keywords Unified set optimization · Linear scalarization · Nonlinear scalarization · Pascoletti–Serafini scalarization

1.1 Introduction

In the last few decades, the study of set optimization problems has received much attention due to its wide applications in many fields, for instance, finance, economics, engineering, game theory and optimal control. For details, we refer to [1] and references therein.

In the case of the scalar optimization problem, the objective function values are compared with the natural ordering. However, due to the absence of total ordering relations for vector and set optimization problems, comparing objective function values is harder than scalar ones. To deal with this disadvantage, one of the extensive

Khushboo
Department of Mathematics, University of Delhi, Delhi 110 007, India
e-mail: thakurkhushboo4@gmail.com

C. S. Lalitha (✉)
Department of Mathematics, University of Delhi South Campus, Benito Juarez Road, New Delhi 110 021, India
e-mail: cslalitha@maths.du.ac.in; cslalitha1@gmail.com

V. Laha et al. (eds.), *Optimization, Variational Analysis and Applications*,
Springer Proceedings in Mathematics & Statistics 355,
https://doi.org/10.1007/978-981-16-1819-2_1

and efficient schemes employed to solve vector and set optimization problems is scalarization. In this scheme, a parametric scalar problem is assigned to the original problem and minimal solutions are characterized through minimal solutions of the scalarized problem. By varying the parameters, one can determine all the minimal solutions to the original problem. Hence, scalarization can be regarded as an algorithm where the same parametric scalar optimization problem is solved for different choices of parameters.

Several linear and nonlinear scalarizations have been developed for vector optimization problems. The most commonly used nonlinear scalar functions are the Gerstewitz function and the oriented distance function, (see [2, 3]). There is an extensive literature dealing with these scalar functions in set optimization as well; for more details, refer to [1, 4–10].

In the literature, optimization problems have been investigated by researchers in a unified setting using general preference relations. In this direction, Rubinov and Gasimov [11] proposed a preference relation using a conic set unifying several well-known ordering relations of vector optimization problems. A unified notion of minimal solution with respect to a preference relation induced by a nonempty proper set was proposed by Flores-Bazán and Hernández [12] and Flores-Bazán et al. [13] which was further investigated in [14] to study scalarization schemes. Later, Khushboo and Lalitha [8] proposed a preference set relation with respect to a nonempty proper set and established scalarization using a generalized Gerstewitz function.

Usually, iterative algorithms lead to an approximate solution rather than an exact solution when applied to solve an optimization problem, which led to growing interest in the study of approximate solutions in the past few years. Scalarization schemes have been employed to characterize approximate minimal solutions of vector optimization problems in the literature (see [12, 13, 15, 16]). Later, Dhingra and Lalitha [9] studied scalarization for approximate minimal solutions using a generalized Gerstewitz function in set optimization.

In this work, we investigate two scalarization schemes for a unified set optimization problem in terms of approximate minimal solutions. We define a unified notion of approximate minimal solutions with respect to a preference set relation considered in [8]. We first establish linear scalar characterizations of the preference set relation and approximate minimal solutions and compare it with an existing one established in [17]. Further, we propose a nonlinear scalarization scheme based on the Pascoletti–Serafini scalarization scheme of vector optimization problem considered in [18].

This chapter is organized as follows. In Sect. 1.2, we introduce a notion of approximate minimal solutions which unify various notions of minimal and approximate minimal solutions considered in the literature. Section 1.3 provides a concise detail of linear and nonlinear scalarization schemes. Section 1.4 deals with linear scalar characterizations of preference set relation and approximate minimal solutions. Section 1.5 is devoted to the study of the Pascoletti–Serafini scalarization scheme for the set optimization problem.

1.2 Preliminaries

Let Y be a real topological linear space and $S \subset Y$ be a nonempty proper set. We denote the family of all nonempty subsets of Y by $\mathcal{P}(Y)$. The topological interior, the topological closure and the complement of a set $A \subseteq Y$ are denoted by $\text{int}A$, $\text{cl}A$ and A^c, respectively. The set A is said to be $(-S)$-closed if $A - S$ is closed and $A^\infty := \{d \in Y \ : \ a + td \in A, \ \text{for all } a \in A, t \in \mathbb{R}_+\}$ is referred to the recession cone of A where $\mathbb{R}_+ = \{t \in \mathbb{R} \ : \ t \geq 0\}$.

In the case of the vector optimization problem, a unified notion of minimal solution is proposed by Flores-Bazán et al. [13], which led to a unified notion of a minimal solution, namely S-minimal solution, based on a preference set relation induced by the set S in [8]. For $A, B \in \mathcal{P}(Y)$

$$A \preceq_S^l B \iff B \subseteq A - S.$$

Generally, the preference relation $\preceq_S^l$ is neither reflexive nor transitive. However, the preference relation $\preceq_S^l$ is reflexive, if $0_Y \in S$ and transitive, if $S + S \subseteq S$. Evidently, a convex cone satisfies both these conditions. Besides cones, there are numerous sets that satisfy these conditions. For instance, $S = \{(y_1, y_2) \in \mathbb{R}^2 \ : \ y_1 + y_2 \geq 0, y_2 \geq 0\} \cup \{(y_1, y_2) \in \mathbb{R}^2 \ : \ y_2 \geq 1\}$.

The following remark investigates that the preference relation $\preceq_S^l$ is a unification of certain well-known preference relations studied by researchers.

Remark 1.1 1. Let $K \subset Y$ be a closed convex pointed cone with nonempty interior. If $S = -K$ $(S = -\text{int}K)$, then the preference relation $\preceq_S^l$ reduces to the lower set order relation $\preceq_K^l (\prec_K^l)$ considered in [19] ([6]).
2. If $S = -E$, where E is an improvement set, then the preference relation $\preceq_S^l$ is the set relation considered in [20].
3. If $A = \{a\}$ and $B = \{b\}$ where $a, b \in Y$, then $\preceq_S^l$ is the preference relation $\preceq_S$ considered in [8] defined as $a \preceq_S b$ iff $a - b \in S$.

We now consider the set-valued optimization problem

$$\text{(P)} \qquad \begin{array}{c} S\text{-Minimize } F(x) \\ \text{subject to } \ x \in X, \end{array}$$

where $F : X \rightrightarrows Y$ is a set-valued map and X is an arbitrary nonempty set.

Throughout the chapter, we assume that $F(x) \neq \emptyset$, for each $x \in X$, and $\epsilon \in \mathbb{R}_+$ unless specified otherwise and $0 \neq q \in Y$.

We define a notion of approximate minimal solutions of (P) extending the notion of approximate l-minimal solutions considered in [9]. This notion is a special case of a notion of approximate minimal solution given in Definition 6.3 of [21].

Definition 1.1 An element $\bar{x} \in X$ is said to be an *ϵ-S-l-minimizer* of (P), if there does not exist any $x \in X$ such that $F(x) + \epsilon q \preceq_S^l F(\bar{x})$.

We denote the set of ϵ-S-l-minimizers of (P) by ϵ-S-l-Mzer. For $\epsilon = 0$ the set of minimizers, namely S-l-minimizers, is denoted by S-l-Mzer. If $-q \in S^\infty$, then it can be easily verified that for $0 < \epsilon_1 < \epsilon_2$,

$$S\text{-}l\text{-Mzer} \subseteq \epsilon_1\text{-}S\text{-}l\text{-Mzer} \subseteq \epsilon_2\text{-}S\text{-}l\text{-Mzer}, \tag{1.1}$$

as in Proposition 6.1 of [21]. Clearly, every S-l-minimizer is an S-l-minimal considered in [8] but the converse is not true; see (Example 3.1 in [8]). Under appropriate assumptions, characterization for S-l-minimizers is given in [8, Lemma 3.1].

Remark 1.2 We have the following observations regarding the notions of ϵ-S-l-minimizers and S-l-minimizers.

1. If $S = -\text{int}K$ and $q \in \text{int}K$, then ϵ-S-l-Mzer reduces to the set ϵ-l-WMzer, where ϵ-l-WMzer denotes the set of ϵ-l-weak minimal solutions considered in [9]. Also, if $S = -\epsilon q - \text{int}K$, then S-l-Mzer reduces to set ϵ-l-WMzer.
2. If $S = -\text{int}K$, then the notion of S-l-minimizer reduces to the notion of weak l-minimal solution considered in [22].
3. If $S = -E$ where E is an improvement set, then under certain assumptions of Proposition 3.3 in [20], the notion of S-l-minimizer reduces to the notion of E-l-minimal solution considered in [20].

1.3 Scalarization

This section provides a brief outline of scalarization techniques for vector and set optimization problems.

We first recall a notion of minimal solutions for vector optimization problem from [23]. Let $K \subset Y$ be a closed convex pointed cone with nonempty interior and $f : X \to Y$. An element $\bar{x} \in X$ is said to be a *weak minimal solution* of minimizing the vector function f over X if

$$f(X) \cap (f(\bar{x}) - \text{int}K) = \emptyset,$$

where $f(X) := \cup_{x \in X} f(x)$. Evidently, if $S = -\text{int}K$ and F is vector-valued map, then the notion of S-l-minimizer reduces to the above notion of weak minimal solution.

Scalarization schemes are mainly based on separation theorems in which a parametric scalar problem is associated with the given problem and minimal solutions are determined in terms of minimal solutions of the scalarized problems. Now, the question that arises is how to choose the objective function of the scalarized problem. From the definition of the weak minimal solution, it is evident that the solution concept of a vector optimization problem is based on the separation of two sets. So, if one is able to determine a functional separating these two sets, then that can be chosen as the objective function of the scalarized problem. By virtue of separation theorems

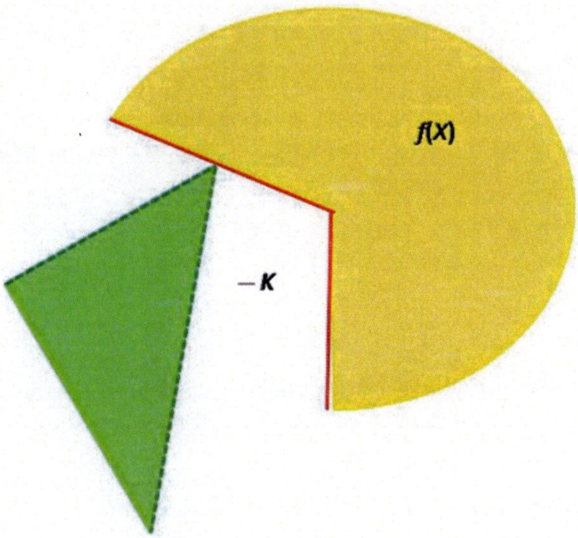

Fig. 1.1 The set of weak minimal solutions of f are marked in red in the figure (Failure of linear scalarization)

for convex sets, linear scalarizations are handy for convex vector optimization problems. However, it can be observed from Fig. 1.1 that in general, linear scalarization fails to hold. To deal with nonconvex optimization problems, Gerth and Weidner [2] introduced a nonlinear scalar function satisfying some nonconvex separation properties. Further, in the literature this function is referred to as the *Gerstewitz function* and used in mathematical finance [24] and mathematical economics [25] under the names of measures of risk and shortage function, respectively.

Let $0 \neq q \in Y$ and $A \in \mathcal{P}(Y)$. The *Gerstewitz function* $\xi_{q,A} : Y \longrightarrow \mathbb{R} \cup \{\pm\infty\}$ is defined as

$$\xi_{q,A}(y) := \inf\{t \in \mathbb{R} \; : \; y \in tq + A\}.$$

We next recall the following nonconvex separation properties of the Gerstewitz function $\xi_{q,A}$ from [13].

Proposition 1.1 [13] *Let* $\lambda \in \mathbb{R}$ *and* $0 \neq q \in Y$ *be such that* $-q \in A^{\infty}$. *Then, the following assertions hold.*

1. $\{y \in Y \; : \; \xi_{q,A}(y) < \lambda\} \subseteq \lambda q + A \subseteq \{y \in Y \; : \; \xi_{q,A}(y) \leq \lambda\}$.
2. If $\mathrm{cl}A - \mathbb{R}_{++}q \subseteq A$, then $\{y \in Y \; : \; \xi_{q,A}(y) \leq \lambda\} = \lambda q + \mathrm{cl}A$.
3. If $A - \mathbb{R}_{++}q \subseteq \mathrm{int}A$, then $\{y \in Y \; : \; \xi_{q,A}(y) < \lambda\} = \lambda q + \mathrm{int}A$.

By virtue of Proposition 1.1, for $\lambda = 0$ it may be noted that for any nonempty set $A \subseteq Y$, the function $\xi_{q,A}$ separates the set A from any set of $B \subseteq A^c$ in the following sense, provided there exists $0 \neq q \in Y$ such that $\mathrm{cl}A - \mathbb{R}_{++}q \subseteq \mathrm{int}A$.

1. $\xi_{q,A}(y) < 0 \iff y \in \text{int}A$.
2. $\xi_{q,A}(y) = 0 \iff y \in \partial A$.
3. $\xi_{q,A}(y) > 0 \iff y \in \text{int}A^c$.

On account of the above separation properties, we provide the following scalarization for the weak minimal solution of a vector optimization problem.

Theorem 1.3.1 *If $f : X \to Y$, $\bar{x} \in X$ and $q \in \text{int}K$, then $\bar{x}$ is a weak minimal solution of f over X if and only if $\bar{x}$ is a strict minimal solution of $\xi_{q,f(\bar{x})-\text{int}K}$ over X, that is, for all $x \in X \backslash \{\bar{x}\}$*

$$\xi_{q,f(\bar{x})-\text{int}K}(f(\bar{x})) < \xi_{q,f(\bar{x})-\text{int}K} f(x).$$

Proof By taking $A = f(\bar{x}) - \text{int}K$, proof follows immediately through the definition of weak minimal solution and the above separation properties. □

In 1984, Pascoletti and Serafini [18] introduced a nonlinear scalarization scheme for the vector-valued optimization problem. This scheme is known to encompass many of the well-known scalairzation schemes such as weighted sum method, ϵ-constraint method and normal boundary intersection method. For more details, refer to the book by Eichfelder [26]. For $p, q \in Y$, Pascoletti–Serafini scalar problem corresponding to (p, q) is given by

$$\begin{aligned} (\text{P}(p,q)) \qquad & \text{Minimize } t \\ \text{subject to} \quad & \\ & p + tq - f(x) = \lambda, \\ & t \in \mathbb{R},\ x \in X,\ \lambda \in K, \end{aligned}$$

where $f : X \to Y$ is a vector-valued map, X is an arbitrary set and $K \subset Y$ is a closed convex pointed cone. Solution of the problem is based on finding the minimal value of t, say $\bar{t}$ for which $(p + \bar{t}q - K) \cap f(X) \neq \emptyset$ and then, the tuple $(\bar{t}, \bar{x}, \bar{\lambda})$ corresponds to the optimal solution of $(\text{P}(p, q))$ where $f(\bar{x}) \in (p + \bar{t}q - K) \cap f(X)$ and $\bar{\lambda} = p + \bar{t}q - f(\bar{x})$. However, the problem can be equivalently formulated as

$$\begin{aligned} (\text{P}(p,q)) \qquad & \text{Minimize } \xi_{q,-K}(f(x) - p) \\ & \text{subject to} \quad x \in X. \end{aligned}$$

Using the Pascoletti–Serafini scalarization scheme, Huong and Yen [27] gave minimal representation formulae for efficient and weakly efficient solution sets and further studied the connectedness of the solution sets of Pascoletti–Serafini's scalar problems. Schematic visualization of Pascoletti–Serafini scheme discussed in Sect. 1.5 is presented in Fig. 1.2.

Another nonlinear scalar function, known as oriented distance function, was proposed by Hiriart-Urruty [3] in the setting of normed linear spaces. Let Y be

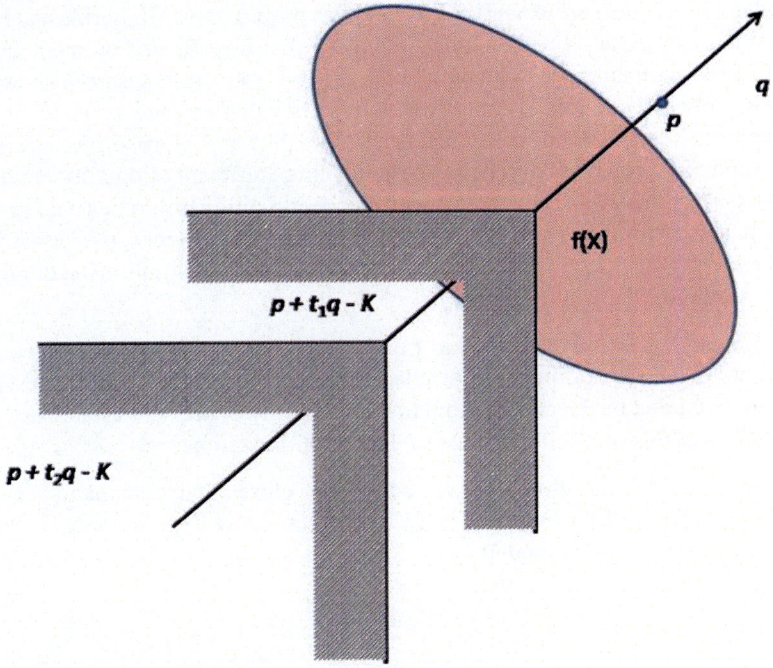

Fig. 1.2 Visualization of Pascoletti–Serafini Scalar problem

a real normed linear space. For $A \subseteq Y$, the *oriented distance function* $\Delta_A : Y \to \mathbb{R} \cup \{\pm\infty\}$ is defined as

$$\Delta_A(y) := d_A(y) - d_{A^c}(y),$$

where $d_A(y) := \inf_{a \in A} \|y - a\|$. Conventionally, for $d_{\emptyset}(y) = +\infty$, any $y \in Y$ and thus $\Delta_{\emptyset}(y) = +\infty$ and $\Delta_Y(y) = -\infty$.

This function was primarily introduced to investigate the geometry of nonsmooth optimization problems and to obtain necessary and sufficient optimality conditions. In comparison to the Gerstewitz function, this function is more compatible in calculation as well as in visualization sense. Analogous to the Gerstewitz function, it has the following separation properties without any restrictions on the set A.

Proposition 1.2 *[28] For $A \in \mathcal{P}(Y)$ and $A \neq Y$, the following assertions hold:*

1. $\Delta_A(y) < 0 \iff y \in \text{int}A$.
2. $\Delta_A(y) = 0 \iff y \in \partial A$.
3. $\Delta_A(y) > 0 \iff y \in \text{int}A^c$.

For vector optimization problems, a scalarization scheme based on an axiomatic approach, mainly on properties known as order (strict order) representing and order (strict order) preserving properties, was proposed by Miglierina and Molho [29]. Amending these properties, this study is further investigated by many authors including set optimization problems as well; see [10, 14, 30]. It has been shown that the Gerstewitz and oriented distance functions satisfy these properties under suitable assumptions. The notions of order and strict order preserving properties correspond to monotonicity properties and help in establishing sufficient optimality conditions using scalarization, while order and strict order representing properties help in establishing necessary optimality conditions using scalarization. Hence, the class of such type scalarizing functions provides complete scalarizations for minimal solutions of a given problem.

Similar to vector optimization problems, scalarization schemes have been broadly used for set optimization problems. In 2007, Hernández and Rodríguez–Marín [6] extended the Gerstewitz function over family of sets and established complete scalarizations of minimal solutions for set optimization.

Definition 1.2 [6, Definition 3.1] Let $K \subset Y$ be a closed convex pointed cone with nonempty interior and $q \in -\text{int}K$. The *generalized Gerstewitz function* $G_q : \mathcal{P}(Y) \times \mathcal{P}(Y) \to \mathbb{R} \cup \{\pm\infty\}$ is defined as

$$G_q(A, B) := \sup_{b \in B} \xi_{q,A}(b).$$

Analogous to the Gerstewitz and oriented distance functions, this nonlinear function has the following separation properties required to establish scalarizations. In this regard, we first recall some notions from [6]. A set $A \in \mathcal{P}(Y)$ is referred to as K-proper if the set $A + K \neq Y$. The set A is referred to as K-compact if for any open cover of A of the form $\{U_\alpha + K \ : \ U_\alpha$ are open sets in $Y\}$ there exists a finite subcover.

Theorem 1.3.2 *Let A and B be K-proper and K-compact sets. Then, the following assertions hold:*

1. $G_q(A, B) < 0 \iff A \prec_K^l B.$
2. $G_q(A, B) = 0 \iff A \preceq_K^l B, A \not\prec_K^l B.$
3. $G_q(A, B) > 0 \iff A \not\preceq_K^l B, A \not\prec_K^l B.$

Using these separation properties, Hernández and Rodríguez–Marín [6] established scalarizations for minimal and weak minimal solutions with respect to lower set order relation. Later, Xu and Li [7] established scalarizations for minimal and weak minimal solutions with respect to upper set order relation using a generalized oriented distance function. In the literature, various scalarization schemes for set optimization problems based on the Gerstewitz function, oriented distance function or their generalizations have been considered; see [1, 5–8, 10]. Recently, Jahn [17] derived linear scalar characterizations for various set order relations defined with respect to closed convex pointed cone. Using these characterizations, he established scalarizations for certain notions of minimal solutions.

1.4 Linear Scalarizations

Inspired by the work of Jahn [17], this section aims to provide linear scalar characterizations for the unified preference set relation $\preceq^l_S$ and the corresponding approximate minimal solution. In this section, we consider Y to be a locally convex space.

We first give a characterization of the preference set relation.

Theorem 1.4.3 *Let $-q \in S^\infty$ and $A, B \in \mathcal{P}(Y)$ be such that $A - S$ is closed and convex. Then, $A + \epsilon q \preceq^l_S B$ if and only if for all $l \in Y^*\backslash\{0_{Y^*}\}$ with $l(q) \leq 0$ we have*

$$\sup_{a\in A} l(a) - \inf_{s\in S} l(s) \geq \sup_{b\in B} l(b) - \epsilon l(q). \tag{1.2}$$

Proof We first assume that $A + \epsilon q \preceq^l_S B$. Suppose on the contrary there exists $l \in Y^*\backslash\{0_{Y^*}\}$ such that $l(q) \leq 0$ but (1.2) does not hold. Then,

$$\sup_{a-s\in A-S} l(a - s + \epsilon q) = \sup_{a\in A} l(a) - \inf_{s\in S} l(s) + \epsilon l(q) < \sup_{b\in B} l(b)$$

which further implies that there exists $\bar{b} \in B$ such that

$$\sup_{a-s\in A-S} l(a - s + \epsilon q) < l(\bar{b}) \leq \sup_{b\in B} l(b).$$

This leads to the fact that $\bar{b} \notin A + \epsilon q - S$, that is, $B \nsubseteq A + \epsilon q - S$ which is a contradiction as $A + \epsilon q \preceq^l_S B$.

Conversely, suppose on the contrary $B \nsubseteq A + \epsilon q - S$. Then, there exists some $b \in B$ such that $b - \epsilon q \notin A - S$. As $A - S$ is closed and convex, therefore by a separation theorem from [31, Theorem 1.1.5], there exists $l \in Y^*\backslash\{0_{Y^*}\}$ and $\alpha \in \mathbb{R}$ such that

$$l(y) \leq \sup_{y\in A-S} l(y) \leq \alpha < l(b - \epsilon q) = l(b) - \epsilon l(q), \tag{1.3}$$

for all $y \in A - S$. Taking $y = a - s + \lambda q \in A - S + \mathbb{R}_{++}q \subseteq A - S$, for some $a \in A$, $s \in S$ and $\lambda > 0$, in (1.3) we have

$$l(a) + l(-s) + \lambda l(q) = l(a - s + \lambda q) \leq \alpha.$$

Since the above relation holds for all $\lambda > 0$, it is clear that $l(q) \leq 0$. Also, from (1.3) we obtain

$$\sup_{a-s\in A-S} l(a - s) = \sup_{a\in A} l(a) - \inf_{s\in S} l(s) < \sup_{b\in B} l(b) - \epsilon l(q)$$

which is a contradiction to (1.2). □

Remark 1.3 It may be observed that the characterization obtained in Theorem 1.2 is different from the one established in [17, Lemma 2.1], for particular choices of $\epsilon = 0$, $S = -K$ and $q \in \text{int}K$ where K is a closed convex pointed cone with a nonempty interior. Also, the above characterization can be derived for any preference set, not necessarily cone, if there exists an element of its recession cone.

The following example justifies that neither closedness nor convexity of the set $A - S$ can be relaxed in the above theorem.

Example 1.1 Let $Y=\mathbb{R}^2$, $A = [0, 1] \times [0, 1]$, $B = \{(1, 0)\}$, $S=\text{int}\mathbb{R}^2_+$, $q = (0, -1)$ and $\epsilon \in [0, 1]$. A linear function $l : \mathbb{R}^2 \to \mathbb{R}$ is of the form $l(y_1, y_2) = c_1 y_1 + c_2 y_2$. Clearly, $l(q) \leq 0$ if and only if $c_1 \in \mathbb{R}$ and $c_2 \geq 0$. Let $(c_1, c_2) \in \{(c_1, c_2) \in \mathbb{R}^2 : c_1 \in \mathbb{R}, c_2 \geq 0\}\backslash\{(0, 0)\}$. Then,

$$\sup_{a\in A} l(a) - \inf_{s\in S} l(s) - \sup_{b\in B} l(b) + \epsilon l(q) = \begin{cases} (1-\epsilon)c_2, & \text{if } c_1 \geq 0, \\ +\infty, & \text{if } c_1 < 0. \end{cases}$$

Clearly, (1.2) is satisfied, $A - S$ is convex but not closed and $B \not\subseteq A + \epsilon q - S$, for every $\epsilon \in [0, 1]$.

However, if we consider $S=\{(y_1, y_2) \in \mathbb{R}^2 : y_1+y_2 \geq 0, y_2 \geq 0\} \cup \{(y_1, y_2) \in \mathbb{R}^2 : y_2 \geq 2\}$ and $B = \{(3, 0)\}$, then

$$\sup_{a\in A} l(a) - \inf_{s\in S} l(s) - \sup_{b\in B} l(b) + \epsilon l(q) = \begin{cases} 0, & \text{if } c_1 = 0, \\ +\infty, & \text{if } c_1 \neq 0, \end{cases}$$

and hence, (1.2) holds. Also, $A - S$ is closed but not convex and $B \not\subseteq A + \epsilon q - S$ for every $\epsilon \in [0, 1]$.

We next present characterization for approximate minimal solutions of (P).

Theorem 1.4.4 *Let $-q \in S^\infty$, $\bar{x} \in X$ and $F(x) - S$ be closed and convex, for each $x \in X$. Then, $\bar{x} \in \epsilon$-S-l-Mzer if and only if for each $x \in X$ there exists $l_x \in Y^*\backslash\{0_{Y^*}\}$ such that $l_x(q) \leq 0$ and*

$$\sup_{y\in F(x)} l_x(y) - \inf_{s\in S} l_x(s) < \sup_{\bar{y}\in F(\bar{x})} l_x(\bar{y}) - \epsilon l_x(q). \tag{1.4}$$

Proof Using Theorem 1.4.3, it is clear that $\bar{x} \in \epsilon$-S-l-Mzer if and only if $F(x) + \epsilon q \not\preceq^l_S F(\bar{x})$, for all $x \in X$, that is, for each $x \in X$ there exists $l_x \in Y^*\backslash\{0_{Y^*}\}$ such that $l_x(q) \leq 0$ and (1.4) holds. □

We now verify the above theorem by means of an example.

Example 1.2 Let $Y = \mathbb{R}^2$, $X = [0, 1]$, $S = \mathbb{R}^2_+$, $q = (0, -1)$, $\epsilon \in (0, 1]$ and $F : X \rightrightarrows Y$ be defined as $F(x) = [0, x] \times [0, x]$, for $x \in X$. Clearly, ϵ-S-l-Mzer $= (1-\epsilon, 1]$. For $\bar{x} = 1$,, it can be easily seen that for all $x \in X$ there exists a linear function $l : \mathbb{R}^2 \to \mathbb{R}$ defined as $l(y_1, y_2) = y_1 + y_2$ such that $l(q) < 0$ and

$$\sup_{y \in F(x)} l(y) - \inf_{s \in S} l(s) - \sup_{\bar{y} \in F(\bar{x})} l(\bar{y}) + \epsilon l(q) = 2(x-1) - \epsilon < 0.$$

Moreover, similar to Example 1.1 it can be seen that for $\bar{x} = 0$ there exists $x = 1$ such that for any $(c_1, c_2) \in \{(c_1, c_2) \in \mathbb{R}^2 : c_1 \in \mathbb{R}, c_2 \geq 0\} \backslash \{(0, 0)\}$,

$$\sup_{y \in F(x)} l(y) - \inf_{s \in S} l(s) - \sup_{\bar{y} \in F(\bar{x})} l(\bar{y}) + \epsilon l(q) = \begin{cases} c_1 + (1-\epsilon)c_2, & \text{if } c_1 \geq 0, \\ +\infty, & \text{if } c_1 < 0 \end{cases}$$

and hence, (1.4) fails to hold.

Remark 1.4 We observe here that the characterization provided in this section also holds for S-l-minimizers by taking $\epsilon = 0$.

1.5 Pascoletti–Serafini Scalarization Scheme

In this section, we extend the Pascoletti–Serafini scalarization scheme to set optimization problem and characterize ϵ-S-l-minimizer in terms of approximate minimal solutions of the Pascoletti–Serafini scalar problem. We observe that the characterizations established in this section, except Theorem 1.5.5, can be obtained only for ϵ-S-l-minimizers but not for S-l-minimizers.

We propose the Pascoletti–Serafini scalar problem $(\mathrm{P}_q(\hat{x}))$ involving set-valued maps for $\hat{x} \in X$ as

$$(\mathrm{P}_q(\hat{x})) \qquad \begin{aligned} &\text{Minimize } t \\ &\text{subject to} \\ &\quad F(x) - tq \preceq^l_S F(\hat{x}), \\ &\quad t \in \mathbb{R},\ x \in X. \end{aligned}$$

We denote the feasible set of $(\mathrm{P}_q(\hat{x}))$ by $\Omega_q(\hat{x})$, that is,

$$\Omega_q(\hat{x}) := \{(t, x) \in \mathbb{R} \times X : F(x) - tq \preceq^l_S F(\hat{x})\}.$$

Remark 1.5 We have the following observations regarding the problem $(\mathrm{P}_q(\hat{x}))$.

1. If $F(\hat{x}) = \{p\}$, $S = -K$, where K is a closed convex pointed cone and F is a vector-valued function, then the problem $(\mathrm{P}_q(\hat{x}))$ reduces to the problem $(\mathrm{P}(p, q))$.
2. The problem $(\mathrm{P}_q(\hat{x}))$ is equivalent to the scalar problem

$$\begin{aligned} &\text{Minimize } G_{q,S}(F(x), F(\hat{x})) \\ &\text{subject to} \quad x \in X, \end{aligned}$$

where $G_{q,S}(F(x), F(\hat{x})) := \inf\{t \in \mathbb{R} : F(x) - tq \preceq^l_S F(\hat{x})\}$.

3. In view of Theorem 4.4 in [8], it is easy to verify that

$$G_{q,S}(F(x), F(\hat{x})) = \sup_{b \in B} \varphi_{q,A}(b),$$

where $\varphi_{q,A}(b) := \inf\{t \in \mathbb{R} : b \in tq + A - S\}$, provided F is $(-S)$-closed valued, $G_{q,S}(F(x), F(\hat{x})) < \infty$, for each $x \in X$ and $\text{cl}S - \mathbb{R}_{++}q \subseteq S$.

s An element $(\bar{t}, \bar{x}) \in \Omega_q(\hat{x})$ is said to be an ϵ-minimal solution of $(\text{P}_q(\hat{x}))$, if $\bar{t} \leq t + \epsilon$, for any $(t, x) \in \Omega_q(\hat{x})$. We observe that $\bar{x}$ may correspond to an ϵ-minimal solution of $(\text{P}_q(\hat{x}))$ with infimum value $-\infty$. We denote the set of ϵ-minimal solutions of $(\text{P}_q(\hat{x}))$ by ϵ-argmin$\text{P}_q(\hat{x})$. If $\epsilon = 0$, then the solution is referred to as an minimal solution and is denoted by argmin$\text{P}_q(\hat{x})$.

We now establish a relationship between ϵ-S-l-minimizer of (P) and ϵ-minimal solution of $(\text{P}_q(\hat{x}))$. In the following theorem, we show that corresponding to every ϵ-S-l-minimizer there exists an ϵ-minimal solution of $(\text{P}_q(x))$, for some $x \in X$.

Theorem 1.5.5 *If* $-q \in S^\infty$ *and* $0_Y \in \text{cl}S$, *then*

$$\epsilon\text{-}S\text{-}l\text{-Mzer} \subseteq \bigcup_{x \in X} \text{proj}_X(\epsilon\text{-argminP}_q(x)), \tag{1.5}$$

where $\text{proj}_X(\epsilon\text{-argminP}_q(x))$ *denotes the projection of* ϵ-argmin$\text{P}_q(x)$ *onto* X.

Proof Let $\bar{x} \in \epsilon$-S-l-Mzer. We claim that $\bar{x} \in \text{proj}_X(\epsilon\text{-argminP}_q(\bar{x}))$. Clearly, $(t, \bar{x}) \in \Omega_q(\bar{x})$, for all $t > 0$ as $0_Y \in \text{cl}S$ and $-q \in S^\infty$ and therefore $G_q(F(\bar{x}), F(\bar{x})) \leq 0$. Suppose on the contrary there exists $(t, x) \in \Omega_q(\bar{x})$ such that $t + \epsilon < 0$. Then, $F(x) - tq \preceq^l_S F(\bar{x})$ which implies that

$$\begin{aligned} F(\bar{x}) &\subseteq -tq + F(x) - S \\ &= -(t + \epsilon)q + \epsilon q + F(x) - S \\ &= \epsilon q + F(x) - S, \end{aligned}$$

as $-q \in S^\infty$. Hence, $F(x) + \epsilon q \preceq^l_S F(\bar{x})$ which contradicts the fact that $\bar{x} \in \epsilon$-S-l-Mzer. □

The following example justifies that the condition $-q \in S^\infty$ cannot be relaxed in the above theorem either when $q \in S$ or $q \notin S$.

Example 1.3 Let $Y = \mathbb{R}^2$, $X = [0, 1]$, $S = \{(0, 0), (0, -1)\}$ and $F : X \rightrightarrows Y$ be defined as

$$F(x) = \begin{cases} \{(0, 0), (0, 1)\}, & \text{if } x = 0, \\ \{(0, 0), (0, 2)\}, & \text{otherwise.} \end{cases}$$

Clearly, $-q \notin S^\infty$, for any $0 \neq q \in Y$. For $q = (0, -1) \in S$ $(q = (0, -2) \notin S)$ it can be seen that ϵ-S-l-Mzer = X and $\text{proj}_X(\epsilon\text{-argminP}_q(x)) = (0, 1]$, for any $x \in X$ and for any $\epsilon > 0$.

Remark 1.6 It may be noted that if $-q \in S^{\infty}$ and $q \in S$, then $0_Y \in \mathrm{cl}S$ and hence, Theorem 1.5.5 holds. However, if $q \notin S$, then the condition $0_Y \in \mathrm{cl}S$ cannot be dropped in Theorem 1.5.5 which is illustrated in the following example.

Example 1.4 Let $Y = \mathbb{R}^2$, $X = [0, 1]$, $S = \{(y_1, y_2) \in \mathbb{R}^2 \ : \ y_1 \geq 1, y_2 \geq 1\}$ and $F : X \rightrightarrows Y$ be defined as

$$F(x) = \begin{cases} \{(0, 0)\}, & \text{if } x = 0, \\ \{(y, y) \ : \ 0 \leq y \leq 1\}, & \text{otherwise.} \end{cases}$$

Let $q = (-1, -1) \notin S$. It can be seen that ϵ-S-l-Mzer $= X$ for any $\epsilon \geq 0$ and $\mathrm{proj}_X(\epsilon\text{-argminP}_q(x)) = \mathrm{proj}_X(\mathrm{argminP}_q(x)) = (0, 1]$, for any $x \in X$. Also, $-q \in S^{\infty}$ but $(0, 0) \notin S$.

The following example justifies that the reverse inclusion in (1.5) may fail to hold.

Example 1.5 Let $Y = \mathbb{R}^2$, $X = [0, 1]$, $q = (-1, -1)$, $S = \mathbb{R}^2_+$ and $F : X \rightrightarrows Y$ be defined as

$$F(x) = \{(y_1, y_2) \in \mathbb{R}^2 \ : \ -1 \leq y_1 \leq x - 1, -1 \leq y_2 \leq x - 1\}.$$

It can be easily verified that for $0 < \epsilon \leq 1$, ϵ-S-l-Mzer $= (1-\epsilon, 1]$ and $\epsilon\text{-argminP}_q(x) = \{(-t + x, t) \ : \ 1 - \epsilon \leq t \leq 1\}$, for all $x \in X$. Then, $1-\epsilon \in \mathrm{proj}_X(\epsilon\text{-argminP}_q(x))$ and $1 - \epsilon \notin \epsilon$-$S$-$l$-Mzer, for all $x \in X$.

Under a suitable assumption, we now establish the reverse inclusion of (1.5).

Theorem 1.5.6 *If* $\epsilon > 0$, $S + S \subseteq S$ *and* $x \in X$ *is such that*

$$|t_1 - t_2| < \epsilon, \tag{1.6}$$

for any $t_1, t_2 \in \mathrm{proj}_{\mathbb{R}}(\epsilon\text{-argminP}_q(x))$, *then*

$$\mathrm{proj}_X(\epsilon\text{-argminP}_q(x)) \subseteq \epsilon\text{-}S\text{-}l\text{-Mzer}.$$

Proof Let $\bar{x} \in \mathrm{proj}_X(\epsilon\text{-argminP}_q(x))$. Then, there exists $\bar{t} \in \mathbb{R}$ such that $(\bar{t}, \bar{x}) \in \epsilon\text{-argminP}_q(x)$. Clearly, $(\bar{t}, \bar{x}) \in \Omega_q(x)$ which implies that

$$F(\bar{x}) \preceq^l_S F(x) + \bar{t}q. \tag{1.7}$$

Suppose on the contrary there exists $x' \in X$ such that

$$F(x') + \epsilon q \preceq^l_S F(\bar{x}). \tag{1.8}$$

Using $S + S \subseteq S$, (1.7) and (1.8), we have

$$F(x') \preceq_S^l F(x) + (\bar{t} - \epsilon)q$$

which implies that $(\bar{t} - \epsilon, x') \in \Omega_q(x)$. Since $(\bar{t}, \bar{x}) \in \epsilon\text{-argmin}\mathrm{P}_q(x)$, therefore for any $(\tilde{t}, \tilde{x}) \in \Omega_q(x)$, we have

$$\bar{t} - \epsilon < \bar{t} \leq \tilde{t} + \epsilon$$

which implies that $(\bar{t} - \epsilon, x') \in \epsilon\text{-argmin}\mathrm{P}_q(x)$. As $\bar{t}$ and $\bar{t} - \epsilon$ are both in $\mathrm{proj}_{\mathbb{R}}(\epsilon\text{-argmin}\mathrm{P}_q(x))$, we get a contradiction to (1.6).

The condition given by (1.6) in the above theorem cannot be relaxed, as evident from Example 1.5. It can be observed that the condition (1.6) does not hold as $\epsilon - 1$ and -1 both belong to $\mathrm{proj}_{\mathbb{R}}(\epsilon\text{-argmin}\mathrm{P}_q(0))$.

Remark 1.7 From Remark 1.6, it is also worth noting that in addition to $-q \in S^\infty$ and $q \in S$ if $S + S \subseteq S$, then for all $\epsilon > 0$ we have ϵ-S-l-Mzer $= \emptyset$ whereas $x \in \mathrm{proj}_X(\epsilon\text{-argmin}\mathrm{P}_q(x))$, for each $x \in X$ with $G_{q,S}(F(x), F(x)) = -\infty$. Hence, we observe that the condition (1.6) is not satisfied in this case and Theorem 1.5.6 fails to hold.

The following theorem presents a complete scalarization for ϵ-S-l-minimizers.

Theorem 1.5.7 *If $\epsilon > 0$, $0_Y \in \mathrm{cl}S$, $-q \in S^\infty$, $S + S \subseteq S$ and for every $x \in X$ we have $|t_1 - t_2| < \epsilon$, for any $t_1, t_2 \in \mathrm{proj}_{\mathbb{R}}(\epsilon\text{-argmin}\mathrm{P}_q(x))$, then*

$$\bigcup_{x \in X} \mathrm{proj}_X(\epsilon\text{-argmin}\mathrm{P}_q(x)) = \epsilon\text{-}S\text{-}l\text{-Mzer}.$$

Proof Proof follows from Theorems 1.5.5 and 1.5.6. □

Remark 1.8 1. In view of Remark 1.7, we note that if there exists $0 \neq q \in Y$ for which all the assumptions of Theorem 1.5.7 holds then q must belong to S^c.

2. In Example 1.5, if we consider $F(1) = \{(-1, -1)\}$, then for $0 < \epsilon < 1$, ϵ-S-l-Mzer $= [1 - \epsilon, 1)$, $\epsilon\text{-argmin}\mathrm{P}_q(1) = \{(-t, t) : 1 - \epsilon \leq t < 1\}$ and $\epsilon\text{-argmin}\mathrm{P}_q(x) = \{(-t + x, t) : 1 - \epsilon \leq t < 1\}$, for all $x \neq 1$ and hence, Theorem 1.5.7 is verified. Note that condition (1.6) holds for all $x \in X$.
3. From Remark 1.5(iii), [6, Theorem 3.6] and [9, Theorem 5.2], it may be noted that if $S = -\mathrm{int}K$ and $q \in \mathrm{int}K$, then ϵ-l-weak minimal solutions considered in [9] can be characterized in terms of ϵ-minimal solution of $(\mathrm{P}_q(x))$, provided the sets $F(X)$ and $F(x)$ are K-compact for each $x \in X$, has a finite subcover. The following example illustrates the fact that Theorem 1.5.7 may hold for ϵ-l-weak minimal solutions in the absence of K-compactness assumption.

Example 1.6 Let $Y = \mathbb{R}^2$, $X = [0, 1] \cup \{2\}$, $q = (1, 1)$, $K = \mathbb{R}^2_+$ and $F : X \rightrightarrows Y$ be defined as

$$F(x) = \begin{cases} \{(y_1, y_2) \in \mathbb{R}^2 : \ y_1 + y_2 \geq 0, y_2 \geq 0\}, & \text{if } x = 2, \\ \{(y_1, y_2) \in \mathbb{R}^2 : \ y_1 + y_2 \geq 0.4, y_2 \geq 0.2\}, & \text{otherwise.} \end{cases}$$

Clearly, $F(2)$ is not K-compact. For $\epsilon \in (0, 0.2)$, ϵ-l-WMzer $= \{2\}$, ϵ-argmin$\text{P}_q(2) = \{(0, 2)\}$ and ϵ-argmin$\text{P}_q(x) = \{(-0.2, 2)\}$ for all $x \in [0, 1]$ and hence, Theorem 1.5.7 is justified. Also, we observe that all the assumptions of Theorem 1.5.7 are satisfied.

In Example 1.5, it was observed that an ϵ-minimal solution of $(\text{P}_q(x))$ may not correspond to an ϵ-S-l-minimizer of (P). However, the following theorem shows that every minimal solution of $(\text{P}_q(x))$ corresponds to an ϵ-S-l-minimizer of (P).

Theorem 1.5.8 *If $\epsilon > 0$ and $S + S \subseteq S$, then*

$$\bigcup_{x \in X} \text{proj}_X(\text{argminP}_q(x)) \subseteq \epsilon\text{-}S\text{-}l\text{-Mzer}.$$

Proof Let $\bar{x} \in \text{proj}_X(\text{argminP}_q(x))$ for some $x \in X$, which implies that there exists $\bar{t} \in \mathbb{R}$ such that $(\bar{t}, \bar{x}) \in \text{argminP}_q(x)$. Proceeding as in Theorem 1.5.6 there exists a feasible solution $(\bar{t} - \epsilon, x')$ of $(\text{P}_q(x))$ which contradicts the optimality of $(\bar{t}, \bar{x})$. □

In the next theorem, we show that if an element is an ϵ-S-l-minimizer of (P) for every $\epsilon > 0$, then it corresponds to a minimal solution of $(\text{P}_q(x))$ for some $x \in X$.

Theorem 1.5.9 *If $\epsilon > 0$ and $0_Y \in S$, then*

$$\bigcap_{\epsilon > 0} \epsilon\text{-}S\text{-}l\text{-Mzer} \subseteq \bigcup_{x \in X} \text{proj}_X(\text{argminP}_q(x)).$$

Proof Let $\bar{x} \in \bigcap_{\epsilon>0} \epsilon$-$S$-$l$-Mzer. Clearly, $(0, \bar{x}) \in \Omega_q(\bar{x})$, as $0_Y \in S$. If possible, there exists $(t, x) \in \Omega_q(\bar{x})$ such that $t < 0$, then $F(x) - tq \preceq^l_S F(\bar{x})$ which is a contradiction as $\bar{x} \in \epsilon$-S-l-Mzer for $\epsilon = -t$. □

The following theorem follows from Theorems 1.5.8 and 1.5.9.

Theorem 1.5.10 *If $\epsilon > 0$, $0_Y \in S$ and $S + S \subseteq S$, then*

$$\bigcap_{\epsilon > 0} \epsilon\text{-}S\text{-}l\text{-Minimizer} = \bigcup_{x \in X} \text{proj}_X(\text{argminP}_q(x)).$$

1.6 Conclusions

Linear and nonlinear scalarization schemes for a unified notion of approximate minimal solutions have been developed for a set optimization problem. It has been shown that the approximate minimal solutions considered in [9] can be characterized by means of the Pasoletti–Serafini scalarization in the absence of K-compactness assumptions as considered by Dhingra and Lalitha [9].

Acknowledgements This research for the first author is supported by CSIR, Senior Research Fellowship, India, National R&D Organization (Ack. No.: 151012/2K18/1), and the second author is supported by the MATRICS scheme of Department of Science and Technology, India. The authors would like to thank the anonymous referees for their careful reading of the manuscript and for providing valuable comments and suggestions.

References

1. Khan, A.A., Tammer, C., Zălinescu, C.: Set-Valued Optimization: An Introduction with Applications. Springer, Berlin (2015)
2. Gerth, C., Weidner, P.: Nonconvex separation theorems and some applications in vector optimization. J. Optim. Theory Appl. **67**(2), 297–320 (1990)
3. Hiriart-Urruty, J.B.: Tangent cones, generalized gradients and mathematical programming in Banach spaces. Math. Oper. Res. **4**(1), 79–97 (1979)
4. Köbis, E., Köbis, M.A.: Treatment of set order relations by means of a nonlinear scalarization functional: a full characterization. Optimization **65**(10), 1805–1827 (2016)
5. Chen, J., Ansari, Q.H., Yao, J.C.: Characterizations of set order relations and constrained set optimization problems via oriented distance function. Optimization **66**(11), 1741–1754 (2017)
6. Hernández, E., Rodríguez-Marín, L.: Nonconvex scalarization in set optimization with setvalued maps. J. Math. Anal. Appl. **325**(1), 1–18 (2007)
7. Xu, Y.D., Li, S.J.: A new nonlinear scalarization function and applications. Optimization **65**(1), 207–231 (2016)
8. Khushboo, Lalitha, C.S.: A unified minimal solution in set optimization. J. Global Optim. **74**(1), 195–211 (2019)
9. Dhingra, M., Lalitha, C.S.: Approximate solutions and scalarization in set-valued optimization. Optimzation **66**(11), 1793–1805 (2017)
10. Gutiérrez, C., Jiménez, B., Miglierina, E., Molho, E.: Scalarization in set optimization with solid and nonsolid ordering cones. J. Global Optim. **61**(3), 525–552 (2015)
11. Rubinov, A.M., Gasimov, R.N.: Scalarization and nonlinear scalar duality for vector optimization with preferences that are not necessarily a pre-order relation. J. Global Optim. **29**(4), 455–477 (2004)
12. Flores-Bazán, F., Hernández, E.: A unified vector optimization problem: complete scalarizations and applications. Optimization **60**(12), 1399–1419 (2011)
13. Flores-Bazán, F., Flores-Bazán, F., Laengle, S.: Characterizing efficiency on infinite dimensional commodity spaces with ordering cones having possibly empty interior. J. Optim. Theory Appl. **164**(2), 455–478 (2015)
14. Khushboo, Lalitha, C.S.: Scalarizations for a unified vector optimization problem based on order representing and order preserving properties. J. Global Optim. **70**(4), 903–916 (2018)
15. Zhao, K., Chen, G., Yang, X.: Approximate proper efficiency in vector optimization. Optimization **64**(8), 1777–1793 (2015)
16. Gutiérrez, C., Jiménez, B., Novo, V.: A unified approach and optimality conditions for approximate solutions of vector optimization problems. SIAM J. Optim. **17**(3), 688–710 (2006)
17. Jahn, J.: Vectorization in set optimization. J. Optim. Theory Appl. **167**(3), 783–795 (2015)
18. Pascoletti, A., Serafini, P.: Scalarizing vector optimization problems. J. Optim. Theory Appl. **42**(4), 499–524 (1984)
19. Kuroiwa, D.: Some duality theorems of set-valued optimization with natural criteria. In: Nonlinear Analysis and Convex Analysis (Niigata, 1998), pp. 221–228. World Scientific Publishing, River Edge, NJ (1999)
20. Dhingra, M., Lalitha, C.S.: Set optimization using improvement sets. Yugosl. J. Oper. Res. **27**(2), 153–167 (2017)

21. Flores-Bazán, F., Gutiérrez, C., Novo, V.: A Brézis-Browder principle on partially ordered spaces and related ordering theorems. J. Math. Anal. Appl. **375**(1), 245–260 (2011)
22. Zhang, W.Y., Li, S.J., Teo, K.L.: Well-posedness for set optimization problems. Nonlinear Anal. **71**(9), 3769–3778 (2009)
23. Luc, D.T.: Theory of Vector Optimization. Lecture Notes in Economics and Mathematical Systems, vol. 319. Springer, Berlin (1989)
24. Hamel, A.H., Heyde, F., Rudloff, B.: Set-valued risk measures for conical market models. Math. Financ. Econom. **5**(1), 1–28 (2011)
25. Makarov, V.L., Levin, M.J., Rubinov, A.M.: Mathematical Economic Theory: Pure and Mixed Types of Economic Mechanisms. Advanced Textbooks in Economics, 33. North-Holland Publishing Co., Amsterdam (1995)
26. Eichfelder, G.: Adaptive Scalarization Methods in Multiobjective Optimization. Springer, Berlin (2008)
27. Huong, N.T.T., Yen, N.D.: The Pascoletti-Serafini scalarization scheme and linear vector optimization. J. Optim. Theory Appl. **162**(2), 559–576 (2014)
28. Zaffaroni, A.: Degrees of efficiency and degrees of minimality. SIAM J. Control Optim. **42**(3), 1071–1086 (2003)
29. Miglierina, E., Molho, E.: Scalarization and stability in vector optimization. J. Optim. Theory Appl. **114**(3), 657–670 (2002)
30. Gutiérrez, C., Jiménez, B., Novo, V.: Optimality conditions via scalarization for a new efficiency concept in vector optimization problems. Eur. J. Oper. Res. **201**(1), 11–22 (2010)
31. Zălinescu, C.: Convex Analysis in General Vector Spaces. World Scientific Publishing, River Edge, NJ (2002)

Chapter 2
A Gradient-Free Method for Multi-objective Optimization Problem

Nantu Kumar Bisui, Samit Mazumder, and Geetanjali Panda

Abstract In this chapter, a gradient-free method is proposed for solving the multi-objective optimization problem in higher dimension. The concept is developed as a modification of the Nelder-Mead simplex technique for the single-objective case. The proposed algorithm is verified and compared with the existing methods with a set of test problems.

Keywords Nelder-Mead simplex method · Multi-objective programming · Gradient-free method · n dimension simplex

2.1 Introduction

A general multi-objective optimization problem is stated as

$$(MOP) \quad \min_{x \in S \subset \mathbb{R}^n} F(x),$$

$F(x) = (f_1(x), f_2(x) \ldots, f_m(x))^T, f_j : \mathbb{R}^n \to \mathbb{R}, j = 1, 2, \ldots, m, m \geq 2$. In practice, (MOP) involves several conflicting and non-commensurate objective functions which have to be optimized simultaneously over $\mathbb{R}^n$. If $x^* \in \mathbb{R}^n$ minimizes all the objective functions simultaneously, then certainly an ideal solution is achieved. But in general, improvement in one criterion results in loss in another criterion, leading to the unlikely existence of an ideal solution. For this reason one has to look for the "best" compromise solution, which is known as an efficient or Pareto optimal

N. K. Bisui · S. Mazumder · G. Panda (✉)
Department of Mathematics, Indian Institute of Technology Kharagpur, Kharagpur 721302, India
e-mail: geetanjali@maths.iitkgp.ac.in

N. K. Bisui
e-mail: nantukrbisuimath@gmail.com

S. Mazumder
e-mail: samitmazumder599@gmail.com

V. Laha et al. (eds.), *Optimization, Variational Analysis and Applications*,
Springer Proceedings in Mathematics & Statistics 355,
https://doi.org/10.1007/978-981-16-1819-2_2

solution. The concept of efficiency arises from a pre-specified partial ordering on $\mathbb{R}^m$. The points satisfying the necessary condition for efficiency are known as critical points. Application of these kinds of problems are found in engineering design, statistics, management science, etc.

Classical methods for solving (MOP) are either scalarization methods or heuristic methods. Scalarization methods reduce the main problem to a single objective optimization problem using predetermined parameters. A widely used scalarization method is due to Geofforion [15], which computes proper efficient solutions. Geofforion's approach has been further developed by several researchers in several directions. Other parameter-free techniques use the concept of order of importance of the objective functions, which have to be specified in advance. Another widely used general solution strategy for multi-objective optimization problems is the ϵ-constrained method [10, 21]. All the above methods are summarized in [1, 8, 19, 23]. These scalarization methods are user-dependent and often have difficulties in finding an approximation to the Pareto front. Heuristic methods [7] do not guarantee the convergence property but usually provide an approximate Pareto front. Some well-known heuristic methods are genetic algorithms, particle swarm optimization, etc. NSGA-II [7] is a well-known genetic algorithm.

Recently, many researchers have developed line search methods for (MOP), which are different from the scalarization process and heuristic approach. These line search numerical techniques are possible extensions of gradient-based line search techniques for the single-objective optimization problem to the multi-objective case. In every gradient-based line search method, the descent direction at an iterative point x is determined by solving a subproblem at x, and a suitable step length α at x in this direction is obtained using the Armijio type condition with respect to each objective function to ensure $f_j(x + \alpha d) < f_j(x)$. Then a descent sequence is generated, which can converge to a critical point. Some recent developments in this direction are summarized below.

The steepest descent method, which is the first line search approach for (MOP), was developed by Fliege and Svaiter [12] in 2000 to find a critical point of (MOP). In this method, descent direction d at every iterating point x is the solution of the following subproblem,

$$\inf_{d \in \mathbb{R}^n} \max_j \nabla f_j(x)^T d,$$

which is same as

$$\begin{aligned} &\min_{t,d} \; t + \frac{1}{2} d^T d \\ \text{subject to } \; &\nabla f_j(x)^T d - t \le 0, \quad j = 1, 2, ..., m \\ &t \in \mathbb{R}, d \in \mathbb{R}^n. \end{aligned}$$

The Newton method for single-objective optimization problem is extended to (MOP) by Fliege et al. [11] in 2009, which uses convexity criteria. Newton direction for (MOP) at x is obtained by solving the following min-max problem, which involves the quadratic approximation of all the objective functions.

$$\inf_{d\in\mathbb{R}^n} \max_{j\in\Lambda_m} \nabla f_j(x)^T d + \frac{1}{2} d^T \nabla^2 f_j(x) d$$

This is equivalent to the following subproblem.

$$\min_{t\in\mathbb{R}, d\in\mathbb{R}^n} \ t$$

$$\text{subject to} \quad \nabla f_j(x)^T d + \frac{1}{2} d^T \nabla^2 f_j(x) d - t \leq 0, \qquad j = 1, 2, ..., m.$$

If every f_j is a strictly convex function, then the above subproblem is a convex programming problem. Using the Karush-Kuhn-Tucker (KKT) optimality condition, the solution of this subproblem becomes the Newton direction $d_N(x)$ as

$$d_N(x) = -[\Sigma_{j=1}^m \lambda_j(x) \nabla^2 f_j(x)]^{-1} \Sigma_{j=1}^m \lambda_j(x) \nabla f_j(x),$$

where $\lambda_j(x)$ are Lagrange multipliers. This iterative process is locally and quadratically convergent for Lipschitz continuous functions.

An extension of the Quasi-Newton method for (MOP) is studied by Qu et al. [27] in 2011 for critical point, which avoids convexity assumptions. The method proposed by Qu et al. [27] uses the approximate Hessian of every objective function. The subproblem in Qu et al. [27] is

$$\min_{t\in\mathbb{R}, d\in\mathbb{R}^n} \ t$$

$$\text{subject to} \quad \nabla f_j(x)^T d + \frac{1}{2} d^T B_j(x) d - t \leq 0 \quad j = 1, 2, ..., m,$$

where $B_j(x)$ is the approximation of $\nabla^2 f_j(x)$.

These individual $B_j(x)$ are replaced by a common positive definite matrix in Ansari and Panda [2] to reduce the complexity of the algorithm. In [2], the descent direction at every iterating point x is determined by solving the following subproblem which involves linear approximation of every objective function along with the common positive definite matrix $B(x)$ in place of individual matrices $B_j(x)$.

$$\begin{aligned} &\min_{t,\ d} \ t + \frac{1}{2} d^T B(x) d \\ &\text{subject to } \nabla f_j(x)^T d - t \leq 0 \ j = 1, 2, ..., m, \\ &\qquad t \in \mathbb{R}, \ \ d \in \mathbb{R}^n. \end{aligned}$$

Here, a sequence of positive definite matrices is generated during the iterative process like the quasi-Newton method for the single-objective case. The Armijo-

Wolfe type line search technique is used to determine the step length. A descent sequence is generated whose accumulation point is a critical point of (MOP) under some reasonable assumptions.

The above line search techniques are restricted to unconstrained multi-objective programming problems, which are further extended to constrained multi-objective problems. A general constrained multi-objective optimization problem is

$$(MOP_C): \begin{cases} \min\limits_{x \in \mathbb{R}^n} \; F(x) \\ \text{subject to } g_i(x) \leq 0, \quad i = 1, 2, ..., p \end{cases}$$

Concept of the line search methods for single objective constrained optimization problems are extended to the multi-objective case in some recent papers; see [3, 4, 13]. A variant of the sequential quadratic programming (SQP) method is developed for inequality constrained (MOP_C) in the light of the SQP method for the single-objective case by Ansari and Panda [4] recently. The following quadratic subproblem is solved to obtain a feasible descent direction at every iterating point x, which involves linear approximations of all functions.

$$\begin{aligned} &\min_{t,\, d} \; t + \frac{1}{2} d^T d \\ &\text{subject to } \nabla f_j(x)^T d \leq t, \quad j = 1, 2, ..., m, \\ &g_i(x) + \nabla g_i(x)^T d \leq t, \quad i = 1, 2, ..., p, \\ &t \in \mathbb{R}, \;\; d \in \mathbb{R}^n. \end{aligned}$$

The same authors consider a different subproblem in [3] which involves quadratic approximations of all the functions, and use the SQCQP technique to develop a descent sequence. This subproblem is

$$\begin{gathered} \min_{t,\, d} \; t \\ \text{subject to } \nabla f_j(x)^T d + \tfrac{1}{2} d^T \nabla^2 f_j(x) d - t \leq 0, \quad j = 1, 2, ..., m \\ g_i(x) + \nabla g_i(x)^T d + \tfrac{1}{2} d^T \nabla^2 g_i(x) d \leq 0, \quad i = 1, 2, ..., p \\ t \in \mathbb{R}, d \in \mathbb{R}^n. \end{gathered}$$

With these subproblems, a non-differentiable penalty function is used to restrict constraint violations. To obtain a feasible descent direction, the penalty function is considered as a merit function with a penalty parameter. The Armijo type line search technique is used to find a suitable step length. Global convergence of these methods is discussed under the Slater constraint qualification.

The above iterative schemes are free from the burden of selection of parameters in advance, and also have the convergence property. These iterative schemes are gradient-based methods, and large-scale problems can be solved efficiently only if the gradient information of the functions is available. Some optimization software packages perform the finite difference gradient evaluation internally. But this is inappropriate when function evaluations are costly and noisy. Hence there is a growing demand for derivative-free optimization methods which neither require derivative information nor approximate the derivatives. The reader may refer to the book Cohn et al. [5] for the recent developments on derivative-free methods for single-objective optimization problem.

Coordinate search is the simplest derivative-free method for the unconstrained single-objective optimization problem. It evaluates the objective function of n variables at $2n$ points around a current iterate defined by displacements along the coordinate directions, their negatives, and a suitable step length. The set of these directions form a positive basis. This method is slow but capable of handling noise and guarantees to converge globally. The implicit filtering algorithm is also a derivative-free line search algorithm that imposes sufficient decrease along a quasi-Newton direction. Here, the true gradient is replaced by the simplex gradient. This method resembles the quasi-Newton approach. The trust region-based derivative-free line search method is also in demand to address noisy functions. In this method, quadratic subproblems are formulated from polynomial interpolation or regression. The implicit filtering is less efficient than the trust region but more capable of capturing noise.

The next choice is the widely cited Nelder-Mead method [24], which is a direct search iterative scheme for single objective optimization problems. This evaluates a finite number of points in every iteration, which takes care of the function values at the vertices of the simplex $\{y^0, y^1, ..., y^n\}$ in n dimension, ordered by increasing values of the objective function which has to be minimized. Action is taken based on simplex operations such as reflections, expansions, and contractions (inside or outside) at every iteration. The Nelder-Mead method attempts to replace the simplex vertex that has the worst function value. In such iterations, the worst vertex y^n is replaced by a point in the line that connects y^n and y^c, where

$$y = y^c + \delta(y^c - y^n), \; y^c = \frac{1}{n}\sum_{i=0}^{n-1} y^i, \quad \delta \in R.$$

$\delta = 1$ indicates a reflection, $\delta = 2$ an expansion, $\delta = 1/2$ an outer contraction, and $\delta = -1/2$ an inside contraction. Nelder-Mead can also perform shrink. Except for the shrinks, the emphasis is on replacing the worse vertex rather than improving the best. The simplices generated by Nelder-Mead may adapt well to the curvature of the function.

In this chapter, a derivative-free iterative scheme is developed for (MOP). The idea of the Nelder-Mead simplex method is imposed in a modified form using the Non-dominated Sorting algorithm to solve (MOP). This algorithm is coded in MATLAB(2019) to generate the Pareto front. The efficiency of this algorithm is justified through a set of test problems, and comparison with a scalarization method and NSGA$-$II is provided in terms of the number of iterations and CPU time.

2.2 Notations and Preliminaries

Consider that $\mathbb{R}^m$ is partially ordered by a binary relation induced by $\mathbb{R}^m_+$, the non-negative orthant of $\mathbb{R}^m$. For $p, q \in \mathbb{R}^m$,

$$p \leqq_{\mathbb{R}^m_+} q \iff q - p \in \mathbb{R}^m_+;$$

$$p \preceq_{\mathbb{R}^m_+} q \iff q - p \in \mathbb{R}^m_+ \setminus \{0\};$$

$$p \prec_{\mathbb{R}^m_+} q \iff q - p \in int\ (\mathbb{R}^m_+).$$

Definition 2.1 A point $x^* \in S$ is called a *weak efficient solution* of the (MOP) if there does not exist $x \in S$ such that $F(x) \prec_{\mathbb{R}^m_+} F(x^*)$. In other words, whenever $x \in S,\ F(x) - F(x^*) \notin -int(\mathbb{R}^m_+)$. In set notation, this becomes $(F(S) - F(x^*)) \cap -int(\mathbb{R}^m_+) = \phi$.

Definition 2.2 A point $x^* \in S$ is called an *efficient solution* of the (MOP) if there does not exist $x \in S$ such that $F(x) \preceq_{\mathbb{R}^m_+} F(x^*)$. In other words, whenever $x \in S,\ F(x) - F(x^*) \notin -(\mathbb{R}^m_+ \setminus \{0\})$. In set notation, this becomes $(F(S) - F(x^*)) \cap (-(\mathbb{R}^m_+ \setminus \{0\})) = \phi$. This solution is also known as the Pareto optimal or non-inferior solution. If X^* is the set of all efficient solutions, then the set $F(X^*)$ is called the Pareto front for (MOP).

Definition 2.3 For $x_1,\ x_2 \in \mathbb{R}^n$, x_1 is said to *dominate* x_2 if and only if $F(x_1) \preceq_{\mathbb{R}^m_+} F(x_2)$, that is, $f_j(x_1) \leq f_j(x_2)$ for all j and $F(x_1) \neq F(x_2)$. x_1 *weakly dominates* x_2 if and only if $F(x_1) \prec_{\mathbb{R}^m_+} F(x_2)$, that is, $f_j(x_1) < f_j(x_2)$ for all j. A point $x_1 \in \mathbb{R}^n$ is said to be *non-dominated* if there does not exist any x_2 such that x_2 dominates x_1.

This concept can also be extended to find a non-dominated set of solutions of a multi-objective programming problem. Consider a set of N points $\{x_1, x_2, ..., x_N\}$, each having $m(> 1)$ objective function values. So $F(x_i) = (f_1(x_i), f_2(x_i), ..., f_m(x_i))$. The following algorithm from Deb [6] can be used to find the non-dominated set of points. This algorithm is used in the next section to order the objective values at every vertex of the simplex.

Algorithm 1[6]
Step 0 : Begin with $i = 1$.
Step 1 : For all $j \neq i$, compare solutions x_i and x_j for domination using Definition 2.3 for all m objectives.
Step 2 : If for any j, x_i is dominated by x_j, mark x_i as "dominated".
Step 3 : If all solutions (that is, when $i = N$ is reached) in the set are considered, go to **Step** 4, else increment i by one and go to **Step** 1.
Step 4 : All solutions that are not marked "dominated" are non-dominated solutions.

2.3 Gradient-Free Method for *MOP*

In this section, we develop a gradient-free algorithm as an extension of the Nelder-Mead simplex method, which is a widely used algorithm for the single-objective case. The dominance property as explained in Algorithm 1 helps to compare $F(x)$ at different points in $\mathbb{R}^m$.

Consider a simplex of $n+1$ vertices in $\mathbb{R}^n$ as $Y = \{y^0, y^1, ..., y^n\}$ ordered by component-wise increasing values of F. To order the vertices component-wise, one can use the "Non-dominated Sorting Algorithm 1". The most common Nelder-Mead iterations for the single-objective case perform a reflection, an expansion, or a contraction (the latter can be inside or outside the simplex). In such iterations, the worst vertex y^n is replaced by a point in the line that connects y^n and y^c,

$$y = y^c + \delta(y^c - y^n), \ \ \delta \in \mathbb{R},$$

where $y^c = \sum\limits_{i=0}^{n-1} \frac{y^i}{n}$ is the centroid of the best n vertices. The value of δ indicates the type of iteration. For instance, when $\delta = 1$ we have a (genuine or isometric) reflection, when $\delta = 2$ an expansion, when $\delta = \frac{1}{2}$ an outside contraction, and when $\delta = -\frac{1}{2}$ an inside contraction. A Nelder-Mead iteration can also perform a simplex shrink, which rarely occurs in practice. When a shrink is performed, all the vertices in Y are thrown away except the best one y^0. Then n new vertices are computed by shrinking the simplex at y^0, that is, by computing, for instance, $y^0 + \frac{1}{2}(y^i - y^0)$, $i = 1, 2, ..., n$. Note that the "shape" of the resulting simplices can change by being stretched or contracted, unless a shrink occurs.

2.3.1 *Modified Nelder-Mead Algorithm*

Choose an initial point of vertices $Y_0 = \{y_0^0, y_0^1, ..., y_0^n\}$. Evaluate F at the points in Y_0. Choose constants:

$$0 < \gamma^s < 1, -1 < \delta^{ic} < 0 < \delta^{oc} < \delta^r < \delta^e.$$

For $k = 0, 1, 2, ...,$ set $Y = Y_k$.

1. Order the $n+1$ vertices of $Y = \{y^0, y^1, ..., y^n\}$ using Algorithm 1 so that

$$F(y^0) \preceq_{\mathbb{R}^m_+} F(y^1) \preceq_{\mathbb{R}^m_+} \cdots \preceq_{\mathbb{R}^m_+} F(y^n).$$

   Denote $F(y^t) = F^t, t = 0, 1, \ldots, n$

2. Reflect the worst vertex y^n over the centroid $y^c = \sum_{i=0}^{n-1} \frac{y^i}{n}$ of the remaining n vertices:

$$y^r = y^c + \delta(y^c - y^n)$$

Evaluate $F^r = F(y^r)$. If F^0 dominates F^r and F^r dominates weakly F^{n-1}, then replace y^n by the reflected point y^r and terminate the iteration:

$$Y_{k+1} = \{y^0, y^1, ..., y^r\}.$$

3. If F^r dominates weakly F^0, then calculate the expansion point

$$y^e = y^c + \delta^r(y^c - y^n)$$

and evaluate $F^e = F(y^e)$. If F^e dominates F^r, replace y^n by the expansion point y^e and terminate the iteration:

$$Y_{k+1} = \{y^0, y^1, ..., y^e\}.$$

Otherwise, replace y^n by the reflected point y^r and terminate the iteration:

$$Y_{k+1} = \{y^0, y^1, ..., y^r\}.$$

4. If F^{n-1} dominates F^r, then a contraction is performed between the best of y^r and y^n.
   (a) If F^r dominates weakly F^n, perform an outside contraction

   $$y^{oc} = y^c + \delta^{oc}(y^c - y^n)$$

   and evaluate $F^{oc} = F(y^{oc})$. If F^{oc} dominates F^r, then replace y^n by the outside contraction point y^{oc} and terminate the iteration:

   $$Y_{k+1} = \{y^0, y^1, ..., y^{oc}\}.$$

   Otherwise, perform a shrink.
   (b) If F^n dominates F^r, perform an inside contraction

   $$y^{ic} = y^c + \delta^{ic}(y^c - y^n)$$

   and evaluate $F^{ic} = F(y^{ic})$. If F^{ic} dominates weakly F^n, then replace y^n by the inside contraction point y^{ic} and terminate the iteration:

   $$Y_{k+1} = \{y^0, y^1, ..., y^{ic}\}.$$

   Otherwise, perform a shrink.

5. Evaluate f at the n points $y^0 + \gamma^s(y^i - y^0)$, $i = 1, ..., n$, and replace $y^1, ..., y^n$ by these points, terminating the iteration:

$$Y_{k+1} = y^0 + \gamma^s(y^i - y^0), i = 0, ..., n.$$

2.4 Numerical Illustrations and Performance Assessment

In this section, the algorithm is executed on some test problems, which are collected from different sources and summarized in Table 2.1 (see Appendix). The results obtained by Algorithm 2.3.1 are compared with the existing methods: the scalarization method (Weighted sum) and NSGA-II. MATLAB code (R2017b) for these three methods is developed. The comparison is provided in Table 2.2 (see Appendix). In this table, "Iter" corresponds to the number of iterations and "CPU time" corresponds to the time for executing the Algorithms. Denote the algorithms in short term as

Algorithm 2.3.1—(NMSM)
Weighted Sum Method—(WSM)
NSGA-II

Pareto front: To generate the Pareto front by Algorithm 2.3.1, the **RAND** strategy is considered for selecting the initial point. 500 uniformly distributed random initial points between lower bound and upper bound are selected. Every test problem is executed 10 times with random initial points. The Pareto front of the test problem "BK1" in NSGA-II, NMSM, and WSM is provided in Fig. 2.1 with red, green, and blue stars, respectively.

2.5 Performance Profile

Performance profile is defined by a cumulative function $\rho(\tau)$ representing a performance ratio with respect to a given metric, for a given set of solvers. Given a set of solvers S and a set of problems P, let $\zeta_{p,s}$ be the performance of solver s on solving problem p. The performance ratio is then defined as $r_{p,s} = \zeta_{p,s}/\min_{s\in S}\zeta_{p,s}$. The cumulative function $\rho_s(\tau)$ $(s \in S)$ is defined as

$$\rho_s(\tau) = \frac{|\{p \in P : r_{p,s} \leq \tau\}|}{|P|}.$$

It is observed that the performance profile is sensitive to the number and types of algorithms considered in the comparison; see [16]. So the algorithms are compared pairwise. In this chapter, the performance profile is compared using purity, Γ, Δ spread metrics.

Purity metric: Let $P_{p,s}$ be the approximated Pareto front of problem p obtained by method s. Then an approximation to the true Pareto front P_p can be built by considering $\bigcup_{s\in S} P_{p,s}$ first and removing the dominated points. The purity metric for algorithms s and problem p is defined by the ratio

$$\bar{t_{p,s}} = |P_{p,s} \cap P_p|/|P_{p,s}|.$$

Clearly, $\bar{t_{p,s}} \in [0, 1]$. When computing the performance profiles of the algorithms for the purity metric, it is required to set $t'_{p,s} = 1/\bar{t_{p,s}}$. $t' = 0$ implies that the algorithm is unable to solve p.

Spread metrics: Two types of spread metrics (Γ and Δ) are used in order to analyze if the points generated by an algorithm are well-distributed in the approximated Pareto front of a given problem. Let $x_1, x_2, \ldots, x_N$ be the set of points obtained by a solver s for problem p and let these points be sorted by objective function j, that is, $f_j(x_i) \leq f_j(x_{i+1})$ $(i = 1, 2, \ldots, N-1)$. Suppose x_0 is the best known approximation of global minimum of f_j and x_{N+1} is the best known global maximum of f_j, computed over all approximated Pareto fronts obtained by different solvers. Define

$$\delta_{i,j} = f_j(x_{i+1}) - f_j(x_i).$$

Then Γ spread metric is defined by

$$\Gamma_{p,s} = \max_{j\in\Lambda_m} \max_{i\in\{0,1,\ldots,N\}} \delta_{i,j}.$$

Define δ_j to be the average of the distances $\delta_{i,j}, i = 1, 2, \ldots, N-1$. For an algorithm s and a problem p, the spread metric Δ is

$$\Delta_{p,s} = \max_{j\in\Lambda_m} \frac{\delta_{0,j} + \delta_{N,j} + \sum_{i=1}^{N-1} |\delta_{i,j} - \bar{\delta_j}|}{\delta_{0,j} + \delta_{N,j} + (N-1)\bar{\delta_j}}.$$

Result Analysis: A deep insight into Figs. 2.2, 2.3, and 2.4 clearly indicates the advantage of the proposed method (NMSM) to the existing methods WSM and NSGA−II. In RAND, NMSM has a better performance ratio in the Γ metric than WSM and NSGA−II and purity and δ metrics than NSGA−II in most of the test problems. Also from the computational details tables, one may observe that NMSM takes less number of iterations and time than WSM and NSGA−II in most of the test problems.

2.6 Conclusions

In this chapter, a Nelder-mead simplex method is developed for solving unconstrained multi-objective optimization problems. This method is modified from the existing Nelder-mead simplex method for single-objective optimization problems. Justification of this iterative process is carried out through numerical computations. This chapter can be further studied for the constrained multi-objective programming problem and for the better spreading technique to generate the Pareto points, which can be considered as the future scope of the present contribution.

Acknowledgements We thank the anonymous reviewers for the valuable comments that greatly helped to improve the content of this chapter.

Appendix

Table 2.1 Multi-objective test problems

Problem	Source	Problem	Source	Problem	Source
BK1	[17]	Jin2	[20]	TKLY1	[26]
Deb41	[6]	Jin3	[20]	LE1	[17]
Deb513	[6]	Jin4	[20]	I1	[17]
Deb521aa	[6]	lovison1	[22]	Far1	[17]
Deb521b	[6]	lovison2	[22]	SK1	[17]
DG01	[17]	lovison3	[22]	SK2	[17]
ex005	[18]	lovison4	[22]	SP1	[17]
ZDT3	[28]	LRS1	[17]	SSFYY1	[17]
Fonseca	[14]	MOP2	[17]	SSFYY2	[17]
Deb53	[9]	MHHM2	[17]	VFM1	[17]
GE5	[9]	MLF1	[17]	ZDT1	[28]
IKK1	[17]	MLF2	[17]	VU1	[17]
ZDT2	[28]	SCH1	[17]	VU2	[17]
Jin1	[20]	MOP1	[17]	KW2	[9]
OKA1	[25]	MOP3	[17]	MOP7	[17]
OKA2	[25]	MOP5	[17]	ZDT4	[28]
QV1	[17]	MOP6	[17]	CEC09_1	[3]
CEC09_2	[3]	CEC09_3	[3]	CEC09_7	[3]
CEC09_8	[3]	CEC09_10	[3]	Deb512a	[6]
Deb512b	[6]	Deb512c	[6]	DTLZ1	[6]
DTLZ1n2	[6]	DTLZ2	[6]	DTLZ2n2	[6]
DTLZ3	[6]	DTLZ3n2	[6]	DTLZ4	[6]
DTLZ4n2	[6]	DTLZ5_a	[6]	DTLZ5n2_a	[6]
DTLZ6	[6]	DTLZ6n2	[6]	FES1	[17]
FES2	[17]	FES3	[17]	IM1	[17]

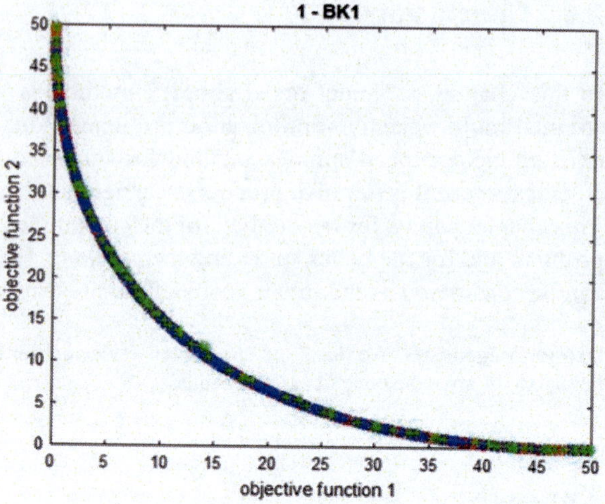

Fig. 2.1 Pareto front of BK1 for Weighted Sum, Nelder-Mead Simplex, and NSGA−II

Table 2.2 Computation details

Problem	NMSM		WSM		NSGA−II	
	Iter	CPU time	Iter	CPU time	Iter	CPU time
BK1	100	175.7102	1000	17.3119	1000	96.1962
CEC09_1	40	40.8337	39185	85.1831	1000	125.0616
CEC09_2	15	14.7780	29659	62.3473	1000	149.2667
CEC09_3	30	43.5034	17463	47.0575	1000	261.1938
CEC09_7	50	58.12	66763	115.43	1000	246.4144
CEC09_8	10	39.5513	11881	24.811	1000	414.9251
CEC09_10	1	7.1234	26521	61.1016	500	424.4517
Deb41	225	46.961	3451	13.1149	200	54.7721
Deb53	1	3.148	4779	16.182	1000	226.228
Deb512a	95	30.8398	6831	16.2914	1000	67.5017
Deb512b	1	1.6304	5665	22.007	500	236.7293
Deb512c	1	1.3482	4169	13.0784	1000	105.0796
Deb513	100	286.7849	1105	8.2348	500	233.5419
Deb521a	200	40.6678	2074	16.2494	500	178.9993
Deb521b	200	34.7894	1832	9.5663	500	86.927
DG01	100	383.6649	3527	10.0337	500	114.6849
DTLZ1	1	8.0783	829	11.3604	500	604.637
DTLZ1n2	50	51.4517	1533	10.5865	500	329.7814

(continued)

Table 2.2 (continued)

Problem	NMSM		WSM		NSGA−II	
	Iter	CPU time	Iter	CPU time	Iter	CPU time
DTLZ2	1	9.1068	4384	12.1977	500	201.0205
DTLZ2n2	30	92.2633	1705	9.2817	500	90.3009
DTLZ3	1	8.5647	546	11.3052	500	180.3397
DTLZ3n2	25	59.364	1489	12.3074	500	837.8227
DTLZ5	5	37.5631	2361	11.2001	500	263.9273
DTLZ5n2	4	19.6248	2417	10.2009	500	277.3301
DTLZ6	100	40.4558	3696	13.6023	500	159.34
DTLZ6n2	150	266.6435	2416	9.9722	500	186.7189
ex005	2	8.6635	2066	9.0123	500	280.0558
Far1	50	54.4216	6633	21.8326	500	122.3895
hline FES1	65	211.8147	22021	69.7231	500	293.3888
FES2	1	3.6761	24335	80.9286	500	146.9458
FES3	1	3.6302	26324	96.1848	500	142.7173
Fonseca	50	83.6762	3872	11.3395	500	95.5476
GE5	1	5.932	2412	9.7671	500	433.6827
I1	4	9.8997	1610	18.8042	500	115.2193
IKK1	1	2.9279	1539	8.1619	500	95.1506
IM1	1	4.971	2353	9.4475	500	185.6666
Jin1	50	35.5375	1091	7.9041	500	83.9951
Jin2	28	5.6778	3834	11.2386	500	152.2696

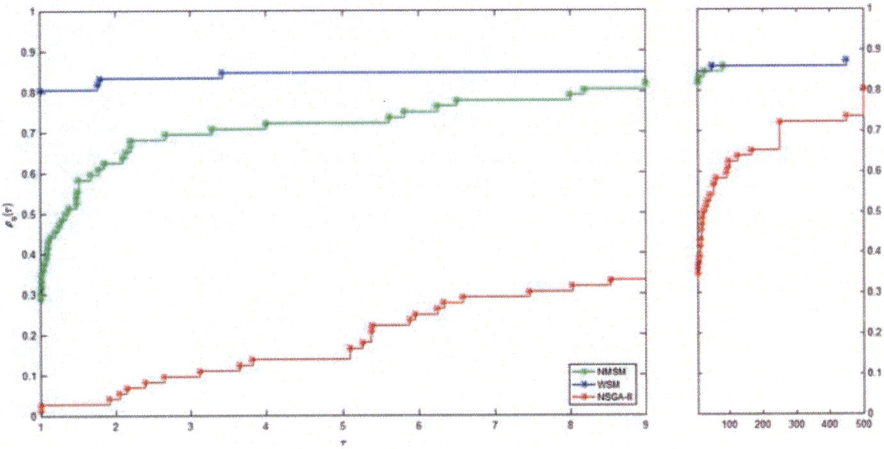

Fig. 2.2 Purity performance profile between NMSM,WSM, and NSGA-II

Table 2.3 Computation details continued

Problem	NMSM		WSM		NSGA−II	
	Iter	CPU time	Iter	CPU time	Iter	CPU time
Jin3	100	15.7487	1565	8.3628	500	196.8463
Jin4	150	24.6867	2815	10.8569	500	111.2916
KW2	50	89.2429	4885	15.7046	500	118.0532
LE1	50	240.3437	9798	19.4636	500	182.9293
Lovison1	1	2.968	1752	9.9745	500	111.5821
Lovison2	150	25.9414	2316	10.4402	500	207.9985
Lovison3	1	2.5782	2389	10.3237	500	206.382
Lovison4	1	4.0833	1949	10.1973	500	279.6307
LRS1	50	336.4876	1002	7.7188	500	142.3477
MHHM2	3	11.0745	3681	11.4456	500	344.4489
MLF1	50	122.7622	1927	9.1655	500	104.0309
MLF2	2	8.4472	4347	11.103	500	358.7373
MOP1	1	5.2652	1006	8.604	500	119.9713
MOP2	50	154.4144	3373	10.4286	500	99.4147
MOP3	2	4.551	4403	12.7004	500	95.9286
MOP5	1	1.2941	4007	10.4313	500	113.9832
MOP6	45	117.3724	1111	8.6546	500	278.6248
MOP7	1	5.1297	3765	12.0097	500	166.5825
OKA1	60	80.5031	5412	17.0589	500	118.3507
OKA2	50	33.6682	5753	16.9394	83	19.2191
QV1	55	251.9818	962	11.4491	500	76.7295
SCH1	50	61.1843	1387	9.228	500	97.6939
SK1	50	53.5915	2171	9.5896	500	98.0009
SK2	50	164.0477	2202	9.9194	500	146.5663
SP1	75	164.2625	3035	10.2907	500	133.1876
SSFYY1	100	323.6209	1002	7.9782	500	91.426
SSFYY2	50	186.4307	2745	10.4761	500	106.6671
TKLY1	55	29.4778	3106	10.9727	500	108.0413
VFM1	1	5.7106	1000	8.5332	500	121.1181
VU1	100	31.6459	1802	8.9375	500	138.3799
VU2	1	3.3975	2116	9.6255	500	400.1829
ZDT1	100	64.8696	4584	19.4772	500	252.9595
ZDT2	100	49.0452	1967	10.6646	500	361.0973
ZDT3	50	44.126	4499	13.469	500	480.667
ZDT4	50	83.5181	8527	19.4524	100	836.2467

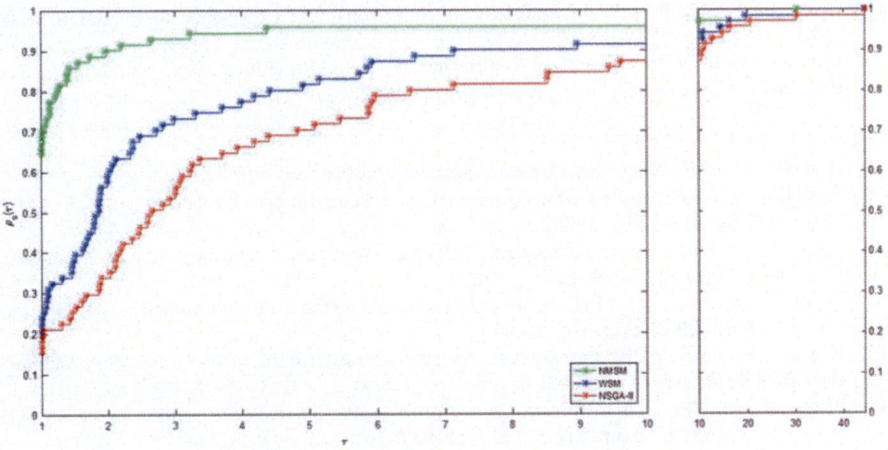

Fig. 2.3 Γ performance profile between NMSM,WSM, and NSGA-II

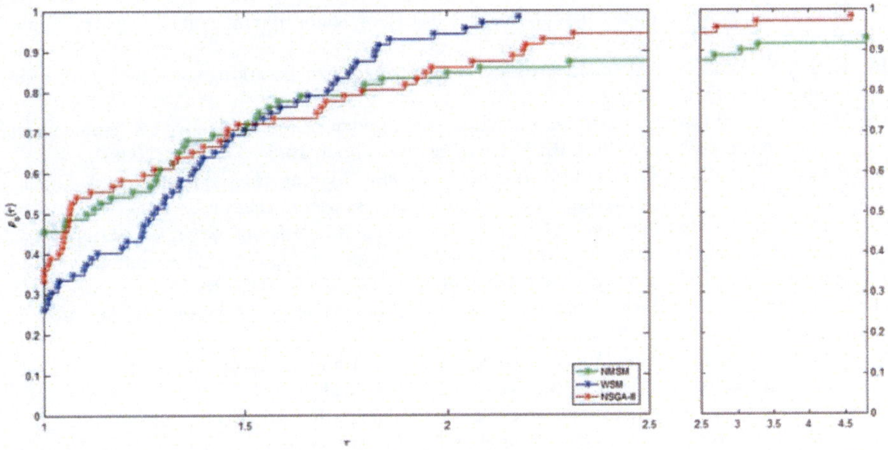

Fig. 2.4 Δ performance profile between NMSM,WSM, and NSGA-II

References

1. Ansari, Q.H., Köbis, E., Yao, J.C.: Vector Variational Inequalities and Vector Optimization. Springer (2018)
2. Ansary, M.A.T., Panda, G.: A modified quasi-newton method for vector optimization problem. Optimization **64**, 2289–2306 (2015)
3. Ansary, M.A.T., Panda, G.: A sequential quadratically constrained quadratic programming technique for a multi-objective optimization problem. Eng. Optim. **51**(1), 22–41 (2019). https://doi.org/10.1080/0305215X.2018.1437154
4. Ansary, M.A.T., Panda, G.: A sequential quadratic programming method for constrained multi-objective optimization problems. J. Appl. Math. Comput. (2020). https://doi.org/10.1007/s12190-020-01359-y

5. Conn, A.R., Scheinberg, K., Vicente, L.N.: Introduction to Derivative-Free Optimization. SIAM (2009)
6. Deb, K.: Multi-Objective Genetic Algorithms: Problem Difficulties and Construction of Test Problems. Wiley India Pvt. Ltd., New Delhi, India (2003)
7. Deb, K.: Multi-objective Optimization Using Evolutionary Algorithms. Wiley India Pvt. Ltd., New Delhi, India (2003)
8. Ehrgott, M.: Multicriteria Optimization. Springer Publication, Berlin (2005)
9. Eichfelder, G.: An adaptive scalarization method in multiobjective optimization. SIAM J. Optim. **19**(4), 1694–1718 (2009)
10. Engau, A., Wiecek, M.M.: Generating ϵ-efficient solutions in multiobjective programming. Eur. J. Oper. Res. **177**, 1566–1579 (2007)
11. Fliege, F., Drummond, L.M.G., Svaiter, B.: Newton's method for multiobjective optimization. SIAM J. Optim **20**(2), 602–626 (2009)
12. Fliege, J., Svaiter, F.V.: Steepest descent methods for multicriteria optimization. Math. Methods Oper. Res. **51**(3), 479–494 (2000)
13. Fliege, J., Vaz, A.I.F.: A method for constrained multiobjective optimization based on SQP techniques. SIAM J. Optim. **26**(4), 2091–2119 (2016)
14. Fonseca, C.M., Fleming, P.J.: Multiobjective optimization and multiple constraint handling with evolutionary algorithms. i. a unified formulation. IEEE Trans. Syst. Man Cybern. Part A: Syst. Hum. **28**(1), 26–37 (1998)
15. Geofforion, A.M.: Proper efficiency and the theory of vector maximization. J. Optim. Theory Appl. **22**, 618–630 (1968)
16. Gould, N., Scott, J.: A note on performance profiles for benchmarking software. ACM Trans. Math. Softw. **43**, 15 (2016)
17. Huband, S., Hingston, P., Barone, L., While, L.: A review of multiobjective test problems and a scalable test problem toolkit. IEEE Trans. Evol. Comput. **10**(5), 477–506 (2006)
18. Hwang, C.L., Yoon, K.: Multiple Attribute Decision Making: Methods and Applications a State-of-the-Art Survey. Springer Science and Business Media (1981)
19. Jahn, J.: Vector Optimization. Theory, Applications, and Extensions. Springer, Berlin (2004)
20. Jin, Y., Olhofer, M., Sendhoff, B.: Dynamic weighted aggregation for evolutionary multi-objective optimization: Why does it work and how? pp. 1042–1049 (2001)
21. Li, Z., Wang, S.: ϵ-approximate solutions in multiobjective optimization. Optimization **34**, 161–174 (1998)
22. Lovison, A.: Singular continuation: generating piecewise linear approximations to pareto sets via global analysis. SIAM J. Optim. **21**(2), 463–490 (2011)
23. Miettinen, K.M.: Nonlinear Multiobjective Optimization. Kluwer, Boston (1999)
24. Nelder, J.A., Mead, R.: A simplex method for function minimization. Comput. J. **7**(4), 308–313 (1965). https://doi.org/10.1093/comjnl/7.4.308
25. Okabe, T., Jin, Y., Olhofer, M., Sendhoff, B.: On test functions for evolutionary multi-objective optimization. In: International Conference on Parallel Problem Solving from Nature, pp. 792–802. Springer (2004)
26. Preuss, M., Naujoks, B., Rudolph, G.: Pareto set and EMOA behavior for simple multimodal multiobjective functions, pp. 513–522 (2006)
27. Qu, S., Goh, M., Chan, F.T.S.: Quasi-newton methods for solving multiobjective optimization. Oper. Res. Lett. **39**, 397–399 (2011)
28. Zitzler, E., Deb, K., Thiele, L.: Comparison of multiobjective evolutionary algorithms: empirical results. Evol. Comput. **8**(2), 173–195 (2000)

Chapter 3
The New Butterfly Relaxation Method for Mathematical Programs with Complementarity Constraints

J.-P. Dussault, M. Haddou, and T. Migot

Abstract We propose a new family of relaxation schemes for mathematical programs with complementarity constraints. We discuss the properties of the sequence of relaxed non-linear programs as well as stationary properties of limiting points. A sub-family of our relaxation schemes has the desired property of converging to an M-stationary point. A stronger convergence result is also proved in the affine case. A comprehensive numerical comparison between existing relaxation methods is performed on the library of test problems MacMPEC which shows promising results for our new method.

Keywords Non-linear programming · MPCC · MPEC · Relaxation methods · Non-linear optimization model · Complementarity

3.1 Introduction

We consider the Mathematical Program with Complementarity Constraints (MPCC)

$$\min_{x \in \mathbb{R}^n} f(x) \text{ s.t. } g(x) \leq 0, h(x) = 0, 0 \leq G(x) \perp H(x) \geq 0, \tag{3.1}$$

where $f : \mathbb{R}^n \to \mathbb{R}$, $g : \mathbb{R}^n \to \mathbb{R}^m$, $h : \mathbb{R}^n \to \mathbb{R}^p$, and $G, H : \mathbb{R}^n \to \mathbb{R}^q$ are assumed continuously differentiable. The notation $0 \leq u \perp v \geq 0$ for two vectors u and v in $\mathbb{R}^q$ is a shortcut for $u \geq 0$, $v \geq 0$ and $u^T v = 0$. This problem has become an active

J.-P. Dussault · T. Migot
Département d'Informatique, faculté des Sciences, Université de Sherbrooke, Sherbrooke, Canada
e-mail: Jean-Pierre.Dussault@USherbrooke.ca

T. Migot
e-mail: tangi.migot@gmail.com

M. Haddou (✉)
INSA Rennes, CNRS, IRMAR - UMR 6625, Univ Rennes, 35000 Rennes, France
e-mail: Mounir.Haddou@insa-rennes.fr

V. Laha et al. (eds.), *Optimization, Variational Analysis and Applications*,
Springer Proceedings in Mathematics & Statistics 355,
https://doi.org/10.1007/978-981-16-1819-2_3

subject in the literature in the last two decades. The wide variety of applications that can be cast as an MPCC is one of the reasons for this popularity.

The MPCC is a non-linear program, but with a special structure since, apart from the usual equality and inequality constraints, they have the additional complementarity constraints, which may be equivalently rewritten as

$$G_i(x) \geq 0,\ H_i(x) \geq 0,\ G_i(x)H_i(x) \leq 0, \quad \forall i \in \{1, \ldots, q\}. \tag{3.2}$$

A popular approach to tackle a non-linear program computes the KKT conditions, which require that some constraint qualification holds at the solution to be an optimality condition. However, it is well-known that these constraint qualifications don't hold in general for (3.1) due to the complementarity constraint. For instance, the classical Mangasarian-Fromowitz constraint qualification is violated at any feasible point [38].

During the past two decades, many researchers introduced necessary optimality conditions such as the Clarke (C-), Mordukhovich (M-), strong (S-), and Bouligand (B-) stationarity conditions for the MPCC; see, e.g., [11, 18, 30, 35–38]. Among these stationarities, the B-stationarity is known to be a good candidate for optimality, but since it is computationally difficult, it is rarely used in algorithmic analysis; the S-stationarity is the strongest and equivalent to the KKT conditions, see, e.g., [10, 15], but its interest is reduced since it does not always hold for the MPCC. The M-stationarity, which has already widely been investigated, see, e.g., [11, 18, 26, 35–38], is the most relevant concept since it is the weakest necessary condition holding, under suitable constraint qualifications, at any local minimizer of the MPCC and is computationally tractable.

The feasible set of the MPCC involves a complementarity constraint equivalent to $G(x) = 0$ OR $H(x) = 0$. This is a *thin* set exhibiting some irregularity when $G(x) = 0$ AND $H(x) = 0$. It is this thinness that makes constraint qualifications fail at any feasible point. In view of the constraint qualification issues that plague the MPCC, the relaxation methods provide an intuitive answer. The complementarity constraint is relaxed using a parameter so that the new feasible domain is not thin anymore. It is assumed here that the classical constraints $g(x) \leq 0$ and $h(x) = 0$ are not more difficult to handle than the complementarity constraint. Finally, as the relaxing parameter is reduced, convergence to the feasible set of (3.1) is obtained similarly to a homotopy technique.

These methods have been suggested in the literature back in 2000 by Scheel and Scholtes in [30, 31]. Their natural approach was later extended by Demiguel, Friedlander, Nogales, and Scholtes in [6]. In [23], Lin and Fukushima improved this relaxation by expressing the same set with two constraints instead of three. This improvement leads to an improved constraint qualification satisfied by the relaxed sub-problem. Even so, the feasible set is not modified; this improved regularity does not come as a surprise, since a constraint qualification measures the way the feasible set is described and not necessarily the geometry of the feasible set itself. In [33], the authors consider a relaxation of the same type but only around the corner $G(x) = H(x) = 0$. In the corresponding papers it has been shown that under classical conditions convergence to some spurious points, called C-stationary points, may still happen, the convergence to M-stationary points being guaranteed only under some second-order condition.

A new perspective for those schemes has been given in [17] relaxing the constraints (3.2) by $t \geq 0$ as

$$(G_i(x) - t)(H_i(x) - t) \leq 0, \quad \forall i \in \{1, \dots, q\}. \tag{3.3}$$

This approximation scheme converges as t decreases to 0 under classical assumptions to M-stationary points without second-order or strict complementarity-type conditions. This is not a relaxation since the feasible domain of (3.1) is not included in the feasible set of the sub-problems. The method has been extended as a relaxation method in [19] through an NCP function ϕ:

$$\phi(G_i(x) - t, H_i(x) - t) \leq 0, \quad \forall i \in \{1, \dots, q\}, \tag{3.4}$$

where $\phi(a, b) := \{ab, \text{ if } a + b \geq 0, -\frac{1}{2}(a^2 + b^2), \text{ otherwise}\}$.

The main aim of this chapter is to continue this discussion and extend the relaxation of Kanzow and Schwartz [19] by introducing the new butterfly relaxation:

$$\phi\left(G_i(x) - t_2\theta_{t_1}(H_i(x)), H_i(x) - t_2\theta_{t_1}(G_i(x))\right) \leq 0, \quad \forall i \in \{1, \dots, q\}.$$

This new method handling two relaxing parameters, t_1 and t_2, allows a non-linear perturbation, θ, of the domain. Thus, we extend the butterfly relaxation introduced in [8] for the mathematical program with vanishing constraints to the case of complementarity constraints.

The following example shows that the butterfly relaxation may improve the relaxations from [17, 19]. Indeed, it illustrates an example where there is no sequence of stationary point[1] that converges to a non-optimal point.

Example 3.1

$$\min_{x \in \mathbb{R}^2} -x_1 \text{ s.t. } x_1 \leq 1, \; 0 \leq x_1 \perp x_2 \geq 0.$$

In this example, there are two stationary points: an S-stationary point $(1, 0)$ that is the global minimum and an M-stationary point $(0, 0)$, which is not a local minimum.[2] Unlike the relaxations (3.3) and (3.4) where for $t_k = \frac{1}{k}$ a sequence $x^k = (t_k, \; 2t_k)^T$, with the Lagrange multiplier associated with the regularized constraint $\lambda^{\Phi,k} = k$, may converge to $(0, 0)$ as k goes to infinity, there is no sequence of stationary points that converges to this undesirable point with the butterfly relaxation.

Our main contributions in this chapter are the following:

1. We prove convergence of the butterfly relaxation scheme to A-stationary points, and to M-stationary points for $t_{2,k} = o(t_{1,k})$.

---

[1] Definitions of stationary points of a non-linear program at the beginning of Sect. 3.2.1.

[2] Definitions of M- and S-stationarity points are given in Definition 3.3.

2. We prove for the affine MPCC that the butterfly relaxation scheme converges to S-stationary points under MPCC-LICQ, thus generalizing the situation of Example 3.1.
3. We prove that the butterfly relaxation scheme computing approximate stationary points at each step converges to an M-stationary point assuming $t_{2,k} = o(t_{1,k})$ and $\epsilon_k = o(\max(G_i(x^k), H_i(x^k))$.
4. We provide extensive numerical results showing that the butterfly relaxation can efficiently solve the MPCC.

In Sect. 3.2, we introduce classical definitions and results from non-linear programming and MPCC literature. In Sect. 3.3, we define the relaxation scheme with the new butterfly relaxation. In Sect. 3.4, we prove theoretical results on convergence and the existence of multipliers of the relaxed sub-problems. We also provide an analysis of the convergence of approximate stationary points. We also generalize the situation of Example 3.1 to illustrate a situation where the non-linear perturbation allows us to escape from undesirable points. In Sect. 3.5, we provide an extensive numerical study by giving details on the implementation and a comparison with other methods. Finally, in Sect. 3.6, we discuss some perspectives of this work.

3.2 Preliminaries

In this section, we introduce classical notations and definitions for non-linear programs and mathematical programs with complementarity constraints used in the sequel.

3.2.1 Non-Linear Programming

Let a general non-linear program be

$$\min_{x \in \mathbb{R}^n} f(x) \text{ s.t. } g(x) \le 0,\ h(x) = 0, \tag{3.5}$$

with $f : \mathbb{R}^n \to \mathbb{R}$, $g : \mathbb{R}^n \to \mathbb{R}^m$, and $h : \mathbb{R}^n \to \mathbb{R}^p$. Denote $\mathcal{F}$ the feasible region of (3.5), and $\mathcal{I}_g(x) := \{i \in \{1, ..., m\} \ : \ g_i(x) = 0\}$ the set of active indices. The Lagrangian function of (3.5) is defined as $\mathcal{L}(x, \lambda) := f(x) + g(x)^T \lambda^g + h(x)^T \lambda^h$, where $\lambda = (\lambda^g, \lambda^h) \in \mathbb{R}^m \times \mathbb{R}^p$ is the vector of Lagrange multipliers.

We call a KKT point a couple (x, λ) with $x \in \mathcal{F}$ such that $\nabla_x \mathcal{L}(x, \lambda) = 0$, $\lambda^g \ge 0$ and $g(x)^T \lambda^g = 0$. We call x a stationary point if there exists λ such that (x, λ) is a KKT point. We remind that the tangent cone of a set $X \subseteq \mathbb{R}^n$ at $x^* \in X$ is a closed cone defined by

$$\mathscr{T}_X(x^*) := \{d \in \mathbb{R}^n \mid \exists \tau_k \ge 0 \text{ and } X \ni x^k \to_{k\to\infty} x^* \text{ s.t. } \tau_k(x^k - x^*) \to_{k\to\infty} d\}.$$

Another useful tool for our study is the linearized cone of (3.5) at $x^* \in \mathcal{F}$ defined by

$$\mathscr{L}(x^*) := \{d \in \mathbb{R}^n \mid \nabla g_i(x)^T d \leq 0, \forall i \in \mathcal{I}_g(x^*),\ \nabla h_i(x)^T d = 0, \forall i = 1, \dots, p\}.$$

In the context of solving non-linear programs, that is finding a local minimum of (3.5), one widely used technique is to compute necessary conditions. The main tool is the Karush-Kuhn-Tucker (KKT) conditions. Let x^* be a local minimum of (3.5) that satisfies a constraint qualification, then there exists a Lagrange multiplier λ^* such that (x^*, λ^*) is a KKT point of (3.5). Constraint qualifications are used to ensure the existence of the multiplier at x^*.

We now define some of the classical constraint qualifications. Note that there exists a wide variety of such notions and we define here those that are essential for our purpose.

Definition 3.1 Let $x^* \in \mathcal{F}$.

(a) *Linear Independence* CQ (LICQ) holds at x^* if the family of gradients $\{\nabla g_i(x^*)\ (i \in \mathcal{I}_g(x^*)),\ \nabla h_i(x^*)\ (i = 1, ..., p)\}$ is linearly independent.
(b) *Mangasarian-Fromovitz* CQ (MFCQ) holds at x^* if the family of gradients $\{\nabla h_i(x^*)\ (i = 1, \dots, p)\}$ is linearly independent and there exists a $d \in \mathbb{R}^n$ such that $\nabla g_i(x^*)^T d < 0\ (i \in \mathcal{I}_g(x^*))$ and $\nabla h_i(x^*)^T d = 0\ (i = 1, \dots, p)$.

Remark 3.1 The definition of MFCQ given here is the most classical. It can be shown using some theorem of the alternative that this definition is equivalent to the family of active gradients being positively linearly independent, so that under MFCQ, the only solution of $\sum_{i\in\mathcal{I}_g(x^*)} \lambda_i^g \nabla g_i(x^*) + \sum_{i=1}^p \lambda_i^h \nabla h_i(x^*) = 0$ with $\lambda_i^g \geq 0, \forall i \in \mathcal{I}_g(x^*)$ is the trivial solution.

A local minimum is characterized by the fact that there is no feasible descent direction for the objective function of (3.5), that is, $-\nabla f(x^*) \in \mathscr{T}_{\mathcal{F}}(x^*)^\circ$, where $\mathscr{T}^\circ$ denotes the polar cone of $\mathscr{T}$. Given a cone $K \subseteq \mathbb{R}^n$, the polar of K is the cone defined by $K^\circ := \{z \in \mathbb{R}^n | z^T x \leq 0, \forall x \in K\}$. On the other hand, the KKT conditions build ∇f using a linearization of the active constraints. In a classical way, we say that a point $x^* \in \mathcal{F}$ satisfies Guignard CQ if $\mathscr{T}_{\mathcal{F}}(x^*)^\circ = \mathscr{L}(x^*)^\circ$ and Abadie CQ if $\mathscr{T}_{\mathcal{F}}(x^*) = \mathscr{L}(x^*)$.

In the context of numerical computations, it is almost never possible to compute stationary points. Hence, it is of interest to consider ϵ-stationary points.

Definition 3.2 Given a general non-linear program (3.5) and $\epsilon \geq 0$. We say that $(x, \lambda) \in \mathbb{R}^n \times \mathbb{R}^{m+p}$ is an ϵ-*KKT point* if it satisfies

$$\begin{aligned}
&\|\nabla_x \mathcal{L}(x, \lambda)\|_\infty \leq \epsilon,\ \|h(x)\|_\infty \leq \epsilon, \\
&g_i(x) \leq \epsilon,\ \lambda_i^g \geq 0,\ \left|\lambda_i^g g_i(x)\right| \leq \epsilon,\ \forall i \in \{1, \dots, m\}.
\end{aligned}$$

We say that x is an ϵ-stationary point if there exists λ such that (x, λ) is an ϵ-KKT point.

3.2.2 *Mathematical Programs with Complementarity Constraints*

We now specialize the general notions above to our specific case of (3.1). Let $\mathcal{Z}$ be the set of feasible points of (3.1). Given $x^* \in \mathcal{Z}$, we denote

$$
\begin{aligned}
\mathcal{I}^{+0} &:= \{i \in \{1,\dots,q\} \mid G_i(x^*) > 0 \text{ and } H_i(x^*) = 0\},\\
\mathcal{I}^{0+} &:= \{i \in \{1,\dots,q\} \mid G_i(x^*) = 0 \text{ and } H_i(x^*) > 0\},\\
\mathcal{I}^{00} &:= \{i \in \{1,\dots,q\} \mid G_i(x^*) = 0 \text{ and } H_i(x^*) = 0\}.
\end{aligned}
$$

In the sequel, we always consider these sets in x^*. In order to derive weaker optimality conditions, we consider an enhanced Lagrangian function. Let $\mathcal{L}_{MPCC}$ be the generalized MPCC-Lagrangian function of (3.1) defined as

$$\mathcal{L}_{MPCC}(x, \lambda) := f(x) + g(x)^T\lambda^g + h(x)^T\lambda^h - G(x)^T\lambda^G - H(x)^T\lambda^H$$

with $\lambda := (\lambda^g, \lambda^h, \lambda^G, \lambda^H) \in \mathbb{R}^m \times \mathbb{R}^p \times \mathbb{R}^q \times \mathbb{R}^q$.

We introduce more stationary concepts as in [24, 26, 27, 30, 35–37]. Those concepts are needed for two reasons:

- unless assuming a restrictive constraint qualification, a local minimizer x^* may fail to be a stationary point, so that optimality conditions need to be weakened in order to obtain a necessary condition;
- when analyzing cluster points of algorithms, other weak stationarity conditions appear naturally.

Definition 3.3 A point $x^* \in \mathcal{Z}$ is said to be

- *W-stationary* if there exists $\lambda \in \mathbb{R}^m \times \mathbb{R}^p \times \mathbb{R}^q \times \mathbb{R}^q$ such that

$$
\begin{aligned}
&\nabla_x \mathcal{L}_{MPCC}(x^*, \lambda) = 0,\\
&\lambda^g \geq 0,\ \lambda_i^g = 0,\ \forall i \notin \mathcal{I}_g,\\
&\lambda_i^G = 0,\ \forall i \in \mathcal{I}^{+0},\ \text{ and, } \lambda_i^H = 0,\ \forall i \in \mathcal{I}^{0+};
\end{aligned}
$$

- *C-stationary*, if it is W-stationary and $\lambda_i^G\lambda_i^H \geq 0,\ \forall i \in \mathcal{I}^{00}$;
- *A-stationary*, if it is W-stationary and $\lambda_i^G \geq 0$ or $\lambda_i^H \geq 0,\ \forall i \in \mathcal{I}^{00}$;
- *M-stationary*, if it is W-stationary and either $\lambda_i^G > 0, \lambda_i^H > 0$ or $\lambda_i^G\lambda_i^H = 0,\ \forall i \in \mathcal{I}^{00}$;
- *S-stationary*, if it is W-stationary and $\lambda_i^G \geq 0,\ \lambda_i^H \geq 0,\ \forall i \in \mathcal{I}^{00}$.

Relations between these definitions are straightforward from the definitions.

As pointed out in [10], strong stationarity is equivalent to the standard KKT conditions of an MPCC. In order to guarantee that a local minimum x^* of (1) is a stationary point in any of the previous senses, one needs to assume that a constraint qualification (CQ) is satisfied in x^*. Since most standard CQs are violated at any feasible point of (3.1), many MPCC-analogues of these CQs have been developed. Here, we mention only those needed later.

Definition 3.4 Let $x^* \in \mathcal{Z}$.

1. *MPCC-LICQ* holds at x^* if the only solution of

$$\sum_{i \in \mathcal{I}_g(x^*)} \lambda_i^g \nabla g_i(x^*) + \sum_{i=1}^{p} \lambda_i^h \nabla h_i(x^*) - \sum_{i \in \mathcal{I}^{0+} \cup \mathcal{I}^{00}} \lambda_i^G \nabla G_i(x^*) - \sum_{i \in \mathcal{I}^{+0} \cup \mathcal{I}^{00}} \lambda_i^H \nabla H_i(x^*) = 0 \tag{3.6}$$

   is the trivial solution.
2. *MPCC-MFCQ* holds at x^* if the only solution of (3.6) with $\lambda_i^g \geq 0, \forall i \in \mathcal{I}_g(x^*)$ is the trivial solution.
3. *MPCC-GMFCQ* holds at x^* if the only solution of (3.6) with $\lambda_i^g \geq 0, \forall i \in \mathcal{I}_g(x^*)$ and either $\lambda_i^G \lambda_i^H = 0$ or $\lambda_i^G > 0, \lambda_i^H > 0, \forall i \in \mathcal{I}^{00}$ is the trivial solution.

Note here that MPCC-MFCQ and MPCC-GMFCQ have been defined using the alternative form of MFCQ mentioned in Remark 3.1. Note that each of these CQs implies that a local minimum is M-stationary, see [9, 36], but only MPCC-LICQ is sufficient to guarantee strong stationarity of a local minimum; see [10, 24, 28]. The MPCC-LICQ is among the first MPCC-tailored constraint qualifications and may already be found in [24, 30]; the MPCC-MFCQ was introduced in [30] and presented in the form above in [16].

3.3 The Butterfly Relaxation Method

Consider a family of continuously differentiable non-decreasing concave functions $\theta : \mathbb{R} \to]-\infty, 1]$ such that

$$\theta(0) = 0, \text{ and, } \lim_{x \to \infty} \theta(x) = 1 \; \forall x \in \mathbb{R}_{++}.$$

Then, for $t_1 > 0$, we introduce $\theta_{t_1}(x) := \theta\left(\frac{x}{t_1}\right)$ if $x \geq 0$, and completed in a smooth way for negative values by considering $\theta_{t_1}(x) = x\theta'(0)/t_1$ if $x < 0$.

Example 3.2 Examples of such functions are

$$\theta_{t_1}^1(x) := \{\frac{x}{x+t_1}, \text{ for } x \geq 0, \quad \frac{x}{t_1}, \text{ for } x < 0.\},$$

and

$$\theta_{t_1}^2(x) := \{1 - \exp^{-\frac{x}{t_1}}, \text{ for } x \geq 0, \quad \frac{x}{t_1}, \text{ for } x < 0.\}.$$

Those functions have already been used in the context of complementarity constraints, for instance, in [1, 2].

To simplify the notation, we denote $t := (t_1, t_2)$. Using this family of functions, we denote

$$F_{1i}(x;t) := H_i(x) - t_2\theta_{t_1}(G_i(x)), \text{ and, } F_{2i}(x;t) := G_i(x) - t_2\theta_{t_1}(H_i(x)).$$

We propose a new family of relaxation of the complementarity constraint with two positive parameters (t_1, t_2) defined such that for all $i \in \{1, \dots, q\}$

$$\Phi_i^B(G(x), H(x); t) := \begin{cases} F_{1i}(x;t)F_{2i}(x;t), & \text{if } F_{1i}(x;t) + F_{2i}(x;t) \geq 0, \\ -\frac{1}{2}\left(F_{1i}(x;t)^2 + F_{2i}(x;t)^2\right) & \text{otherwise.} \end{cases} \tag{3.7}$$

This new relaxation uses two parameters t_1 and t_2 chosen such that

$$t_2\theta'(0) < t_1. \tag{3.8}$$

This condition ensures that the intersection of the sets $\{x \in \mathbb{R}^n \mid F_1(x; t_1, t_2) = 0\}$ and $\{x \in \mathbb{R}^n \mid F_2(x; t_1, t_2) = 0\}$ is reduced to the origin. In other words, the two branches of the relaxation do not cross each other. A typical choice will be to take $t_2 = o(t_1)$ motivated by strong convergence properties as discussed in Sect. 3.4.1.

The parametric non-linear program related to the butterfly relaxation of the complementarity constraints defined in (3.7), and augmented with a regularization of the non-negativity constraints parametrized by $\bar{t}$, is given by

$$\min_{x \in \mathbb{R}^n} f(x) \text{ s.t } x \in \mathcal{X}_{t,\bar{t}}^B, \tag{$R_{t,\bar{t}}^B$}$$

with

$$\mathcal{X}_{t,\bar{t}}^B := \{x \in \mathbb{R}^n \ : g(x) \leq 0, h(x) = 0,$$
$$G(x) \geq -\bar{t}e, H(x) \geq -\bar{t}e, \Phi^B(G(x), H(x); t) \leq 0\},$$

where e denotes the vector of all ones.

This method is similar to the methods (3.3) from [17] and (3.4) from [19] in the sense that they can also be written as a product of two functions. The main difference is that handling two parameters allows bringing the two "wings" of the relaxation closer. A comparison of the feasible set of these relaxations can be seen in Fig. 3.1.

The sets of indices used in the sequel are defined in the following way:

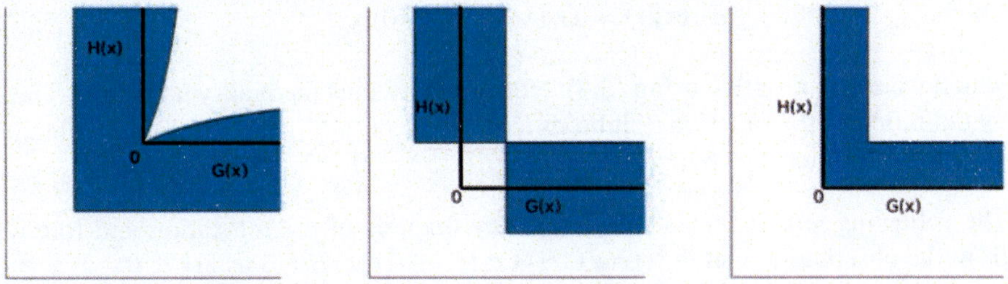

Fig. 3.1 The feasible set of the butterfly relaxation, the approximation from [17] and the relaxation from [19]

$$
\begin{aligned}
\mathcal{I}_G(x;\bar{t}) &:= \{i = 1,\dots,q \mid G_i(x) + \bar{t} = 0\},\\
\mathcal{I}_H(x;\bar{t}) &:= \{i = 1,\dots,q \mid H_i(x) + \bar{t} = 0\},\\
\mathcal{I}_{GH}(x;t) &:= \{i = 1,\dots,q \mid \Phi_i^B(G(x),H(x);t) = 0\},\\
\mathcal{I}_{GH}^{0+}(x;t) &:= \{i \in \mathcal{I}_{GH}(x;t) \mid F_{1i}(x;t) = 0,\ F_{2i}(x;t) > 0\},\\
\mathcal{I}_{GH}^{+0}(x;t) &:= \{i \in \mathcal{I}_{GH}(x;t) \mid F_{1i}(x;t) > 0,\ F_{2i}(x;t) = 0\},\\
\mathcal{I}_{GH}^{00}(x;t) &:= \{i \in \mathcal{I}_{GH}(x;t) \mid F_{1i}(x;t) = F_{2i}(x;t) = 0\}.
\end{aligned}
$$

Several relations between these sets follow directly from the definition of the relaxation. For instance, it holds that

$$\mathcal{I}_G \cap \mathcal{I}_{GH} = \mathcal{I}_H \cap \mathcal{I}_{GH} = \emptyset.$$

Additionally, by definition of the relaxation mapping, it holds

$$\Phi_i^B(G(x),H(x);t) = 0 \Longrightarrow F_{1i}(x;t) + F_{2i}(x;t) \geq 0.$$

The following two lemmas give more insights on the relaxation.

Lemma 3.3.1 *Let* $x \in \mathcal{X}_{t,\bar{t}}^B$, *then it is true for the relaxation* (3.7) *that*

(a) $\{i \in \mathcal{I}_{GH}(x;t) \mid F_{1i}(x;t) = 0,\ F_{2i}(x;t) < 0\} = \{i \in \mathcal{I}_{GH}(x;t) \mid F_{1i}(x;t) < 0,\ F_{2i}(x;t) = 0\} = \emptyset$;

(b) $i \in \mathcal{I}_{GH}(x;t) \Longrightarrow G_i(x) \geq 0,\ H_i(x) \geq 0.$

Proof Case (a) is directly considering that $\Phi_i^B(G(x),H(x);t_1,t_2) \neq 0$ for $F_{1i}(x;t) + F_{2i}(x;t) < 0$.

By symmetry of the relaxation, it is sufficient to assume that $F_{1i}(x;t) = H_i(x) - t_2\theta_{t_1}(G_i(x)) = 0$ for some $i = 1,\dots,q$. Then, by definition of $F_{2i}(x;t)$, it holds that

$$F_{2i}(x;t) = G_i(x) - t_2\theta_{t_1}(H_i(x)) = G_i(x) - t_2\theta_{t_1}(t_2\theta_{t_1}(G_i(x))),$$

so $G_i(x) \geq 0$ since in the other case, i.e., $G_i(x) < 0$, it would follow that

$$F_{2i}(x;t) = G_i(x)(1 - (\theta'(0)t_2/t_1)^2),$$

which would be negative using (3.8). Note that previous inequality holds true since, by definition of the function θ, it holds that $t_2\theta_{t_1}(z) = t_2\theta'(0)z/t_1$ for $z \leq 0$. Finally, $G_i(x) \geq 0$ implies that $H_i(x) \geq 0$ since $F_{1i}(x;t) = 0$.

The following lemma states two of the key features of the relaxation and follows from the observation that $\Phi_i^B(G(x), H(x); t) \to G_i(x)H_i(x)$ as $\|t\| \downarrow 0$.

Lemma 3.3.2 *The set $\mathcal{X}_{t,\bar{t}}^B$ satisfies the following properties:*

1. $\lim_{\|t,\bar{t}\|\to 0} \mathcal{X}_{t,\bar{t}}^B = \mathcal{Z}$ *where the limit is assumed pointwise*[3]*;*
2. $\bigcap_{t,\bar{t}\geq 0} \mathcal{X}_{t,\bar{t}}^B = \mathcal{Z}$.

If the feasible set of the (3.1) is non-empty, then the feasible sets of the relaxed sub-problems are also non-empty for all $t \geq 0$. If for some parameter $t \geq 0$ the set $\mathcal{X}_{t,\bar{t}}^B$ is empty, then it implies that $\mathcal{Z}$ is empty. Finally, if a local minimum of $R_{t,\bar{t}}^B$ already belongs to $\mathcal{Z}$, then it is a local minimum of the (3.1).

Finally, numerical results will be presented in Sect. 3.5 and we will show that these new methods are very competitive compared to existing methods.

Before moving to our main statements regarding convergence and regularity properties of the butterfly relaxation, we provide two technical lemmas. Direct computation gives the gradient of $\Phi^B(G(x), H(x); t)$ in the following lemma.

Lemma 3.3.3 *For all $i \in \{1, \ldots, q\}$, the gradient of $\Phi_i^B(G(x), H(x); t)$ w.r.t. x for the relaxation (3.7) is given by*

$$\nabla_x \Phi_i^B(G(x), H(x); t) = \begin{cases} \left(F_{1i}(x;t) - t_2\theta'_{t_1}(G_i(x))F_{2i}(x;t)\right)\nabla G_i(x) \\ +\left(F_{2i}(x;t) - t_2\theta'_{t_1}(H_i(x))F_{1i}(x;t)\right)\nabla H_i(x), \\ \text{if } F_{1i}(x;t) \geq -F_{2i}(x;t), \\ \left(t_2\theta'_{t_1}(G_i(x))F_{1i}(x;t) - F_{2i}(x;t)\right)\nabla G_i(x) \\ +\left(t_2\theta'_{t_1}(H_i(x))F_{2i}(x;t) - F_{1i}(x;t)\right)\nabla H_i(x), \\ \text{if } F_{1i}(x;t) < -F_{2i}(x;t). \end{cases}$$

The following lemma illustrates the behavior of functions θ_{t_1} and their derivatives.

Lemma 3.3.4 *Given two sequences $\{t_{1,k}\}$ and $\{t_{2,k}\}$, which converge to 0 as k goes to infinity and $\forall k \in \mathbb{N}$, $(t_{1,k}, t_{2,k}) \in \mathbb{R}^2_{++}$. It holds true that*

$$\lim_{k\to\infty} t_{2,k}\theta'_{t_{1,k}}(z) = 0 \;\forall z \in \mathbb{R}_{++}.$$

---

[3] $\lim_{k\to\infty} A^k = A$ pointwise means that for all sequences $\{x^k\}$ with $x^k \in A^k$ for all k implies $\lim_{k\to\infty} x^k \in A$ and for any $x^* \in A$ there exists a sequence x^k with $x^k \in A^k$ such that $\lim_{k\to\infty} x^k = x^*$.

Furthermore, assuming that $t_{2,k} = o(t_{1,k})$, *yields*

$$\lim_{k \to \infty} t_{2,k} \theta'_{t_{1,k}}(z) = 0 \, \forall z \in \mathbb{R}_+.$$

Proof The first part of the lemma follows from the definition of functions $\theta_{t_{1,k}}$. Indeed, it holds for all $x \in \mathbb{R}_+$ that $\theta_{t_{1,k}}(x) \in [0, 1]$. Therefore, $\lim_{k \to \infty} t_{2,k} \theta_{t_{1,k}}(x_k) = 0$.

The second part of the lemma uses the fact that functions $\theta_{t_{1,k}}$ are defined as perspective functions, that is, for all $z \in \mathbb{R}_+$

$$\theta_{t_{1,k}}(z) = \theta\left(\frac{z}{r_k}\right),$$

and so, computing the derivative gives

$$t_{2,k} \theta'_{t_{1,k}}(z) = \frac{t_{2,k}}{t_{1,k}} \theta'\left(\frac{z}{t_{1,k}}\right).$$

So, for k sufficiently large $t_{1,k} \leq z$, and by concavity of θ_r, we get

$$0 \leq \lim_{k \to \infty} t_{2,k} \theta'_{t_{1,k}}(z) \leq \lim_{k \to \infty} t_{2,k} \theta'_{t_{1,k}}(t_{1,k}) = \lim_{k \to \infty} \frac{t_{1,k}}{t_{1,k}} \theta'(1),$$

and the result follows. □

We focus on the sequel on the properties of these new relaxation schemes. We prove that the method converges to an A-stationary point in Theorem 3.4.1 and to an M-stationary point, Theorem 3.4.2, with some relation between the sequences $\{t_{2,k}\}$ and $\{t_{1,k}\}$. Furthermore, we prove in the affine case convergence to S-stationary point under MPCC-LICQ (Theorem 3.4.3).

The main motivation to consider relaxation methods for (3.1) is to solve a sequence of regular problems. Under classical assumptions, the butterfly relaxed non-linear programs satisfy the Guignard CQ, as proved in Theorem 3.4.4.

3.4 Theoretical Properties

The study of the theoretical properties of the butterfly relaxation method is split into three parts: convergence of the sequence of stationary points, the existence of Lagrange multipliers for the relaxed non-linear program, and convergence of the sequence of approximate stationary points.

3.4.1 Convergence

In this section, we focus on the convergence properties of the butterfly relaxation method and the constraint qualifications guaranteeing convergence of the sequence of stationary points generated by the method. Our aim is to compute an M-stationary point or, at least, provide a certificate if we converge to an undesirable point.

We prove in Theorem 3.4.1 that the butterfly relaxation converges to an A-stationary point. This result is improved to convergence to M-stationary points for some choices on the parameters t_2 and t_1 in Theorem 3.4.2.

Finally, we prove convergence to the S-stationary point in a specific situation in Theorem 3.4.3. To the best of our knowledge, this kind of result without the second-order condition is new from the literature and allows the butterfly relaxation to escape from spurious points.

Theorem 3.4.1 *Given two sequences $\{t_k\}$ and $\{\bar{t}_k\}$ of positive parameters satisfying (3.8) and decreasing to zero. Let $\{x^k, \eta^{g,k}, \eta^{h,k}, \eta^{G,k}, \eta^{H,k}, \eta^{\Phi,k}\}$ be a sequence of KKT points of $(R^B_{t,\bar{t}})$ with $x^k \to x^*$ such that MPCC-MFCQ holds at x^*. Then, x^* is an A-stationary point.*

Proof First, we identify the expressions of the multipliers of the complementarity constraint in Definition 3.3 through the stationary points of $(R^B_{t,\bar{t}})$. The representation of $\nabla\Phi^B$ immediately gives $\nabla\Phi^B_i(G(x^k), H(x^k); t_k) = 0,\ \forall i \in \mathcal{I}^{00}_{GH}(x^k; t_k)$. Thus, we can write

$$-\nabla f(x^k)=\sum_{i=1}^{m}\lambda_i^{g,k}\nabla g_i(x^k)+\sum_{i=1}^{p}\lambda_i^{h,k}\nabla h_i(x^k)-\sum_{i=1}^{q}\lambda_i^{G,k}\nabla G_i(x^k)-\sum_{i=1}^{q}\lambda_i^{H,k}\nabla H_i(x^k), \tag{3.9}$$

where $\lambda^{g,k} = \eta^{g,k}$, $\lambda^{h,k} = \eta^{h,k}$ and

$$\lambda_i^{G,k} = \begin{cases} \eta_i^{G,k}, & \text{if } i \in \mathcal{I}_G(x^k; \bar{t}_k), \\ \eta_i^{\Phi,k} t_{2,k}\theta'_{t_{1,k}}(G_i(x^k))F_{2i}(x^k; t_k), & \text{if } i \in \mathcal{I}^{0+}_{GH}(x^k; t_k), \\ -\eta_i^{\Phi,k} F_{1i}(x^k; t_k), & \text{if } i \in \mathcal{I}^{+0}_{GH}(x^k; t_k), \\ 0, & \text{otherwise}, \end{cases}$$

$$\lambda_i^{H,k} = \begin{cases} \eta_i^{H,k}, & \text{if } i \in \mathcal{I}_H(x^k; \bar{t}_k), \\ \eta_i^{\Phi,k} t_{2,k}\theta'_{t_{1,k}}(H_i(x^k))F_{1i}(x^k; t_k), & \text{if } i \in \mathcal{I}^{+0}_{GH}(x^k; t_k), \\ -\eta_i^{\Phi,k} F_{2i}(x^k; t_k), & \text{if } i \in \mathcal{I}^{0+}_{GH}(x^k; t_k), \\ 0, & \text{otherwise}. \end{cases}$$

First, by (3.9), it holds that $\nabla\mathcal{L}_{MPCC}(x^k, \lambda^k) = 0$ for all k. Thus, the first condition of the W-stationary conditions is satisfied. Moreover, by definition of $\{\lambda^{g,k}\}$ it holds that $\mathcal{I}_g(x^k) \subseteq \mathcal{I}_g(x^*)$ and so $\lim_{k\to\infty} \lambda_i^{g,k} = 0, \forall i \notin \mathcal{I}_g(x^*)$.

Denote $\|\lambda^k\|_\infty := \|\lambda^{g,k}, \lambda^{h,k}, \lambda^{G,k}, \lambda^{H,k}\|_\infty$. Using the definition of $\lambda^{G,k}$ and $\lambda^{H,k}$ in (3.9) and since by (3.8) it holds that $t_k\theta'_{t_{1,k}}(G_i(x^k)) \leq t_k\theta'_{t_{1,k}}(0) < 1$ for all $i \in \mathcal{I}_{GH}(x^k; t_k)$, it can be observed that[4]

$$\|\lambda^k\|_\infty = \|\eta^{g,k}, \eta^{h,k}, \eta^{G,k}, \eta^{H,k}, \eta^{\Phi,k} \circ F_2(x^k; t_k), \eta^{\Phi,k} \circ F_1(x^k; t_k))\|_\infty, \quad (3.10)$$

where $\circ$ denotes the component-wise product of two vectors.

We now verify that $\lambda_i^{G,k}/\|\lambda^k\|_\infty \to 0$ for indices $i \in \mathcal{I}^{+0}$. By symmetry, it would follow that $\lambda_i^{H,k}/\|\lambda^k\|_\infty \to 0$ for indices $i \in \mathcal{I}^{0+}$.

Let $i \in \mathcal{I}^{+0}$. Clearly, $i \in \mathcal{I}_{GH}^{0+}(x^k; t_k)$ as otherwise $G_i(x^k) = \bar{t}_k$ for $i \in \mathcal{I}_G(x^k; t_k)$ or $G_i(x^k) = t_{2,k}\theta_{t_{1,k}}(H_i(x^k))$ for $i \in \mathcal{I}_{GH}^{+0}(x^k; t_k)$ which in both cases, for k sufficiently large, contradicts the fact $G_i(x^k) \to G_i(x^*) > 0$. Now, $i \in \mathcal{I}_{GH}^{0+}(x^k; t_k)$ yields

$$\lambda_i^{G,k} = \eta_i^{\Phi,k} t_{2,k}\theta'_{t_{1,k}}(G_i(x^k))F_{2i}(x^k; t_k).$$

Moreover, $\|\lambda^k\|_\infty \geq |\eta_i^{\Phi,k} F_{2i}(x^k; t_k)|$ by (3.10), thus

$$\frac{\lambda_i^{G,k}}{\|\lambda^k\|_\infty} \leq \frac{\eta_i^{\Phi,k} t_{2,k}\theta'_{t_{1,k}}(G_i(x^k))F_{2i}(x^k; t_k)}{|\eta_i^{\Phi,k} F_{2i}(x^k; t_k)|} = t_{2,k}\theta'_{t_{1,k}}(G_i(x^k)) \to 0$$

since $G_i(x^k) \to G(x^*) > 0$ and using Lemma 3.3.4.

Now, let us prove that the sequence $\{\lambda^k\}$ is bounded. Assume by contradiction that is not bounded, then the sequence $\{\lambda^k/\|\lambda^k\|_\infty\}$ is bounded and converges, up to a subsequence to a non-trivial limit $\hat{\lambda}$. Therefore, dividing (3.9) by $\|\lambda^k\|_\infty$ and passing to the limit gives

$$\sum_{i\in\mathcal{I}_g(x^*)} \hat{\lambda^g}_i \nabla g_i(x^*) + \sum_{i=1}^{p} \hat{\lambda^h}_i \nabla h_i(x^*) - \sum_{i\in\mathcal{I}^{0+}\cup\mathcal{I}^{00}} \hat{\lambda^G}_i \nabla G_i(x^*) - \sum_{i\in\mathcal{I}^{+0}\cup\mathcal{I}^{00}} \hat{\lambda^H}_i \nabla H_i(x^*) = 0,$$

which leads to a contradiction since x^* satisfies MPCC-MFCQ.

So, the sequence $\{\lambda^k\}$ is bounded, hence $\lambda_i^{G,k} \to 0, \forall i \in \mathcal{I}^{+0}$ and $\lambda_i^{H,k} \to 0, \forall i \in \mathcal{I}^{0+}$. Therefore, x^* is a W-stationary point of the MPCC.

Finally, let us now verify that x^* is an A-stationary point. Denote λ^* the limit, up to a subsequence, of the sequence $\{\lambda^k\}$. Let $i \in \mathcal{I}^{00}$. Assume without loss of generality that $\lambda_i^{G,*} < 0$ (the other case will be similar by symmetry) and we show that $\lambda_i^{H,*} \geq 0$. $\lambda_i^{G,*} < 0$ implies that $i \in \mathcal{I}_{GH}^{+0}(x^k; t_k)$ for k sufficiently large by definition of $\lambda_i^{G,k}$. So, $\lambda_i^{H,k} = \eta_i^{\Phi,k} t_{2,k}\theta'_{t_{1,k}}(H_i(x^k))F_{1i}(x^k; t_k)$, which is non-negative. So x^* is an A-stationary point. □

[4] For indices $i \in \mathcal{I}_{GH}^{0+}(x^k; t_k)$ (symmetry for indices $i \in \mathcal{I}_{GH}^{+0}(x^k; t_k)$), then $\lambda_i^{G,k} = \eta_i^{\Phi,k} t_{2,k}\theta'_{t_{1,k}}(G_i(x^k))F_{2i}(x^k; t_k)$ and $\lambda_i^{H,k} = \eta_i^{\Phi,k} F_{2i}(x^k; t_k)$. Therefore, considering that $t_k\theta'_{t_{1,k}}(G_i(x^k)) < 1$, we get $\lambda_i^{G,k} < \lambda_i^{H,k}$. All in all the infinite norm is not obtained at these components.

The following example shows that the result of Theorem 3.4.1 is sharp since convergence cannot be ensured, assuming only that MPCC-GMFCQ holds at the limit point.

Example 3.3 Consider the following two-dimensional example:

$$\min_{x \in \mathbb{R}^2} x_2 \text{ s.t. } 0 \le x_1 + x_2^2 \perp x_1 \ge 0.$$

MPCC-GMFCQ holds at $(0,0)^T$. The point $(0,0)^T$ is even not a W-stationary point. In this case, there exists a sequence of stationary points of the relaxation such that $\{x^k\}$ converges to the origin. Given a sequence $\{x^k\}$, with $\{1\} \in \mathcal{I}_{GH}(x^k; t_k)$, such that $x^k \to (0,0)^T$ then $\eta^{G,k} = \eta^{H,k} = 0$ and we can choose $\eta^{\Phi,k}$ that satisfies

$$\lambda^{G,k} = -\lambda^{H,k} = \frac{1}{2x_2^k}.$$

The sequence $\{x^k\}$ converges to an undesirable point.

The result of Theorem 3.4.1 can be tightened if we consider a particular choice of parameter. It is an essential result since it shows that a subfamily of the butterfly relaxation has the desired property to converge to an M-stationary point.

Theorem 3.4.2 *Given two sequences $\{t_k\}$ and $\{\bar{t}_k\}$ of positive parameters satisfying (3.8) and decreasing to zero. Let $\{x^k, \eta^{g,k}, \eta^{h,k}, \eta^{G,k}, \eta^{H,k}, \eta^{\Phi,k}\}$ be a sequence of KKT points of $(R^B_{t,\bar{t}})$ with $x^k \to x^*$ such that MPCC-GMFCQ holds at x^*. If $t_{2,k} = o(t_{1,k})$, then, x^* is an M-stationary point.*

Proof In the proof of Theorem 3.4.1, we already showed that $\nabla \mathcal{L}_{MPCC}(x^k, \lambda^k) = 0$ for all k, $\lim_{k\to\infty} \lambda_i^{g,k} = 0, \forall i \notin \mathcal{I}_g(x^*)$, $\lim_{k\to\infty} \lambda_i^{G,k}/\|\lambda^k\|_\infty = 0, \forall i \in \mathcal{I}^{+0}$, and $\lim_{k\to\infty} \lambda_i^{H,k}/\|\lambda^k\|_\infty = 0, \forall i \in \mathcal{I}^{0+}$.

Let us now check that either $\lim_{k\to\infty} \lambda_i^{G,k}\lambda_i^{H,k}/\|\lambda^k\|_\infty^2 = 0$ or $\lim_{k\to\infty} \lambda_i^{G,k}/\|\lambda^k\|_\infty > 0$, $\lim_{k\to\infty} \lambda_i^{H,k}/\|\lambda^k\|_\infty > 0$ using the contrapositive, i.e.,

$$\lim_{k\to\infty} \lambda_i^{G,k}/\|\lambda^k\|_\infty < 0 \Longrightarrow \lim_{k\to\infty} \lambda_i^{H,k}/\|\lambda^k\|_\infty = 0,$$

and the other case will be similar by symmetry.

Let $i \in \mathcal{I}^{00}$. $\lim_{k\to\infty} \lambda_i^{G,k}/\|\lambda^k\|_\infty < 0$ implies that $i \in \mathcal{I}_{GH}^{+0}(x^k; t_k)$ for k sufficiently large by definition of $\lambda_i^{G,k}$ as the function θ is non-decreasing. So, $\lambda_i^{H,k} = \eta_i^{\Phi,k} t_{2,k}\theta'_{t_{1,k}}(H_i(x^k))F_{1i}(x^k; t_k)$. Moreover, $\lim_{k\to\infty} t_{2,k}\theta'_{t_{1,k}}(H_i(x^k)) = 0$ by Lemma 3.3.4 with $t_{2,k} = o(t_{1,k})$ and

$$\lim_{k\to\infty} \eta_i^{\Phi,k} F_{1i}(x^k; t_k)/\|\lambda^k\|_\infty = \lim_{k\to\infty} -\lambda_i^{G,k}/\|\lambda^k\|_\infty < \infty.$$

Thus, $\lim_{k\to\infty} \lambda_i^{H,k}/\|\lambda^k\|_\infty = 0$.

Finally, following the same reasoning as in the proof of Theorem 3.4.1, using MPCC-GMFCQ, the sequence $\{\lambda^k\}$ is bounded, and x^* is an M-stationary point. □

The following example shows that this result is sharp, since it illustrates a situation where MPCC-GMFCQ does not hold and the method converges to an undesirable W-stationary point. This phenomenon only happens if the sequence of multipliers defined in (3.9) is unbounded.

Example 3.4 Consider the problem

$$\min_{x\in\mathbb{R}^2} x_2^2 \text{ s.t. } 0 \leq x_1^2 \perp x_1 + x_2^2 \geq 0.$$

The feasible set is $\mathcal{Z} = \{(x_1, x_2)^T \in \mathbb{R}^2 \mid x_1 = 0\} \cup \{(x_1, x_2)^T \in \mathbb{R}^2 \mid x_1 = -x_2^2\}$. $(0, 0)^T$ is the unique M-stationary, with $(\lambda^G, \lambda^H = 0)$.

MPCC-GMFCQ fails to hold at any point $(0, a \in \mathbb{R})^T$ since the gradient of x_1^2 is non-zero only for $x \neq 0$.

Consider a sequence such that for $(t_{1,k}, t_{2,k})$ sufficiently small $F_2(x^k; t_k) = 0$ and

$$x_1^k = t_{2,k}\theta'_{t_{1,k}}(x_1^k + a^2),\ x_2^k = a,\ \eta^{\Phi,k} F_{1i}(x^k; t_k) = \frac{1}{-t_{2,k}\theta'_{t_{1,k}}(x_1^k + a^2)}.$$

Obviously, the sequence x^k goes to $x^* = (0, a \neq 0)^T$, which is not a W-stationary point. Indeed, we have

$$\lambda^{G,k} = \frac{1}{t_{2,k}\theta'_{t_{1,k}}(x_1^k + a^2)} \to \infty \text{ and } \lambda^{H,k} = -1 \neq 0.$$

The following result motivated by Example 3.1 shows that the butterfly relaxation may improve its behavior in some specific cases. Example 3.1 also indicates that this cannot be expected with the other relaxations. In the sequel, we denote $\text{supp}(z) := \{i \mid z_i \neq 0\}$ the non-zero indices of z.

Theorem 3.4.3 *Assume that f, g, h, G, H are affine functions. Given two sequences $\{t_k\}$ and $\{\bar{t}_k\}$ of positive parameters satisfying (3.8) and decreasing to zero as k goes to infinity. Let $\{x^k, \eta^{g,k}, \eta^{h,k}, \eta^{G,k}, \eta^{H,k}, \eta^{\Phi,k}\}$ be a sequence of KKT points of $(R^B_{t,\bar{t}})$ with $x^k \to x^*$ such that MPCC-LICQ holds at x^*. If $t_{2,k} = o(t_{1,k})$, and, for all k sufficiently large*

$$\text{supp}(\eta^{\Phi,k}) \cap \left(\mathcal{I}^{+0} \cup \mathcal{I}^{0+}\right) = \emptyset, \tag{3.11}$$

then, x^* is an S-stationary point.

Proof Theorem 3.4.2 already proves that x^* is an M-stationary point. Assume by contradiction that x^* is not an S-stationary point. Then, it holds that this point cannot be a stationary point of $(R^B_{t,\bar{t}})$.

We already mention in the proof of Theorem 3.4.1 that for all k it holds

$$-\nabla f = \sum_{i=1}^{m} \lambda_i^{g,k} \nabla g_i + \sum_{i=1}^{p} \lambda_i^{h,k} \nabla h_i - \sum_{i=1}^{q} \lambda_i^{G,k} \nabla G_i - \sum_{i=1}^{q} \lambda_i^{H,k} \nabla H_i,$$

where we omit the dependence in k in the expression of the gradients, since they are constant by linear/affine assumption. Clearly, for k sufficiently large, it holds that $\operatorname{supp}(\lambda^{g,k}) \subseteq \mathcal{I}_g(x^*)$, $\operatorname{supp}(\lambda^{G,k}) \subset \mathcal{I}^{0+} \cup \mathcal{I}^{00}$, and $\operatorname{supp}(\lambda^{H,k}) \subseteq \mathcal{I}^{00} \cup \mathcal{I}^{+0}$ by (3.11).

Now, by continuity, linear independence of these gradients holds in a neighborhood of x^*. So, we get finite convergence of the λ^k, and for k sufficiently large it holds

$$\lambda^{g,k} = \lambda^{g,\infty},\ \lambda^{h,k} = \lambda^{h,\infty},\ \lambda^{G,k} = \lambda^{G,\infty},\ \lambda^{H,k} = \lambda^{H,\infty}. \tag{3.12}$$

Let $i \in \mathcal{I}^{00} \cap \operatorname{supp}(\eta^{\Phi,\infty})$, where we remind that $\operatorname{supp}(\eta^{\Phi,\infty}) \subseteq \operatorname{supp}(\eta^{\Phi,k}) \subseteq \mathcal{I}_{GH}(x^k; t_k)$. If no such index exists, then for all k sufficiently large $\eta^{\Phi,k}$ is zero and x^* is S-stationary.

By stationarity assumption on x^*, we assume that $\lambda_i^{G,\infty} < 0$ (the case $\lambda_i^{H,\infty}$ will be symmetrical). It implies that $i \in \mathcal{I}_{GH}^{+0}$ by definition of the multipliers in (3.9) and so

$$\lambda_i^{G,k} = -\eta_i^{\Phi,k} F_{1i}(x^k; t_k), \text{ and, } \lambda_i^{H,k} = \eta_i^{\Phi,k} t_{2,k} \theta'_{t_{1,k}}(H_i(x^k)) F_{1i}(x^k; t_k).$$

We proved in Theorem 3.4.2 that $\lambda^{G,k}$ and $\lambda^{H,k}$ have bounded limits, so by Lemma 3.3.4 with $t_{2,k} = o(t_{1,k})$ we have $\lim_{k\to\infty} \lambda_i^{H,k} = 0$. By (3.12), we get $\eta_i^{\Phi,k} = 0$ for all k sufficiently large, which contradicts $\lambda_i^{G,\infty} < 0$. □

3.4.2 *Existence of Lagrange Multipliers for the Relaxed Sub-Problems*

In this section, we study some regularity properties of the relaxed non-linear programs. Indeed, to guarantee the existence of a sequence of stationary points, the relaxed non-linear programs must satisfy some constraint qualifications in the neighborhood of the limit point.

Theorem 3.4.4 *Let $x^* \in \mathcal{Z}$, satisfying MPCC-LICQ. Then, there exists $t^* > 0$ and a neighborhood $U(x^*)$ of x^* such that*

$$\forall t \in (0, t^*] : \ x \in U(x^*) \cap \mathcal{X}_{t,\bar{t}}^{B} \Longrightarrow \text{standard GCQ holds at } x \text{ for } (R_{t,\bar{t}}^{B}).$$

Proof Let $x \in U(x^*) \cap \mathcal{X}_{t,\bar{t}}^{B}$. We know that $\mathscr{L}_{\mathcal{X}_{t,\bar{t}}^{B}}(x)^\circ \subseteq \mathscr{T}_{\mathcal{X}_{t,\bar{t}}^{B}}(x)^\circ$. So, it is sufficient to show the converse inclusion.

The linearized cone of $(R_{t,\bar{t}}^{B})$ is given by

$$\begin{aligned}\mathscr{L}_{\mathcal{X}^B_{t,\bar{t}}}(x) = \{d \in \mathbb{R}^n \mid \nabla g_i(x)^T d \le 0, \forall i \in \mathcal{I}_g(x),\ \nabla h_i(x)^T d = 0, \forall i = 1, \dots, p,\\ \nabla G_i(x)^T d \ge 0, \forall i \in \mathcal{I}_G(x; \bar{t}),\ \nabla H_i(x)^T d \ge 0, \forall i \in \mathcal{I}_H(x; \bar{t}),\\ \nabla \Phi^B_i(G(x), H(x); t)^T d \le 0, \forall i \in \mathcal{I}^{0+}_{GH}(x; t) \cup \mathcal{I}^{+0}_{GH}(x; t)\Big\},\end{aligned}$$

using that $\nabla\Phi^B_i(G(x), H(x); t) = 0$ for all $i \in \mathcal{I}^{00}_{GH}(x, t)$.

Let us compute the polar of the tangent cone. Consider the following set of non-linear constraints parametrized by $z \in \mathcal{X}^B_{t,\bar{t}}$ and a partition (I, I^c, I^-) of $\mathcal{I}^{00}_{GH}(z; t)$,[5] defined by

$$\begin{aligned}\mathrm{S}_{(I,I^c,I^-)}(z) := \{x \in \mathbb{R}^n \mid g(x) \le 0, h(x) = 0, G(x) \ge -\bar{t}e,\ H(x) \ge -\bar{t}e,\\ \Phi^B_i(G(x), H(x); t) \le 0,\ i \notin \mathcal{I}^{00}_{GH}(z; t),\\ F_{1i}(x; t) \le 0,\ F_{2i}(x; t) \ge 0,\ i \in I,\\ F_{1i}(x; t) \ge 0,\ F_{2i}(x; t) \le 0,\ i \in I^c,\\ F_{1i}(x; t) \le 0,\ F_{2i}(x; t) \le 0,\ i \in I^-\}.\end{aligned} \tag{3.13}$$

Since $z \in \mathcal{X}^B_{t,\bar{t}}$, it is obvious that $z \in \mathrm{S}_{(I,I^c,I^-)}(z)$.

By construction of $U(x^*)$ and t^*, the gradients $\{\nabla g_i(x^*)\ \ (i \in \mathcal{I}_g(x^*)),$ $\nabla h_i(x^*)\ \ (i = 1, \dots, m), \nabla G_i(x^*)\ \ (i \in \mathcal{I}^{00} \cup \mathcal{I}^{0+}), \nabla H_i(x^*)\ \ (i \in \mathcal{I}^{+0} \cup \mathcal{I}^{00})\}$ remain linearly independent for all $x \in U(x^*)$ by continuity of the gradients, and we have

$$\begin{aligned}&\mathcal{I}_g(x) \subseteq \mathcal{I}_g(x^*),\ \mathcal{I}_G(x; \bar{t}) \subseteq \mathcal{I}^{00} \cup \mathcal{I}^{0+},\ \mathcal{I}_H(x; \bar{t}) \subseteq \mathcal{I}^{+0} \cup \mathcal{I}^{00},\\ &\mathcal{I}^{00}_{GH}(x; t) \cup \mathcal{I}^{+0}_{GH}(x; t) \subseteq \mathcal{I}^{00} \cup \mathcal{I}^{0+},\\ &\mathcal{I}^{00}_{GH}(x; t) \cup \mathcal{I}^{0+}_{GH}(x; t) \subseteq \mathcal{I}^{+0} \cup \mathcal{I}^{00}.\end{aligned} \tag{3.14}$$

Therefore, by Lemma 3.7.6, LICQ holds for (3.13) at x. Furthermore, by [32, Lemma 8.10], and since LICQ in particular implies Abadie CQ it follows that

$$\mathscr{T}_{\mathcal{X}^B_{t,\bar{t}}}(x) = \bigcup_{\forall (I,I^c,I^-)} \mathscr{T}_{\mathrm{S}_{(I,I^c,I^-)}(x)}(x) = \bigcup_{\forall (I,I^c,I^-)} \mathscr{L}_{\mathrm{S}_{(I,I^c,I^-)}(x)}(x).$$

By [5, Theorem 3.1.9], passing to the polar, we get

$$\mathscr{T}_{\mathcal{X}^B_{t,\bar{t}}}(x)^\circ = \bigcap_{\forall (I,I^c,I^-)} \mathscr{L}_{\mathrm{S}_{(I,I^c,I^-)}(x)}(x)^\circ.$$

By [5, Theorem 3.2.2], we know that

[5] (I, I^c, I^-) is a partition of $\mathcal{I}^{00}_{GH}(z; t)$ means that $I \cup I^c \cup I^- = \mathcal{I}^{00}_{GH}(z; t)$ and $I \cap I^c = I \cap I^- = I^c \cap I^- = \emptyset$.

$$\mathcal{L}_{\mathrm{S}_{(I,I^c,I^-)}(x)}(x)^\circ = \{ \sum_{i\in\mathcal{I}_g(x)} \eta_i^g \nabla g_i(x) + \sum_{i=1}^{p} \eta_i^h \nabla h_i(x)$$
$$-\sum_{i\in\mathcal{I}_G(x;\bar{t})} \eta_i^G \nabla G_i(x) - \sum_{i\in\mathcal{I}_H(x;\bar{t})} \eta_i^H \nabla H_i(x)$$
$$+ \sum_{i\in\mathcal{I}_{GH}^{+0}(x;t)\cup\mathcal{I}_{GH}^{0+}(x;t)} \eta_i^\Phi \nabla\Phi_i^B(G(x), H(x); t)$$
$$-\sum_{i\in I} \eta_i^G \nabla G_i(x) + \sum_{i\in I^c} \eta_i^G \nabla G_i(x)$$
$$+\sum_{i\in I} \eta_i^H \nabla H_i(x) - \sum_{i\in I^c} \eta_i^H \nabla H_i(x)$$
$$+\sum_{i\in I^-} \delta_i^G \nabla G_i(x) + \sum_{i\in I^-} \delta_i^H \nabla H_i(x) : (\eta^g, \eta^G, \eta^H, \eta^\Phi) \geq 0\}.$$

For $v \in \mathcal{T}_{X^B_{t,\bar{t}}}(x)^\circ$, we have $v \in \mathcal{L}_{\mathrm{S}_{(I,I^c,I^-)}(x)}(x)^\circ$ for any partition (I, I^c, I^-) of $\mathcal{I}_{GH}^{00}(x;t)$. If we fix I and set $I^- = \emptyset$, then there exists some multipliers η^h and $\eta^g, \eta^G, \eta^H, \eta^\Phi \geq 0$ so that

$$v = \sum_{i\in\mathcal{I}_g(x)} \eta_i^g \nabla g_i(x) + \sum_{i=1}^{p} \eta_i^h \nabla h_i(x) - \sum_{i\in\mathcal{I}_G(x;\bar{t})} \eta_i^G \nabla G_i(x) - \sum_{i\in\mathcal{I}_H(x;\bar{t})} \eta_i^H \nabla H_i(x)$$
$$+ \sum_{i\in\mathcal{I}_{GH}^{+0}(x;t)\cup\mathcal{I}_{GH}^{0+}(x;t)} \eta_i^\Phi \nabla\Phi_i^B(G(x), H(x); t)$$
$$-\sum_{i\in I} \eta_i^G \nabla G_i(x) + \sum_{i\in I^c} \eta_i^G \nabla G_i(x) - \sum_{i\in I} \eta_i^H \nabla H_i(x) + \sum_{i\in I^c} \eta_i^H \nabla H_i(x).$$

Now, it also holds that $v \in \mathcal{L}_{\mathrm{S}_{(I^c,I,I^-)}(x)}(x)^\circ$ and so there exists some multipliers η^h and $\eta^g, \eta^G, \eta^H, \eta^\Phi \geq 0$ such that

$$v = \sum_{i\in\mathcal{I}_g(x)} \eta_i^g \nabla g_i(x) + \sum_{i=1}^{p} \eta_i^h \nabla h_i(x) - \sum_{i\in\mathcal{I}_G(x;\bar{t})} \eta_i^G \nabla G_i(x) - \sum_{i\in\mathcal{I}_H(x;\bar{t})} \eta_i^H \nabla H_i(x)$$
$$+ \sum_{i\in\mathcal{I}_{GH}^{+0}(x;t)\cup\mathcal{I}_{GH}^{0+}(x;t)} \eta_i^\Phi \nabla\Phi_i^B(G(x), H(x); t)$$
$$+\sum_{i\in I} \eta_i^G \nabla G_i(x) - \sum_{i\in I^c} \eta_i^G \nabla G_i(x) + \sum_{i\in I} \eta_i^H \nabla H_i(x) - \sum_{i\in I^c} \eta_i^H \nabla H_i(x).$$

By the construction of t^* and $U(x^*)$, the gradients involved here are linearly independent and so the multipliers in both previous equations must be equal. Thus, the multipliers η_i^G and η_i^H with indices i in $I \cup I^c$ vanish.

Therefore, $v \in \mathscr{L}_{\mathcal{X}^B_{t,\bar{t}}}(x)^\circ$ and as v has been chosen arbitrarily then $\mathscr{T}_{\mathcal{X}^B_{t,\bar{t}}}(x)^\circ \subseteq \mathscr{L}_{\mathcal{X}^B_{t,\bar{t}}}(x)^\circ$, which concludes the proof. □

This result is sharp, as shown by the following example since Abadie CQ does not hold.

Example 3.5 Consider the problem

$$\min_{x \in \mathbb{R}^2} f(x) \text{ s.t. } 0 \leq x_1 \perp x_2 \geq 0.$$

At $x^* = (0, 0)^T$ it holds that $\nabla \Phi^B(G(x), H(x); t) = (0, 0)^T$ and so $\mathscr{L}_{\mathcal{X}^B_{t,\bar{t}}}(x^*) = \mathbb{R}^2$, which is obviously different from the tangent cone at x^* for $t_2\theta'(0) < t_1$ and $\bar{t} > 0$.

The following example shows that we cannot have a similar result using MPCC-GMFCQ.

Example 3.6 Consider the set

$$C := \{(x_1, x_2)^T \mid 0 \leq x_1 + x_2^2 \perp x_1 \geq 0\}.$$

MPCC-GMFCQ holds at $x^* = (0, 0)^T$, since the gradients are linearly dependent but only with coefficients $\lambda^G = -\lambda^H$ that does not satisfy the condition given in Definition 3.4.

Now, we can choose a sequence of points such that $x^k \to x^*$ and

$$F_2(x^k; t_k) = 0, -t_{2,k}\theta'_{t_{1,k}}(H(x^k)) \to -1.$$

Since $\nabla G(x^*) = \nabla H(x^*)$, it holds that $\nabla F_2(x^*; 0) = (0\ 0)^T$ and so MFCQ does not hold for (3.13).

It is disappointing to require MPCC-LICQ to obtain the only GCQ, but when $\mathcal{I}^{00}_{GH}$ is empty, we get the stronger LICQ.

Theorem 3.4.5 *Let $x^* \in \mathcal{Z}$, satisfying MPCC-LICQ. Then, there exists $t^* > 0$ and a neighborhood $U(x^*)$ of x^* such that*

$$\forall t \in (0, t^*] : x \in U(x^*) \cap \mathcal{X}^B_{t,\bar{t}} \text{ and } \mathcal{I}^{00}_{GH}(x; t) = \emptyset \Longrightarrow \text{LICQ holds at } x \text{ for } (R^B_{t,\bar{t}}).$$

Proof Let $x \in U(x^*) \cap \mathcal{X}^B_{t,\bar{t}}$ and t sufficiently small. We prove that the gradients of the constraints involved in $(R^B_{t,\bar{t}})$ are linearly independent, by verifying that the trivial solution is the only solution to the following equation:

$$0 = \sum_{i \in \mathcal{I}_g(x)} \eta_i^g \nabla g_i(x) + \sum_{i=1}^{p} \eta_i^h \nabla h_i(x) + \sum_{i \in \mathcal{I}_G(x;\bar{t})} \nabla G_i(x)\eta_i^G + \sum_{i \in \mathcal{I}_H(x;\bar{t})} \nabla H_i(x)\eta_i^H$$
$$+ \sum_{i \in \mathcal{I}_{GH}^{+0}(x;t)} \nabla G_i(x) \left(\eta_i^{\Phi}(F_{1i}(x;t) - F_{2i}(x;t)t_2\theta'_{t_1}(G_i(x)))\right)$$
$$+ \sum_{i \in \mathcal{I}_{GH}^{0+}(x;t)} \nabla H_i(x) \left(\eta_i^{\Phi}(F_{2i}(x;t) - F_{1i}(x;t)t_2\theta'_{t_1}(H_i(x)))\right).$$

MPCC-LICQ and the inclusions (3.14) give that the only solution is the trivial one. □

3.4.3 Convergence of the Epsilon-Stationary Points

Non-linear programming algorithms usually compute sequences of approximate stationary points or ϵ-stationary points (see Definition 3.2). We present below in relations (3.16)–(3.21) our specific definition and hypothesis of ϵ-stationary points. This approach, which has become an active subject recently, can significantly alter the convergence analysis of relaxation methods, as shown in [17, 20, 21, 29].

Previous results in [21] prove convergence to C-stationary point for the relaxation from Scheel and Scholtes [31] and the one from Lin and Fukushima [23], under some hypotheses on the sequence ϵ_k, respectively $\epsilon_k = O(t_k)$ and $\epsilon_k = o(t_k^2)$. Furthermore, the authors in [21] also provide a counter-example with a sequence converging to a W-stationary point if these conditions do not hold. Additionally, the authors in [21], prove that relaxations (3.3) and (3.4) converge only to a W-stationary point, and they require more hypotheses on the sequences ϵ_k and x_k to prove the convergence to a C- or an M-stationary point.

In the same way as in Theorem 3.4.1, we consider through this section a sequence of multipliers that should verify the stationary conditions. We denote for all $i \in \{1, \ldots, q\}$

$$\lambda_i^{G,k} := \begin{cases} \eta_i^{G,k} + \eta_i^{\Phi,k}\left(t_{2,k}\theta'_{t_{1,k}}(G_i(x^k))F_{2i}(x^k;t_k) - F_{1i}(x^k;t_k)\right), \\ \text{if } F_{1i}(x^k;t_k) \geq -F_{2i}(x^k;t_k) \\ \eta_i^{G,k} + \eta_i^{\Phi,k}\left(F_{2i}(x^k;t_k) - t_{2,k}\theta'_{t_{1,k}}(G_i(x^k))F_{1i}(x^k;t_k)\right), \\ \text{if } F_{1i}(x^k;t_k) < -F_{2i}(x^k;t_k), \end{cases}$$
$$\lambda_i^{H,k} := \begin{cases} \eta_i^{H,k} + \eta_i^{\Phi,k}\left(t_{2,k}\theta'_{t_{1,k}}(H_i(x^k))F_{1i}(x^k;t_k) - F_{2i}(x^k;t_k)\right), \\ \text{if } F_{1i}(x^k;t_k) \geq -F_{2i}(x^k;t_k) \\ \eta_i^{H,k} + \eta_i^{\Phi,k}\left(F_{1i}(x^k;t_k) - t_{2,k}\theta'_{t_{1,k}}(H_i(x^k))F_{2i}(x^k;t_k)\right), \\ \text{if } F_{1i}(x^k;t_k) < -F_{2i}(x^k;t_k). \end{cases} \tag{3.15}$$

The representation of $\nabla\Phi_i^B(G(x^k), H(x^k); t_k)$ immediately gives for all $i \in \mathcal{I}_{GH}^{00}(x^k; t_k)$ and all k that $\nabla\Phi_i^B(G(x^k), H(x^k); t_k) = 0$. Thus, x^k being a ϵ_k-stationary point for $(R_{t,\bar{t}}^B)$ satisfies

$$\left\|\mathcal{L}_{MPCC}(x^k, \lambda^k)\right\|_\infty \leq \epsilon_k,$$

with $(\lambda^{g,k}, \lambda^{h,k}) = (\eta^{g,k}, \eta^{h,k})$ and $\lambda^{G,k}, \lambda^{H,k}$ defined in (3.15), and

$$|h_i(x^k)| \leq \epsilon_k, \forall i = 1, \ldots, p, \tag{3.16}$$

$$g_i(x^k) \leq \epsilon_k,\ \eta_i^{g,k} \geq 0,\ \left|\eta_i^{g,k} g_i(x^k)\right| \leq \epsilon_k, \forall i = 1, \ldots, m, \tag{3.17}$$

$$G_i(x^k) + \bar{t}_k \geq -\epsilon_k,\ \eta_i^{G,k} \geq 0,\ \left|\eta_i^{G,k}(G_i(x^k) + \bar{t}_k)\right| \leq \epsilon_k, \forall i=1, \ldots, q, \tag{3.18}$$

$$H_i(x^k) + \bar{t}_k \geq -\epsilon_k,\ \eta_i^{H,k} \geq 0,\ \left|\eta_i^{H,k}(H_i(x^k) + \bar{t}_k)\right| \leq \epsilon_k, \forall i=1, \ldots, q, \tag{3.19}$$

$$\Phi_i^B(G(x^k), H(x^k); t_k) \leq \epsilon_k, \eta_i^{\Phi,k} \geq 0, \tag{3.20}$$

$$\left|\eta_i^{\Phi,k}\Phi_i^B(G(x^k), H(x^k); t_k)\right| \leq \epsilon_k, \forall i = 1, \ldots, q. \tag{3.21}$$

In order to prove our main convergence theorem, we first prove a technical lemma.

Lemma 3.4.5 *Consider the same assumptions as Theorem 3.4.6 below. Additionally, assume that for $i \in \mathcal{I}^{+0} \cup \mathcal{I}^{00}$, $\lim_{k\to\infty} \eta_i^{G,k} = \lim_{k\to\infty} \eta_i^{\Phi,k} F_{1i}(x^k; t_k) = 0$ and for k sufficiently large $F_{1i}(x^k; t_k) \geq -F_{2i}(x^k; t_k)$. Then,*

$$\lim_{k\to\infty} \frac{|\eta_i^{\Phi,k} t_{2,k}\theta'_{t_{1,k}}(G_i(x^k))F_{2i}(x^k; t_k)|}{\|\lambda^k\|_\infty} = 0.$$

As a consequence, $\lim_{k\to\infty} \frac{\lambda_i^{G,k}}{\|\lambda^k\|} = 0.$

Proof Without loss of generality, let us assume that $\lim_{k\to\infty} \eta_i^{\Phi,k} F_{2i}(x^k; t_k) \neq 0$, otherwise we are done. Since $\|\lambda^k\|_\infty \geq |\lambda_i^{H,k}|$, by (3.15) and $\lim_{k\to\infty} \eta_i^{\Phi,k} F_{1i}(x^k; t_k) = 0$, it is sufficient to show that

$$\lim_{k\to\infty} \frac{|\eta_i^{\Phi,k} t_{2,k}\theta'_{t_{1,k}}(G_i(x^k))F_{2i}(x^k; t_k)|}{|\eta_i^{H,k} - \eta_i^{\Phi,k} F_{2i}(x^k; t_k)|} = 0. \tag{3.22}$$

We now consider two cases: either $\lim_{k\to\infty} \eta_i^{H,k} = 0$ or $\lim_{k\to\infty} \eta_i^{H,k} \neq 0$.

- If $\lim_{k\to\infty} \eta_i^{H,k} = 0$. Then, the left-hand side in (3.22) is equal to $\lim_{k\to\infty} t_{2,k}\theta'_{t_{1,k}}(G_i(x^k))$, which goes to zero by Lemma 3.3.4 as $t_{2,k} = o(t_{1,k})$.

- Consider the case, $\lim_{k\to\infty} \eta_i^{H,k} \neq 0$. Dividing by $\bar{t}_k$ in the complementarity condition in (3.19) implies $H_i(x^k) \sim -\bar{t}_k$ as $\epsilon_k = o(\bar{t}_k)$. Thus, $H_i(x^k) < 0$ for k sufficiently large.
  We prove that $\lim_{k\to\infty} \eta_i^{\Phi,k} F_{2i}(x^k; t_k) = 0$. Dividing by $H_i(x^k)$ in the complementarity condition in (3.20) gives for $H_i(x^k) \sim -\bar{t}_k$ that

$$\left| \eta_i^{\Phi,k} F_{2i}(x^k; t_k) \left(1 - \frac{t_{2,k}\theta_{t_{1,k}}(G_i(x^k))}{H_i(x^k)}\right) \right| \leq \frac{\epsilon_k}{|H_i(x^k)|} \to 0, \tag{3.23}$$

as $\epsilon_k = o(\bar{t}_k)$. However, $\lim_{k\to\infty} \frac{t_{2,k}\theta_{t_{1,k}}(G_i(x^k))}{H_i(x^k)} \neq 1$, otherwise $G_i(x^k) \leq 0$ and $|G_i(x^k)| \geq |H_i(x^k)|$ would yield $\lim_{k\to\infty} \frac{t_{2,k}\theta_{t_{1,k}}(G_i(x^k))}{H_i(x^k)} \leq \lim_{k\to\infty} \frac{t_{2,k}\theta_{t_{1,k}}(H_i(x^k))}{H_i(x^k)} = 0$ as θ is non-decreasing and $t_{2,k} = o(t_{1,k})$. Therefore, (3.22) follows as (3.23) implies $\lim_{k\to\infty} \eta_i^{\Phi,k} F_{2i}(x^k; t_k) = 0$. □

The following result proves convergence of the butterfly relaxation in this context.

Theorem 3.4.6 *Given the three sequences $\{t_k\}, \{\bar{t}_k\}, \{\epsilon_k\}$ decreasing to zero and satisfying (3.8). Assume that $\epsilon_k = o(\max(|G(x^k)|, |H(x^k)|))$, $\epsilon_k = o(\bar{t}_k)$, and $t_{2,k} = o(t_{1,k})$. Let $\{x^k, \eta^{g,k}, \eta^{h,k}, \eta^{G,k}, \eta^{H,k}, \eta^{\Phi,k}\}$ be a sequence of ϵ_k-KKT points of $(R^B_{t,\bar{t}})$ with $x^k \to x^*$ such that MPCC-GMFCQ holds at x^*. Then, x^* is an M-stationary point.*

The notation $\epsilon_k = o(\max(|G(x^k)|, |H(x^k)|))$ means here that for all $i = 1, \ldots, q$, $\epsilon_k = o(\max(|G_i(x^k)|, |H_i(x^k)|))$. For two sequences $\{g_k\}, \{h_k\}$ with the same signs for k sufficiently large, we also denote $g_k \sim h_k$ whenever $\lim_{k\to\infty} g_k/h_k = 1$.

Proof Proceeding in the same way as Theorem 3.4.2, we verify that

(i) $x^* \in \mathcal{Z}$, $\lim_{k\to\infty} \nabla\mathcal{L}_{MPCC}(x^k, \lambda^k) = 0$, $\lim_{k\to\infty} \lambda_i^{g,k} = 0, \forall i \notin \mathcal{I}_g(x^*)$,
(ii) $\lim_{k\to\infty} \lambda_i^{G,k}/\|\lambda^k\|_\infty = 0, \forall i \in \mathcal{I}^{+0}$, $\lim_{k\to\infty} \lambda_i^{H,k}/\|\lambda^k\|_\infty = 0, \forall i \in \mathcal{I}^{0+}$,
(iii) $\lim_{k\to\infty} \lambda_i^{G,k}\lambda_i^{H,k}/\|\lambda^k\|_\infty^2 = 0$ or $\lim_{k\to\infty} \lambda_i^{G,k}/\|\lambda^k\|_\infty > 0$, $\lim_{k\to\infty} \lambda_i^{H,k}/\|\lambda^k\|_\infty > 0, \forall i \in \mathcal{I}^{00}$.

Clearly (i) follows from the stationarity of x^k as $\epsilon_k \downarrow 0$.

Let us now show that for indices $i \in \mathcal{I}^{+0}$, $\lim_{k\to\infty} \lambda_i^{G,k}/\|\lambda^k\|_\infty = 0$. The opposite case for indices $i \in \mathcal{I}^{0+}$ would follow in a completely similar way. So, let i be in $\mathcal{I}^{+0}$.

The complementarity condition in (3.18) gives that $\lim_{k\to\infty} \eta_i^{G,k} = 0$, since $\epsilon_k \downarrow 0$ and $G_i(x^k) + \bar{t}_k \to G_i(x^*) > 0$.

Note that we are necessarily in the case $F_{1i}(x^k;t_k)+F_{2i}(x^k;t_k)\geq 0$, as $F_{1i}(x^k;t_k)+F_{2i}(x^k;t_k)\to G_i(x^*)>0$.[6] In this case, we get $\lim_{k\to\infty} F_{1i}(x^k;t_k)\eta_i^{\Phi,k}=0$ since $\left|\eta_i^{\Phi,k}\Phi_i^B(G(x^k),H(x^k);t_k)\right|\leq\epsilon_k$ by (3.20) and $\lim_{k\to\infty} F_{2i}(x^k;t_k)>0$.

Since $\lim_{k\to\infty} F_{1i}(x^k;t_k)\eta_i^{\Phi,k}=\lim_{k\to\infty}\eta_i^{G,k}=0$, applying Lemma 3.4.5, we obtain that $\lim_{k\to\infty}\lambda_i^{G,k}/\|\lambda^k\|_\infty=0$ for $i\in\mathcal{I}^{+0}$.

We now consider indices $i\in\mathcal{I}^{00}$. Without loss of generality suppose that $\max(|G_i(x^k)|,|H_i(x^k)|)=|G_i(x^k)|$, and so $\lim_{k\to\infty}\frac{\epsilon_k}{|G_i(x^k)|}=0$. Let α (possibly infinite) be such that

$$\alpha:=\lim_{k\to\infty}\frac{|G_i(x^k)|}{|t_{2,k}\theta_{t_{1,k}}(H_i(x^k))|}. \tag{3.24}$$

It should be noticed that $\alpha>1$, otherwise for k sufficiently large there would exist a constant C such that $|G_i(x^k)|\leq C|t_{2,k}\theta_{t_{1,k}}(H_i(x^k))|$, which is a contradiction with $|G_i(x^k)|\geq|H_i(x^k)|$ and $t_{2,k}=o(t_{1,k})$.

Another consequence is that $F_{2i}(x^k;t_k)\sim G_i(x^k)$, since $F_{2i}(x^k;t_k)\leq G_i(x^k)+t_{2,k}\theta_{t_{1,k}}(|G_i(x^k)|)$ and by definition of functions θs.

We consider separately the two cases: (a) $F_{1i}(x^k;t_k)+F_{2i}(x^k;t_k)\geq 0$, and (b) $F_{1i}(x^k;t_k)+F_{2i}(x^k;t_k)<0$.

(a) When $F_{1i}(x^k;t_k)+F_{2i}(x^k;t_k)\geq 0$, dividing by $|G_i(x^k)|$ in the complementarity condition in (3.20) yields

$$\frac{\epsilon_k}{|G_i(x^k)|}\geq\left|\eta_i^{\Phi,k}F_{1i}(x^k;t_k)\left(1-\frac{t_{2,k}\theta_{t_{1,k}}(H_i(x^k))}{G_i(x^k)}\right)\right|,$$

so $\eta_i^{\Phi,k}F_{1i}(x^k;t_k)\to 0$ since $\alpha>1$.

Now, consider two cases either $\{\eta_i^{G,k}\}$ tends to zero or not. In the former case, the conclusion of case a) would follow by applying Lemma 3.4.5.

So, let $\lim_{k\to\infty}\lambda_i^{G,k}\neq 0$. Dividing by $G_i(x^k)$ in the complementarity condition in (3.18) gives $|\eta_i^{G,k}(1+\bar{t}_k/G_i(x^k))|\leq\epsilon_k/|G_i(x^k)|$ and so $G_i(x^k)\sim-\bar{t}_k$. Besides, it can be noted that for k sufficiently large there is no constant $C>0$ such that $H_i(x^k)\leq C\epsilon_k$ as this would lead to a contradiction with $F_{1i}(x^k;t_k)+F_{2i}(x^k;t_k)\geq 0$. Indeed, as $H_i(x^k)\geq G_i(x^k)$, we would obtain

$$F_{1i}(x^k;t_k)+F_{2i}(x^k;t_k)\leq G_i(x^k)+C\epsilon_k-2t_{2,k}\theta_{t_{1,k}}(G_i(x^k)),$$

which is negative for k sufficiently large by definition of θ and $\epsilon_k=o(G_i(x^k))$. So, $\epsilon_k=o(H_i(x^k))$ and $H_i(x^k)>0$. Thus, dividing by $H_i(x^k)$ in the complementarity condition (3.20), we obtain $\lim_{k\to\infty}\eta_i^{\Phi,k}F_{2i}(x^k;t_k)=0$. This con-

[6] We remind that $F_{1i}(x^k;t_k)=H_i(x)-t_{2,k}\theta_{t_{1,k}}(G_i(x))$ and $F_{2i}(x^k;t_k)=G_i(x)-t_{2,k}\theta_{t_{1,k}}(H_i(x))$. Thus, $\lim_{k\to\infty}(F_{2i}(x^k;t_k),F_{1i}(x^k;t_k))=(G_i(x^*),0)$ and $G_i(x^*)>0$.

cludes case a), since $\lim_{k\to\infty} \eta_i^{\Phi,k} F_{2i}(x^k;t_k) = \lim_{k\to\infty} \eta_i^{\Phi,k} F_{1i}(x^k;t_k) = 0$ gives that $(\lambda_i^{G,*}, \lambda_i^{H,*}) = \lim_{k\to\infty} (\eta^{G,k}, \eta^{H,k}) \geq 0$.

(b) When $F_{1i}(x^k;t_k) + F_{2i}(x^k;t_k) < 0$, the complementarity condition in (3.20) gives $\left|\eta_i^{\Phi,k} F_{2i}(x^k;t_k)^2\right| \leq 2\epsilon_k$, and dividing by $|G_i(x^k)|$ yields

$$\left|\eta_i^{\Phi,k} F_{2i}(x^k;t_k)\left(1 - \frac{t_{2,k}\theta_{t_{1,k}}(H_i(x^k))}{G_i(x^k)}\right)\right| \leq \frac{2\epsilon_k}{|G_i(x^k)|}.$$

This implies that $\lim_{k\to\infty} \eta_i^{\Phi,k} F_{2i}(x^k;t_k) = 0$, by assumption on ϵ_k and $\alpha > 1$. Now, by definition of functions θs and the triangle inequality, we get

$$\left|F_{1i}(x^k;t_k) + F_{2i}(x^k;t_k)\right| \leq 2|G_i(x^k)| + 2t_{2,k}\theta_{t_{1,k}}(|G_i(x^k)|) \sim 2|G_i(x^k)|. \quad (3.25)$$

Using that $F_{2i}(x^k;t_k) \sim G_i(x^k)$ as noticed in the beginning of case (iii), we obtain that $\lim_{k\to\infty} \eta_i^{\Phi,k} F_{2i}(x^k;t_k) = \lim_{k\to\infty} \eta_i^{\Phi,k} G_i(x^k) = 0$. So, multiplying by $\eta_i^{\Phi,k}$ and going to the limit in (3.25) yields $\lim_{k\to\infty} \eta_i^{\Phi,k}\left(F_{1i}(x^k;t_k) + F_{2i}(x^k;t_k)\right) = 0$. As a consequence, it holds that $\lim_{k\to\infty} (\lambda_i^{G,k}, \lambda_i^{H,k})/\|\lambda^k\|_\infty = \lim_{k\to\infty} (\eta^{G,k}, \eta^{H,k}) \geq 0$.

All in all, we completed cases (a) and (b), so (iii) is satisfied.

Finally, since (i)–(iii) are satisfied, we conclude as in Theorem 3.4.2 so that under MPCC-GMFCQ the sequence $\{\lambda^k\}$ is bounded and x^* is an M-stationary point. □

The assumption in Theorem 3.4.6 is not entirely satisfactory since the sequence of parameter ϵ_k depends on the iterates. However, this is in the same vein as the existing results in [7, 21]. Further research may try to exploit this weak point to propose more adequate conditions.

Another benefit of considering approximate stationary points is that they may exist even though the assumptions presented in the previous section are not satisfied; see [3, 4].

The following example, from [19], shows that the butterfly relaxation with $t_{2,k} = o(t_{1,k})$ may converge to an undesirable A-stationary point without the hypothesis that $\epsilon_k = o(\max(|G(x^k)|, |H(x^k)|))$.

Example 3.7 Consider the problem

$$\min_{x\in\mathbb{R}^2} x_2 - x_1 \text{ s.t. } 0 \leq x_1 \perp x_2 \geq 0.$$

Let $t_{2,k} = {t_{1,k}}^2$ and choose any positive sequences $\{t_{1,k}\}$ and $\{\epsilon_k\}$ such that $t_{1,k}, \epsilon_k \to 0$. Consider the following ϵ-stationary sequence

$$x^k = (\epsilon_k, \epsilon_k/2)^T,\ \eta^{G,k} = 0,\ \eta^{H,k} = 1 - \eta^{\Phi,k}({t_{1,k}}^2\theta_{t_{1,k}}\left(\frac{\epsilon_k}{2}\right) F_1(x^k;t_k) - F_2(x^k;t_k))$$

and

$$\eta^{\Phi,k} = \frac{1}{t_{1,k}{}^2\theta_{t_{1,k}}(\epsilon_k)\,F_2(x^k;t_k) - F_1(x^k;t_k)}.$$

This sequence converges to $x^* = (0,0)$, which is an A-stationary point.

The ϵ-feasible set of the butterfly relaxation is similar to the relaxation from [31]. Therefore, it is not surprising that we can only expect to converge to a C-stationary point without strong hypotheses. Those issues clearly deserve a specific study that is left for further research.

3.5 Numerical Results

In this section, we focus on the numerical implementation of the butterfly relaxation. Our aim is to compare the new method with the existing ones in the literature and to show some of its features. This comparison uses the collection of test problems MacMPEC [22]. This collection has been widely used in the literature to compare relaxation methods as in [16, 17, 33]. The test problems included in MacMPEC are extracted from the literature and real-world applications.

3.5.1 On the Implementation of the Butterfly Relaxation

Practical implementation could consider a slightly different model, by skipping the relaxation of the positivity constraint and adding a new parameter t_3 in order to shift the intersection of both wings to the point $(G(x), H(x)) = (t_3, t_3)$. This can be done by redefining $F_1(x; t_1, t_2, t_3)$ and $F_2(x; t_1, t_2, t_3)$ such that

$$F_{1i}(x;t_1,t_2,t_3) = H_i(x) - t_3 - t_2\theta_{t_1}(G_i(x) - t_3),$$
$$F_{2i}(x;t_1,t_2,t_3) = G_i(x) - t_3 - t_2\theta_{t_1}(H_i(x) - t_3).$$

Even if we did not give any theoretical proof regarding this modified system, this modification does not alter the behavior of the butterfly relaxation. This formulation is clearly an extension of the relaxation (3.4).

The numerical comparison of the butterfly relaxation with other existing methods considers three schemes:

1. $B_{(t_2=t_1)}$: $t_3 = 0, t_2 = t_1$;
2. $B_{(t_2=t_1{}^{3/2})}$: $t_3 = 0, t_2 = t_1{}^{3/2}$;
3. $B_{(t_3=t_2,2t_2=t_1)}$: $t_3 = t_2, 2t_2 = t_1$.

In all these tests, we fixed $\bar{t} = 0$. Our tests concern many variants, not all of which were covered by our analysis, but they give a broader insight into the new relaxations.

3.5.2 Comparison of the Relaxation Methods

We provide in this section and Algorithm 1 some more details on the implementation and the comparison between relaxation methods. It is to be noted that we aim to compare the methods and so no attempt to optimize any method has been carried out. We use 101 test problems from MacMPEC, which omit the problems that exceed the limit of 300 variables or constraints and some problems with the evaluation error of the objective function or the constraints.

Algorithm 1 is coded in Matlab and uses the AMPL API. R_{t_k} denotes the relaxed non-linear program associated with a generic relaxation, where except the butterfly methods, the parameter $t_{1,k}$ does not play any role. At each step, we compute x^{k+1} as a solution of R_{t_k} starting from x^k. Therefore, at each step, the initial point is more likely to be infeasible for R_{t_k}. The iterative process stops when $t_{2,k}$ and $t_{1,k}$ are smaller than some tolerance, denoted by $p_{\min}$ which is set as 10^{-15} here, or when the solution x^{k+1} of R_{t_k} is considered an ϵ-solution of (3.1). To consider x^{k+1} as a ϵ-solution, with ϵ set as 10^{-7}, we check three criteria:

(a) Feasibility of the last relaxed non-linear program:

$$\nu_f(x) := \max(-g(x), |h(x)|, -\Phi(x));$$

(b) Feasibility of the complementarity constraint: $\nu_{comp}(x) := \min(G(x), H(x))^2$;
(c) The complementarity between the Lagrange multipliers and the constraints of the last relaxed non-linear program:

$$\nu_c(x) := \left\| g(x) \circ \eta^g, h(x) \circ \eta^h, G(x) \circ \eta^G, H(x) \circ \eta^H, \Phi^B(x) \circ \eta^\Phi \right\|_\infty .$$

Obviously, it is hard to ask a tighter condition on the complementarity constraint since the feasibility only guarantees that the product component-wise is less than ϵ. Using these criteria, we define a measure of optimality

$$optimal(x) := \max\left(\nu_f(x), \nu_{comp}(x), \nu_c(x)\right).$$

A fourth criterion could be the dual feasibility, which is the norm of the gradient of the Lagrangian. However, solvers like SNOPT or MINOS do not use this criterion as a stopping criterion, but use the gradient of the Lagrangian scaled by the norm of the Lagrange multiplier. One reason among others to discard such a criterion could be numerical issues implied by the degeneracy in the KKT conditions. In the case of an infeasible or unbounded sub-problem R_{t_k}, the algorithm stops and returns a certificate.

Data:
starting vector x^0; initial relaxation parameter t_0; update parameter $(\sigma_{t_1}, \sigma_{t_2}) \in (0, 1)^2$; p_{min} the minimum parameter value; ϵ the precision tolerance ;
1 **Begin** ;
2 Set $k := 0$;
3 **while** $\max(t_{2,k}, t_{1,k}) > p_{\min}$ ***and*** $optimal(x^k) > \epsilon$ **do**
4 | x^{k+1} solution of $R_{t_{1,k}, t_{2,k}}$ with x^k initial point;
5 | $(t_{1k+1}, t_{2k+1}) := (t_{1,k}\sigma_{t_1}, t_{2,k}\sigma_{t_2})$;
6 **return**: f_{opt} the optimal value at the solution x_{opt} or a decision of infeasibility or unboundedness.

Algorithm 1: Basic Relaxation methods for (3.1), with a relaxed non-linear program R_{t_k}.

Step 4 in Algorithm 1 is performed using three different solvers accessible through AMPL [13], which are SNOPT 7.2-8 [14], MINOS 5.51 [25], and IPOPT 3.12.4 [34] with their default parameters. A previous similar comparison in the literature in [16] only considered SNOPT to solve the sub-problems. We compare the butterfly schemes with the most popular relaxations SS from [30] and (3.4). We also take into account the results of the non-linear programming solver without specific MPCC tuning denoted by NL.

In order to compare the various relaxation methods, we need to have a coherent use of the parameters. In a similar way as in [16], we consider the value of the "intersection between G and H", which is (t, t) for (3.4) and (3.7), $(\sqrt{t}, \sqrt{t})$ for SS. Then, we run a sensitivity analysis on several values of the parameters $T \in \{100, 25, 10, 5, 1, 0.5, 0.05\}$ and $S \in \{0.1, 0.075, 0.05, 0.025, 0.01\}$, which corresponds to t_0 and σ_t as described in Table 3.1. In [16], the authors consider as a stopping criterion the feasibility of the last non-linear parametric program in particular by considering the complementarity constraint by the minimum component-wise. Table 3.2 provides our result with this criterion. We report elementary statistics by considering the percentage of success for each set of parameters. A problem is considered solved if criteria (a) and (b) are satisfied.

First, we see that the method NL is giving decent results. It is not a surprise, as was pointed out in [12]. Practical implementation of relaxation methods would select the best choice of parameters so that we focus most of our attention on the line "best". In all cases, the relaxations manage to improve or at least equal the number of problems solved by NL. By using SNOPT, KS, and butterfly with $t_2 = t_1{}^{3/2}$ methods, we get 1% of improvement, and with IPOPT, the method butterfly with $t_2 = t_1{}^{3/2}$ is

Table 3.1 Parameter links among the methods

Relaxation	NL	SS	KS	Butterfly
t_0	none	T^2	T	T
σ_t	none	S^2	S	S

Table 3.2 Sensitivity analysis for MacMPEC test problems considering the feasibility of (3.1). Results are a percentage of success. Best: percentage of success with the best set of parameters (independent of the problem); worst: percentage of success with the worst set of parameters; average: average percentage of success among the distribution of (T, s); std: standard deviation

Solver SNOPT	NL	SS	KS	$B_{(t_2=t_1)}$	$B_{(t_3=t_2, 2t_2=t_1)}$	$B_{(t_2={t_1}^{3/2})}$
Best	97.03	97.03	98.02	97.03	97.03	98.02
Average	97.03	95.02	94.71	95.39	93.89	94.88
Worst	97.03	91.09	91.09	92.08	91.09	91.09
Std	0	1.64	2.09	1.50	1.97	2.42
Solver MINOS	NL	SS	KS	$B_{(t_2=t_1)}$	$B_{(t_3=t_2, 2t_2=t_1)}$	$B_{(t_2={t_1}^{3/2})}$
Best	89.11	94.06	93.07	90.10	95.05	89.11
Average	89.11	91.20	90.89	83.54	91.06	81.92
Worst	89.11	87.13	87.13	77.23	86.14	76.24
Std	0	1.50	1.44	2.81	2.15	2.89
Solver IPOPT	NL	SS	KS	$B_{(t_2=t_1)}$	$B_{(t_3=t_2, 2t_2=t_1)}$	$B_{(t_2={t_1}^{3/2})}$
Best	98.02	99.01	98.02	99.01	98.02	100
Average	98.02	98.16	96.38	94.03	93.89	94.79
Worst	98.02	95.05	93.07	89.11	88.12	88.12
Std	0	0.97	1.99	2.62	2.80	3.60

the only one that attains 100%. The relaxation methods seem to make a significant improvement over NL with MINOS. In this case, it is clear that the butterfly methods benefit from the introduction of the parameter s, and the method with $t_3 = t_2$, $2t_2 = t_1$ is very competitive.

Our goal by solving (3.1) is to compute a local minimum. The results using the local minimum criterion defined above as a measure of success are given in Table 3.3. Once again, we provide percentages of success.

In comparison with Table 3.2, this new criterion appears to be more selective. Independent of the solver, the relaxation methods with some correct choices of parameters provide improved results. Using SNOPT as a solver, the methods KS and butterfly give the highest number of results. The method butterfly with $t_2 = {t_1}^{3/2}$ even improved the number of problems that SNOPT alone solved on average. Similarly, as in the previous experiment, the butterfly method benefits from the introduction of the parameter s when using MINOS as a solver.

Table 3.3 Sensitivity analysis for MacMPEC test problems considering the optimality of (3.1). The results are percentages of success. Best: percentage of success with the best set of parameters; worst: percentage of success with the worst set of parameters; average: average percentage of success among the distribution of (T, s); std: standard deviation

Solver SNOPT	NL	SS	KS	$B_{(t_2=t_1)}$	$B_{(t_3=t_2,2t_2=t_1)}$	$B_{(t_2=t_1{}^{3/2})}$
Best	92.08	94.06	96.04	96.04	97.03	96.04
Average	92.08	90.78	91.17	92.08	90.04	92.33
Worst	92.08	83.17	86.14	87.13	82.18	87.13
Std	0	3.15	2.59	2.45	2.86	2.77
Solver MINOS	NL	SS	KS	$B_{(t_2=t_1)}$	$B_{(t_3=t_2,2t_2=t_1)}$	$B_{(t_2=t_1{}^{3/2})}$
Best	85.15	94.06	93.07	88.11	94.06	87.13
Average	85.15	90.94	90.18	81.92	90.04	80.11
Worst	85.15	87.13	86.14	76.23	85.15	74.26
Std	0	1.50	1.62	2.65	2.31	2.95
Solver IPOPT	NL	SS	KS	$B_{(t_2=t_1)}$	$B_{(t_3=t_2,2t_2=t_1)}$	$B_{(t_2=t_1{}^{3/2})}$
Best	91.09	93.07	93.07	94.06	93.07	94.06
Average	91.09	91.82	89.84	89.05	88.80	89.02
Worst	91.09	90.10	86.14	84.16	84.16	81.19
Std	0	1.14	2.19	3.09	2.72	3.86

3.6 Concluding Remarks

This chapter proposes a new family of relaxation schemes for the mathematical program with complementarity constraints. We prove convergence of the method in the general case and show that a specific relation between the parameters allows the method to converge to the desired M-stationary point. Additionally, in the particular case where MPCC-LICQ holds, S-stationary conditions can be expected to hold at a local minimum. We prove that in the affine case, the butterfly relaxation method converges to such a point without assuming any second-order conditions or strict complementarity-type conditions, which is an improvement over other methods.

We provide a complete numerical study with remarks regarding the implementation as well as a comparison with existing methods in the literature. These numerical experiments show that the butterfly schemes are very competitive.

Future research will focus on the main difficulty regarding relaxation schemes that are the convergence of approximate stationary sequences. A discussion regarding the above problem has been initiated in [7, 21] and appeal for further study.

Acknowledgements This research was partially supported by the NSERC grantand partially by a french grant from “l’Ecole des Docteurs de l’UBL” and “le Conseil Régional de Bretagne”. The authors would like to thank the referees for their help and valuable comments.

Appendix

3.7 Proof of a Technical Lemma

In the proof of Theorem 3.4.4 and Theorem 3.4.5, we use the following lemma that links the gradients of G and H with the gradients of $F_1(x; t)$ and $F_2(x; t)$.

Lemma 3.7.6 *Let (I, I^c, I^-) be any partition of $\mathcal{I}_{GH}^{00}(x; t)$. Assume that the gradients*

$$\begin{aligned}&\{\nabla g_i(x)\ (i \in \mathcal{I}_g(x)),\ \nabla h_i(x)\ (i = 1, \ldots, p),\\ &\nabla G_i(x)\ (i \in \mathcal{I}_G(x; \bar{t}) \cup \mathcal{I}_{GH}^{00}(x; t) \cup \mathcal{I}_{GH}^{+0}(x; t)),\\ &\nabla H_i(x)\ (i \in \mathcal{I}_H(x; \bar{t}) \cup \mathcal{I}_{GH}^{00}(x; t) \cup \mathcal{I}_{GH}^{0+}(x; t))\}\end{aligned}$$

are linearly independent. Then, LICQ holds at x for (3.13).

Proof We show that the gradients of the constraints of (3.13) are positively linearly independent. For this purpose, we prove that the trivial solution is the only solution to the equation

$$\begin{aligned}0 = &\sum_{i \in \mathcal{I}_g(x)} \eta_i^g \nabla g_i(x) + \sum_{i=1}^{p} \eta_i^h \nabla h_i(x) - \sum_{i \in \mathcal{I}_G(x;\bar{t})} \eta_i^G \nabla G_i(x) - \sum_{i \in \mathcal{I}_H(x;\bar{t})} \eta_i^H \nabla H_i(x)\\ &+ \sum_{i \in \mathcal{I}_{GH}^{+0}(x;t) \cup \mathcal{I}_{GH}^{0+}(x;t)} \eta_i^\Phi \nabla \Phi_i^B(G(x), H(x); t)\\ &+ \sum_{i \in \mathcal{I}_{GH}^{00}(x;t)} \left(\nu_i^{F_1(x;t)} - \mu_i^{F_1(x;t)} + \delta_i^{F_1(x;t)}\right) \nabla F_{1i}(x; t)\\ &+ \left(-\nu_i^{F_2(x;t)} + \mu_i^{F_2(x;t)} + \delta_i^{F_2(x;t)}\right) \nabla F_{2i}(x; t),\end{aligned}$$

where $\operatorname{supp}(\eta^g) \subseteq \mathcal{I}_g(x)$, $\operatorname{supp}(\eta^G) \subseteq \mathcal{I}_G(x; \bar{t})$, $\operatorname{supp}(\eta^H) \subseteq \mathcal{I}_H(x; \bar{t})$, $\operatorname{supp}(\eta^\Phi) \subseteq \mathcal{I}_{GH}^{+0}(x; t) \cup \mathcal{I}_{GH}^{0+}(x; t)$, $\operatorname{supp}(\nu^{F_1(x;t)}) \subseteq I$, $\operatorname{supp}(\nu^{F_2(x;t)}) \subseteq I$, $\operatorname{supp}(\mu^{F_1(x;t)}) \subseteq I^c$, $\operatorname{supp}(\mu^{F_2(x;t)}) \subseteq I^c$, $\operatorname{supp}(\delta^{F_1(x;t)}) \subseteq I^-$, and $\operatorname{supp}(\delta^{F_2(x;t)}) \subseteq I^-$ where $I \cup I^c \cup I^- = \mathcal{I}_{GH}^{00}(x; t)$ and I, I^c, I^- have a two-by-two empty intersection.

By definition of $F_1(x; t)$ and $F_2(x; t)$, it holds that

$$\begin{aligned}\nabla F_{1i}(x; t) &= \nabla H_i(x) - t_2 \theta'_{t_1}(G_i(x)) \nabla G_i(x),\\ \nabla F_{2i}(x; t) &= \nabla G_i(x) - t_2 \theta'_{t_1}(H_i(x)) \nabla H_i(x).\end{aligned}$$

The gradient of $\Phi^B(G(x), H(x); t)$ is given by Lemma 3.3.3.

We now replace those gradients in the equation above

$$0 = \sum_{i \in \mathcal{I}_g(x)} \lambda_i^g \nabla g_i(x) + \sum_{i=1}^{p} \lambda_i^h \nabla h_i(x) + \sum_{i=1}^{q} \lambda_i^G \nabla G_i(x) + \sum_{i=1}^{q} \lambda_i^H \nabla H_i(x),$$

with

$$\lambda_i^G = -\eta_i^G + \eta_i^\Phi F_{1i}(x;t) - \left(\eta_i^\Phi F_{2i}(x;t) + \nu_i^{F_1(x;t)} - \mu_i^{F_1(x;t)} + \delta_i^{F_1(x;t)}\right) t_2\theta'_{t_1}(G_i(x))$$
$$- \nu_i^{F_2(x;t)} + \mu_i^{F_2(x;t)} + \delta_i^{F_2(x;t)},$$
$$\lambda_i^H = -\eta_i^H + \eta_i^\Phi F_{2i}(x;t) - \left(\eta_i^\Phi F_{1i}(x;t) - \nu_i^{F_2(x;t)} + \mu_i^{F_2(x;t)} + \delta_i^{F_2(x;t)}\right) t_2\theta'_{t_1}(H_i(x))$$
$$+ \nu_i^{F_1(x;t)} - \mu_i^{F_1(x;t)} + \delta_i^{F_1(x;t)}.$$

By linear independence assumption, we obtain

$$\eta^g = 0, \eta^h = 0, \eta^G = 0, \eta^H = 0, \eta_i^\Phi = 0\, \forall i \in \mathcal{I}_{GH}^{0+}(x;t) \cup \mathcal{I}_{GH}^{+0}(x;t),$$
$$-\nu_i^{F_1(x;t)} t_2\theta'_{t_1}(G_i(x)) - \nu_i^{F_2(x;t)} = 0 \text{ and } \nu_i^{F_2(x;t)} t_2\theta'_{t_1}(H_i(x)) + \nu_i^{F_1(x;t)} = 0, \forall i \in I,$$
$$\mu_i^{F_1(x;t)} t_2\theta'_{t_1}(G_i(x)) + \mu_i^{F_2(x;t)} = 0 \text{ and } -\mu_i^{F_2(x;t)} t_2\theta'_{t_1}(H_i(x)) - \mu_i^{F_1(x;t)} = 0, \forall i \in I^c,$$
$$-\delta_i^{F_1(x;t)} t_2\theta'_{t_1}(G_i(x)) + \delta_i^{F_2(x;t)} = 0 \text{ and } -\delta_i^{F_2(x;t)} t_2\theta'_{t_1}(H_i(x)) + \delta_i^{F_1(x;t)} = 0, \forall i \in I^-.$$

So, it follows for $i \in I^-$ that

$$\delta_i^{F_2(x;t)} = \delta_i^{F_1(x;t)} t_2\theta'_{t_1}(G_i(x)) \text{ and } \delta_i^{F_1(x;t)} = \delta_i^{F_2(x;t)} t_2\theta'_{t_1}(H_i(x)).$$

So $\delta_i^{F_1(x;t)} = \delta_i^{F_2(x;t)} = 0$, since $i \in \mathcal{I}_{GH}^{00}(x;t)$ gives

$$t_2\theta'_{t_1}(G_i(x)) t_2\theta'_{t_1}(H_i(x)) = t_2\theta'_{t_1}(0) t_2\theta'_{t_1}(0) < 1$$

by properties of θ and (3.8). Similarly, we get $\mu_i^{F_1(x;t)} = \mu_i^{F_2(x;t)} = \nu_i^{F_2(x;t)} = \nu_i^{F_1(x;t)} = 0$. □

References

1. Abdallah, L., Haddou, M., Migot, T.: Solving absolute value equation using complementarity and smoothing functions. J. Comput. Appl. Math. **327**, 196–207 (2018)
2. Abdallah, L., Haddou, M., Migot, T.: A sub-additive DC approach to the complementarity problem. Comput. Optim. Appl. **73**(2), 509–534 (2019)
3. Andreani, R., Haeser, G., Secchin, L.D., Silva, P.J.S.: New sequential optimality conditions for mathematical programs with complementarity constraints and algorithmic consequences. SIAM J. Optim. **29**(4), 3201–3230 (2019)
4. Andreani, R., Martínez, J.M., Svaiter, B.F.: A new sequential optimality condition for constrained optimization and algorithmic consequences. SIAM J. Optim. **20**(6), 3533–3554 (2010)
5. Bazaraa, M.S., Shetty, C.M.: Foundations of Optimization, vol. 122. Springer Science & Business Media (2012)
6. DeMiguel, V., Friedlander, M.P., Nogales, F.J., Scholtes, S.: A two-sided relaxation scheme for mathematical programs with equilibrium constraints. SIAM J. Optim. **16**(2), 587–609 (2005)

7. Dussault, J.-P., Haddou, M., Kadrani, A., Migot, T.: On approximate stationary points of the regularized mathematical program with complementarity constraints. J. Optim. Theory Appl. **186**(2), 504–522 (2020)
8. Dussault, J.-P., Haddou, M., Migot, T.: Mathematical programs with vanishing constraints: constraint qualifications, their applications, and a new regularization method. Optimization **68**(2–3), 509–538 (2019)
9. Flegel, M.L., Kanzow, C.: Abadie-type constraint qualification for mathematical programs with equilibrium constraints. J. Optim. Theory Appl. **124**(3), 595–614 (2005)
10. Flegel, M.L., Kanzow, C.: On the Guignard constraint qualification for mathematical programs with equilibrium constraints. Optimization **54**(6), 517–534 (2005)
11. Flegel, M.L., Kanzow, C.: A direct proof for M-stationarity under MPEC-GCQ for mathematical programs with equilibrium constraints. Springer (2006)
12. Fletcher, R., Leyffer, S.: Solving mathematical programs with complementarity constraints as nonlinear programs. Optim. Methods Softw. **19**(1), 15–40 (2004)
13. Fourer, R., Gay, D., Kernighan, B.: Ampl, vol. 119. Boyd & Fraser (1993)
14. Gill, P.E., Murray, W., Saunders, M.A.: SNOPT: an SQP algorithm for large-scale constrained optimization. SIAM Rev. **47**, 99–131 (2005)
15. Guo, L., Lin, G.-H., Ye, J.J.: Solving mathematical programs with equilibrium constraints. J. Optim. Theory Appl. **166**(1), 234–256 (2015)
16. Hoheisel, T., Kanzow, C., Schwartz, A.: Theoretical and numerical comparison of relaxation methods for mathematical programs with complementarity constraints. Math. Program. **137**(1–2), 257–288 (2013)
17. Kadrani, A., Dussault, J.-P., Benchakroun, A.: A new regularization scheme for mathematical programs with complementarity constraints. SIAM J. Optim. **20**(1), 78–103 (2009)
18. Kanzow, C., Schwartz, A.: Mathematical programs with equilibrium constraints: enhanced fritz john-conditions, new constraint qualifications, and improved exact penalty results. SIAM J. Optim. **20**(5), 2730–2753 (2010)
19. Kanzow, C., Schwartz, A.: A new regularization method for mathematical programs with complementarity constraints with strong convergence properties. SIAM J. Optim. **23**(2), 770–798 (2013)
20. Kanzow, C., Schwartz, A.: Convergence properties of the inexact Lin-Fukushima relaxation method for mathematical programs with complementarity constraints. Comput. Optim. Appl. **59**(1–2), 249–262 (2014)
21. Kanzow, C., Schwartz, A.: The price of inexactness: convergence properties of relaxation methods for mathematical programs with complementarity constraints revisited. Math. Oper. Res. **40**(2), 253–275 (2015)
22. Leyffer, S.: Macmpec: Ampl collection of mpecs. Argonne National Laboratory. www.mcs.anl.gov/leyfier/MacMPEC (2000)
23. Lin, G.-H., Fukushima, M.: A modified relaxation scheme for mathematical programs with complementarity constraints. Ann. Oper. Res. **133**(1–4), 63–84 (2005)
24. Luo, Z.-Q., Pang, J.-S., Ralph, D.: Mathematical programs with equilibrium constraints. Cambridge University Press (1996)
25. Murtagh, B.A., Saunders, M.A.: Minos 5.0 user's guide. Technical report, DTIC Document (1983)
26. Outrata, J.V.: Optimality conditions for a class of mathematical programs with equilibrium constraints. Math. Oper. Res. **24**(3), 627–644 (1999)
27. Outrata, J.V.: A generalized mathematical program with equilibrium constraints. SIAM J. Control Optim. **38**(5), 1623–1638 (2000)
28. Pang, J.-S., Fukushima, M.: Complementarity constraint qualifications and simplified b-stationarity conditions for mathematical programs with equilibrium constraints. Comput. Optim. Appl. **13**(1), 111–136 (1999)
29. Ramos, A.: Mathematical programs with equilibrium constraints: a sequential optimality condition, new constraint qualifications and algorithmic consequences. Optim. Meth. Softw. **36**(1), 45–81 (2021)

30. Scheel, H., Scholtes, S.: Mathematical programs with complementarity constraints: stationarity, optimality, and sensitivity. Math. Oper. Res. **25**(1), 1–22 (2000)
31. Scholtes, S.: Convergence properties of a regularization scheme for mathematical programs with complementarity constraints. SIAM J. Optim. **11**(4), 918–936 (2001)
32. Schwartz, A.: Mathematical programs with complementarity constraints: theory, methods, and applications. PhD thesis, Ph. D. dissertation, Institute of Applied Mathematics and Statistics, University of Würzburg (2011)
33. Steffensen, S., Ulbrich, M.: A new relaxation scheme for mathematical programs with equilibrium constraints. SIAM J. Optim. **20**(5), 2504–2539 (2010)
34. Wächter, A., Biegler, L.T.: On the implementation of an interior-point filter line-search algorithm for large-scale nonlinear programming. Math. program. **106**(1), 25–57 (2006)
35. Ye, J.J.: Constraint qualifications and necessary optimality conditions for optimization problems with variational inequality constraints. SIAM J. Optim. **10**(4), 943–962 (2000)
36. Ye, J.J.: Necessary and sufficient optimality conditions for mathematical programs with equilibrium constraints. J. Math. Anal. Appl. **307**(1), 350–369 (2005)
37. Ye, J.J., Ye, X.Y.: Necessary optimality conditions for optimization problems with variational inequality constraints. Math. Oper. Res. **22**(4), 977–997 (1997)
38. Ye, J.J., Zhu, D.L., Zhu, Q.J.: Exact penalization and necessary optimality conditions for generalized bilevel programming problems. SIAM J. Optim. **7**(2), 481–507 (1997)

Chapter 4
Copositive Optimization and Its Applications in Graph Theory

S. K. Neogy and Vatsalkumar N. Mer

Abstract Recently, copositive optimization has received a lot of attention to the Operational Research community, and it is rapidly expanding and becoming a fertile field of research. In this chapter, we demonstrate the diversity of copositive formulations in different domains of optimization: continuous, discrete, and stochastic optimization problems. Further, we discuss the role of copositivity for local and global optimality conditions. Finally, we talk about some applications of copositive optimization in graph theory and game theory.

Keywords Nonconvex quadratic program · Completely positive program · Fractional quadratic optimization · Lifted problem · Graph theory · Maximum weight clique problem

4.1 Introduction

Copositive matrices appear in various applications in mathematics, and especially, in the characterization of the solution set of constrained optimization problems and the linear complementarity problem. Recently, Copositive optimization has been an object of research because many NP-hard combinatorial problems have a representation in this domain. Copositive optimization deals with minimizing a linear function in matrix variables subject to linear constraints and the constraint that the matrix should be in the convex cone of copositive matrices. In what follows, we make use of the following notations.

S. K. Neogy (✉) · V. N. Mer
Indian Statistical Institute, 7, S.J.S Sansanwal Marg, New Delhi 110016, India
e-mail: skn@isid.ac.in

V. N. Mer
e-mail: vnm232657@gmail.com

V. Laha et al. (eds.), *Optimization, Variational Analysis and Applications*,
Springer Proceedings in Mathematics & Statistics 355,
https://doi.org/10.1007/978-981-16-1819-2_4

$\mathcal{S}^n$	Set of symmetric matrices of order n
$\mathcal{C}^n$	Set of copositive matrices of order n
$\mathcal{C}^{*n}$	Set of comp. positive matrices of order n
E	Matrix of all-ones
conv($\mathcal{M}$)	Convex hull of a set $\mathcal{M}$
Copositive cone $\mathcal{C}^n$	$\{A \in \mathcal{S}^n \mid x^T Ax \geq 0 \; \forall x \in R^n\}$
Completely positive cone $\mathcal{C}^{*n}$	$\{BB^T \in \mathcal{S}^n \mid B \geq 0\}$
For an arbitrary givn cone $\mathcal{K} \subseteq \mathcal{S}$, dual cone $\mathcal{K}^*$	$= \{\sum_{i=1}^n {a_i}^T a_i \mid a_i \in R_+^n \; \forall i\}$ $\{A \in \mathcal{S} \mid \langle A, B\rangle \geq 0, \; \forall \; B \in \mathcal{K}\}$
Recession cone of A,	$rec(A) := \{y \in \mathbb{R}^n : \forall x \in A, \forall \lambda \geq 0 : \; x + \lambda y \in A\}$
Inner Product $\langle A, B\rangle$	trace$\langle B, A\rangle = \sum_{i,j=1}^n a_{ij}b_{ij}$

Consider the standard quadratic problem (stQP)

$$\min x^T Qx \text{ s.t. } e^T x = 1, x \geq 0.$$

where e denotes the all-ones vector. This optimization problem asks for the minimum of a (not necessarily convex) quadratic function over the standard simplex $\Delta = \{x \in \mathbb{R}_+^n : e^T x = 1\}$.

Note that $\langle x^T Qx\rangle = \langle Q, xx^T\rangle$ E is square matrix consisting entirely of unit entries, so that $x^T Ex = (e^T x)^2 = 1$ on Δ.

So $e^T x = 1$ transforms to $\langle E, xx^T\rangle = 1$.

Hence, the problem stQP can be written as

$$\min\langle Q, X\rangle$$

$$\text{s. t. } \langle E, X\rangle = 1,$$

$$X \in \mathcal{C}^{*n}.$$

More generally, a primal-dual pair in copositive optimization (COP) is of the following form:

$$\begin{aligned} &\min\langle C, X\rangle \\ &\text{s.t. } \langle A_i, X\rangle = b_i \; (i = 1, \ldots, m) \\ &X \in \mathcal{C}^n, \end{aligned} \tag{4.1}$$

where $\mathcal{C}^n = \{A \in \mathcal{S}^n : x^T Ax \geq 0 \; \forall \;\; x \in \mathbb{R}_+^n\}$ is the cone of copositive matrices. Bundfuss and Dür [8] developed an efficient algorithm to solve the optimization problem (4.1) over the copositive cone using iteratively polyhedral inner and outer approximations of the copositive cone.

Associated with problem (4.1), there is a dual problem which involves the constraint that the dual variable lies in the dual cone of $\mathcal{C}^n$, that is, the convex cone $\mathcal{C}^{*n}$ of completely positive matrices: $\mathcal{C}^{*n} = conv\{xx^T : x \in \mathbb{R}_+^n\}$

The dual of (4.1) is

$$\max \sum_{i=1}^{m} b_i y_i \tag{4.2}$$
$$\text{s.t. } C - \sum_{i=1}^{m} y_i A_i \in C^{*n},\ y_i \in \mathbb{R},$$

where $C^{*n} = \text{conv}\{xx^T : x \in \mathbb{R}^n_+\}$ is the cone of completely positive matrices. Clearly, (4.1) and (4.2) are convex optimization problems since both C^n and C^{*n} are convex cones. Note that KKT optimality conditions hold if Slater's condition is satisfied and imposing a constraint qualification guarantees strong duality, i.e., equality of the optimal values of (4.1) and (4.2). It is well known that most common constraint qualification assume that both problems are feasible and one of them strictly feasible.

Copositive programming can be visualized as a convexification approach for nonconvex quadratic programs. In many cases, nonconvex optimization problems admit exact copositive formulation. In this chapter, we show that some nonconvex quadratic programming problems that arise in graph theory can be converted into a convex quadratic problem. The first account of copositive optimization goes back to [4], which established a copositive representation of a subclass of particular interest, namely, in standard quadratic optimization (StQP).

4.1.1 *Quadratic Programming Problem with Binary and Continuous Variables*

Burer [6] considered an extremely large class of nonconvex quadratic programs with a mixture of binary and continuous variables, and showed that they can be expressed as completely positive program (CPPs).

We consider the following problem:

$$\begin{aligned} \min\ & x^T Qx + 2c^T x \\ \text{s.t. }\ & a_i^T x = b_i\ (i = 1, \ldots, m) \\ & x \geq 0, \\ & x_j \in \{0, 1\}\ \forall\ j \in B \text{ where } B \subseteq \{1, \ldots, n\}. \end{aligned} \tag{4.3}$$

Burer [6] showed that (4.3) is equivalent to the following linear problem over the cone of completely positive matrices.

$$\begin{aligned} &\min \ \langle Q, X\rangle + 2c^T x \\ &\text{s.t.}\ a_i^T x \ = b_i \ (i = 1, \ldots, m) \\ &\qquad \langle a_i a_i^T, X\rangle \ = b_i^2 \ (i = 1, \ldots, m) \\ &\qquad x_j = X_{jj} \ (j \in B) \\ &\qquad \begin{bmatrix} 1 & x \\ x & X \end{bmatrix} \in C^{*n+1}. \end{aligned} \tag{4.4}$$

This is a nice result since a nonconvex quadratic integer problem is equivalently written as a linear problem over a convex cone. Note that the dual problem of a completely positive program is an optimization problem over the cone of copositive matrices. Clearly, both problem classes are NP-hard since they are equivalent to an integer programming problem. Bundfuss and Dür [8] posed an open question whether problems with general quadratic constraints can similarly be restated as completely positive problems. Bomze [2] demonstrated the diversity of copositive formulations in various domains of optimization, namely, continuous and discrete, deterministic and stochastic.

4.1.2 Fractional Quadratic Optimization Problem

Consider the fractional quadratic optimization problem

$$\min_x f(x) = \min_x \frac{x^T Cx + 2c^T x + \gamma}{x^T Bx + 2b^T x + \beta} : Ax = a, x \in \mathbb{R}^n_+, \tag{4.5}$$

where B is positive semidefinite matrix and $C = C^T \in \mathbb{R}^{n\times n}$, $\{b, c\} \subset \mathbb{R}^n$, $A \in \mathbb{R}^{m\times n}$, $a \in \mathbb{R}^m$ and $\beta, \gamma \in \mathbb{R}$.

Now define the symmetric $(n+1) \times (n+1)$ matrices $\tilde{A} = \begin{bmatrix} a^T a & -a^t A \\ -A^T a & A^T A \end{bmatrix}$, $\tilde{B} = \begin{bmatrix} \beta & b^T \\ b & B \end{bmatrix}$, $\tilde{C} = \begin{bmatrix} \gamma & c^t \\ c & C \end{bmatrix}$. We further assume that the problem in (4.5) is well defined. Amral et al. [1] observed that the problem (4.5) can be written as the completely positive problem:

$$\min\{< \tilde{C}, X >: < \tilde{B}, X >= 1, \ < \tilde{A}, X >= 0, \ X \in C^{*n+1}\}.$$

The above problem occurs in many engineering applications. For further details, see [2] and references therein.

4.1.3 More on Nonconvex Quadratic Programming Problems

Burer [7] generalized the sign constraints $x \in \mathbb{R}^n_+$ to arbitrary cone constraints $x \in \mathcal{K}$, where $\mathcal{K}$ is a closed, convex cone, and studied the following (nonconvex) quadratic cone-constrained problem.

$$\begin{aligned} \min\ & x^T Qx + 2c^T x \\ \text{s.t.}\ & Ax = b \\ & x \in \mathcal{K}. \end{aligned} \tag{4.6}$$

Note that the dimension of the problem is increased by one by passing from the cone $\mathcal{K} \subseteq \mathbb{R}^n$ to the cone $\hat{\mathcal{K}} = \mathbb{R}_+ \times \mathcal{K}$. Let $C_{\hat{\mathcal{K}}} = conv\{zz^T : z \in \mathcal{K}\}$, the dual cone $C^*_{\hat{\mathcal{K}}}$ of all $\hat{\mathcal{K}}$-copositive $(n+1) \times (n+1)$ matrices.

In [2, 7, 13], it has been shown that (4.6) is equivalent to the (generalized) completely positive problem of the following form.

$$\begin{aligned} \min\ & \langle \tilde{Q}, Y\rangle + 2c^T x \\ \text{s.t.}\ & Ax = b \\ & (AXA^T)_{ii} = b_i^2 \ (i = 1, \ldots, m) \\ & Y = \begin{bmatrix} 1 & x \\ x & X \end{bmatrix} \in C^*_{\hat{\mathcal{K}}}, \end{aligned} \tag{4.7}$$

where $\tilde{Q} = \begin{bmatrix} 0 & c^T \\ c & Q \end{bmatrix}$.

4.1.4 Quadratic Optimization Problem and the Concept of Lifted Problem

Nguyen [17] presents a general concept of lifting a nonconvex quadratic optimization problem into an equivalent convex optimization problem with matrix variables. Further, they apply this lifting concept to a class of quadratic optimization problem with linear inequality and mixed binary constraints.

Nguyen [17] consider the following quadratic optimization problem (QP)

$$\begin{aligned} \min\ & x^T Qx \\ \text{s.t.}\ & x \in F(QP), \end{aligned} \tag{4.8}$$

where $Q \in \mathcal{S}^n$ and $F(QP)$ is some non-empty feasible set in R^n.

Consider the following subsets of $\mathcal{S}^n$.

$$C := \text{conv}\{xx^T : x \in F(QP)\},$$
$$\mathcal{R} := \text{conv}\{yy^T : y \in \text{rec}F(QP)\}.$$

The optimization problem

$$\begin{aligned} &\min \langle Q, X\rangle \\ &\text{s.t. } X \in C + \mathcal{R}, \end{aligned} \tag{4.9}$$

is called the lifted problem according to the original quadratic problem (4.8).

Proposition 4.1 (Proposition 2.2, [17]) *Assume that an optimal solution of* (4.8) *exists. Then the problem* (4.8) *and* (4.9) *are equivalent in the sense that they have the same optimal value, and any optimal solution of* (4.9) *is a convex combination of matrices* $x^i(x^i)^T$*, where* x^i *are optimal solution of* (4.8).

Note that the original Problem (4.8) of minimizing is not necessarily a convex quadratic function over a not necessarily convex set. However, the lifted problem (4.9) is a convex optimization problem. Therefore, as every local optimal solution obtained by solving (4.9) is a global one, we can obtain global optimal solutions for (4.8) by computing local optimal solutions of (4.9).

4.1.5 *Quadratic Optimization Problem and the Role of Special Matrix Classes*

In this section, we discuss about some matrix classes that plays a role in quadratic optimization problem. Consider QP(q, A) : $[\min x^T(Ax + q); x \geq 0,\ Ax + q \geq 0]$. We denote by $S^1(q, A)$, the set of optimal solutions of QP(q, A) and feasible solutions by $F(q, A) = \{x : Ax + q \geq 0, x \geq 0\}$. Applying the Farkas-Lemma, the feasibility is equivalent to the following condition:

$$x \geq 0, A^T x \leq 0 \Rightarrow q^T x \geq 0.$$

Let us consider the polyhedral convex cone

$$C_A = \{x \geq 0 \,|\, A^T x \leq 0\}$$

and its polar cone

$$C_A^* = \{x^* \,|\, x^T x^* \leq 0\ \forall x \in C_A\}.$$

Thus QP(q, A) is feasible iff $-q \in C_A^*$.

$$S^1(q, A) \neq \emptyset \text{ if and only if } -q \in C_A^*.$$

Assume that $x^* \in S^1(q, A)$. Then in view of KKT-condition for optimality there exist $u, v \in \mathbb{R}^n$ such that

$$(A + A^T)x^* + q - A^T u - v = 0, \tag{4.10}$$

$$x^*, u, v, Ax^* + q \geq 0, \tag{4.11}$$

$$x^{*T} v = u^T (Ax^* + q) = 0. \tag{4.12}$$

We denote by $S^2(q, A)$ the set of points for which such u and v exist. $S^2(q, A)$ is called the set of KKT-stationary points. We are interested in conditions implying that $S^2(q, A) = S^1(q, A)$.

In what follows, we introduce the following matrix classes. A is said to be *column sufficient* if for all $x \in \mathbb{R}^n$ the following implication holds:

$$x_i(Ax)_i \leq 0 \, \forall \, i \;\Rightarrow\; x_i(Ax)_i = 0 \, \forall \, i.$$

A is said to be *row sufficient* if A^T is column sufficient. A is *sufficient* if A and A^T are both column sufficient. We say that A is *positive semidefinite* (PSD) if $x^T Ax \geq 0 \, \forall \, x \in \mathbb{R}^n$ and A is *positive definite* (PD) if $x^T Ax > 0 \, \forall \, 0 \neq x \in \mathbb{R}^n$. A matrix $A \in \mathbb{R}^{n \times n}$ is a positive subdefinite (PSBD) matrix if for all $x \in \mathbb{R}^n$

$$x^T Ax < 0 \text{ implies either } A^T x \leq 0 \text{ or } A^T x \geq 0.$$

A matrix $A \in \mathbb{R}^{n \times n}$ is said to be *generalized positive subdefinite matrix* (GPSBD) if there exist two nonnegative diagonal matrices S and T with $S + T = I$ such that

$$\forall \, z \in \mathbb{R}^n, \quad z^t Az < 0 \Rightarrow \begin{cases} \text{either} & -Sz + TA^T z \geq 0 \\ \text{or} & -Sz + TA^T z \leq 0. \end{cases} \tag{4.13}$$

A matrix A is called *merely GPSBD matrix* (MGPSBD) if it is not a PSBD matrix. For details on these classes, see [9, 10, 16]. We now state the following theorem.

Theorem 4.1.1 *Assume any one of the following conditions hold:*

(i) *A is a copositive PSBD matrix with $rank(A) \geq 2$.*
(ii) *A is a copositive MGPSBD matrix with $0 < t_i < 1$ for all i.*

Then A is a row sufficient matrix.

The following result is an immediate consequence of the above theorem.

Lemma 4.1.1 *Suppose A is a copositive PSBD matrix with $rank(A) \geq 2$ or a copositive MGPSBD matrix with $0 < t_i < 1$ for all i. For each vector $q \in \mathbb{R}^n$, if $(\tilde{x}, \tilde{u})$ is a Karush–Kuhn–Tucker (KKT) pair of the quadratic program $QP(q, A)$: $[\min x^T(Ax + q); x \geq 0,\ Ax + q \geq 0]$, then $\tilde{x}$ solves $QP(q, A)$.*

4.2 Applications of Copositive Optimization in Graph Theory

We discuss the connection between nonconvex quadratic optimization and copositive optimization that allows the reformulation of nonconvex quadratic problems as convex ones in a unified way. Copositive optimization is a new approach for analyzing the specific, difficult case of optimizing a general nonconvex quadratic function over a polyhedron $\{x : Ax = b, x \geq 0\}$. In this section, we consider graph theoretic problems and reformulate stQP discussed in Sect. 4.1 as a convex quadratic optimization problem. We begin with some preliminaries on graph theory which will be used throughout the section. A graph G is a set of points $V(G)$ called vertices along with a set of line segments $E(G)$ called edges joining pairs of vertices. We say that two vertices are adjacent if there is an edge joining them. The set of vertices adjacent to v is the neighborhood of v which we denote as $N(v)$. If e is an edge joining v to one of its neighbors, we say e is incident to v. The degree of a vertex v, denoted $\deg(v)$, is the number of vertices adjacent to v. A graph is connected if there exists a path between every pair of distinct vertices. A closed walk is a walk with the same starting and ending vertex. An open walk is a walk in which the start and end vertices differ. A path is a walk in which no vertex is repeated. The distance between two vertices v and w is the length of the shortest path between v and w. A cycle is a closed walk in which no vertex is repeated (except that the starting and ending vertices are the same). The diameter of a connected graph G, denoted $diam(G)$ is the greatest distance between any two vertices of G. A tree is a connected graph that contains no cycles. A pendant vertex is a vertex whose degree is one. A tree on n vertices has $n - 1$ edges. Let $G = (V, E)$ be a connected graph with vertex set $V(G)$ and edge set $E(G)$. Let $c : V(G) \to \mathbb{R}^+$ be a nonnegative vertex weight function such that the total weight of the vertices is $N = \sum_{v \in V(G)} c(v)$.

Suppose $d_G(u, v)$ (or simply $d(u, v)$) denotes the usual distance (the length of the shortest path) between u and v in G. Then the total distance of G with respect to c, is defined by

$$d_c(G) = \sum_{\{u,v\} \subseteq V(G)} c(u)c(v)d_G(u, v).$$

Among all nonnegative weight functions c of given weight N, we seek to find one that maximizes $d_c(G)$.

Let G be a graph with vertices $\{1, 2, \ldots, n\}$. The distance matrix of G is defined as $D = [d_{ij}]$ where d_{ij} (which we also denote as $d(i, j)$) is the distance between vertices i and j. As an example, consider the tree

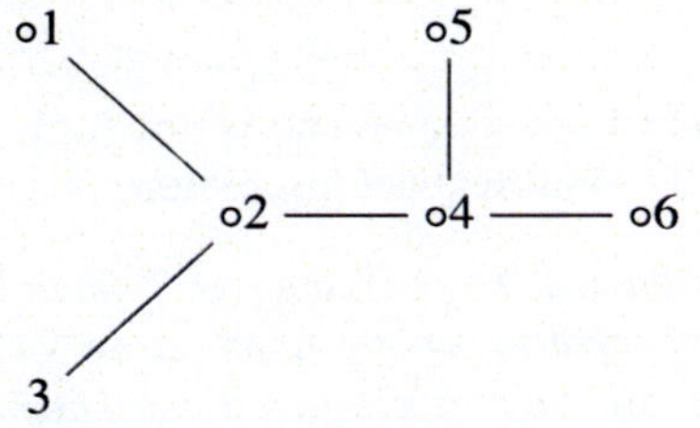

The distance matrix of the tree is given by

$$\begin{bmatrix} 0&1&2&2&3&3\\ 1&0&1&1&2&2\\ 2&1&0&2&3&3\\ 2&1&2&0&1&1\\ 3&2&3&1&0&2\\ 3&2&3&1&2&0 \end{bmatrix}$$

Let $\mathcal{D}$ be the distance matrix of a tree with n vertices. Let $\Delta = \{x \in \mathbb{R}^n_+ : e^T x = 1\}$. We consider the problem:

Problem I $\max x^T \mathcal{D} x$ subject to $x \in \Delta$.

If T is a tree on n vertices with distance matrix $\mathcal{D}$, then clearly, Problem I is equivalent to maximizing $d_c(T)$ over all nonnegative weight functions with given fixed weight N.

Note that Problem I and more general versions of it have occurred in the literature in different contexts. Apart from graph theory literature (see [11] and the references therein) there are at least two other areas where the problem has been considered. These areas are: (i) a generalized notion of diameter of finite metric space and (ii) Nash equilibria of symmetric bimatrix games associated with the distance matrix involving tree and resistance distance.

Theorem 4.2.2 *Let T be a tree with vertex set $\{1, \ldots, n\}$ and let $\mathcal{D}$ be the distance matrix of T. Then, there exists α_0 such that for all $\alpha > \alpha_0$, the matrix $\alpha\mathcal{E} - \mathcal{D}$ is positive definite, where $\mathcal{E}$ is a $m \times m$ matrix of all-ones.*

Note that D is a copositive matrix and the Problem I is a nonconvex quadratic Programming (NQP) problem, and we may write the equivalent convex quadratic programming (CQP) problem. By Theorem 4.2.2 there exists k such that $\tilde{\mathcal{D}} = k\mathcal{E} - \mathcal{D}$ is positive definite. We remark that to construct $\tilde{\mathcal{D}}$, it is sufficient to find the diameter (length of the longest path) of the tree. This can be done in polynomial time. Note that the maximum of $\frac{1}{2}x^T \mathcal{D} x$ over all $x \in \Delta$ is attained at x^* if and only if the minimum of $\frac{1}{2}x^T\tilde{\mathcal{D}}x$ over all $x \in \Delta$ is attained at x^*. Therefore, we solve Problem II.

Problem II: min $\frac{1}{2}x^T\tilde{\mathcal{D}}x$ subject to $Ax \geq b$ and $x \geq 0$ where $A = \begin{bmatrix} e_n^T \\ -e_n^T \end{bmatrix}$ and and $b = \begin{bmatrix} 1 \\ -1 \end{bmatrix}$.

A vertex of a tree of degree 1 is called an end vertex (or a pendant vertex) of T. The following result is useful for subsequent discussion.

Lemma 4.2.2 *[11, Proposition 2, p. 15] Given a tree T on at least two vertices and a real $N > 0$. Let c be a nonnegative weight function on $V(T)$ of total weight N that maximizes $d_c(T)$ among all such weight functions. Then, $c(v) > 0$ only if v is an end vertex of T*

In view of Lemma 4.2.2, we may replace $\tilde{D}$ in Problem II by the principal submatrix $\tilde{D}_p$ of $\tilde{\mathcal{D}}$ corresponding to the end vertices of the tree. The matrix A will be modified to A_p by replacing e_n by e_p, where p is the number of pendant vertices. We denote this problem as

Problem III: min $\frac{1}{2}y^T\tilde{\mathcal{D}}_p y$ subject to $A_p y \geq b$ and $y \in R^p_+$ where $A_p = \begin{bmatrix} e_p^T \\ -e_p^T \end{bmatrix}$ and and $b = \begin{bmatrix} 1 \\ -1 \end{bmatrix}$.

Lemma 4.2.3 *Problem II has a unique solution if and only if Problem III has a unique solution.*

We will write PD for positive definite and PSD for positive semidefinite. We may rewrite Problem II or III as a linear complementarity problem (denoted as LCP(q, M)) which is defined as follows. Given a real square matrix $A \in \mathbb{R}^{n\times n}$ and a vector $q \in \mathbb{R}^n$, the linear complementarity problem is to find $w, z \in \mathbb{R}^n$ such that $w - Mz = q,\ w \geq 0,\ z \geq 0$ and $w^T z = 0$.

The Karush–Kuhn–Tucker (KKT) necessary and sufficient optimality conditions specialized to Problem III yields the linear complementarity problem LCP(q, M) with $M = \begin{bmatrix} \tilde{\mathcal{D}}_p & -A_p^T \\ A_p & 0 \end{bmatrix}$, $q = \begin{bmatrix} 0 \\ -b \end{bmatrix}$. If (w, z) solves LCP(q, M) where $w = \begin{bmatrix} u \\ v \end{bmatrix}$ and $z = \begin{bmatrix} x \\ y \end{bmatrix}$ then x solves Problem III. It is easy to see that M is a PSD matrix.

Granot and Skorin-Kapov [14] extend Tardos' results and present a polynomial algorithm for solving strictly convex quadratic programming problems, in which, the number of arithmetic steps is independent of the size of the numbers on the right-hand side and the linear cost coefficients. Under the assumption that M is positive semidefinite, Kojima et al. [15] present a polynomial time algorithm that solves LCP(q, M) in $O(n^3L)$ arithmetic operations.

Remark 4.1 Dubey and Neogy [12] consider the question of solving the quadratic programming problem of finding maximum of $x^T\mathcal{R}x$ subject to $x \in \Delta = \{x \in \mathbb{R}^n_+ :$

$e^T x = 1\}$ and observe that this problem can be solved in polynomial time for the class of simple graphs with resistance distance matrix ($\mathcal{R}$) which are not necessarily a tree by reformulating this problem as a strictly convex quadratic programming problem.

4.2.1 Maximum Weight Clique Problem

We consider a copositive reformulation for the maximum weight clique problem. Consider an undirected graph $G = (V, E)$ with n nodes. A clique $\mathcal{S}$ is a subset of the node set V which corresponds to a complete subgraph of G (i.e., any pair of nodes in $\mathcal{S}$ is an edge in $\mathcal{E}$, the edge set). A clique $\mathcal{S}$ is said to be maximal if there is no larger clique containing $\mathcal{S}$.

Let A_G denotes the adjacency matrix of the graph G. Let f^* denotes the optimal value of the standard quadratic optimization problem max $f(x), x \in \Delta$ where $f(x) = x^T A_G x$. Then $\frac{1}{(1-f^*)}$ is the size of a maximum clique. This approach has served as the basis of many clique-finding algorithms and to determine theoretical bounds on the maximum clique size.

In [3], this problem was reformulated as a standard quadratic optimization problem and in [4] standard quadratic optimization problems were, in turn, reformulated as a copositive optimization problems. Therefore, the maximum weight clique problem is equivalent to copositive optimization problems.

4.3 The Notion of Transfinite Diameter in a Finite Metric Space and Copositive Optimization Problem

Let $M = (X, d)$ be a finite metric space, where $X = \{x_1, \ldots, x_n\}$. The distance matrix D of the metric space is the $n \times n$ matrix $D = [d_{ij}]$, where $d_{ij} = d(x_i, x_j)$. The metric space is completely described by its distance matrix. As a generalization of the diameter, the notion of transfinite diameter has been introduced. The notion of transfinite diameter (the maximal average distance in a multiset of points placed in the space), is a natural generalization of the diameter. The ∞-extender is the load vectors realizing the transfinite diameter provide strong structural information about metric spaces. It is, therefore, natural to study conditions under which ∞-extender are unique. The transfinite diameter of M equals the maximum of $x^T Dx$ over $x \in \Delta$. The vector that attains the maximum has been called ∞-extender of M can be posed as a copositive optimization problem. In what follows, we need the following definition to state a result related to a unique ∞-extender. The matrix A is said to be conditionally negative definite (c.n.d.) if $x^T Ax \leq 0$ for all $x \in \mathbb{R}^n$ such that $\sum_{i=1}^{n} x_i = 0$. Furthermore, a c.n.d. matrix is said to be strictly c.n.d. if $x^T Ax = 0$

only for $x = 0$. The matrix space M is said to be of negative type if D is c.n.d., while it is of strictly negative type if D is strictly c.n.d. Now we have the following theorem.

Theorem 4.3.3 *Let (X, d) be a finite metric space. If (X, d) is of strictly negative type, then (X, d) has a unique ∞-extender.*

4.4 Symmetric Bimatrix Game as a Copositive Optimization Problem

A bimatrix game is a noncooperative two-person game described by a pair $(\mathcal{A}, \mathcal{B})$ of $m \times n$ matrices. There are two players, Player 1 and Player 2, with m and n pure strategies respectively. If Player 1 chooses the i-th strategy and Player 2 chooses the j-th strategy, then a_{ij} and b_{ij} are the payoffs to Players 1 and 2, respectively. The mixed strategy spaces of Players 1 and 2 are Δ_m and Δ_n, respectively. A pair of strategies $(x^*, y^*) \in \Delta_m \times \Delta_n$ is a Nash equilibrium if $x^T \mathcal{A} y^* \leq x^{*T} \mathcal{A} y^*$ and $x^{*T} \mathcal{B} y \leq x^{*T} \mathcal{B} y^*$, for all $x \in \Delta_m$, $y \in \Delta_n$.

The celebrated theorem of Nash guarantees the existence of an equilibrium pair in any bimatrix game.

A bimatrix game is said to be symmetric if there is symmetry in strategies and payoffs, that is, if $m = n$ and $\mathcal{B} = \mathcal{A}^T$. A symmetric bimatrix game there is at least one symmetric Nash equilibrium, that is, an equilibrium of the form $(x^*, x^*) \in \Delta_n \times \Delta_n$. It can be seen that (x^*, x^*) is a symmetric Nash equilibrium of $(\mathcal{A}, \mathcal{A}^T)$ if and only if $(\mathcal{A}x^*)_i \leq x^{*T} \mathcal{A} x^*$, $i = 1, \ldots, n$; or equivalently, x^* maximizes $x^T \mathcal{A} x$ over $x \in \Delta_n$. In what follows, we consider symmetric bimatrix game associated with a tree.

Let T be a tree with n vertices and let $\mathcal{D}$ be the distance matrix of T. Consider the symmetric bimatrix game $(\mathcal{D}, \mathcal{D})$ in [5]. This game is interpreted as follows. Players 1 and 2 both choose a vertex each of the trees and tries to be as away from each other as possible. In view of the preceding discussion, $(x^*, x^*) \in \Delta_n \times \Delta_n$ is a symmetric Nash equilibrium of the game $(\mathcal{D}, \mathcal{D})$ if and only if x^* is a solution of Problem I. Note that the game $(\mathcal{D}, \mathcal{D})$ has a unique symmetric Nash equilibrium. The symmetric bimatrix game associated with a tree is extended by Dubey and Neogy [12] for resistance matrix as payoff matrix.

Let G be a connected graph with vertex set $\{1, \ldots, n\}$ and $\mathcal{R}$ be the resistance matrix where $\mathcal{R} = [r_{ij}]$ with its (i, j)-entry r_{ij} equal to the resistance distance between the i-th and the j-th vertices. In [12], Dubey and Neogy consider the symmetric bimatrix game $(\mathcal{R}, \mathcal{R})$. $(\tilde{x}, \tilde{x}) \in \Delta_n \times \Delta_n$ is a symmetric Nash equilibrium of the game $(\mathcal{R}, \mathcal{R})$ if and only if $\tilde{x}$ is a solution of Problem I. By using the same argument, it is easy to see that the game $(\mathcal{R}, \mathcal{R})$ has a unique symmetric Nash equilibrium.

Acknowledgements The authors would like to thank the anonymous referees for their constructive suggestions which considerably improve the overall presentation of the chapter.

References

1. Amaral, P.A., Bomze, I.M., Júdice, J.: Copositivity and constrained fractional quadratic problems. J. Math. Program. **146**, 325–350 (2014)
2. Bomze, I.M.: Copositive optimization - recent developments and applications. Eur. J. Oper. Res. **216**, 509–520 (2012)
3. Bomze, I.M.: On standard quadratic optimization problems. J. Global Optim. **13**(4), 369–387 (1998)
4. Bomze, I.M., Dür, M., de Klerk, E., Roos, C., Quist, A.J., Terlaky, T.: On copositive programming and standard quadratic optimization problems. J. Global Optim. **18**, 301–320 (2000)
5. Bapat, R.B., Neogy, S.K.: On a quadratic programming problem involving distances in trees. Ann. Oper. Res. **243**, 365–373 (2016)
6. Burer, S.: On the copositive representation of binary and continuous nonconvex quadratic programs. Math. Program. **120**, 479–495 (2009)
7. Burer, S.: Copositive programming. In: Anjos, M.F., Lasserre, J.B. (eds.) Handbook of Semidefinite, Cone and Polynomial Optimization: Theory, Algorithms, Software and Applications, International Series in Operations Research and Management Science. Springer, New York (2012)
8. Bundfuss, S., Dür, M.: An adaptive linear approximation algorithm for copositive programs. SIAM J. Opt. **20**, 30–53 (2009)
9. Crouzeix, J.-P., Hassouni, A., Lahlou, A., Schaible, S.: Positive subdefinite matrices, generalized monotonicity and linear complementarity problems. SIAM J. Matrix Anal. Appl. **22**, 66–85 (2000)
10. Crouzeix, J.-P., Komlósi, S.: The linear complementarity problem and the class of generalized positive subdefinite matrices. Optimization Theory (Matrahaza, 1999), Applied Optimization 59. Kluwer Academic Publisher, Dordrecht, pp. 45–63 (2001)
11. Dankelmann, P.: Average distance in weighted graphs. Discrete Math. **312**, 12–20 (2012)
12. Dubey, D., Neogy, S.K.: On solving a non-convex quadratic programming problem involving resistance distances in graphs. Ann. O.R. **287**, 643–651 (2020)
13. Eichfelder, G., Povh, J.: On the set-semidefinite representation of nonconvex quadratic programs over arbitrary feasible sets. Optim. Lett. **7**, 1373–1386 (2013)
14. Granot, F., Skorin-Kapov, J.: Towards a strongly polynomial algorithm for strictly convex quadratic programs: an extension of Tardos' algorithm. Math. Program. **46**, 225–236 (1990)
15. Kojima, M., Mizuno, S., Yoshise, A.: A polynomial-time algorithm for a class of linear complementarity problems. Math. Program. **44**, 1–26 (1989)
16. Mohan, S.R., Neogy, S.K., Das, A.K.: More on positive subdefinite matrices and the linear complementarity problem. Linear Algebra Appl. **338**, 275–285 (2001)
17. Nguyen, D.V.: Completely positive and copositive program modelling for quadratic optimization problems. Optimization, https://doi.org/10.1080/02331934.2020.1712392

Chapter 5
Hermite–Hadamard Type Inequalities For Functions Whose Derivatives Are Strongly η-Convex Via Fractional Integrals

Nidhi Sharma, Jaya Bisht, and S. K. Mishra

Abstract In this chapter, we establish some Hermite–Hadamard and Féjer type inequalities for strongly η-convex functions. We derive fractional integral inequalities for strongly η-convex functions. Further, some applications of these results to special means of real numbers are also discussed. Moreover, our results include several new and known results in particular cases.

Keywords Convex functions · Strongly η-convex functions · Hermite-Hadamard Féjer inequalities · Hölder's inequality

Mathematics Subject Classification (2010): 26A51, 26D15

5.1 Introduction

In 1969, Karamardian [8] introduced a strongly convex function and established a relationship between the generalized convexity of the function and the concepts of monotonicity of its gradient functions. Karamardian [8] also showed that every bidifferentiable function is strongly convex if and only if its Hessian matrix is strongly positive definite. For more details, one can refer to [10, 12].

Işcan [6] established Hermite–Hadamard–Féjer inequality for fractional integrals. Further, Park [14] obtained new estimates on the generalization of Hermite–Hadamard–Féjer type inequalities for differentiable functions whose derivatives in absolute value at certain powers are convex. Gordji et al. [4] introduced the new

N. Sharma · J. Bisht · S. K. Mishra (✉)
Department of Mathematics, Institute of Science, Banaras Hindu University, Varanasi 221005, India
e-mail: bhu.skmishra@gmail.com

N. Sharma
e-mail: sharmanidhirock@gmail.com

J. Bisht
e-mail: bishtjaya782@gmail.com

V. Laha et al. (eds.), *Optimization, Variational Analysis and Applications*,
Springer Proceedings in Mathematics & Statistics 355,
https://doi.org/10.1007/978-981-16-1819-2_5

class of convex functions known as η-convex functions and investigated the Jensen and Hermite–Hadamard type inequalities related to η-convex functions. Gordji et al. [4] showed that if $\xi : I \subseteq \mathbb{R} \to \mathbb{R}$ is φ-convex and φ is bounded from above on $\xi(I) \times \xi(I)$ with M_φ as an upper bound. Then ξ satisfies the Lipschitz condition on any closed interval $[c, d]$ contained in the interior I^0 of I. Hence, ξ is absolutely continuous on $[c, d]$ and continuous on I^0.

Awan et al. [1] introduced the notion of strongly η-convex functions and obtained some new integral inequalities of Hermite–Hadamard and Hermite–Hadamard–Féjer type for strongly η-convex functions. In 2019, Mishra and Sharma [11] introduced the concept of strongly η-convex functions of higher order, as a generalization of the strongly η-convex functions and investigated the Hermite–Hadamard and Hermite–Hadamard–Féjer type inequalities for strongly η-convex functions of higher order. For more details on Hermite–Hadamard inequalities, we refer the interested reader [2, 3, 7, 15].

The fractional inequalities play an important role in calculating different means for generalized convexity, so researchers are attracting to develop fractional integral inequalities for generalized convexity. Recently, Kwun et al. [9] established Hermite–Hadamard and Féjer type inequalities and derived fractional integral inequalities for η-convex functions. Further, Yang [17] investigated some Hermite–Hadamard type fractional integral inequalities for generalized h-convex functions. However, fractional integral inequalities for strongly η-convex functions have not been studied. Thus, the purpose of this chapter is to establish some Hermite–Hadamard and Féjer type inequalities for strongly η-convex functions. We derive some fractional integral inequalities for strongly η-convex functions. Further, we discuss some applications to special means of real numbers with the help of these results.

5.2 Preliminaries

Throughout this chapter, let I be an interval in real line $\mathbb{R}$ and I^0 denotes the interior of I.

Let $\xi : [c, d] \to \mathbb{R}$ be a convex function with $c < d$. Then the following double inequality is known as Hermite–Hadamard inequality in the literature.

$$\xi\left(\frac{c+d}{2}\right) \leq \frac{1}{d-c}\int_c^d \xi(x)dx \leq \frac{\xi(c)+\xi(d)}{2}. \tag{5.1}$$

Definition 5.1 [5] A function $\xi : I \subseteq \mathbb{R} \longrightarrow \mathbb{R}$ is said to be *η-convex* function with respect to $\eta : \mathbb{R} \times \mathbb{R} \longrightarrow \mathbb{R}$, if

$$\xi(\delta x + (1-\delta)y) \leq \xi(y) + \delta\eta(\xi(x), \xi(y)), \quad \forall x, y \in I, \quad \delta \in [0, 1].$$

Definition 5.2 [1] A function $\xi : I \subseteq \mathbb{R} \to \mathbb{R}$ is said to be *strongly η-convex* function with respect to $\eta : \mathbb{R} \times \mathbb{R} \to \mathbb{R}$ and modulus $\mu > 0$ if

$$\xi(\delta x + (1-\delta)y) \leq \xi(y) + \delta\eta(\xi(x), \xi(y)) - \mu\delta(1-\delta)(x-y)^2, \ \ \forall\, x, y \in I, \ \delta \in [0, 1]. \tag{5.2}$$

Lemma 5.2.1 *[16] Let $\xi : I \subseteq \mathbb{R} \to \mathbb{R}$ be a differentiable function on I^0 such that $\xi' \in L^1[c, d]$, where $c, d \in I$ with $c < d$. If $\alpha, \beta \in \mathbb{R}$, then*

$$\begin{aligned}&\frac{\alpha\xi(c) + \beta\xi(d)}{2} + \frac{2-\alpha-\beta}{2}\xi\left(\frac{c+d}{2}\right) - \frac{1}{d-c}\int_c^d \xi(x)dx \\ &= \frac{d-c}{4}\int_0^1 \left[(1-\alpha-\delta)\xi'\left(\delta c + (1-\delta)\frac{c+d}{2}\right)\right. \\ &\left. + (\beta-\delta)\xi'\left(\delta\frac{c+d}{2} + (1-\delta)d\right)\right] d\delta.\end{aligned}$$

Lemma 5.2.2 *[16] For $m > 0$ and $0 \leq \rho \leq 1$, we have*

$$\int_0^1 |\rho - \delta|^m d\delta = \frac{\rho^{m+1} + (1-\rho)^{m+1}}{m+1}$$

and

$$\int_0^1 \delta|\rho - \delta|^m d\delta = \frac{\rho^{m+2} + (m+\rho+1)(1-\rho)^{m+1}}{(m+1)(m+2)}.$$

Lemma 5.2.3 *For $m > 0$ and $0 \leq \rho \leq 1$, we have*

$$\int_0^1 \delta^2|\rho - \delta|^m d\delta = \frac{2\rho^{m+3} + (1-\rho)^{m+3}}{m+3} + \frac{2\rho(1-\rho)^{m+2}}{m+2} + \frac{\rho^2(1-\rho)^{m+1}}{m+1}.$$

Lemma 5.2.4 *[9] Let $\xi : I \subseteq \mathbb{R} \to \mathbb{R}$ be a differentiable mapping on I^0 with $\xi'' \in L^1[c, d]$, where $c, d \in I$ and $c < d$. Then*

$$\begin{aligned}\frac{1}{d-c}\int_c^d \xi(x)dx - \xi\left(\frac{c+d}{2}\right) = \frac{(d-c)^2}{16}&\left[\int_0^1 \delta^2\xi''\left(\delta\frac{c+d}{2} + (1-\delta)c\right) d\delta\right. \\ &\left. + \int_0^1 (\delta-1)^2\xi''\left(\delta d + (1-\delta)\frac{c+d}{2}\right) d\delta\right].\end{aligned}$$

Theorem 5.2.1 *[9] Let $\xi : I \subseteq \mathbb{R} \to \mathbb{R}$ be an η-convex function with $\xi \in L^1[c, d]$, where $c, d \in I$ with $c < d$. Then*

$$\xi\left(\frac{c+d}{2}\right)-\frac{1}{2(d-c)}\int_c^d \eta(\xi(c+d-x),\xi(x))dx$$
$$\leq \frac{1}{d-c}\int_c^d \xi(x)dx \leq \xi(d)+\frac{1}{2}\eta(\xi(c),\xi(d)).$$

5.3 Main Results

In this section, first, we prove Hermite–Hadamard and Féjer type inequalities for strongly η-convex functions.

Theorem 5.3.2 *Let $\xi : I \subset \mathbb{R} \to \mathbb{R}$ be an strongly η-convex function with modulus μ and $\xi \in L^1[c,d]$ where $c, d \in I$ with $c < d$, Then*

$$\xi\left(\frac{c+d}{2}\right)-\frac{1}{2(d-c)}\int_c^d \eta(\xi(c+d-x),\xi(x))dx+\frac{\mu}{12}(c-d)^2$$
$$\leq \frac{1}{(d-c)}\int_c^d \xi(x)dx$$
$$\leq \xi(d)+\frac{1}{2}\eta(\xi(c),\xi(d))-\frac{\mu}{6}(c-d)^2.$$

Proof From the definition of strong η-convexity, we have

$$\xi(\delta x+(1-\delta)y)\leq \xi(y)+\delta\eta(\xi(x),\xi(y))-\mu\delta(1-\delta)(x-y)^2,\quad \forall x, y \in I,\ \delta \in [0,1].$$

Using $x = tc + (1-t)d$, $y = (1-t)c + td$ and $\delta = \frac{1}{2}$, we get

$$\xi\left(\frac{1}{2}(tc+(1-t)d)+\frac{1}{2}((1-t)c+td)\right)$$
$$\leq \xi((1-t)c+td)+\frac{1}{2}\eta(\xi(tc+(1-t)d),\xi((1-t)c+td))-\frac{\mu}{4}((2t-1)(c-d))^2.$$

This implies

$$\xi\left(\frac{c+d}{2}\right)\leq \xi((1-t)c+td)+\frac{1}{2}\eta(\xi(tc+(1-t)d),\xi((1-t)c+td))$$
$$-\frac{\mu}{4}((2t-1)(c-d))^2.$$

Integrating above inequality from 0 to 1 with respect to t on both sides, we have

$$\xi\left(\frac{c+d}{2}\right) \leq \int_0^1 \xi((1-t)c+td)dt + \frac{1}{2}\int_0^1 \eta(\xi(tc+(1-t)d), \xi((1-t)c+td))dt$$
$$-\frac{\mu}{4}(c-d)^2\int_0^1 (2t-1)^2 dt.$$

Using the change of variable technique in above inequality, we have

$$\xi\left(\frac{c+d}{2}\right) \leq \frac{1}{d-c}\int_c^d \xi(x)dx + \frac{1}{2(d-c)}\int_c^d \eta(\xi(c+d-x), \xi(x))dx$$
$$-\frac{\mu}{12}(c-d)^2,$$

that is,

$$\xi\left(\frac{c+d}{2}\right) - \frac{1}{2(d-c)}\int_c^d \eta(\xi(c+d-x), \xi(x))dx + \frac{\mu}{12}(c-d)^2 \leq \frac{1}{d-c}\int_c^d \xi(x)dx. \tag{5.3}$$

We now prove the second pair of inequality. Using $x = c$ and $y = d$ in the definition of strong η-convexity, then we have

$$\xi(\delta c + (1-\delta)d) \leq \xi(d) + \delta\eta(\xi(c), \xi(d)) - \mu\delta(1-\delta)(c-d)^2, \quad \delta \in [0, 1].$$

Integrating above inequality from 0 to 1 on both sides with respect to δ, we get

$$\int_0^1 \xi(\delta c + (1-\delta)d)d\delta \leq \xi(d)\int_0^1 d\delta + \eta(\xi(c), \xi(d))\int_0^1 \delta d\delta - \mu(c-d)^2\int_0^1 \delta(1-\delta)d\delta.$$

This implies

$$\frac{1}{d-c}\int_c^d \xi(x)dx \leq \xi(d) + \frac{1}{2}\eta(\xi(c), \xi(d)) - \frac{\mu}{6}(c-d)^2. \tag{5.4}$$

From (5.3) and (5.4), we have

$$\xi\left(\frac{c+d}{2}\right) - \frac{1}{2(d-c)}\int_c^d \eta(\xi(c+d-x), \xi(x))dx + \frac{\mu}{12}(c-d)^2$$
$$\leq \frac{1}{(d-c)}\int_c^d \xi(x)dx$$
$$\leq \xi(d) + \frac{1}{2}\eta(\xi(c), \xi(d)) - \frac{\mu}{6}(c-d)^2.$$

This completes the proof. □

Remark 5.1 When $\mu = 0$, then above theorem reduces to Theorem 2.1 of [9]. If $\mu = 0$ and $\eta(x, y) = x - y$, then above theorem reduces to (5.1).

Theorem 5.3.3 *Let ξ and ϕ be nonnegative strongly η-convex functions with modulus μ_1 and μ_2, respectively, and $\xi\phi \in L^1[c, d]$, where $c, d \in I$, $c < d$. Then*

$$\frac{1}{(d-c)}\int_c^d \xi(x)\phi(x)dx \leq P(c, d),$$

where

$$\begin{aligned}P(c,d) &= \xi(d)\phi(d) + \frac{1}{2}[\xi(d)\eta(\phi(c),\phi(d)) + \phi(d)\eta(\xi(c),\xi(d))]\\&+\frac{1}{3}\eta(\xi(c),\xi(d))\eta(\phi(c),\phi(d)) - \frac{(c-d)^2}{12}(\mu_1\eta(\phi(c),\phi(d)) + \mu_2\eta(\xi(c),\xi(d)))\\&-\frac{(c-d)^2}{6}(\mu_1\phi(d)+\mu_2\xi(d)) + \frac{\mu_1\mu_2}{30}(c-d)^4.\end{aligned}$$

Proof Since ξ and ϕ are strongly η-convex functions with modulus μ_1 and μ_2, respectively, therefore

$$\xi(\delta c + (1-\delta)d) \leq \xi(d) + \delta\eta(\xi(c),\xi(d)) - \mu_1\delta(1-\delta)(c-d)^2, \ \forall\, \delta \in [0,1] \tag{5.5}$$

and

$$\phi(\delta c + (1-\delta)d) \leq \phi(d) + \delta\eta(\phi(c),\phi(d)) - \mu_2\delta(1-\delta)(c-d)^2, \ \forall\, \delta \in [0,1]. \tag{5.6}$$

From (5.5) and (5.6), we obtain

$$\begin{aligned}\xi(\delta c + (1-\delta)d)\phi(\delta c + (1-\delta)d) &\leq \xi(d)\phi(d) + \delta(\xi(d)\eta(\phi(c),\phi(d))\\&+\phi(d)\eta(\xi(c),\xi(d))) + \delta^2\eta(\xi(c),\xi(d))\eta(\phi(c),\phi(d))\\&-\delta^2(1-\delta)(c-d)^2(\mu_1\eta(\phi(c),\phi(d)) + \mu_2\eta(\xi(c),\xi(d)))\\&-\delta(1-\delta)(c-d)^2(\mu_1\phi(d)+\mu_2\xi(d)) + \delta^2(1-\delta)^2\mu_1\mu_2(c-d)^4.\end{aligned}$$

Integrating above inequality from 0 to 1 on both sides with respect to δ, we have

$$\begin{aligned}\int_0^1 \xi(\delta c + (1-\delta)d)\phi(\delta c + (1-\delta)d)d\delta &\leq \xi(d)\phi(d)\\&+\frac{1}{2}[\xi(d)\eta(\phi(c),\phi(d)) + \phi(d)\eta(\xi(c),\xi(d))] + \frac{1}{3}\eta(\xi(c),\xi(d))\eta(\phi(c),\phi(d))\\&-\frac{(c-d)^2}{12}(\mu_1\eta(\phi(c),\phi(d)) + \mu_2\eta(\xi(c),\xi(d))) - \frac{(c-d)^2}{6}(\mu_1\phi(d)+\mu_2\xi(d))\\&+\frac{\mu_1\mu_2}{30}(c-d)^4.\end{aligned}$$

This implies

$$\frac{1}{(d-c)}\int_c^d \xi(x)\phi(x)dx \leq P(c,d).$$

This completes the proof. □

Remark 5.2 When $\mu = 0$, then above theorem reduces to Theorem 2.2 of [9]. If $\mu = 0$ and $\eta(x, y) = x - y$, then above theorem reduces to Theorem 1 of [13].

Theorem 5.3.4 *Let ξ be an strongly η-convex function with modulus μ and $\xi \in L^1[c, d]$, where $c, d \in I$, $c < d$ and $\phi : [c, d] \to \mathbb{R}$ be nonnegative, integrable, and symmetric about $\left(\frac{c+d}{2}\right)$. Then*

$$\int_c^d \xi(x)\phi(x)dx \leq \left(\xi(d) + \frac{1}{2}\eta(\xi(c), \xi(d))\right)\int_c^d \phi(x)dx - \mu\int_c^d (x-c)(d-x)\phi(x)dx.$$

Proof Since ξ be an strongly η-convex function with modulus μ, and ϕ nonnegative, integrable, and symmetric about $\left(\frac{c+d}{2}\right)$, therefore, we have

$$\begin{aligned}
\int_c^d \xi(x)\phi(x)dx &= \frac{1}{2}\left[\int_c^d \xi(x)\phi(x)dx + \int_c^d \xi(c+d-x)\phi(c+d-x)dx\right]\\
&= \frac{1}{2}\left[\int_c^d \xi(x)\phi(x)dx + \int_c^d \xi(c+d-x)\phi(x)dx\right]\\
&= \frac{1}{2}\int_c^d \left[\xi\left(\frac{d-x}{d-c}c + \frac{x-c}{d-c}d\right) + \xi\left(\frac{x-c}{d-c}c + \frac{d-x}{d-c}d\right)\right]\phi(x)dx\\
&\leq \frac{1}{2}\int_c^d \left[\left(\xi(d) + \left(\frac{d-x}{d-c}\right)\eta(\xi(c), \xi(d))\right.\right.\\
&\quad \left. - \mu\left(\frac{x-c}{d-c}\right)\left(\frac{d-x}{d-c}\right)(c-d)^2\right) + \left(\xi(d) + \left(\frac{x-c}{d-c}\right)\eta(\xi(c), \xi(d))\right.\\
&\quad \left.\left. - \mu\left(\frac{x-c}{d-c}\right)\left(\frac{d-x}{d-c}\right)(c-d)^2\right)\right]\phi(x)dx\\
&= \left(\xi(d) + \frac{1}{2}\eta(\xi(c), \xi(d))\right)\int_c^d \phi(x)dx - \mu\int_c^d (x-c)(d-x)\phi(x)dx.
\end{aligned}$$

This completes the proof. □

Remark 5.3 When $\mu = 0$, then above theorem reduces to Theorem 2.3 of [9]. If $\mu = 0$, $\eta(x, y) = x - y$ and $\phi(x) = 1$, then above theorem reduces to second inequality of (5.1).

Now we establish the results on fractional integral inequalities for strongly η-convex functions.

Theorem 5.3.5 *Let $\xi : I \subseteq \mathbb{R} \to \mathbb{R}$, be a differentiable mapping on I^0 with $\xi' \in L^1[c, d]$, where $c, d \in I$, $c < d$. If $|\xi'(x)|^q$ for $q \geq 1$ is strongly η-convex with modulus μ on $[c, d]$ and $0 \leq \alpha, \beta \leq 1$, then*

$$\left|\frac{\alpha\xi(c)+\beta\xi(d)}{2}+\frac{2-\alpha-\beta}{2}\xi\left(\frac{c+d}{2}\right)-\frac{1}{d-c}\int_c^d\xi(x)dx\right|$$
$$\leq\left(\frac{d-c}{8}\right)\left(\frac{1}{24}\right)^{1/q}\Big[(2\alpha^2-2\alpha+1)^{1-\frac{1}{q}}(24(2\alpha^2-2\alpha+1)|\xi'(d)|^q$$
$$+4(-2\alpha^3+12\alpha^2-9\alpha+4)\eta(|\xi'(c)|^q,|\xi'(d)|^q)$$
$$-\mu(-7\alpha^4+28\alpha^3-30\alpha^2+12\alpha)(c-d)^2)^{1/q}$$
$$+(2\beta^2-2\beta+1)^{1-\frac{1}{q}}(24(2\beta^2-2\beta+1)|\xi'(d)|^q$$
$$+4(2\beta^3-3\beta+2)\eta(|\xi'(c)|^q,|\xi'(d)|^q)$$
$$-\mu(-7\beta^4+8\beta^3-8\beta+5)(c-d)^2)^{1/q}\Big].$$

Proof Recall Lemma 5.2.1;

$$\frac{\alpha\xi(c)+\beta\xi(d)}{2}+\frac{2-\alpha-\beta}{2}\xi\left(\frac{c+d}{2}\right)-\frac{1}{d-c}\int_c^d\xi(x)dx$$
$$=\frac{d-c}{4}\int_0^1\Big[(1-\alpha-\delta)\xi'\left(\delta c+(1-\delta)\frac{c+d}{2}\right)$$
$$+(\beta-\delta)\xi'\left(\delta\frac{c+d}{2}+(1-\delta)d\right)\Big]d\delta.$$

This implies

$$\left|\frac{\alpha\xi(c)+\beta\xi(d)}{2}+\frac{2-\alpha-\beta}{2}\xi\left(\frac{c+d}{2}\right)-\frac{1}{d-c}\int_c^d\xi(x)dx\right|$$
$$\leq\frac{d-c}{4}\Big[\int_0^1|1-\alpha-\delta|\left|\xi'\left(\delta c+(1-\delta)\frac{c+d}{2}\right)\right|d\delta$$
$$+\int_0^1|\beta-\delta|\left|\xi'\left(\delta\frac{c+d}{2}+(1-\delta)d\right)\right|d\delta\Big]. \tag{5.7}$$

Applying Hölder's inequality and the definition of strong η-convexity in (5.7), we have

$$
\begin{aligned}
&\left|\frac{\alpha\xi(c)+\beta\xi(d)}{2}+\frac{2-\alpha-\beta}{2}\xi\left(\frac{c+d}{2}\right)-\frac{1}{d-c}\int_c^d \xi(x)dx\right| \\
&\leq \frac{d-c}{4}\left[\left(\int_0^1 |1-\alpha-\delta|d\delta\right)^{1-\frac{1}{q}}\left(\int_0^1 |1-\alpha-\delta|\left(|\xi'(d)|^q\right.\right.\right. \\
&+\left.\left(\frac{1+\delta}{2}\right)\eta(|\xi'(c)|^q,|\xi'(d)|^q)-\mu\left(\frac{1+\delta}{2}\right)\left(\frac{1-\delta}{2}\right)(c-d)^2\right)d\delta\Bigg)^{1/q} \\
&+\left(\int_0^1 |\beta-\delta|d\delta\right)^{1-\frac{1}{q}}\left(\int_0^1 |\beta-\delta|\left(|\xi'(d)|^q+\left(\frac{\delta}{2}\right)\eta(|\xi'(c)|^q,|\xi'(d)|^q)\right.\right. \\
&-\left.\left.\left.\mu\left(\frac{\delta}{2}\right)\left(1-\frac{\delta}{2}\right)(c-d)^2\right)d\delta\right)^{1/q}\right]. \qquad (5.8)
\end{aligned}
$$

Using Lemmas 5.2.2 and 5.2.3, we calculate

$$
\begin{aligned}
&\int_0^1 |1-\alpha-\delta|\left(|\xi'(d)|^q+\left(\frac{1+\delta}{2}\right)\eta(|\xi'(c)|^q,|\xi'(d)|^q)-\frac{\mu}{4}(1-\delta^2)(c-d)^2\right)d\delta \\
&=\left(|\xi'(d)|^q+\frac{1}{2}\eta(|\xi'(c)|^q,|\xi'(d)|^q)-\frac{\mu}{4}(c-d)^2\right)\int_0^1 |1-\alpha-\delta|d\delta \\
&+\frac{1}{2}\eta(|\xi'(c)|^q,|\xi'(d)|^q)\int_0^1 \delta|1-\alpha-\delta|d\delta+\frac{\mu}{4}(c-d)^2\int_0^1 \delta^2|1-\alpha-\delta|d\delta \\
&=\frac{1}{2}(2\alpha^2-2\alpha+1)|\xi'(d)|^q+\frac{1}{12}(-2\alpha^3+12\alpha^2-9\alpha+4)\eta(|\xi'(c)|^q,|\xi'(d)|^q) \\
&-\frac{\mu}{48}(-7\alpha^4+28\alpha^3-30\alpha^2+12\alpha)(c-d)^2 \qquad (5.9)
\end{aligned}
$$

and

$$
\begin{aligned}
&\int_0^1 |\beta-\delta|\left(|\xi'(d)|^q+\left(\frac{\delta}{2}\right)\eta(|\xi'(c)|^q,|\xi'(d)|^q)-\mu\left(\frac{\delta}{2}\right)\left(1-\frac{\delta}{2}\right)(c-d)^2\right)d\delta \\
&=|\xi'(d)|^q\int_0^1 |\beta-\delta|d\delta+\left(\frac{1}{2}\eta(|\xi'(c)|^q,|\xi'(d)|^q)-\frac{\mu}{2}(c-d)^2\right)\int_0^1 \delta|\beta-\delta|d\delta \\
&+\frac{\mu}{4}(c-d)^2\int_0^1 \delta^2|\beta-\delta|d\delta \\
&=\frac{1}{2}(2\beta^2-2\beta+1)|\xi'(d)|^q+\frac{1}{12}(2\beta^3-3\beta+2)\eta(|\xi'(c)|^q,|\xi'(d)|^q) \\
&-\frac{\mu}{48}(-7\beta^4+8\beta^3-8\beta+5)(c-d)^2. \qquad (5.10)
\end{aligned}
$$

From (5.8)–(5.10) and Lemma 5.2.2, we have

$$\left|\frac{\alpha\xi(c)+\beta\xi(d)}{2}+\frac{2-\alpha-\beta}{2}\xi\left(\frac{c+d}{2}\right)-\frac{1}{d-c}\int_c^d \xi(x)dx\right|$$
$$\leq\left(\frac{d-c}{8}\right)\left(\frac{1}{24}\right)^{1/q}\Big[(2\alpha^2-2\alpha+1)^{1-\frac{1}{q}}(24(2\alpha^2-2\alpha+1)|\xi'(d)|^q$$
$$+4(-2\alpha^3+12\alpha^2-9\alpha+4)\eta(|\xi'(c)|^q,|\xi'(d)|^q)$$
$$-\mu(-7\alpha^4+28\alpha^3-30\alpha^2+12\alpha)(c-d)^2)^{1/q}$$
$$+(2\beta^2-2\beta+1)^{1-\frac{1}{q}}(24(2\beta^2-2\beta+1)|\xi'(d)|^q$$
$$+4(2\beta^3-3\beta+2)\eta(|\xi'(c)|^q,|\xi'(d)|^q)-\mu(-7\beta^4+8\beta^3-8\beta+5)(c-d)^2)^{1/q}\Big].$$

This completes the proof. □

Remark 5.4 When $\mu=0$, then above theorem reduces to Theorem 3.1 of [9].

Corollary 5.1 *If $\alpha=\beta$ in above theorem, then*

$$\left|\frac{\alpha}{2}(\xi(c)+\xi(d))+(1-\alpha)\xi\left(\frac{c+d}{2}\right)-\frac{1}{d-c}\int_c^d \xi(x)dx\right|$$
$$\leq\left(\frac{d-c}{8}\right)\left(\frac{1}{24}\right)^{1/q}(2\alpha^2-2\alpha+1)^{1-\frac{1}{q}}$$
$$\times[(24(2\alpha^2-2\alpha+1)|\xi'(d)|^q+4(-2\alpha^3+12\alpha^2-9\alpha+4)\eta(|\xi'(c)|^q,|\xi'(d)|^q)$$
$$-\mu(-7\alpha^4+28\alpha^3-30\alpha^2+12\alpha)(c-d)^2)^{1/q}+(24(2\alpha^2-2\alpha+1)|\xi'(d)|^q$$
$$+4(2\alpha^3-3\alpha+2)\eta(|\xi'(c)|^q,|\xi'(d)|^q)-\mu(-7\alpha^4+8\alpha^3-8\alpha+5)(c-d)^2)^{1/q}].$$

Corollary 5.2 *If $\alpha=\beta=\frac{1}{2}$ in Corollary 5.1, then*

$$\left|\frac{1}{2}\left[\frac{\xi(c)+\xi(d)}{2}+\xi\left(\frac{c+d}{2}\right)\right]-\frac{1}{d-c}\int_c^d \xi(x)dx\right|$$
$$\leq\left(\frac{d-c}{16}\right)\left(\frac{1}{192}\right)^{1/q}[(192|\xi'(d)|^q+144\eta(|\xi'(c)|^q,|\xi'(d)|^q)-25\mu(c-d)^2)^{1/q}$$
$$+(192|\xi'(d)|^q+48\eta(|\xi'(c)|^q,|\xi'(d)|^q)-25\mu(c-d)^2)^{1/q}].$$

Corollary 5.3 *If $q=1$ in Corollary 5.2, then*

$$\left|\frac{1}{2}\left[\frac{\xi(c)+\xi(d)}{2}+\xi\left(\frac{c+d}{2}\right)\right]-\frac{1}{d-c}\int_c^d \xi(x)dx\right|$$
$$\leq\left(\frac{d-c}{1536}\right)[192|\xi'(d)|+96\eta(|\xi'(c)|,|\xi'(d)|)-25\mu(c-d)^2].$$

Theorem 5.3.6 *Let $\xi: I\subseteq\mathbb{R}\to\mathbb{R}$, be a differentiable mapping on I^0 with $\xi'\in L^1[c,d]$, where $c,d\in I$, $c<d$. If $|\xi'(x)|^q$ for $q\geq 1$ is strongly η-convex with modulus μ on $[c,d]$ and $0\leq\alpha,\beta\leq 1$, then*

$$\left|\frac{\alpha\xi(c)+\beta\xi(d)}{2}+\frac{2-\alpha-\beta}{2}\xi\left(\frac{c+d}{2}\right)-\frac{1}{d-c}\int_c^d \xi(x)dx\right|$$
$$\leq \frac{d-c}{4}\left[\left(\left(\frac{(1-\alpha)^{q+1}+\alpha^{q+1}}{q+1}\right)|\xi'(d)|^q\right.\right.$$
$$+\left(\frac{(q+2)((1-\alpha)^{q+1}+2\alpha^{q+1})+(1-\alpha)^{q+2}-\alpha^{q+2}}{2(q+1)(q+2)}\right)\eta(|\xi'(c)|^q,|\xi'(d)|^q)$$
$$\left.-\frac{\mu}{4}(c-d)^2\left(\frac{(1-\alpha)^{q+1}-\alpha^{q+3}+2\alpha^{q+2}}{q+1}-\frac{2(1-\alpha)\alpha^{q+2}}{q+2}-\frac{2(1-\alpha)^{q+3}\alpha^{q+3}}{q+3}\right)\right)^{1/q}$$
$$+\left(\left(\frac{(1-\beta)^{q+1}+\beta^{q+1}}{q+1}\right)|\xi'(d)|^q+\left(\frac{(q+\beta+1)(1-\beta)^{q+1}+\beta^{q+2}}{2(q+1)(q+2)}\right)\right.$$
$$\times\ \eta(|\xi'(c)|^q,|\xi'(d)|^q)-\frac{\mu}{4}(c-d)^2$$
$$\times\left(\frac{2(q+\beta+1)(1-\beta)^{q+1}-2(q+1)\beta(1-\beta)^{q+2}-(q+2)\beta^2(1-\beta)^{q+1}+2\beta^{q+2}}{(q+1)(q+2)}\right.$$
$$\left.\left.\left.-\frac{2\beta^{q+3}+(1-\beta)^{q+3}}{q+3}\right)\right)^{1/q}\right].$$

Proof From Lemma 5.2.1, we have

$$\left|\frac{\alpha\xi(c)+\beta\xi(d)}{2}+\frac{2-\alpha-\beta}{2}\xi\left(\frac{c+d}{2}\right)-\frac{1}{d-c}\int_c^d \xi(x)dx\right|$$
$$\leq \frac{d-c}{4}\left[\int_0^1|1-\alpha-\delta|\left|\xi'\left(\delta c+(1-\delta)\frac{c+d}{2}\right)\right|d\delta\right.$$
$$\left.+\int_0^1|\beta-\delta|\left|\xi'\left(\delta\frac{c+d}{2}+(1-\delta)d\right)\right|d\delta\right].$$

Using Hölder's inequality and the definition of strong η-convexity, we have

$$\left|\frac{\alpha\xi(c)+\beta\xi(d)}{2}+\frac{2-\alpha-\beta}{2}\xi\left(\frac{c+d}{2}\right)-\frac{1}{d-c}\int_c^d \xi(x)dx\right|$$
$$\leq \frac{d-c}{4}\left[\left(\int_0^1 d\delta\right)^{1-\frac{1}{q}}\left(\int_0^1|1-\alpha-\delta|^q\left|\xi'\left(\delta c+(1-\delta)\frac{c+d}{2}\right)\right|^q d\delta\right)^{1/q}\right.$$
$$\left.+\left(\int_0^1 d\delta\right)^{1-\frac{1}{q}}\left(\int_0^1|\beta-\delta|^q\left|\xi'\left(\delta\frac{c+d}{2}+(1-\delta)d\right)\right|^q d\delta\right)^{1/q}\right]$$
$$\leq \frac{d-c}{4}\left[\left(\int_0^1|1-\alpha-\delta|^q\left(|\xi'(d)|^q+\left(\frac{1+\delta}{2}\right)\eta(|\xi'(c)|^q,|\xi'(d)|^q)\right.\right.\right.$$
$$\left.\left.-\frac{\mu}{4}(1-\delta^2)(c-d)^2\right)d\delta\right)^{1/q}+\left(\int_0^1|\beta-\delta|^q\left(|\xi'(d)|^q\right.\right.$$
$$\left.\left.\left.+\left(\frac{\delta}{2}\right)\eta(|\xi'(c)|^q,|\xi'(d)|^q)-\mu\left(\frac{\delta}{2}\right)\left(1-\frac{\delta}{2}\right)(c-d)^2\right)d\delta\right)^{1/q}\right]$$

$$= \frac{d-c}{4}\left[\left(\left(|\xi'(d)|^q + \frac{1}{2}\eta(|\xi'(c)|^q, |\xi'(d)|^q) - \frac{\mu}{4}(c-d)^2\right)\int_0^1 |1-\alpha-\delta|^q d\delta\right.\right.$$

$$\left. + \frac{1}{2}\eta(|\xi'(c)|^q, |\xi'(d)|^q)\int_0^1 \delta|1-\alpha-\delta|^q d\delta + \frac{\mu}{4}(c-d)^2\int_0^1 \delta^2|1-\alpha-\delta|^q d\delta\right)^{1/q}$$

$$+ \left(|\xi'(d)|^q \int_0^1 |\beta-\delta|^q d\delta + \left(\frac{1}{2}\eta(|\xi'(c)|^q, |\xi'(d)|^q) - \frac{\mu}{2}(c-d)^2\right)\int_0^1 \delta|\beta-\delta|^q d\delta\right.$$

$$\left.\left. + \frac{\mu}{4}(c-d)^2\int_0^1 \delta^2|\beta-\delta|^q d\delta\right)^{1/q}\right].$$

Applying Lemmas 5.2.2 and 5.2.3, we obtain

$$\left|\frac{\alpha\xi(c)+\beta\xi(d)}{2} + \frac{2-\alpha-\beta}{2}\xi\left(\frac{c+d}{2}\right) - \frac{1}{d-c}\int_c^d \xi(x)dx\right|$$

$$\leq \frac{d-c}{4}\left[\left(\left(\frac{(1-\alpha)^{q+1}+\alpha^{q+1}}{q+1}\right)|\xi'(d)|^q\right.\right.$$

$$+ \left(\frac{(q+2)((1-\alpha)^{q+1}+2\alpha^{q+1}) + (1-\alpha)^{q+2} - \alpha^{q+2}}{2(q+1)(q+2)}\right)\eta(|\xi'(c)|^q, |\xi'(d)|^q)$$

$$- \frac{\mu}{4}(c-d)^2\left(\frac{(1-\alpha)^{q+1} - \alpha^{q+3} + 2\alpha^{q+2}}{q+1} - \frac{2(1-\alpha)\alpha^{q+2}}{q+2}\right.$$

$$\left.\left. - \frac{2(1-\alpha)^{q+3}+\alpha^{q+3}}{q+3}\right)\right)^{1/q}$$

$$+ \left(\left(\frac{(1-\beta)^{q+1}+\beta^{q+1}}{q+1}\right)|\xi'(d)|^q + \left(\frac{(q+\beta+1)(1-\beta)^{q+1}+\beta^{q+2}}{2(q+1)(q+2)}\right)\right.$$

$$\times\ \eta(|\xi'(c)|^q, |\xi'(d)|^q) - \frac{\mu}{4}(c-d)^2$$

$$\times \left(\frac{2(q+\beta+1)(1-\beta)^{q+1} - 2(q+1)\beta(1-\beta)^{q+2} - (q+2)\beta^2(1-\beta)^{q+1} + 2\beta^{q+2}}{(q+1)(q+2)}\right.$$

$$\left.\left.\left. - \frac{2\beta^{q+3}+(1-\beta)^{q+3}}{q+3}\right)\right)^{1/q}\right].$$

This completes the proof. □

Remark 5.5 When $\mu = 0$, then above theorem reduces to [9, Theorem 3.2].

Corollary 5.4 *If* $\alpha = \beta$ *in above theorem, then*

$$\left|\frac{\alpha}{2}(\xi(c)+\xi(d))+(1-\alpha)\xi\left(\frac{c+d}{2}\right)-\frac{1}{d-c}\int_c^d \xi(x)dx\right|$$
$$\leq \frac{d-c}{4}\left[\left(\left(\frac{(1-\alpha)^{q+1}+\alpha^{q+1}}{q+1}\right)|\xi'(d)|^q\right.\right.$$
$$+\left(\frac{(q+2)((1-\alpha)^{q+1}+2\alpha^{q+1})+(1-\alpha)^{q+2}-\alpha^{q+2}}{2(q+1)(q+2)}\right)\eta(|\xi'(c)|^q,|\xi'(d)|^q)$$
$$\left.-\frac{\mu}{4}(c-d)^2\left(\frac{(1-\alpha)^{q+1}-\alpha^{q+3}+2\alpha^{q+2}}{q+1}-\frac{2(1-\alpha)\alpha^{q+2}}{q+2}-\frac{2(1-\alpha)^{q+3}\alpha^{q+3}}{q+3}\right)\right)^{1/q}$$
$$+\left(\left(\frac{(1-\alpha)^{q+1}+\alpha^{q+1}}{q+1}\right)|\xi'(d)|^q\right.$$
$$+\left(\frac{(q+\alpha+1)(1-\alpha)^{q+1}+\alpha^{q+2}}{2(q+1)(q+2)}\right)\eta(|\xi'(c)|^q,|\xi'(d)|^q)-\frac{\mu}{4}(c-d)^2$$
$$\times\left(\frac{2(q+\alpha+1)(1-\alpha)^{q+1}-2(q+1)\alpha(1-\alpha)^{q+2}-(q+2)\alpha^2(1-\alpha)^{q+1}+2\alpha^{q+2}}{(q+1)(q+2)}\right.$$
$$\left.\left.\left.-\frac{2\alpha^{q+3}+(1-\alpha)^{q+3}}{q+3}\right)\right)^{1/q}\right].$$

Theorem 5.3.7 *Let $\xi : I \subset [0,\infty) \to \mathbb{R}$ be a differentiable mapping on I^0 with $\xi'' \in L^1[c,d]$, where $c, d \in I$ and $c < d$. If $|\xi''|$ is strongly η-convex with modulus μ on $[c,d]$, then*

$$\left|\xi\left(\frac{c+d}{2}\right)-\frac{1}{d-c}\int_c^d \xi(x)dx\right|$$
$$\leq \frac{(d-c)^2}{16}\left[\frac{1}{3}\left(|\xi''(c)|+\left|\xi''\left(\frac{c+d}{2}\right)\right|\right)\right.$$
$$\left.+\frac{1}{4}\left(\eta\left(\left|\xi''\left(\frac{c+d}{2}\right)\right|,|\xi''(c)|\right)+\frac{1}{3}\eta\left(|\xi''(d)|,\left|\xi''\left(\frac{c+d}{2}\right)\right|\right)\right)-\frac{\mu}{40}(d-c)^2\right].$$

Proof Recall Lemma 5.2.4, we have

$$\left|\xi\left(\frac{c+d}{2}\right)-\frac{1}{d-c}\int_c^d \xi(x)dx\right| \leq \frac{(d-c)^2}{16}\left[\int_0^1 \delta^2\left|\xi''\left(\delta\frac{c+d}{2}+(1-\delta)c\right)\right|d\delta\right.$$
$$\left.+\int_0^1(\delta-1)^2\left|\xi''\left(\delta d+(1-\delta)\frac{c+d}{2}\right)\right|d\delta\right].$$

Using the definition of strong η-convexity, we obtain

$$\left|\xi\left(\frac{c+d}{2}\right)-\frac{1}{d-c}\int_c^d \xi(x)dx\right|$$
$$\leq \frac{(d-c)^2}{16}\left[\int_0^1 \delta^2\left(|\xi''(c)|+\delta\eta\left(\left|\xi''\left(\frac{c+d}{2}\right)\right|,|\xi''(c)|\right)-\mu\delta(1-\delta)\left(\frac{d-c}{2}\right)^2\right)d\delta\right.$$
$$+\int_0^1(\delta-1)^2\left(\left|\xi''\left(\frac{c+d}{2}\right)\right|+\delta\eta\left(|\xi''(d)|,\left|\xi''\left(\frac{c+d}{2}\right)\right|\right)\right.$$
$$\left.\left.-\mu\delta(1-\delta)\left(\frac{d-c}{2}\right)^2\right)d\delta\right]$$
$$=\frac{(d-c)^2}{16}\left[\left(\frac{1}{3}|\xi''(c)|+\frac{1}{4}\eta\left(\left|\xi''\left(\frac{c+d}{2}\right)\right|,|\xi''(c)|\right)-\frac{\mu}{20}\left(\frac{d-c}{2}\right)^2\right)\right.$$
$$\left.+\left(\frac{1}{3}\left|\xi''\left(\frac{c+d}{2}\right)\right|+\frac{1}{12}\eta\left(|\xi''(d)|,\left|\xi''\left(\frac{c+d}{2}\right)\right|\right)-\frac{\mu}{20}\left(\frac{d-c}{2}\right)^2\right)\right]$$
$$=\frac{(d-c)^2}{16}\left[\frac{1}{3}\left(|\xi''(c)|+\left|\xi''\left(\frac{c+d}{2}\right)\right|\right)\right.$$
$$\left.+\frac{1}{4}\left(\eta\left(\left|\xi''\left(\frac{c+d}{2}\right)\right|,|\xi''(c)|\right)+\frac{1}{3}\eta\left(|\xi''(d)|,\left|\xi''\left(\frac{c+d}{2}\right)\right|\right)\right)-\frac{\mu}{40}(d-c)^2\right].$$

This completes the proof. □

Remark 5.6 When $\mu = 0$, then above theorem reduces to Theorem 3.3 of [9].

Theorem 5.3.8 *Let $\xi : I \subset [0, \infty) \to \mathbb{R}$ be a differentiable mapping on I^0 with $\xi'' \in L^1[c, d]$, where $c, d \in I$ and $c < d$. If $|\xi''|^q$ for $q \geq 1$ with $\frac{1}{p}+\frac{1}{q}=1$ is strongly η-convex with modulus μ on $[c, d]$, then*

$$\left|\xi\left(\frac{c+d}{2}\right)-\frac{1}{d-c}\int_c^d \xi(x)dx\right|$$
$$\leq \frac{(d-c)^2}{16}\left(\frac{1}{3}\right)^{\frac{1}{p}}\left[\left(\frac{1}{3}|\xi''(c)|^q+\frac{1}{4}\eta\left(\left|\xi''\left(\frac{c+d}{2}\right)\right|^q,|\xi''(c)|^q\right)\right.\right.$$
$$\left.-\frac{\mu}{80}(d-c)^2\right)^{\frac{1}{q}}+\left(\frac{1}{3}\left|\xi''\left(\frac{c+d}{2}\right)\right|^q\right.$$
$$\left.\left.+\frac{1}{12}\eta\left(|\xi''(d)|^q,\left|\xi''\left(\frac{c+d}{2}\right)\right|^q\right)-\frac{\mu}{80}(d-c)^2\right)^{\frac{1}{q}}\right].$$

Proof From Lemma 5.2.4, we have

$$\left|\xi\left(\frac{c+d}{2}\right)-\frac{1}{d-c}\int_c^d \xi(x)dx\right| \leq \frac{(d-c)^2}{16}\left[\int_0^1 \delta^2\left|\xi''\left(\delta\frac{c+d}{2}+(1-\delta)c\right)\right|d\delta\right.$$
$$\left.+\int_0^1(\delta-1)^2\left|\xi''\left(\delta d+(1-\delta)\frac{c+d}{2}\right)\right|d\delta\right].$$

Using Hölder's inequality, we have

$$\left|\xi\left(\frac{c+d}{2}\right)-\frac{1}{d-c}\int_c^d \xi(x)dx\right|$$
$$\leq \frac{(d-c)^2}{16}\left[\int_0^1 \delta^{\frac{2}{p}}\delta^{\frac{2}{q}}\left|\xi''\left(\delta\frac{c+d}{2}+(1-\delta)c\right)\right|d\delta\right.$$
$$\left.+\int_0^1 (\delta-1)^{\frac{2}{p}}(\delta-1)^{\frac{2}{q}}\left|\xi''\left(\delta d+(1-\delta)\frac{c+d}{2}\right)\right|d\delta\right]$$
$$\leq \frac{(d-c)^2}{16}\left[\left(\int_0^1 \delta^2 d\delta\right)^{\frac{1}{p}}\left(\int_0^1 \delta^2\left|\xi''\left(\delta\frac{c+d}{2}+(1-\delta)c\right)\right|^q d\delta\right)^{\frac{1}{q}}\right.$$
$$\left.+\left(\int_0^1 (\delta-1)^2 d\delta\right)^{\frac{1}{p}}\left(\int_0^1 (\delta-1)^2\left|\xi''\left(\delta d+(1-\delta)\frac{c+d}{2}\right)\right|^q d\delta\right)^{\frac{1}{q}}\right].$$

Since $|\xi''|$ is strongly η-convex function with modulus $\mu>0$, therefore

$$\left|\xi\left(\frac{c+d}{2}\right)-\frac{1}{d-c}\int_c^d \xi(x)dx\right|$$
$$\leq \frac{(d-c)^2}{16}\left[\left(\int_0^1 \delta^2 d\delta\right)^{\frac{1}{p}}\left(\int_0^1 \delta^2\left(|\xi''(c)|^q+\delta\eta\left(\left|\xi''\left(\frac{c+d}{2}\right)\right|^q,|\xi''(c)|^q\right)\right.\right.\right.$$
$$\left.\left.-\frac{\mu}{4}\delta(1-\delta)(d-c)^2\right)d\delta\right)^{\frac{1}{q}}+\left(\int_0^1 (\delta-1)^2 d\delta\right)^{\frac{1}{p}}\left(\int_0^1 (\delta-1)^2\left(\left|\xi''\left(\frac{c+d}{2}\right)\right|^q\right.\right.$$
$$\left.\left.\left.+\delta\eta\left(|\xi''(d)|^q,\left|\xi''\left(\frac{c+d}{2}\right)\right|^q\right)-\frac{\mu}{4}\delta(1-\delta)(d-c)^2\right)d\delta\right)^{\frac{1}{q}}\right]$$
$$=\frac{(d-c)^2}{16}\left[\left(\frac{1}{3}\right)^{\frac{1}{p}}\left(\frac{1}{3}|\xi''(c)|^q+\frac{1}{4}\eta\left(\left|\xi''\left(\frac{c+d}{2}\right)\right|^q,|\xi''(c)|^q\right)\right.\right.$$
$$\left.-\frac{\mu}{80}(d-c)^2\right)^{\frac{1}{q}}+\left(\frac{1}{3}\right)^{\frac{1}{p}}\left(\frac{1}{3}\left|\xi''\left(\frac{c+d}{2}\right)\right|^q\right.$$
$$\left.\left.+\frac{1}{12}\eta\left(|\xi''(d)|^q,\left|\xi''\left(\frac{c+d}{2}\right)\right|^q\right)-\frac{\mu}{80}(d-c)^2\right)^{\frac{1}{q}}\right].$$

This completes the proof. □

Remark 5.7 When $\mu=0$, then above theorem reduces to Theorem 3.4 of [9].

5.3.1 Application to Means

Now, we consider the following special means for positive real numbers $c, d > 0$:

- Arithmetic mean: $A(c,d) = \frac{c+d}{2}$.
- Geometric mean: $G(c,d) = \sqrt{cd}$.
- Harmonic mean: $H(c,d) = \frac{2}{\frac{1}{c}+\frac{1}{d}}$.
- Generalized logarithmic mean: $L(c,d) = \begin{cases} \left[\frac{d^{m+1}-c^{m+1}}{(m+1)(d-c)}\right]^{1/m}, & \text{if } c \neq d, \\ c, & \text{if } c = d. \end{cases}$
- Identric mean: $I(c,d) = \begin{cases} \frac{1}{e}\left(\frac{d^d}{c^c}\right)^{1/(d-c)}, & \text{if } c \neq d, \\ c, & \text{if } c = d. \end{cases}$
- Heronian mean: $H_{w,m}(c,d) = \begin{cases} \left[\frac{c^m + w(cd)^{\frac{m}{2}} + d^m}{(w+2)}\right]^{1/m}, & \text{if } m \neq 0, \\ \sqrt{cd}, & \text{if } m = 0, \end{cases}$

  for $0 \leq w < \infty$.

Now, using the above results in previous theorems, we have some applications to the special means of positive real numbers.

Theorem 5.3.9 *Let $c, d > 0$, $c \neq d$, $q \geq 1$, and either $m > 1$ and $(m-1)q \geq 1$ or $m < 0$. Then*

$$\begin{aligned}
&\left|A(\alpha c^m, \beta d^m) + \frac{2-\alpha-\beta}{2} A^m(c,d) - L^m(c,d)\right| \\
&\leq \left(\frac{d-c}{8}\right)\left(\frac{1}{24}\right)^{1/q} \Big[(2\alpha^2 - 2\alpha + 1)^{1-\frac{1}{q}} (24(2\alpha^2 - 2\alpha + 1)|md^{m-1}|^q \\
&+ 4(-2\alpha^3 + 12\alpha^2 - 9\alpha + 4)\eta(|mc^{m-1}|^q, |md^{m-1})|^q) \\
&- \mu(-7\alpha^4 + 28\alpha^3 - 30\alpha^2 + 12\alpha)(c-d)^2)^{1/q} + (2\beta^2 - 2\beta + 1)^{1-\frac{1}{q}} \\
&\times (24(2\beta^2 - 2\beta + 1)|md^{m-1})|^q + 4(2\beta^3 - 3\beta + 2)\eta(|mc^{m-1}|^q, |md^{m-1}|^q) \\
&- \mu(-7\beta^4 + 8\beta^3 - 8\beta + 5)(c-d)^2)^{1/q}\Big].
\end{aligned}$$

Proof Applying Theorem 5.3.5 with $\xi(x) = x^m$. Then we obtain the result immediately. □

Example 5.1 Let $\xi(x) = x^2$, $\eta(x,y) = x + y + (x-y)^2$, $\mu = 1$, $\alpha = \beta = 1$, $c = 1$, $d = 2$, $q = 1$. Then above theorem is verified.

Theorem 5.3.10 *Let $c, d > 0$, $c \neq d$, $q \geq 1$, Then*

$$\left|\frac{\ln G^2(c^\alpha, d^\beta)}{2} + \frac{2-\alpha-\beta}{2}\ln A(c,d) - \ln I(c,d)\right|$$
$$\leq \left(\frac{d-c}{8}\right)\left(\frac{1}{24}\right)^{1/q}\Bigg[(2\alpha^2-2\alpha+1)^{1-\frac{1}{q}}\Bigg(24(2\alpha^2-2\alpha+1)\left(\frac{1}{d}\right)^q$$
$$+\ 4(-2\alpha^3+12\alpha^2-9\alpha+4)\eta\left(\left(\frac{1}{c}\right)^q, \left(\frac{1}{d}\right)^q\right)$$
$$-\ \mu(-7\alpha^4+28\alpha^3-30\alpha^2+12\alpha)(c-d)^2\Bigg)^{1/q}$$
$$+\ (2\beta^2-2\beta+1)^{1-\frac{1}{q}}\Bigg(24(2\beta^2-2\beta+1)\left(\frac{1}{d}\right)^q + 4(2\beta^3-3\beta+2)\eta\left(\left(\frac{1}{c}\right)^q, \left(\frac{1}{d}\right)^q\right)$$
$$-\ \mu(-7\beta^4+8\beta^3-8\beta+5)(c-d)^2\Bigg)^{1/q}\Bigg].$$

Proof Applying Theorem 5.3.5 with $\xi(x) = \ln x$. Then we obtain the result immediately. □

Theorem 5.3.11 *For $d > c > 0$, $c \neq d$, $w \geq 0$, and $s \geq 4$ or $0 \neq s < 1$, we have*

$$\left|\frac{1}{2}\left[\frac{H^m_{w,m}(c,d)}{H(c^m,d^m)} + H^m_{w,m}\left(\frac{c}{d}+\frac{d}{c}, 1\right)\right] - H^m_{w,m}\left(L\left(\frac{c}{d}, \frac{d}{c}\right), 1\right)\right|$$
$$\leq \frac{(d-c)A(c,d)}{768\ G^2(c,d)}\Bigg[\frac{192|m|}{w+2}\left(G^{2(m-1)}\left(d, \frac{1}{c}\right) + \frac{w}{2}G^{2(\frac{m}{2}-1)}\left(d, \frac{1}{c}\right)\right)$$
$$+\ 96\eta\Bigg(\frac{|m|}{w+2}\left(G^{2(m-1)}\left(c, \frac{1}{d}\right) + \frac{w}{2}G^{2(\frac{m}{2}-1)}\left(c, \frac{1}{d}\right)\right),$$
$$\frac{|m|}{w+2}\left(G^{2(m-1)}\left(d, \frac{1}{c}\right) + \frac{w}{2}G^{2(\frac{m}{2}-1)}\left(d, \frac{1}{c}\right)\right)\Bigg) - \frac{100\mu(d-c)^2A^2(c,d)}{G^2(c^2,d^2)}\Bigg].$$

Proof From Corollary 5.3, we have

$$\left|\frac{1}{2}\left[\frac{\xi\left(\frac{c}{d}\right)+\xi\left(\frac{d}{c}\right)}{2} + \xi\left(\frac{\frac{c}{d}+\frac{d}{c}}{2}\right)\right] - \frac{1}{\frac{d}{c}-\frac{c}{d}}\int_{\frac{c}{d}}^{\frac{d}{c}}\xi(x)dx\right|$$
$$\leq \left(\frac{\frac{d}{c}-\frac{c}{d}}{1536}\right)\left[192\left|\xi'\left(\frac{d}{c}\right)\right| + 96\eta\left(\left|\xi'\left(\frac{c}{d}\right)\right|, \left|\xi'\left(\frac{d}{c}\right)\right|\right) - 25\mu\left(\frac{c}{d}-\frac{d}{c}\right)^2\right]. \tag{5.11}$$

Applying $\xi(x) = \frac{x^m + wx^{\frac{m}{2}}+1}{w+2}$ for $x > 0$ and $m \notin (1, 4)$ in above inequality, we obtain

$$\frac{\xi\left(\frac{c}{d}\right)+\xi\left(\frac{d}{c}\right)}{2}=\frac{\left(\frac{c}{d}\right)^m+w\left(\frac{c}{d}\right)^{\frac{m}{2}}+1}{2(w+2)}+\frac{\left(\frac{d}{c}\right)^m+w\left(\frac{d}{c}\right)^{\frac{m}{2}}+1}{2(w+2)}$$
$$=\frac{1}{2(w+2)}\left[\frac{c^{2m}+wc^m(cd)^{\frac{m}{2}}+2c^md^m+wd^m(cd)^{\frac{m}{2}}+d^{2m}}{c^md^m}\right]$$
$$=\frac{1}{2(w+2)}\left[\frac{(c^m+w(cd)^{\frac{m}{2}}+d^m)(c^m+d^m)}{c^md^m}\right]$$
$$=\frac{H^m_{w,m}(c,d)}{H(c^m,d^m)}, \tag{5.12}$$

$$\xi\left(\frac{\frac{c}{d}+\frac{d}{c}}{2}\right)=\frac{\left(\frac{\frac{c}{d}+\frac{d}{c}}{2}\right)^m+w\left(\frac{\frac{c}{d}+\frac{d}{c}}{2}\right)^{\frac{m}{2}}+1}{(w+2)}=H^m_{w,m}\left(\frac{\frac{c}{d}+\frac{d}{c}}{2},1\right), \tag{5.13}$$

$$\frac{1}{\frac{d}{c}-\frac{c}{d}}\int_{\frac{c}{d}}^{\frac{d}{c}}\xi(x)dx=\frac{1}{(w+2)}\left[\left\{\frac{(\frac{d}{c})^{m+1}-(\frac{c}{d})^{m+1}}{(m+1)(\frac{d}{c}-\frac{c}{d})}\right\}+w\left\{\frac{(\frac{d}{c})^{\frac{m}{2}+1}-(\frac{c}{d})^{\frac{m}{2}+1}}{(\frac{m}{2}+1)(\frac{d}{c}-\frac{c}{d})}\right\}+1\right]$$
$$=H^m_{w,m}\left(L\left(\frac{c}{d},\frac{d}{c}\right),1\right), \tag{5.14}$$

and

$$\left(\frac{\frac{d}{c}-\frac{c}{d}}{1536}\right)\left[192\left|\xi'\left(\frac{d}{c}\right)\right|+96\eta\left(\left|\xi'\left(\frac{c}{d}\right)\right|,\left|\xi'\left(\frac{d}{c}\right)\right|\right)-25\mu\left(\frac{c}{d}-\frac{d}{c}\right)^2\right]$$
$$=\frac{(d-c)A(c,d)}{768\,G^2(c,d)}\left[\frac{192|m|}{w+2}\left(G^{2(m-1)}\left(d,\frac{1}{c}\right)+\frac{w}{2}G^{2(\frac{m}{2}-1)}\left(d,\frac{1}{c}\right)\right)\right.$$
$$+96\eta\left(\frac{|m|}{w+2}\left(G^{2(m-1)}\left(c,\frac{1}{d}\right)+\frac{w}{2}G^{2(\frac{m}{2}-1)}\left(c,\frac{1}{d}\right)\right),\right.$$
$$\left.\frac{|m|}{w+2}\left(G^{2(m-1)}\left(d,\frac{1}{c}\right)+\frac{w}{2}G^{2(\frac{m}{2}-1)}\left(d,\frac{1}{c}\right)\right)\right)-\left.\frac{100\mu(d-c)^2A^2(c,d)}{G^2(c^2,d^2)}\right]. \tag{5.15}$$

Applying (5.12)–(5.15) in (5.11), we have

$$\left|\frac{1}{2}\left[\frac{H^m_{w,m}(c,d)}{H(c^m,d^m)}+H^m_{w,m}\left(\frac{c}{d}+\frac{d}{c},1\right)\right]-H^m_{w,m}\left(L\left(\frac{c}{d},\frac{d}{c}\right),1\right)\right|$$
$$\leq\frac{(d-c)A(c,d)}{768\,G^2(c,d)}\left[\frac{192|m|}{w+2}\left(G^{2(m-1)}\left(d,\frac{1}{c}\right)+\frac{w}{2}G^{2(\frac{m}{2}-1)}\left(d,\frac{1}{c}\right)\right)\right.$$
$$+96\eta\left(\frac{|m|}{w+2}\left(G^{2(m-1)}\left(c,\frac{1}{d}\right)+\frac{w}{2}G^{2(\frac{m}{2}-1)}\left(c,\frac{1}{d}\right)\right),\right.$$
$$\left.\frac{|m|}{w+2}\left(G^{2(m-1)}\left(d,\frac{1}{c}\right)+\frac{w}{2}G^{2(\frac{m}{2}-1)}\left(d,\frac{1}{c}\right)\right)\right)-\left.\frac{100\mu(d-c)^2A^2(c,d)}{G^2(c^2,d^2)}\right].$$

5.4 Conclusion

In this chapter, we derived some Hermite–Hadamard type and fractional integral inequalities for strongly η-convex functions. The results obtained in this chapter are generalizations of the previously known results. Our results may have further applications in future research work.

Acknowledgements The authors are grateful to the referees for valuable comments and suggestions that helped us improve this chapter.

References

1. Awan, M.U., Noor, M.A., Noor, K.I., Safdar, F.: On strongly generalized convex functions. Filomat **31**(18), 5783–5790 (2017)
2. Delavar, M.R., Sen, M.D.L.: Some Hermite–Hadamard–Féjer type integral inequalities for differentiable η-convex functions with applications. J. Math. **2017** (2017)
3. Dragomir, S.S., Pearce, C.E.M.: Selected Topics on Hermite–Hadamard Inequalities and Applications. Victoria University, Australia (2000)
4. Gordji, M.E., Delavar, M.R., Dragomir, S.S.: Some inequalities related to η-convex functions. RGMIA **18** (2015)
5. Gordji, M.E., Delavar, M.R., Sen, M.D.L.: On φ-convex functions. J. Math. Inequalities **10**(1), 173–183 (2016)
6. Işcan, I.: Hermite–Hadamard Féjer type inequalities for convex functions via fractional integrals. Stud. Univ. Babes-Bolyai Math. **60**(3), 355–366 (2015)
7. Jiang, W.D., Niu, D.W., Hua, Y., Qi, F.: Generalizations of Hermite–Hadamard inequality to n-time differentiable functions which are s-convex in the second sense. Analysis (Munich) **32**(3), 209–220 (2012)
8. Karamardian, S.: The nonlinear complementarity problem with applications. Part 2. J. Optim. Theory Appl. **4**(3), 167–181 (1969)
9. Kwun, Y.C., Saleem, M.S., Ghafoor, M., Nazeer, W., Kang, S.M.: Hermite–Hadamard type inequalities for functions whose derivatives are η-convex via fractional integrals. J. Inequalities Appl. **2019**(1), 1–16 (2019)
10. Merentes, N., Nikodem, K.: Remarks on strongly convex functions. Aequ. Math. **80**, 193–199 (2010)
11. Mishra, S.K., Sharma, N.: On strongly generalized convex functions of higher order. Math. Inequalities Appl. **22**(1), 111–121 (2019)

12. Nikodem, K., Páles, Z.: Characterizations of inner product spaces by strongly convex functions. Banach J. Math. Anal. **5**(1), 83–87 (2011)
13. Pachpatte, B.G.: On some inequalities for convex functions. RGMIA Res. Rep. Coll. **6**(1), 1–9 (2003)
14. Park, J.: Inequalities of Hermite–Hadamard–Féjer type for convex functions via fractional integrals. Int. J. Math. Anal. **8**(59), 2927–2937 (2014)
15. Sharma, N., Mishra, S.K., Hamdi, A.: Weighted version of Hermite–Hadamard type inequalities for strongly GA-convex functions. Int. J. Adv. Appl. Sci. **7**(3), 113–118 (2020)
16. Xi, B.Y., Qi, F.: Some integral inequalities of Hermite–Hadamard type for convex functions with applications to means. J. Funct. Spaces Appl. Artical ID 980438 (2012)
17. Yang, Y., Saleem, M.S., Ghafoor, M., Qureshi, M.I.: Fractional integral inequalities of Hermite–Hadamard type for differentiable generalized h-convex functions. **2020**, 13 (2020)

Chapter 6
Set Order Relations, Set Optimization, and Ekeland's Variational Principle

Qamrul Hasan Ansari and Pradeep Kumar Sharma

Abstract This chapter provides a brief survey on different kinds of set order relations which are used to compare the objective values of set-valued maps and play a key role to study set optimization problems. The solution concepts of set optimization problems and their relationships with respect to different kinds of set order relations are provided. The nonlinear scalarization functions for vector-valued maps as well as for set-valued maps are very useful to study the optimality solutions of vector optimization/set optimization problems. A survey of such nonlinear scalarization functions for vector-valued maps/set-valued maps is given. We give some new results on the existence of optimal solutions of set optimization problems. In the end, we gather some recent results, namely, Ekeland's variational principle and some equivalent variational principle for set-valued maps with respect to different kinds of set order relations.

Mathematics Subject Classification (2010): 49J53, 90C29, 90C30, 90C46, 58E30

6.1 Introduction

An optimization problem whose objective function is a set-valued map is known as set optimization or set-valued optimization problem.

Let S be a nonempty subset of a vector space X, Y be a topological vector space, and $F : S \rightrightarrows Y$ be a set-valued map with nonempty values. The *set optimization problem* is defined as follows:

Q. H. Ansari (✉)
Department of Mathematics, Aligarh Muslim University, Aligarh 202002, India
e-mail: qhansari@gmail.com

P. K. Sharma
Department of Mathematics, University of Delhi South Campus, New Delhi 110021, India
e-mail: sharmapradeepmsc@gmail.com

V. Laha et al. (eds.), *Optimization, Variational Analysis and Applications*,
Springer Proceedings in Mathematics & Statistics 355,
https://doi.org/10.1007/978-981-16-1819-2_6

$$\begin{aligned} &\min F(x) \\ &\text{subject to } x \in S. \end{aligned} \tag{SOP}$$

The study of such problems is known as set optimization or set-valued optimization. Since the set-valued maps include single-valued maps as well as vector-valued maps, the set optimization can be considered as an extension of scalar optimization and/or vector optimization. Since the middle of eighties, the theory of set optimization has received increasing interest in the optimization community and many authors have studied and investigated set optimization problems due to its extensive applications in different branches of applied mathematics, engineering, economics, finance, and medical sciences. Note that several problems from game theory [73], multivariate statistics [70, 106], radiotherapy treatment (medical image registration, intensity-modulated radiation therapy) [42, 104, 120], uncertain optimization [7, 105], welfare economics [16], socio-economics [132], mathematical finance [36, 69], optimal control [75], etc. can be written in the form of mathematical formulation of set optimization problems. Not only this, the robust optimization problems and stochastic/fuzzy programming problems can also be modeled as set optimization problems. For an overview and further detailed investigations, we refer to the books [68, 97].

Let us consider and distinguish simple examples of scalar optimization problem, vector optimization problem, and set optimization problem.

- To find the fastest bowler from a set of cricket players is the scalar optimization problem (maximization problem) where the objective function gives the speed of a player.
- To find the bowler(s) from a set of cricket players in such a way that he/she (they) is (are) having several qualities, namely, speed, in swing/out swing, etc., is a vector optimization problem. Consider an objective function from a set of players to the set of all such qualities, that is, the value of the objective function can be regarded as a vector whose coordinates consist of one's ability, speed, in swing/out swing, etc. In other words, the objective function is vector-valued.
- Consider the objective function whose values are teams and assume that a team is a set of players and each player is regarded as a vector whose coordinates consist of one's ability, speed, in swing/out swing, popularity, and so on. Then one can formulate the problem of choosing a good team for a cricket league in the form of set optimization problem with the objective function defined as above.

For a set optimization problem, it seems natural the first thing that has to be done is to decide how to define the solution of a set optimization problem. There are two popular approaches to define the solution concepts of a set optimization problem: one is the vector approach and another is the set approach.

In vector approach, one directly generalizes the concepts known from vector optimization to set optimization, that is, we try to find the best element, in some sense, of the union of all image sets of the set-valued objective map over the feasible set. In other words, in vector approach, a minimizer $(\bar{x}, \bar{y})$ depends on only certain special element $\bar{y}$ of $F(\bar{x})$ and other elements of $F(\bar{x})$ are ignored. That is, an element $\bar{x} \in S$

for which there exists at least one element $\bar{y} \in F(\bar{x})$ which is Pareto minimal point of the image set of F even if there exists many bad elements in $F(\bar{x})$ is a solution of the set optimization problem (SOP). The set optimization problems with vector approach have been studied and investigated by Corley [33, 34]; Luc [128]; Lin [123]; Jahn and Rauh [93]; Chen and Jahn [29]; Götz and Jahn [55]; Li [122]; Crespi, Ginchev, and Rocca [35], Alonso and Rodríguez-Marín [2], Hernández, Rodríguez-Marín and Sama [84], Hernández [79], etc. For more detail, we refer to the books [90, 97], the survey papers [41, 64, 79] and the references therein. Of course, vector optimization problems provide a very important special case of set optimization with numerous applications. Moreover, the answer to certain problems in vector optimization can be found, if the vector optimization problem is considered in a set-valued framework, see [67]. Note that the solution concept based on vector approach is of mathematical interest but it can not be often used in practice. This solution concept is not suitable to deal with the set optimization problem defined in the above example. For example, we can see that a team which has at least one good player is a solution, though most of the members of such teams are useless. Is it true that such team can achieve good results?

These solutions must be almost invalid and improper. This drawback gave birth to the set approach which is based on the comparison of values (sets) of objective set-valued map, that is, using the set approach, the sets $F(x)$ are compared by using some kinds of set order relations with the aim to choose the best one in some sense. The credit for the birth of set approach goes to Kuroiwa [109]. To resolve this problem, Kuroiwa [109] introduced six kinds of set order relations which are further studied and investigated in [1, 25, 72, 79–82, 84, 110, 111, 115, 132] and the references therein. Note that these set order relations were independently introduced in different fields, for example, in terms of algebraic structures by Young [150] in 1931, in the theory of fixed points of monotonic operators by Nishnianidze [134] in 1984, in interval arithmetic by Chiriaev and Walster [31] in 1998, and in theoretical computer science by Brink [22] in 1993. In 2011, Jahn and Ha [92] introduced the, so-called, minmax set order relations to deal with the solutions of the problem (SOP) where the above mentioned six kinds of set order relations fail. Since the notion of the set approach was introduced, there has been rapid growth in this field. On the contrary, the main disadvantage of the set approach over the vector approach is the loss of lineal structure. Hamel [67] studied the structure of the power set of Y by introducing a conlineal space. In order to avoid such a problem, several authors have considered specializations of F or tools to study the problem (SOP) via a structure well known or simpler than a conlineal space. For instance, Hernández [80] characterized the solutions of the problem (SOP) via nonlinear scalarization, see also [13, 129]. Kuroiwa and Nuriya [114] constructed an embedding vector space. Maeda's [130] work on n-dimensional Euclidean spaces shows that whenever the set-valued map is rectangle-valued (SOP), then it is equivalent to a pair of vector-valued optimization problems.

In general, there is no relation among solutions of the problem (SOP) obtained by vector approach and solutions obtained by set approach. Moreover, the existence of solutions by one approach does not imply the existence of solutions of the other approach, see [1, 63, 84, 99] and the references therein. Even though both criteria are different but under certain assumptions, the relation among solution concepts of the problem (SOP) with vector approach and set approach has been studied in [1, 79, 84, 129, 130] and the references therein.

In 2017, Chen et al. [30] introduced a set order relation called weighted set order relation. This weighted set order relation is the combination of Kuroiwa's [109] upper and lower set order relations. So, under some assumptions, this new set order relation is more general than Kuroiwa's upper and lower set order relations. It is useful for formulating solution concepts for researchers who do not specifically rely on either the upper or lower set order relation. Recently, Ansari et al. [6] studied Ekeland's variational principle and some equivalent results for set-valued maps by using weighted set order relations and gave some applications to order intervals.

In 2018, Karaman et al. [96] introduced set order relations on the family of sets based on the Minkowski difference. In comparison to Kuroiwa's set order relations, these set order relations are partial ordered on the family of bounded sets, and hence provide a new approach to study set optimization problems. Khushboo and Lalitha [99] studied the relationship among different kinds of solution sets of set optimization problems defined by means of Kuroiwa's set order relations and Karaman's set order relations. They also investigated that the solution sets of a set optimization problem defined by different kinds of set order relations are different. Therefore, it is interesting and important to investigate the set optimization problems by using Karaman's set order relations. Very recently, the set optimization problems have been investigated and studied in [12, 94–96, 99, 139] by using Karaman's set order relations.

Besides the set order relations with fixed ordering cone, the interest in set order relations with variable cone has increased during the last years due to some applications in different problems, see [7, 10, 20, 40, 42–44, 101, 102, 104, 105, 119, 120] and references therein. Therefore, in the order relations defined by convex cone to compare sets, the cone is replaced by a variable domination structure. This variable domination structure is defined by a set-valued map, called ordering map, which associates with each element of the space an individual set of preferred or dominated directions. In 2016, Eichfelder and Pilecka [42, 44] introduced the set order relations equipped with variable domination structures. They provided scalarization results for obtaining optimality conditions for the solutions of the problem (SOP). Further, Köbis [101, 102] introduced new set order relations equipped with variable domination structures and differentiated between a concept of domination and preference. In the recent years, set optimization problems with respect to variable domination structures have been studied and investigated in [7, 10, 20, 42, 44, 101, 102, 104, 119] and the references therein.

In the recent years, the set order relations has played an important role to deal with several problems from nonlinear analysis and optimization with set-valued maps, for instance, Ekeland's variational principle and related results [12, 13, 63, 72], continuity and convexity of set-valued maps [115, 117], minimax theorem for set-valued maps [116], well-posedness [62], stabilty [77], connectedness [78], concepts of efficiency for uncertain multi-objective optimiztion [88], optimality notions for (SOP) [1, 84, 94], and so on.

There are various techniques to deal with the set optimization problems, for instance, scalarization, vectorization, etc., see [13, 14, 72, 80, 91, 99, 151] and the references therein. One of the most and widely used techniques to deal with set optimization problems is the scalarization by which we can convert a set optimization problem into a scalar optimization problem, that is, by using scalarization, set optimization problem is replaced by a family of scalar optimization problems which allow to relate the solutions of both problems and solve the set optimization problem by a numerical method applicable for the scalar problems. To study set optimization problems, scalarization functions are one of the most essential tools from a theoretical as well as computational point of view. Several scalarization techniques for set optimization problems are available in the literature. Most of them are based on Gerstewitz function [54], oriented distance function [86], or their extensions [11, 13, 14, 72, 80, 96, 99, 151]. The original idea of the nonlinear scalarization functions was given by Krasnosel'skij [107] and Rubinov [140]. Krasnosel'skij [107] used them in order to establish necessary and sufficient conditions for a cone to be normal. Also, these types of functionals have been used in theoretical investigations within the framework of ordered linear spaces, see the book [38] by M. M. Day as an elegant tool for proof of the fact that the Hahn–Banach extension and a linear closure property imply the interpolation property. Furthermore, Feldman [48] and Rubinov [140] investigated the dual properties of such kinds of functionals, namely, their so-called support sets. The nonlinear scalarization functional for vector optimization with its concrete definition was given by Tammer (Gerstewitz) [52] in 1983 and applied to study separation theorems for not necessary convex sets by Tammer (Gerstewitz) and Iwanow [53] in 1985. Such nonlinear scalarization functions are now known as Gerstewitz nonlinear scalarization functional. Luc [125–127] also gave early contributions to this topic. On the other hand, Hiriart-Urruty [86] introduced the notion of oriented distance function to study optimality conditions of nonsmooth optimization problems from the geometric point of view. For more details on oriented distance function and their extensions, we refer [7, 8, 35, 60, 99, 151, 152] and the references therein.

The idea of nonlinear scalarization for sets was first investigated in 2000 by Tanaka–Georgiev [51]. In 2006, Hamel and Löhne [72] extended the above functions to two different functions on a power set of Y corresponding to the set order relations. Further, Hernández and Rodríguez-Marín, [80] investigated nonlinear scalarizing functions for sets by introducing cone-topological concepts, see [9]. Furthermore, in 2009, Knwano–Tanaka–Yamada [118] introduced a unified approach for such scalarizations for sets using Kuroiwa's set order relations. In the recent past, Araya [13, 14] investigated six types of nonlinear scalarizing functions for set-valued maps and their relationships. In the literature, expressions using inf–sup of the Gerstewitz function can be found in [60, 61] which were used to study necessary and sufficient optimality conditions in set optimization problems with set order relations. Khoshkhabar-amiranloo et al. [98] and Sach [141] also introduced slightly different nonlinear scalarization functions to study set optimization problmes. Recently,

Karaman et al. [96] introduced nonlinear scalarization functions by using the set order relations defined by the Minkowski difference and studied optimality notions for (SOP). Very recently, Ansari et al. [6] introduced the notions of nonlinear scalarization functions by using weighted set order relations. Several applications of the nonlinear scalarization functions can be found in the literature, for instance, to study Ekeland's variational principle and related variational principle [6, 12, 13, 63, 65, 72]; nonconvex separation type theorems [13, 14, 54]; Gordan's type alternative theorems [13, 135]; equilibrium problems [9, 56]; minimax theorems [116]; vector variational inequalities [9, 56]; robustness and stochastic programming [100]; and stability and well-posedness [62, 77]. In the recent years, several authors have studied and investigated nonlinear scalarizing technique for set optimization problem, see [6, 12, 60, 61, 72, 97, 98, 141] and their references therein.

In recent years, scalarization functions with variable domination structures also gained increasing interest in the optimization community. Eichfelder and Pilecka [44] introduced a nonlinear scalarization function when the images of ordering maps are Bishop Phelps cones. Further, Köbis et al. [104] and Ansari et al. [7] introduced nonlinear scalarizing methods to characterize several set order relations and minimal solutions for set optimization problems equipped with variable domination structures with their applications in medical image registration and uncertain multi-objective optimization and to derive necessary optimality conditions for solutions of set optimization problems with respect to variable domination structures. Very recently, Kobis et al. [105] introduced a new nonlinear scalarization functional in set optimization equipped with variable domination structures, which are further studied by Ansari and Sharma [10] to obtain Ekeland's variational principle. For more details on scalarization functions with respect to variable domination structures, we refer [7, 10, 44, 104, 105] and the references therein.

The present chapter is organized as follows: In the next section, we recall some definitions and concepts which will be used in the sequel. In Sect. 6.3, we gather different kinds for set order relations with their properties. The relationships among these set order relations are provided along with theoretical and geometrical illustrations. In Sect. 6.4, a survey of nonlinear scalarization functions for vector-valued maps/set-valued maps is given. Such nonlinear scalarization functions for vector-valued maps as well as for set-valued maps are very useful to study the optimality solutions of vector optimization/set optimization problems and to study some set order relations. In Sect. 6.5, solution concepts for set optimization problems based on vector approach and set approach and relations among them are given. Several examples are given to illustrate each type of solution concept. Some new results on the existence of optimal solutions of set optimization problems are given in Sect. 6.6. In the last section, we investigate Ekeland's variational principle for set-valued maps in different settings and also by using different kinds of set order relations. Further, we investigate some other equivalent variational principles, namely minimal element theorem, Takahashi minimization theorem, and Caristi fixed point theorem for set-valued maps.

6.2 Preliminaries

Throughout the chapter, all vector spaces are assumed to be defined over the field of real numbers, and we adopt the following notations, unless otherwise specified.

We denote by $\mathbb{N}$, $\mathbb{Q}$, and $\mathbb{R}$ the set of all natural numbers, the set of all rational numbers, and the set of all real numbers, respectively, and $\mathbb{R}_+ = [0, \infty)$. We denote by $\mathbb{R}^n$ the n-dimensional Euclidean space and by $\mathbb{R}^n_+$ the nonnegative orthant in $\mathbb{R}^n$. The zero element in a vector space will be denoted by $\mathbf{0}$. Let Y be a topological vector space with its topological dual Y^*. We denote by 2^Y (respectively, $P(Y)$) and $\mathcal{B}(Y)$ the family of all (respectively, nonempty) subsets of Y and the family of all nonempty bounded subsets of Y, respectively. For a set $A \subseteq Y$, we denote by $\text{int}A$, $\overline{A}$ or clA, ∂A, and A^c, the interior, the closure, the boundary, and the complement of A, respectively.

For arbitrary nonempty sets X and Y, we denote by P_X and P_Y, the projection of $X \times Y$ onto X and Y, respectively, that is,

$$P_X(x, y) = x \quad \text{and} \quad P_Y(x, y) = y, \quad \text{for all } (x, y) \in X \times Y.$$

A function $F : X \to 2^Y$ is said to be a *set-valued map*, and it is denoted by $F : X \rightrightarrows Y$. For the set-valued map $F : X \rightrightarrows Y$, the *image* of F at $x \in X$ is a subset $F(x)$ of Y. The *domain* of F is

$$\text{dom}\, F = \{x \in X : F(x) \neq \emptyset\},$$

and the *image* of F is

$$\text{Im}\, F = \{y \in Y : \text{ there exists } x \in X \text{ such that } y \in F(x)\}.$$

The set-valued map $F : X \rightrightarrows Y$ can be identified by its *graph* which is defined as

$$\text{graph}\, F = \{(x, y) \in X \times Y : x \in X,\ y \in F(x)\}.$$

The *image* of the set $S \subseteq X$ under F is

$$F(S) := \bigcup_{x \in S} F(x),$$

so, $\text{Im}\, F = F(X)$. The set

$$\text{Graph}\, F = \{(x, V) \in X \times P(Y) : V = F(x)\}$$

is designated as *graph* of F by Hamel and Löhne [72].

A subset C of a vector space Y is said to be a *cone* if for all $x \in C$ and $\lambda \geq 0$, we have $\lambda y \in C$. The set C of Y is called a *convex cone* if it is convex and a cone, that is, for all $x, y \in C$ and $\lambda, \mu \geq 0$, we have $\lambda x + \mu y \in C$.

Definition 6.1 A cone C in Y is said to be

(a) *solid* if it has nonempty interior, that is, $\text{int}C \neq \emptyset$;
(b) *nontrivial* or *proper* if $C \neq \{\mathbf{0}\}$ and $C \neq Y$;
(c) *reproducing* if $C - C = Y$;
(d) *pointed* if for $\mathbf{0} \neq x \in C$, we have $-x \notin C$, that is, $C \cap (-C) = \{\mathbf{0}\}$;
(e) *closed cone* if it is also closed.

The *dual* of a cone $C \subseteq Y$ is defined by

$$C^* := \{y^* \in Y^* : \langle y^*, y \rangle \geq 0 \text{ for all } y \in C\},$$

where $\langle y^*, y \rangle$ denotes the value of the functional y^* at y.

The convex cone $C \subseteq Y$ induces an ordering on Y as

$$x \preceq_C y \Leftrightarrow y - x \in C, \quad \text{for all } x, y \in Y.$$

If $\text{int}C \neq \emptyset$, then we have

$$x \prec_C y \Leftrightarrow y - x \in \text{int}C, \quad \text{for all } x, y \in Y.$$

Further, if C is pointed, then the ordering $\preceq_C$ is a partial ordering on Y.

Note that there is a one-to-one correspondence between an ordering and a convex cone (see [9]).

Definition 6.2 Let $A, B \in P(Y)$.

- The *algebraic sum* of A and B is defined as

$$A + B := \{a + b : a \in A,\ b \in B\}.$$

- The *algebraic difference* of A and B is defined as

$$A - B := \{a - b : a \in A,\ b \in B\}.$$

- The *Minkowski* (*Pontryagin*) *difference* of A and B is defined as

$$A \dot{-} B := \{y \in Y : y + B \subseteq A\} = \bigcap_{b \in B} (A - b).$$

- For $\lambda \in \mathbb{R}$, $\lambda A := \{\lambda x : x \in A\}$.

It is worth to mention that the set equation $A + A = 2A$ does not hold in general for a nonempty subset A of a vector space. The Minkowski difference of a set and a vector coincides with their algebraic difference, that is, $A \dot{-} a = A - a$ for all $A \in P(Y)$ and $a \in Y$.

Note that the *Minkowski (Pontryagin) difference* plays a very important role in many applications such as robot motion planning [124], morphological image analysis [143], and computer-aided design and manufacturing [121]. For further details on *Minkowski (Pontryagin) difference*, we refer to the book [136].

The following example illustrates different types of set addition and set difference.

Example 6.1 Let $A = [-1, 1] \times [-1, 1]$ and $B = [-1, 0] \times [-1, 0]$. Then,

$$A + B = [-2, 1] \times [-2, 1], \quad A - B = [-1, 2] \times [-1, 2], \quad \text{and} \quad A \dot{-} B = [0, 1] \times [0, 1].$$

See, Fig. 6.1 for an illustration of the sets A, B, $A + B$, $A - B$, and $A \dot{-} B$.

We present some basic properties of the *Minkowski (Pontryagin) difference*.

Proposition 6.1 [96] *Let Y be a normed space, $A, B \in P(Y)$, and $\alpha \in Y$. The following assertions hold.*

(a) $(\alpha + A) \dot{-} B = \alpha + (A \dot{-} B)$.
(b) $A \dot{-} (\alpha + B) = -\alpha + (A \dot{-} B)$.
(c) *If A is closed, then $A \dot{-} B$ is also closed.*
(d) *If A is bounded, then $A \dot{-} A = \{\mathbf{0}\}$.*

Definition 6.3 [9, 127] Let C be a closed convex cone in Y. A nonempty subset A of Y is said to be

(a) *C-proper* if $A + C \neq Y$;
(b) *C-closed* if $A + C$ is a closed set;
(c) *C-bounded* if, for each neighborhood U of $\mathbf{0} \in Y$, there is a positive number t such that $A \subset tU + C$;
(d) *C-compact* if each cover of A of the form $\{U_\lambda + C : U_\lambda \text{ is an open set}, \lambda \in \Lambda\}$ admits a finite subcover, where Λ denotes the index set.

Clearly, if C is a closed convex cone, so is $-C$. The replacement of C by $-C$ in the above definition produces $(-C)$-closed, $(-C)$-bounded, etc. For more detail and examples on cone-topological concepts, we refer to [9, 127].

We denote by Ω_C the family of all C-proper subsets of Y and by Ω_C^{cb} the family of all nonempty, C-proper, closed, and bounded subsets of Y.

6.3 Set Order Relations

This section deals with different kinds of set order relations to study set optimization problems.

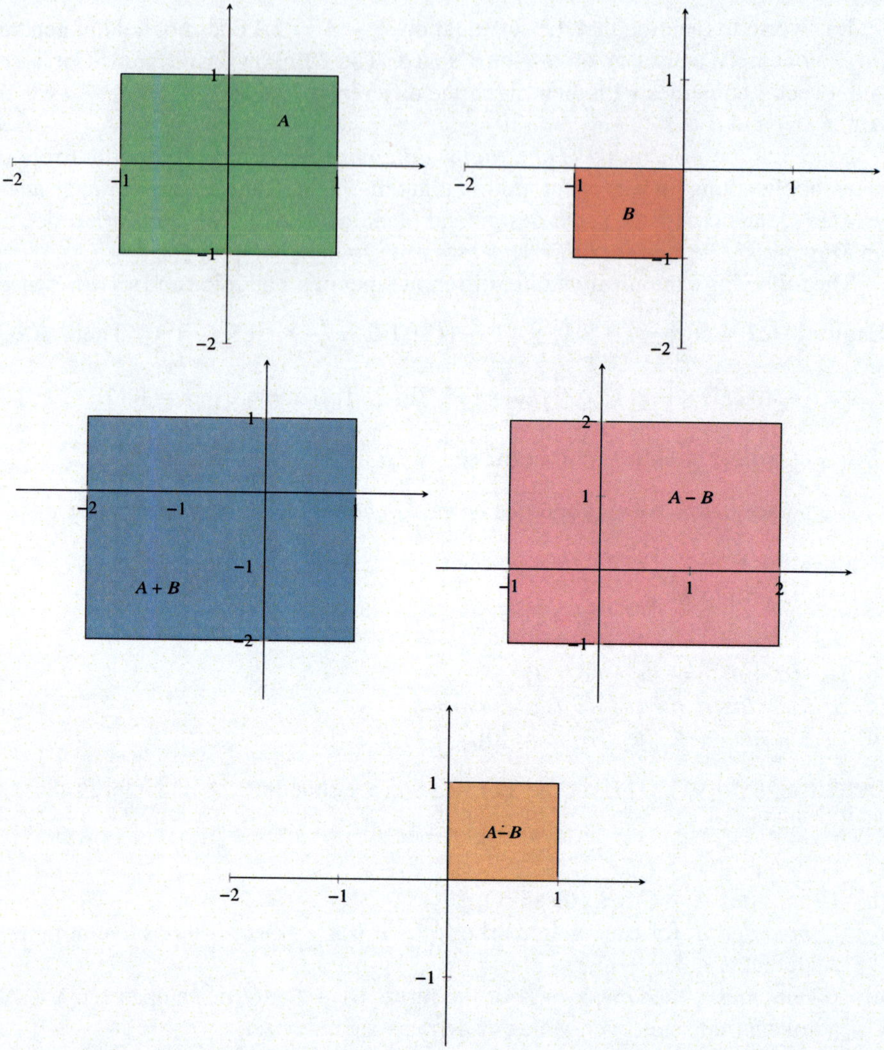

Fig. 6.1 Visualization of sets A, B, $A + B$, $A - B$, and $A \dot{-} B$

Definition 6.4 [92, 115] Let Y be a topological vector space, $A, B \in P(Y)$, and C be a proper convex cone in Y. The *set order relations* on $P(Y)$ with respect to C are defined as follows:

(a) The *lower set less order relation* $\preceq_C^l$ is defined by

$$A \preceq_C^l B \quad \Leftrightarrow \quad B \subseteq A + C,$$

or equivalently, for all $b \in B$, there exists $a \in A$ such that $a \preceq_C b$.

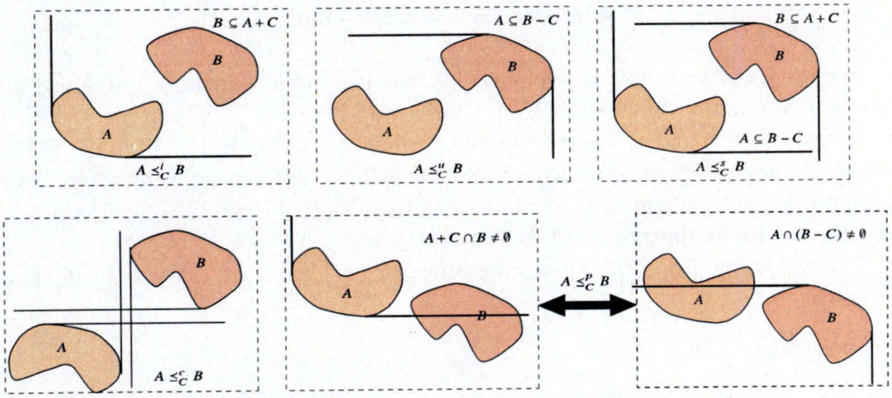

Fig. 6.2 Illustration of set order relations in $\mathbb{R}^2$ with $C = \mathbb{R}^2_+$

(b) The *upper set less order relation* $\preceq_C^u$ is defined by

$$A \preceq_C^u B \quad \Leftrightarrow \quad A \subseteq B - C,$$

or equivalently, for all $a \in A$, there exists $b \in B$ such that $a \preceq_C b$.

(c) The *set less order relation* $\preceq_C^s$ is defined by

$$A \preceq_C^s B \quad \Leftrightarrow \quad B \subseteq A + C \text{ and } A \subseteq B - C,$$

or equivalently, for all $b \in B$, there exists $a \in A$ such that $a \preceq_C b$, and for all $a \in A$, there exists $b \in B$ such that $a \preceq_C b$.

(d) The *certainly set less order relation* $\preceq_C^c$ is defined by

$$A \preceq_C^c B \Leftrightarrow (A = B) \text{ or } (A \neq B, \text{ for all } b \in B, \text{ for all } a \in A \text{ such that } a \preceq_C b),$$

or equivalently, $A = B$, or $B - A \subset C$ whenever $A \neq B$.

(e) The *possibly set less order relation* $\preceq_C^p$ is defined by

$$A \preceq_C^p B \quad \Leftrightarrow \quad \text{there exists } b \in B, \text{ there exists } a \in A \text{ such that } a \preceq_C b,$$

which is equivalent to $A \cap (B - C) \neq \emptyset$ or $B \cap (A + C) \neq \emptyset$.

When the cone C has nonempty interior, that is, $\text{int}C \neq \emptyset$, then we can define the corresponding weak set order relations $\prec_C^\alpha$, $\alpha \in \{l, u, s, c, p\}$ as the relations $\preceq_C^\alpha$, $\alpha \in \{l, u, s, c, p\}$ by replacing C with $\text{int}C$.

For instance, the weak lower set less order relation $\prec_C^l$ is defined as

$$A \prec_C^l B \quad \Leftrightarrow \quad B \subseteq A + \text{int}C,$$

and the weak upper set less order relation $\prec_C^u$ is defined as

$$A \prec^u_C B \quad \Leftrightarrow \quad A \subseteq B - \text{int}C.$$

Note that the set less order relation $\preceq^s_C$ has been independently introduced by Young [150] and Nishnianidze [134]. Chiriaev and Walster [31] used the set order relations $\preceq^s_C$, $\preceq^c_C$, and $\preceq^p_C$ in the interval arithmetic and implemented in the FORTRAN compiler f95 of SUN Microsystems [146]. These set order relations have been presented by Kuroiwa [115] in the modified form as defined in Definition 6.4. See Fig. 6.2 for an illustration of these set order relations.

The following proposition gives the properties of the set order relations defined as above.

Proposition 6.2 [92]

(a) *The set order relations* $\preceq^l_C$, $\preceq^u_C$, *and* $\preceq^s_C$ *are pre-order and compatible with respect to addition and scalar multiplication on* $P(Y)$.
(b) *The set order relation* $\preceq^c_C$ *is a pre-order and compatible with respect to addition and scalar multiplication on* $P(Y)$. *If the ordering cone* C *is pointed, then the set order relation* $\preceq^c_C$ *is antisymmetric and hence, a partial order relation.*
(c) *The set order relation* $\preceq^p_C$ *is reflexive and compatible with respect to addition and scalar multiplication on* $P(Y)$. *In general, it is not transitive and not antisymmetric.*
(d) *In general, the set order relations* $\preceq^l_C$, $\preceq^u_C$, *and* $\preceq^s_C$ *are not antisymmetric. More precisely, for arbitrary sets* $A, B \in P(Y)$, *we have*

    (1) $(A \preceq^l_C B \text{ and } B \preceq^l_C A) \Leftrightarrow A + C = B + C;$
    (2) $(A \preceq^u_C B \text{ and } B \preceq^u_C A) \Leftrightarrow A - C = B - C;$
    (3) $(A \preceq^s_C B \text{ and } B \preceq^s_C A) \Leftrightarrow (A + C = B + C \text{ and } A - C = B - C).$

Remark 6.1 The pointedness of the cone C in Proposition 6.2(b) cannot be relaxed. Indeed, let $Y = \mathbb{R}^2$ and $C = \mathbb{R} \times \{\mathbf{0}\}$. Then C is not pointed. For $A = [-1, 1] \times \{\mathbf{0}\}$ and $B = [3, 5] \times \{\mathbf{0}\}$, we have $A \preceq^c_C B$ and $B \preceq^c_C A$ but $A \neq B$.

The following example shows that the set order relation $\preceq^p_C$ is in fact not transitive and not antisymmetric.

Example 6.2 Let $Y = \mathbb{R}^2$ and $C = \mathbb{R}^2_+$. Consider the sets

$$\begin{aligned}
A_1 &= \{(y_1, y_2) \in \mathbb{R}^2 : y_1^2 + y_2^2 \leq 2^2, y_1 \geq 0, y_2 \geq 0\},\\
A_2 &= \{(y_1, y_2) \in \mathbb{R}^2 : (y_1 - 3)^2 + (y_2 + 1)^2 \leq 1\},\\
A_3 &= \text{conv}\{(4, -2), (6, -2), (6, -4)\},
\end{aligned}$$

where conv denotes the convex hull. One can easily see from Fig. 6.3 that

$$A_1 \preceq^p_C A_2 \quad \text{and} \quad A_2 \preceq^p_C A_1 \quad \text{but} \quad A_1 \neq A_2,$$

and

$$A_1 \preceq^p_C A_2 \quad \text{and} \quad A_2 \preceq^p_C A_3 \quad \text{but} \quad A_1 \npreceq^p_C A_3.$$

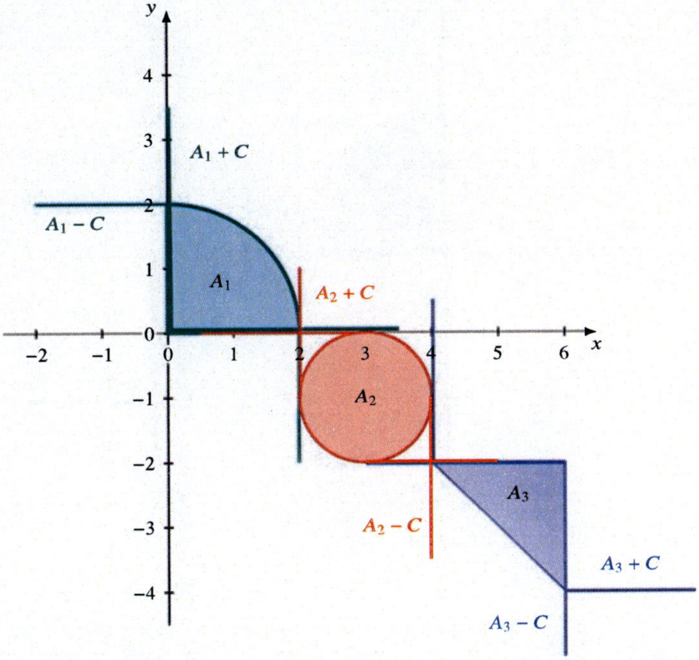

Fig. 6.3 Visualization of Example 6.2 with $C = \mathbb{R}^2_+$

We have the following relation between the lower set less order relation $\preceq^l_C$ and the upper set less order relation $\preceq^u_C$:

$$A \preceq^l_C B \Leftrightarrow B \subseteq A + C \Leftrightarrow B \subseteq A - (-C) \Leftrightarrow B \preceq^u_{-C} A \Leftrightarrow (-B) \preceq^u_C (-A).$$

Similarly,

$$A \prec^l_C B \Leftrightarrow B \subseteq A + \text{int}C \Leftrightarrow B \subseteq A - (-\text{int}C) \Leftrightarrow B \prec^u_{-C} A \Leftrightarrow (-B) \prec^u_C (-A).$$

Proposition 6.3 [92] *Let $A, B \in P(Y)$ with $A \neq B$. Then,*

(a) $A \preceq^s_C B \Rightarrow A \preceq^l_C B \Rightarrow A \preceq^p_C B$;
(b) $A \preceq^s_C B \Rightarrow A \preceq^u_C B \Rightarrow A \preceq^p_C B$;
(c) *$A \preceq^l_C B$ does not always imply $A \preceq^u_C B$, and $A \preceq^u_C B$ does not always imply $A \preceq^l_C B$.*

The following example shows that the implications in Proposition 6.3 are strict, that is, the converse implications do not hold.

Example 6.3 Let $Y = \mathbb{R}^2$ and $C = \mathbb{R}^2_+$. Consider the sets

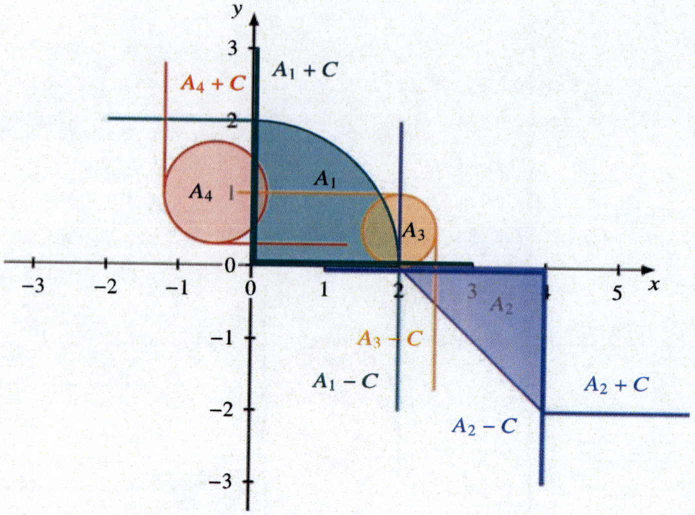

Fig. 6.4 Visualization of Example 6.3 with $C = \mathbb{R}^2_+$

$$A_1 = \{(y_1, y_2) \in \mathbb{R}^2 : y_1^2 + y_2^2 \leq 2^2, y_1 \geq 0, y_2 \geq 0\},$$
$$A_2 = \text{conv}\{(2, 0), (4, 0), (4, -2)\},$$
$$A_3 = \{(y_1, y_2) \in \mathbb{R}^2 : (y_1 - 2)^2 + (y_2 - 0.5)^2 \leq 0.5^2\},$$
$$A_4 = \{(y_1, y_2) \in \mathbb{R}^2 : (y_1 + 0.5)^2 + (y_2 - 1)^2 \leq 0.7^2\}.$$

From Fig. 6.4, it can be easily visualized that

$$A_1 \preceq^p_C A_2 \quad \text{but} \quad A_1 \not\preceq^l_C A_2 \quad \text{and} \quad A_1 \not\preceq^u_C A_2,$$

$$A_1 \preceq^l_C A_3 \quad \text{but} \quad A_1 \not\preceq^u_C A_3 \quad \text{and hence} \quad A_1 \not\preceq^s_C A_3,$$

and

$$A_4 \preceq^u_C A_1 \quad \text{but} \quad A_4 \not\preceq^l_C A_1 \quad \text{and hence} \quad A_4 \not\preceq^s_C A_1.$$

Let us illustrate the set order relations by the following example with order intervals.

Example 6.4 [92] Let $a_1, a_2, b_1, b_2 \in Y$ be arbitrarily given with $a_1 \preceq_C a_2$ and $b_1 \preceq_C b_2$, and consider the intervals

$$A = [a_1, a_2] := \{y \in \mathbb{R} : a_1 \preceq_C y \preceq_C a_2\}$$

and

$$B = [b_1, b_2] := \{y \in \mathbb{R} : b_1 \preceq_C y \preceq_C b_2\}.$$

(a) $[a_1, a_2] \preceq_C^s [b_1, b_2] \Leftrightarrow a_1 \preceq_C b_1$ and $a_2 \preceq_C b_2$.
(b) $[a_1, a_2] \preceq_C^c [b_1, b_2] \Leftrightarrow a_2 \preceq_C b_1$.
(c) $[a_1, a_2] \preceq_C^p [b_1, b_2] \Leftrightarrow a_1 \preceq_C b_2$.
(d)

$$A \preceq_C^s B \Leftrightarrow \begin{cases} \min A \in \min B - C, & \min B \in \min A + C, \\ \max A \in \max B - C, & \max B \in \max A + C, \end{cases}$$
$$\Leftrightarrow \quad \min B - \min A \in C \text{ and } \max B - \max B \in C.$$

and

$$A \preceq_C^c B \Leftrightarrow \min B \in \max A + C \text{ and } \max A \in \min B - C,$$

where $\min A := \{a \in A : A \cap (a - C) = \{a\}\}$ and $\max A := \{a \in A : A \cap (a + C) = \{a\}\}$ are the sets of minimal elements and maximal elements, respectively, with respect to the convex pointed cone C.

This observation was one of the motivations to Jahn and Ha [92] to introduce new set order relations involving minimal and maximal elements.

From a practical point of view, the set order relations $\preceq_C^s$ and $\preceq_C^c$ are more appropriate in applications than the other set order relations. In the case of order intervals, the set order relations $\preceq_C^s$ and $\preceq_C^c$ are described by a pre-order of the minimal and maximal elements of these intervals. But for general nonempty sets A and B, which possess minimal elements and maximal elements, this property may not be fulfilled. The following figure illustrates two sets $A, B \in P(Y)$ with $A \preceq_C^s B$ and the properties $\max A \subseteq \max B - C$ but $\max B \not\subseteq \max A + C$. This means that there may be elements $b \in \max B$ and $a \in \max A$ which are not comparable with respect to the pre-order (see Fig. 6.5). In order to avoid this drawback, Jahn and Ha [92] defined new set order relations involving the minimal and maximal elements of a set. This leads to various definitions of "minmax less" set order relations. For further details, see [92].

We denote by $\Xi := \{A \in P(Y) : \min A \neq \emptyset \text{ and } \max A \neq \emptyset\}$, where $\min A := \{a \in A : A \cap (a - C) = \{a\}\}$ and $\max A := \{a \in A : A \cap (a + C) = \{a\}\}$ are the sets of minimal elements and maximal elements, respectively, with respect to the convex pointed cone C in a topological vector space Y.

Definition 6.5 [92] Let $A, B \in \Xi$ and C be a proper, convex, and pointed cone in a toplogical vector space Y. The minmax set order relations on Ξ with respect to C are defined as follows:

(a) The *minmax set less order relation* $\preceq_C^m$ is defined by

$$A \preceq_C^m B \quad \Leftrightarrow \quad \min A \preceq_C^s \min B \text{ and } \max A \preceq_C^s \max B.$$

(b) The *minmax certainly set less order relation* $\preceq_C^{mc}$ is defined by

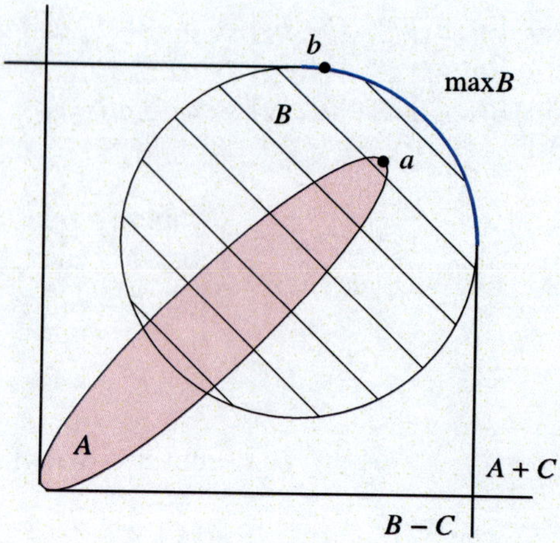

Fig. 6.5 Illustration of two sets $A,\ B \in P(Y)$ with $A \preceq^s_C B$, and $a \in \max A$ and $b \in \max B$ with $a \not\preceq_C b$ and $b \not\preceq_C a$, and $C = \mathbb{R}^2_+$

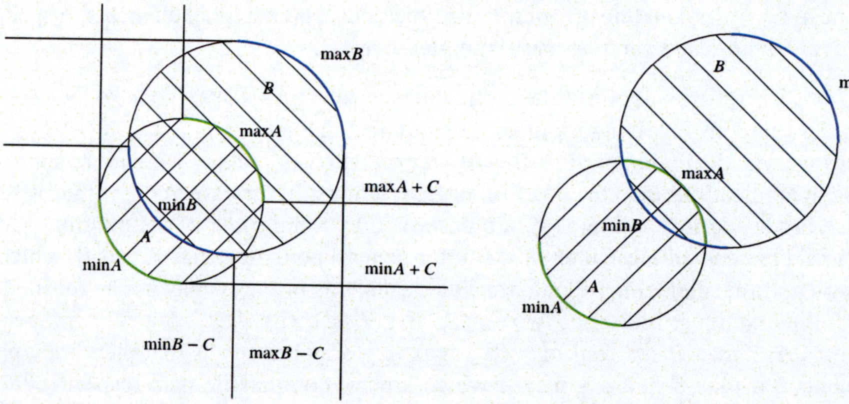

Fig. 6.6 Illustration of two sets $A,\ B \in \Xi$ with $A \preceq^m_C B$ and $A \preceq^{mc}_C B$, and $C = \mathbb{R}^2_+$

$$A \preceq^{mc}_C B \quad \Leftrightarrow \quad (A = B) \text{ or } (A \neq B,\ \min A \preceq^c_C \min B \text{ and } \max A \preceq^c_C \max B).$$

(c) The *minmax certainly nondominated set less order relation* $\preceq^{mn}_C$ is defined by

$$A \preceq^{mn}_C B \quad \Leftrightarrow \quad (A = B) \text{ or } (A \neq B,\ \max A \preceq^s_C \min B).$$

Neukel [132, 133] used set order relations defined in Definitions 6.4 and 6.5 to deal with the building conflict situation in the surroundings of the Frankfurt airport and cryptanalysis of substitution ciphers. See Fig. 6.2 and Fig. 6.6 for an illustration of these set order relations.

Definition 6.6 [92] A set $A \in \Xi$ is said to have the *quasi domination property* if and only if the following equivalent conditions hold:

(a) $\min A + C = A + C$ and $\max A - C = A - C$.

(b) $A \subseteq \min A + C$ and $A \subseteq \max A - C$.

Proposition 6.4 [92]

(a) *The set order relations $\preceq_C^m$ and $\preceq_C^{mc}$ are pre-order on Ξ and compatible with respect to the scalar multiplication with nonnegative real numbers. In general, they are not antisymmetric.*

(b) *Let $A, B \in \Xi$ have the quasi domination property. The set order relation $\preceq_C^{mn}$ is pre-order on Ξ and compatible with respect to the scalar multiplication with nonnegative real numbers. If the ordering cone C is pointed, then the set order relation $\preceq_C^{mn}$ is antisymmetric.*

Remark 6.2 The pointedness of the cone C in Proposition 6.4(b) cannot be dropped. From Remark 6.1, it is easy to see that for the sets A and B, we have $A \preceq_C^{mn} B$ and $B \preceq_C^{mn} A$ but $A \neq B$.

More precisely, for any $A, B \in \Xi$, we have

(a) $(A \preceq_C^m B$ and $B \preceq_C^m A) \Leftrightarrow (\min A + C = \min B + C,\ \min A - C = \min B - C,\ \max A + C = \max B + C,$ and $\max A - C = \max B - C)$.

(b) If C is pointed, then

$$(A \preceq_C^{mc} B \text{ and } B \preceq_C^{mc} A) \quad \Leftrightarrow \quad (\min A = \min B \text{ and } \max A = \max B),$$

(c) If C is pointed and A, B have the quasi domination property, then

$$(A \preceq_C^{m} B \text{ and } B \preceq_C^{m} A) \quad \Leftrightarrow \quad (\min A = \min B \text{ and } \max A = \max B).$$

The following result provides the relation among different kinds of set order relations.

Proposition 6.5 [92] *Let $A, B \in \Xi$ with $A \neq B$. Suppose that A and B have the quasi domination property. Then,*

(a) $A \preceq_C^c B \Rightarrow A \preceq_C^{mc} B \Rightarrow A \preceq_C^m B \Rightarrow A \preceq_C^s B$;

(b) $A \preceq_C^c B \Rightarrow A \preceq_C^{mn} B \Rightarrow A \preceq_C^m B$;

(c) $A \preceq_C^{mn} B$ *does not always imply* $A \preceq_C^{mc} B$ *and* $A \preceq_C^{mc} B$ *does not always imply* $A \preceq_C^{mn} B$.

The following example illustrates that the implications in the above proposition are strict, that is, the converse implications do not hold.

Example 6.5 [92] Let $Y = \mathbb{R}^2$ and $C = \mathbb{R}^2_+$. Consider the sets

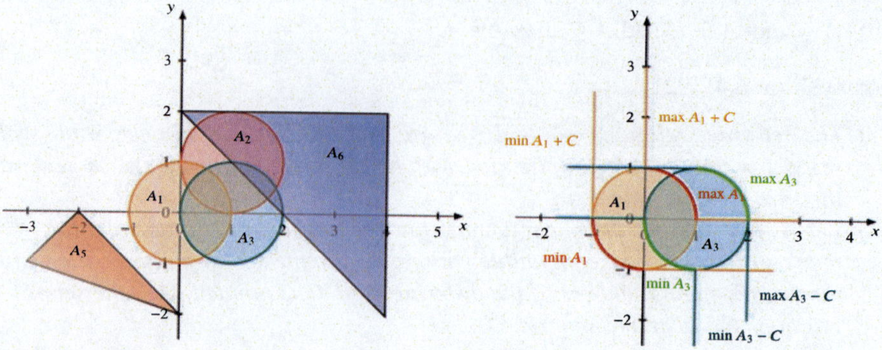

Fig. 6.7 Visualization of Example 6.5 with $C = \mathbb{R}^2_+$

$$
\begin{aligned}
A_1 &= \{(y_1, y_2) \in \mathbb{R}^2 : y_1^2 + y_2^2 \leq 1\},\\
A_2 &= \{(y_1, y_2) \in \mathbb{R}^2 : (y_1 - 1)^2 + (y_2 - 1)^2 \leq 1\},\\
A_3 &= \{(y_1, y_2) \in \mathbb{R}^2 : (y_1 - 1)^2 + y_2^2 \leq 1\},\\
A_4 &= \{(y_1, y_2) \in \mathbb{R}^2 : (y_1 - 1)^2 + (y_2 - 1)^2 \leq 1,\, y_1^2 + y_2^2 \geq 1\},\\
A_5 &= \operatorname{conv}\{(-2, 0), (-3, -1), (0, -2)\},\\
A_6 &= \operatorname{conv}\{(4, 2), (0, 2), (4, -2)\}.
\end{aligned}
$$

From Fig. 6.7, one can easily visualize that

$$A_1 \preceq_C^{mc} A_2 \quad \text{but} \quad A_1 \not\preceq_C^{mn} A_2 \quad \text{and} \quad A_1 \not\preceq_C^{c} A_2,$$

$$A_1 \preceq_C^{m} A_3 \quad \text{but} \quad A_1 \not\preceq_C^{mn} A_3 \quad \text{and hence} \quad A_1 \not\preceq_C^{mc} A_3,$$

$$A_1 \preceq_C^{mn} A_4 \quad \text{but} \quad A_1 \not\preceq_C^{c} A_4,$$

$$A_5 \preceq_C^{mn} A_6 \quad \text{but} \quad A_5 \not\preceq_C^{mc} A_6.$$

Remark 6.3 From Propositions 6.2 and 6.5, it is clear that the set order relation $\preceq_C^p$ is the weakest one and the set order relation $\preceq_C^c$ is the strongest one. Furthermore, in contrast to the set order relations $\preceq_C^c$ and $\preceq_C^{mn}$, the set order relations $\preceq_C^\alpha$, $\alpha \in \{l, u, s, m, mc\}$ are generally not antisymmetric. To see this, it suffices to consider the case with the set order relation $\preceq_C^{mc}$ because this set order relation is the strongest one among all other set order relations.

Let $Y = \mathbb{R}^2$ and $C = \mathbb{R}^2_+$. Consider the sets

$$
\begin{aligned}
A_1 &= \{(y_1, y_2) \in \mathbb{R}^2 : y_1^2 + y_2^2 \leq 1\},\\
A_2 &= \{(y_1, y_2) \in \mathbb{R}^2 : y_1^2 + y_2^2 \leq 1, -1 \leq y_1 - y_2 \leq 1\}.
\end{aligned}
$$

Then we can see that $A \neq B$, $A \preceq_C^{mc} B$, and $B \preceq_C^{mc} A$.

6.3.1 Set Order Relations in Terms of the Minkowski Difference

Recently, Karaman et al. [96] introduced the following set order relations on the family of sets by using the Minkowski difference.

Definition 6.7 [96] Let Y be a normed space and $A, B, K \in P(Y)$.

(a) The *m-upper set less order relation*, denoted by $\preceq_K^{mu}$, is defined as

$$A \preceq_K^{mu} B \quad \Leftrightarrow \quad (B \dot{-} A) \cap K \neq \emptyset.$$

(b) The *m-lower set less order relation*, denoted by $\preceq_K^{ml}$, is defined as

$$A \preceq_K^{ml} B \quad \Leftrightarrow \quad (A \dot{-} B) \cap (-K) \neq \emptyset.$$

If A and B are bounded and $A \dot{-} B \neq \emptyset$, $B \dot{-} A \neq \emptyset$, then $A \preceq_K^{mu} B$ if and only if $A \preceq_K^{ml} B$. If A and B are singleton sets and K is a convex and pointed cone with $\mathbf{0} \in K$, then $\preceq_K^{mu}$ and $\preceq_K^{ml}$ coincide with the vector order relation $\preceq_C$ on Y, that is, for any $a, b \in Y$, we have

$$\{a\} \preceq_K^{mu} \{b\} \quad \Leftrightarrow \quad \{a\} \preceq_K^{ml} \{b\} \quad \Leftrightarrow \quad a \preceq_K b.$$

It is pointed out in [96] that

(a) if K is a convex cone in Y and $\mathbf{0} \in K$, then $\preceq_K^{mu}$ and $\preceq_K^{ml}$ are pre-order on $P(Y)$;
(b) if K is a pointed convex cone in Y with $\mathbf{0} \in K$, then $\preceq_K^{mu}$ and $\preceq_K^{ml}$ are partial order on $\mathcal{B}(Y)$;
(c) $\preceq_K^{mu}$ and $\preceq_K^{ml}$ are compatible with addition;
(d) $\preceq_K^{mu}$ and $\preceq_K^{ml}$ are compatible with scalar multiplication if and only if K is a cone.

From now onward, we consider the ordering cone C on Y instead of K, then $\preceq_K^{mu}$ and $\preceq_K^{ml}$ turn to $\preceq_C^{mu}$ and $\preceq_C^{ml}$.

The set order relations $\preceq_C^{mu}$ and $\preceq_C^{ml}$ and the set order relations $\preceq_C^{u}$ and $\preceq_C^{l}$ have the following relations: For any $A, B \in P(Y)$,

$$A \preceq_C^{mu} B \Rightarrow A \preceq_C^{u} B \quad \text{and} \quad A \preceq_C^{ml} B \Rightarrow A \preceq_C^{l} B,$$

but the converse of the above implications may not be true.

The following example illustrates that the set order relation $\preceq_C^{u}$ does not imply $\preceq_C^{mu}$.

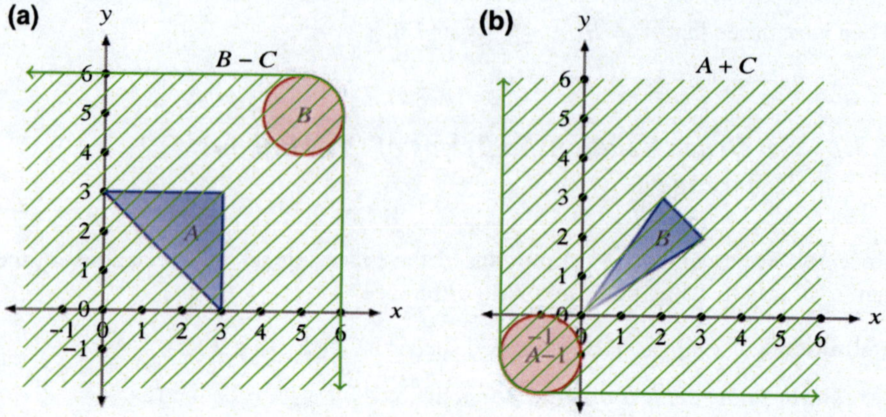

Fig. 6.8 **(a)** Illustration of sets in Example 6.6. **(b)** Illustration of sets in Example 6.7

Example 6.6 Let $Y = \mathbb{R}^2$ and $C = \mathbb{R}^2_+$. Consider the sets

$$A = \text{conv}\{(2,0), (3,3), (0,2)\}$$

and

$$B = \{(y_1, y_2) \in \mathbb{R}^2 : (y_1 - 5)^2 + (y_2 - 5)^2 \leq 1\}.$$

As in Fig. 6.8 (a), $A \subseteq B - C$ which gives us $A \preceq^u_C B$. On the other hand, there does not exist any $x \in \mathbb{R}^2$ such that $x + A \subseteq B$. Hence, we have $(B \dot{-} A) \cap C = \emptyset$, that is, $A \npreceq^{mu}_C B$.

The following example shows that the set order relation $\preceq^l_C$ does not imply the set order relation $\preceq^{ml}_C$.

Example 6.7 Let $Y = \mathbb{R}^2$ and $C = \mathbb{R}^2_+$. Consider the sets

$$A = \{(y_1, y_2) \in \mathbb{R}^2 : (y_1 + 1)^2 + (y_2 + 1)^2 \leq 1\}$$

and

$$B = \text{conv}\{(0,0), (3,2), (2,3)\}.$$

As in Fig. 6.8 (b), $B \subseteq A + C$ which gives us $A \preceq^l_C B$. On the other hand, there does not exist any $x \in \mathbb{R}^2$ such that $x + B \subseteq A$. Hence, we have $(A \dot{-} B) \cap (-C) = \emptyset$, that is, $A \npreceq^{ml}_C B$.

The strict version of $\preceq^{mu}_C$ and $\preceq^{ml}_C$ is defined as follows:

Definition 6.8 [96] Let Y be a normed space, $A, B \in P(Y)$, and C be a convex cone in Y with $\text{int}C \neq \emptyset$.

(a) The *strictly m-upper set less order relation*, denoted by $\prec_C^{mu}$, is defined as

$$A \prec_C^{mu} B \quad \Leftrightarrow \quad (B \dot{-} A) \cap \text{int} C \neq \emptyset.$$

(b) The *strictly m-lower set less order relation*, denoted by $\prec_C^{ml}$, is defined as

$$A \prec_C^{ml} B \quad \Leftrightarrow \quad (A \dot{-} B) \cap \text{int}(-C) \neq \emptyset.$$

Remark 6.4 Let $\alpha \in \{mu, ml\}$ and $A, B \in P(Y)$. If $A \prec_C^{\alpha} B$, then $A \preceq_C^{\alpha} B$.

It is pointed out in [96] that

(a) $\prec_C^{mu}$ and $\prec_C^{ml}$ are compatible with addition;
(b) $\prec_C^{mu}$ and $\prec_C^{ml}$ are compatible with scalar multiplication.

If C is a pointed convex cone, even then the relations $\prec_C^{mu}$ and $\prec_C^{ml}$ are not reflexive unless $C = Y$, and hence, $\prec_C^{mu}$ and $\prec_C^{ml}$ are not partial order. The following example clarifies this fact.

Example 6.8 Let $Y=\mathbb{R}^2$, $C=\mathbb{R}^2_+$, and $A = \{(x, y) : x^2 + y^2 \leq 1\}$. Since $A \dot{-} A = \{\mathbf{0}\}$, we have $\{\mathbf{0}\} \cap \text{int} C = \emptyset$ and $A \not\prec_C^{mu} A$. Similarly, since $(A \dot{-} A) \cap \text{int}(-C) \neq \emptyset$, we obtain $A \not\prec_C^{ml} A$.

6.3.2 Set Order Relations with Respect to Variable Domination Structures

In the recent past, Eichfelder and Pilacka [42, 44] and Köbis et al. [7, 101, 102] are among major contributors to study set optimization problems with respect to variable ordering structures with applications to different real-world problems. The importance of incorporating variable ordering structures for intensity-modulated radiation therapy (IMRT) in order to allow an improved modeling of the decision-making problem is already discussed in [40, Chap. 10]. Another significant application of set optimization problems with respect to variable domination structures can be found in the theory of consumer demand [119], medical image registration [104], and uncertain optimization [7].

To study set optimization problems with respect to variable domination structures by using a set approach, we recall the following six kinds of generalized variable set order relations to compare sets in a topological vector space Y.

Definition 6.9 [104] Let $A, B \in P(Y)$ and $\mathcal{K} : Y \rightrightarrows Y$ be a set-valued map. The following binary relations on $P(Y)$ with respect to $\mathcal{K}$ are defined as follows:

(a) The *variable generalized lower set less order relation* $\preceq_l^{\mathcal{K}}$ is defined by

$$A \preceq_l^{\mathcal{K}} B \quad \Leftrightarrow \quad B \subseteq \bigcup_{a \in A} (a + \mathcal{K}(a)).$$

(b) The *variable generalized upper set less order relation* $\preceq_u^{\mathcal{K}}$ is defined by

$$A \preceq_u^{\mathcal{K}} B \quad \Leftrightarrow \quad A \subseteq \bigcup_{b \in B} (b - \mathcal{K}(b)).$$

(c) The *variable generalized certainly lower set less order relation* $\preceq_{cl}^{\mathcal{K}}$ is defined by

$$A \preceq_{cl}^{\mathcal{K}} B \quad \Leftrightarrow \quad B \subseteq \bigcap_{a \in A} (a + \mathcal{K}(a)).$$

(d) The *variable generalized certainly upper set less order relation* $\preceq_{cu}^{\mathcal{K}}$ is defined by

$$A \preceq_{cu}^{\mathcal{K}} B \quad \Leftrightarrow \quad A \subseteq \bigcap_{b \in B} (b - \mathcal{K}(b)).$$

(e) The *variable generalized possible lower set less order relation* $\preceq_{pl}^{\mathcal{K}}$ is defined by

$$A \preceq_{pl}^{\mathcal{K}} B \quad \Leftrightarrow \quad B \cap \bigcup_{a \in A} (a + \mathcal{K}(a)) \neq \emptyset.$$

(f) The *variable generalized possible upper set less order relation* $\preceq_{pu}^{\mathcal{K}}$ is defined by

$$A \preceq_{pu}^{\mathcal{K}} B \quad \Leftrightarrow \quad A \cap \bigcup_{b \in B} (b - \mathcal{K}(b)) \neq \emptyset.$$

Remark 6.5 For all $y \in Y$, if $\mathcal{K}(y) = C$ is a convex cone with $\text{int}C \neq \emptyset$ in Y, then the set order relations $\preceq_l^{\mathcal{K}}$ and $\preceq_u^{\mathcal{K}}$ reduce to the set order relations $\preceq_C^l$ and $\preceq_C^u$, respectively. See Fig. 6.9 for illustration of variable generalized set order relations.

Proposition 6.6 [119] *Let $A, B \in P(Y)$. Then, the following assertions hold:*

(a) $A \preceq_u^{\mathcal{K}} B \Leftrightarrow B \preceq_l^{-\mathcal{K}} A$;
(b) $A \preceq_{cu}^{\mathcal{K}} B \Leftrightarrow B \preceq_{cl}^{-\mathcal{K}} A$;
(c) $A \preceq_{pu}^{\mathcal{K}} B \Leftrightarrow B \preceq_{pl}^{-\mathcal{K}} A$;
(d) $A \preceq_{cl}^{\mathcal{K}} B \Rightarrow A \preceq_l^{\mathcal{K}} B \Rightarrow A \preceq_{pl}^{\mathcal{K}} B$;
(e) $A \preceq_{cu}^{\mathcal{K}} B \Rightarrow A \preceq_u^{\mathcal{K}} B \Rightarrow A \preceq_{pu}^{\mathcal{K}} B$.

Köbis et al. [104] established the following useful properties of the set order relations $\preceq_t^{\mathcal{K}}, t \in \{l, u, cl, cu, pl, pu\}$

Proposition 6.7 [104] *Let $\mathcal{K} : Y \rightrightarrows Y$ be a set-valued map. The following statements hold:*

(a) *If $\mathbf{0} \in \mathcal{K}(y)$ for all $y \in Y$, then the set order relations $\preceq_l^{\mathcal{K}}$, $\preceq_u^{\mathcal{K}}$, $\preceq_{pl}^{\mathcal{K}}$, and $\preceq_{pu}^{\mathcal{K}}$ are reflexive.*

Fig. 6.9 Visualization of variable generalized set order relations defined in Definition 6.9

(b) *If* $\mathcal{K}(y) + \mathcal{K}(y) \subseteq \mathcal{K}(y)$ *for all* $y \in Y$ *and* $\mathcal{K}(y + d) \subseteq \mathcal{K}(y)$ *for all* $y \in Y$ *and all* $d \in \mathcal{K}(y)$*, then the set order relations* $\preceq_l^{\mathcal{K}}$ *and* $\preceq_{cl}^{\mathcal{K}}$ *are transitive.*

(c) *If* $\mathcal{K}(y) + \mathcal{K}(y) \subseteq \mathcal{K}(y)$ *for all* $y \in Y$ *and* $\mathcal{K}(y - d) \subseteq \mathcal{K}(y)$ *for all* $y \in Y$ *and all* $d \in \mathcal{K}(y)$*, then the set order relations* $\preceq_u^{\mathcal{K}}$ *and* $\preceq_{cu}^{\mathcal{K}}$ *are transitive.*

(d) *If* $\mathcal{K}(y) \cap (-\mathcal{K}(z)) = \{0\}$ *for all* $y, z \in Y$*, then the set order relations* $\preceq_{cl}^{\mathcal{K}}$ *and* $\preceq_{cu}^{\mathcal{K}}$ *are antisymmetric.*

6.4 Nonlinear Scalarization Functions

We first recall the linear scalarization method for vectors. The most representative example of linear scalarizing functions is an inner product. For any $y, k \in Y$, in case of vector, the linear scalarizing function is defined by

$$h_k(y) := \langle y, k \rangle. \tag{6.1}$$

Based on this scalarization, we can consider the following scalarizing functions for a set $A \subseteq Y$ defined by

$$\varphi_k(A) := \inf_{y \in A} \langle y, k \rangle \quad \text{and} \quad \phi_k(A) := \sup_{y \in A} \langle y, k \rangle.$$

Rest of this section, we assume that C is a proper, solid, closed convex cone in a topological vector space Y and $k \in \text{int}C$. The nonlinear scalarization functional $\varphi_{C,k} : Y \to (-\infty, \infty]$ is defined by

$$\varphi_{C,k}(y) = \inf\{t \in \mathbb{R} : y \preceq_C tk\} = \inf\{t \in \mathbb{R} : y \in tk - C\}, \quad \text{for all } y \in Y. \tag{6.2}$$

As mentioned in [103], Fig. 6.10 visualizes the functional $\varphi_{C,k}$ with $C = \mathbb{R}^2_+$ and $k \in \text{int}C$. We can see that the set $-C$ is moved along the line $\mathbb{R} \cdot k$ up until y belongs to $tk - C$. The functional $\varphi_{C,k}$ assigns the smallest value t such that the property $y \in tk - C$ is fulfilled.

It can be shown that all minimal elements of a vector optimization problem can be found by means of $\varphi_{C,k}$ if $k \in C \setminus \{\mathbf{0}\}$, and all weakly minimal elements of a vector optimization problem can be determined if $k \in \text{int}C$ (see [54]). In Fig. 6.10, we can easily see that for the given cone $C = \mathbb{R}^2_+$, by a variation of the vector $k \in C \setminus \{\mathbf{0}\}$, all minimal elements of the vector optimization problem without any convexity assumptions can be found. The scalarizing functional $\varphi_{C,k}$ was used in [54] to prove nonconvex separation theorems and has applications in coherent risk measures in financial mathematics (see, for instance, [66, 85]).

We note that the set $\{t \in \mathbb{R} : y \in tk - C\}$ may be empty, and in this case $\varphi_{C,k}$ will take $+\infty$ as by convention $\inf \emptyset = +\infty$. For further details, see [56]. On the other hand, if $k \in C$, then the lower level set of $\varphi_{C,k}$ at each height t coincides with a parallel translation of $-C$ at offset tk, that is,

$$\{y \in Y : \varphi_{C,k}(y) \leq t\} = tk - C,$$

and hence $\varphi_{C,k}$ is the smallest strictly monotonic function with respect to the ordering cone C in case $k \in \text{int}C$. Also, this scalarization function has a dual form as follows:

$$-\varphi_{C,k}(-y) = \sup\{t \in \mathbb{R} : tk \preceq_C y\} = \sup\{t \in \mathbb{R} : y \in tk + C\}, \quad \text{for all } y \in Y.$$

The importance of this function is due to the fact that it characterizes, under some appropriate assumptions, the relation $\preceq_C$ as

Fig. 6.10 Illustration of the functional (6.2) with $C = \mathbb{R}^2_+$

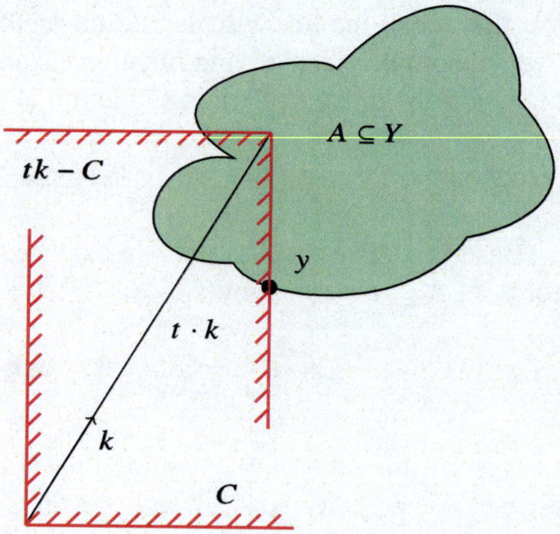

$$y_1 \preceq_C y_2 \quad \Leftrightarrow \quad \varphi_{C,k}(y_1 - y_2) \leq 0.$$

Another essential feature of this function is the so-called translativity property (see [56] for details), that is,

$$\text{for all } y \in Y \text{ and all } \alpha \in \mathbb{R}\colon \varphi_{C,k}(y + \alpha k) = \varphi_{C,k}(y) + \alpha.$$

In [148], functionals of type (6.2) have been applied in order to obtain vector-valued variants of Ekeland's variational principle. For this topic, see also [59] and [65]. Note that the originality of the approach in [54, 148] relies on the fact that the set C defining a functional via (6.2) was assumed neither to be a cone nor convex. In some papers, this functional has been treated and regarded as a generalization of the Chebyshev scalarization, see the books [9, 56, 127]. Essentially, it is equivalent to the smallest strictly monotonic function with respect to int$C$ defined by Luc in [127].

Recently, Köbis et al. [103] characterized the upper and lower set less order relations defined in Definition 6.4(a) and (b) by using the scalarization functional $\varphi_{C,k}$ as follows.

Theorem 6.4.1 [103, Theorems 3.3 and 3.8] *Let C be a proper closed convex cone in a topological vector space Y and $A, B \in P(Y)$.*

(a) *If $k_0 \in C\backslash\{\mathbf{0}\}$ is such that $\inf_{b\in B} \varphi_{C,k_0}(a - b)$ is attained for all $a \in A$, then*

$$\sup_{a\in A} \inf_{b\in B} \varphi_{C,k_0}(a - b) \leq 0 \quad \Leftrightarrow \quad A \subseteq B - C.$$

(b) *If $k_1 \in C\backslash\{\mathbf{0}\}$ is such that $\inf_{a\in A} \varphi_{C,k_1}(a - b)$ is attained for all $b \in B$, then*

$$\sup_{b\in B} \inf_{a\in A} \varphi_{C,k_1}(a - b) \leq 0 \quad \Leftrightarrow \quad B \subseteq A + C.$$

Recently, Köbis et al. [105] and Ansari et al. [10] studied and investigated new nonlinear scalarization functions for the relations $\preceq_l^{\mathcal{K}}$ and $\preceq_u^{\mathcal{K}}$ and discussed some of its properties.

Let $A, B \in P(Y)$ and $\mathcal{K} : Y \rightrightarrows Y$ be a set-valued map. For each $k \in Y \setminus \{\mathbf{0}\}$, let

$$[0, +\infty)k + \mathcal{K}(y) \subseteq \mathcal{K}(y), \quad \text{for all } y \in Y. \tag{6.3}$$

Let $B \in P(Y)$ be arbitrary but fixed. Consider the scalarization functionals $\varphi_{k,B} : P(Y) \to \mathbb{R} \cup \{+\infty\}$ and $\phi_{k,B} : P(Y) \to \mathbb{R} \cup \{+\infty\}$ defined by

$$\varphi_{k,B}(A) := \inf\{t \geq 0 \,:\, A \preceq_u^{\mathcal{K}} tk + B\}, \quad \text{for all } A \in P(Y), \tag{6.4}$$

and

$$\phi_{k,B}(A) := \inf\{t \geq 0 \,:\, A \preceq_l^{\mathcal{K}} tk + B\}, \quad \text{for all } A \in P(Y), \tag{6.5}$$

respectively.

If we consider $B = \{0\}$, then the scalarization functionals $\varphi_{k,B}$ and $\phi_{k,B}$ defined by (6.4) and (6.5) can be written as $h^u_k : P(Y) \to \mathbb{R} \cup \{+\infty\}$ and $h^l_k : P(Y) \to \mathbb{R} \cup \{+\infty\}$, respectively, and we have $h^u_k : P(Y) \to \mathbb{R} \cup \{+\infty\}$ and $h^l_k : P(Y) \to \mathbb{R} \cup \{+\infty\}$, respectively, and we have

$$h^u_k(A) = \inf\{t \geq 0 : A \preceq^{\mathcal{K}}_u tk\}, \tag{6.6}$$

and

$$h^l_k(A) = \inf\{t \geq 0 : A \preceq^{\mathcal{K}}_l tk\}, \tag{6.7}$$

for all $A \in P(Y)$.

For more details, we refer to [10, 105].

6.4.1 Weighted Set Order Relations

Rest of this subsection, we assume the following assumption.

Assumption 1 The ordering cone $C \neq Y$ is solid, closed, and convex in a Hausdorff topological vector space Y and $k \in \text{int}C$ is such that $\inf_{b\in B} \varphi_{C,k}(a - b)$ is attained for all $a \in A$ and $\inf_{a\in A} \varphi_{C,k}(a - b)$ is attained for all $b \in B$ whenever A and B are closed and bounded sets in Y.

By using the characterization of the set order relations $\preceq^l_C$ and $\preceq^u_C$ given in Theorem 6.4.1, Chen et al. [30] introduced the so-called weighted set order relations as follows.

Definition 6.10 Let $A, B \in \Omega^{cb}$ and $\lambda \in [0, 1]$. The *weighted set order relation* $\preceq^\lambda_C$ for sets $A, B \in P(Y)$ is defined by

$$A \preceq^\lambda_C B \quad \Leftrightarrow \quad \lambda g^u(A, B) + (1 - \lambda) g^l(A, B) \leq 0,$$

where

$$g^u(A, B) := \sup_{a\in A} \inf_{b\in B} \varphi_{C,k}(a - b) \quad \text{and} \quad g^l(A, B) := \sup_{b\in B} \inf_{a\in A} \varphi_{C,k}(a - b).$$

Remark 6.6 [30] For any $\lambda \in [0, 1]$, the relation $\preceq^\lambda_C$ is reflexive and transitive, that is, $\preceq^\lambda_C$ is a pre-order. Moreover, the relation $\preceq^\lambda_C$ is compatible with nonnegative scalar multiplication, that is, for any $A, B \in P(Y)$ and $\alpha \geq 0$, one has

$$A \preceq^\lambda_C B \quad \Rightarrow \quad \alpha A \preceq^\lambda_C \alpha B.$$

Remark 6.7 For $\lambda = 1$, $\preceq^{\lambda}_{C}$ reduces to $\preceq^{u}_{C}$, and for $\lambda = 0$, $\preceq^{\lambda}_{C}$ reduces to $\preceq^{l}_{C}$. If $\preceq^{u}_{C}$ and $\preceq^{l}_{C}$ hold, then $\preceq^{\lambda}_{C}$ is true for all $\lambda \in [0, 1]$, but the converse is not true and this was exactly the intention of introducing $\preceq^{\lambda}_{C}$.

Note that the parameter λ serves as a weight vector which indicates the importance of either of the two relations $\preceq^{u}_{C}$ and $\preceq^{l}_{C}$. The relation which is more important should be associated with a higher weight factor. For instance, if $g^{u}(A, B) \leq 0$ and $g^{l}(A, B) > 0$, then, for large enough λ, $A \preceq^{\lambda}_{C} B$ can hold and the $A \preceq^{u}_{C} B$ "outweighs" the effects of $A \not\preceq^{l}_{C} B$.

Remark 6.8 Chen et al. [30] gave the definition of the weighted set order relation under the assumption of Y being a quasicompact topological space. Recalling that usually a Hausdorff (separated) topological space is called compact if it is quasicompact, one may realize that [30, Definition 2.5] is basically empty (as well as the following results, for example, [30, Proposition 2.9]): Up to trivial examples, there are no (quasi) compact topological linear spaces; not even the real line with the usual topology satisfies this assumption. Therefore, we modified Assumption 2.4 in [30] to the version above: It is certainly satisfied in any finite-dimensional space when the usual topology since every closed bounded set in such a space is compact and the function $\varphi_{C,k}$ is continuous for $k \in \text{int}C$.

We provide an example below to illustrate the weighted set order relations $\preceq^{\lambda}_{C}$ and discuss the role of the parameter λ.

Example 6.9 [30] Let $A = [a, c]$ and $B = [b, d]$ be compact sets in $\mathbb{R}$. We choose $C = \mathbb{R}_{+}$ and $k = 1$. Then,

$$
\begin{aligned}
g^{u}(A, B) &= \sup_{a\in A} \inf_{b\in B} \varphi_{C,k}(a - b) = \sup_{a\in A} \inf_{b\in B} \inf\{t \in \mathbb{R} : a - b \leq t\} \\
&= \sup_{a\in A} \inf_{b\in B} (a - b) = \sup_{a\in A} a - \sup_{b\in B} b = c - d, \\
g^{u}(A, B) &= a - b,\ g^{u}(B, A) = d - c,\ g^{l}(B, A) = b - a.
\end{aligned}
$$

Consider $a = 5, c = 10, b = 0, and\, d = 11$. Then $B \not\preceq^{u}_{C} A$, but $B \preceq^{l}_{C} A$. Also, $A \preceq^{u}_{C} B$, but $A \not\preceq^{l}_{C} B$. However, we can see that the "amount" of B that is bigger than the supremum of A is very small compared to how the lower bound of B is smaller than the lower bound of A. In that sense, when a decision-maker has no clear understanding of how to choose a set, the weighted set order relation $\preceq^{\lambda}_{C}$ can be helpful. We have $g^{u}(A, B) = -1$, $g^{l}(A, B) = 5$. So, in order for $A \preceq^{\lambda}_{C} B$ to hold, $\lambda \in [\frac{5}{6}, 1]$. Similarly, as $g^{u}(B, A) = 1$, $g^{l}(B, A) = -5$, $\lambda \in [0, \frac{5}{6}]$ for $B \preceq^{\lambda}_{C} A$ to hold true.

6.5 Solution Concepts in Set Optimization

This section deals with the solution concepts of the set optimization problem (SOP) with vector approach and set approach. The solution concept based on the vector approach is of mathematical interest but it cannot be often used in practice.

In the vector approach, an element $\bar{x} \in S$ for which there exists at least one element $\bar{y} \in F(\bar{x})$ which is Pareto minimal point of the image set of F is a solution of the set optimization problem (SOP). In the past, solution concepts based on vector approach has been studied and investigated in [1, 2, 29, 33–35, 41, 55, 59, 61, 64, 80, 93, 122, 123] and the references therein.

Definition 6.11 [1, 80] An element $\bar{x} \in S$ is said to be

(a) a *minimal solution* of the problem (SOP) if there exists $\bar{y} \in F(\bar{x})$ such that $\bar{y}$ is a minimal element of image set $F(S)$, that is,

$$(\{\bar{y}\} - C) \cap F(S) = \{\bar{y}\}.$$

(b) a *weak minimal solution* of the problem (SOP) if there exists $\bar{y} \in F(\bar{x})$ such that $\bar{y}$ is a weak minimal element of image set $F(S)$, that is,

$$(\{\bar{y}\} - \text{int}C) \cap F(S) = \emptyset.$$

We denote the set of minimal and weak minimal elements of (SOP) by $\text{Min}(F, S)$ and $\text{WMin}(F, S)$, respectively.

Recall that $\min A := \{a \in A : A \cap (a - C) = \{a\}\}$ and $\text{wmin}A := \{a \in A : A \cap (a - \text{int}C) = \emptyset\}$ are the sets of minimal elements and weak minimal elements, respectively, with respect to the convex pointed cone C in a topological vector space Y.

Note that Definition 6.11 can also be written as follows:
An element $\bar{x} \in S$ is said to be

(a) a *minimal solution* [99] of the problem (SOP) if there exists $\bar{y} \in F(\bar{x})$ such that

$$(F(S) - \{\bar{y}\}) \cap (-C) = \{\mathbf{0}\};$$

(b) a *weak minimal solution* [99] of the problem (SOP) if there exists $\bar{y} \in F(\bar{x})$ such that

$$(F(S) - \{\bar{y}\}) \cap (-\text{int}C) = \emptyset.$$

Another form of Definition 6.11 can also be written as follows:
An element $\bar{x} \in S$ is said to be

(a) a *minimal solution* [96] of the problem (SOP) if $F(\bar{x}) \cap \min F(S) \neq \emptyset$;
(b) a *weak minimal solution* [96] of the problem (SOP) if $F(\bar{x}) \cap \text{wmin}F(S) \neq \emptyset$.

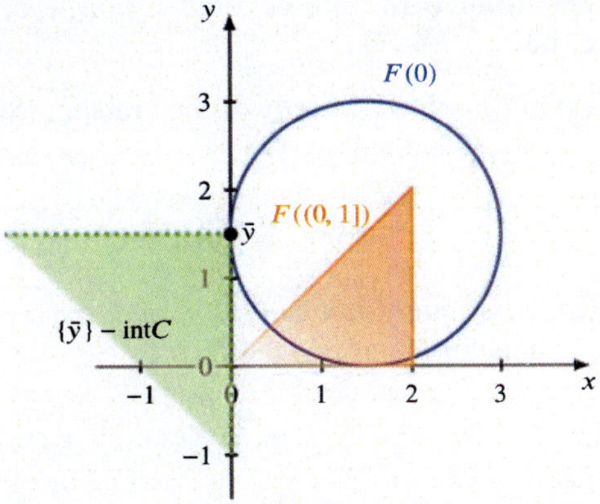

Fig. 6.11 Illustration of Example 6.10 with $C = \mathbb{R}^2_+$

It is clear that $\text{Min}(F, S) \subseteq \text{WMin}(F, S)$. However, the reverse inclusion may not hold.

Example 6.10 Let $X=\mathbb{R}$, $S=[0, 1]$, $Y = \mathbb{R}^2$, $C = \mathbb{R}^2_+$, and $F : S \rightrightarrows Y$ be defined by

$$F(x) = \begin{cases} \{(y_1, y_2) \in \mathbb{R}^2 : (y_1 - 3/2)^2 + (y_2 - 3/2)^2 = (3/2)^2\}, & \text{if } x = 0, \\ \text{conv}\{(0, 0), (2, 0), (2, 2)\}, & \text{otherwise.} \end{cases}$$

From Fig. 6.11, we can see that there exists $\bar{y} \in F(0)$ such that $(\{\bar{y}\} - \text{int}C) \cap F(S) = \emptyset$ and hence $0 \in \text{WMin}(F, S)$ but there does not exist any $\bar{y} \in F(0)$ such that $(\{\bar{y}\} - C) \cap F(S) = \{\bar{y}\}$, so $0 \notin \text{Min}(F, S)$.

As we have seen above, in the vector approach, we consider only a minimal element $\bar{y}$ of the image set $F(S)$. However, only one minimal element does not imply that the whole set $F(\bar{x})$ be in a certain sense minimal with respect to all sets $F(x)$ with $x \in S$. To overcome this drawback, the solution concepts based on the set approach are very helpful and important. In the set approach, solution concepts are defined by using different kinds of set order relations, and these solutions are based on the comparison of values of set-valued objective map using set order relations. In the recent past, solution concepts based on set approach have been studied and investigated in [1, 8, 25, 30, 67, 79–84, 92, 96, 109, 110, 112, 115] and the references therein.

As the set order relations $\preceq^l_C$, $\preceq^u_C$, $\preceq^s_C$, $\preceq^c_C$ on $P(Y)$; $\preceq^m_C$, $\preceq^{mc}_C$, $\preceq^{mn}_C$ on Ξ; and $\preceq^\lambda_C$ on Ω^{cb} are pre-order, we can define optimal solutions with respect to the pre-order $\preceq^t_C$, where $t \in \{l, u, s, c, m, mc, mn, \lambda\}$. For the set order relation $\preceq^t_C$, we assume the following condition:

$$F \text{ takes values on } \begin{cases} P(Y), & \text{if } t \in \{l, u, s, c\}, \\ \Xi, & \text{if } t \in \{m, mc, mn\}, \\ \Omega^{cb}, & \text{if } t = \lambda. \end{cases}$$

Definition 6.12 [30, 92] Let $t \in \{l, u, s, c, m, mc, mn, \lambda\}$. An element $\bar{x} \in S$ is said to be

(a) a *t-minimal solution* of the problem (SOP) with respect to the set order relation $\preceq_C^t$ if and only if

$$F(x) \preceq_C^t F(\bar{x}) \text{ for some } x \in S \quad \Rightarrow \quad F(\bar{x}) \preceq_C^t F(x);$$

(b) a *t-strongly minimal solution* of the problem (SOP) with respect to the set order relation $\preceq_C^t$ if and only if

$$F(\bar{x}) \preceq_C^t F(x), \quad \text{for all } x \in S;$$

(c) a *t-weak minimal solution* of the problem (SOP) with respect to the set order relation $\prec_C^t$, $t \neq \lambda$ if and only if

$$F(x) \prec_C^t F(\bar{x}) \text{ for some } x \in S \;\Rightarrow\; F(\bar{x}) \prec_C^t F(x).$$

We denote the family of t-minimal, t-strongly minimal, and t-weak minimal elements of S by $t-\text{Min}(F, S)$, $t-\text{SMin}(F, S)$, and $t-\text{WMin}(F, S)$, respectively, where $t \in \{l, u, s, c, m, mc, mn, \lambda\}$.

It is clear that $t-\text{Min}(F, S) \subseteq t-\text{WMin}(F, S)$ for $t \in \{l, u\}$. However, the reverse inclusion may not hold.

Example 6.11 Let $X = \mathbb{R}$, $S = [0, 1]$, $Y = \mathbb{R}^2$, $C = \mathbb{R}^2_+$, and $F : S \rightrightarrows Y$ be defined by

$$F(x) = \begin{cases} [(-1, -1), (1, 1)], & \text{if } x = 0, \\ \{(u, v) \in \mathbb{R}^2 : u^2 + v^2 \leq x^2\}, & \text{otherwise.} \end{cases}$$

Then it can be easily seen that $l-\text{Min}(F, S) = \{0\}$ and $l-\text{WMin}(F, S) = \{0, 1\}$.

Example 6.12 Let $X = \mathbb{R}$, $S = [-1, 0]$, $Y = \mathbb{R}^2$, $C = \mathbb{R}^2_+$, and $F : S \rightrightarrows Y$ be defined by

$$F(x) = \begin{cases} [0, -2] \times [0, -2], & \text{if } x = 0, \\ [0, -3) \times (0, -3), & \text{otherwise.} \end{cases}$$

Then we see that $0 \in u-\text{WMin}(F, S)$ but $0 \notin u-\text{Min}(F, S)$.

Definition 6.13 [77, 78] Let S be a nonempty convex subset of X. A set-valued map $F : S \rightrightarrows Y$ is said to be

(a) *strictly natural l-type C-quasi-convex* on S if for all $x_1, x_2 \in S$ with $x_1 \neq x_2$ and all $t \in (0, 1)$, there exists $\lambda \in [0, 1]$ such that

$$F(tx_1 + (1-t)x_2) \prec_C^l \lambda F(x_1) + (1-\lambda)F(x_2);$$

(b) *strictly natural u-type C-quasi-convex* on S if for all $x_1, x_2 \in S$ with $x_1 \neq x_2$ and all $t \in (0, 1)$, there exists $\lambda \in [0, 1]$ such that

$$F(tx_1 + (1-t)x_2) \prec_C^u \lambda F(x_1) + (1-\lambda)F(x_2).$$

Proposition 6.8 [77, 78] *Assume that S is a convex subset of X, $F : S \rightrightarrows Y$ is a strictly natural l-type C-quasi-convex map on S with nonempty compact values. Then, $l - \mathrm{Min}(F, S) = l - \mathrm{WMin}(F, S)$.*

Proposition 6.9 [77, 78] *Assume that S is a convex subset of X, $F : S \rightrightarrows Y$ is a strictly natural u-type C-quasi-convex map on S with nonempty compact values. Then, $u - \mathrm{Min}(F, S) = u - \mathrm{WMin}(F, S)$.*

The following examples show that there is no relation between minimal and l-minimal solutions.

Example 6.13 [79] Let $X = \mathbb{R}$, $S = \mathbb{R}_+$, $Y = \mathbb{R}^2$, $C = \mathbb{R}^2_+$, and $F : S \rightrightarrows Y$ be defined by

$$F(x) = \begin{cases} \{(0, 0)\}, & \text{if } x = 0, \\ [(0, 0), (-x, \frac{1}{x})], & \text{otherwise.} \end{cases}$$

Then we can easily obtain $\mathrm{Min}(F, S) = S$ and $l - \mathrm{Min}(F, S) = \emptyset$.

Example 6.14 [79] Let $X = \mathbb{R}$, $S = [-1, 0]$, $Y = \mathbb{R}^2$, $C = \mathbb{R}^2_+$, and $F : S \rightrightarrows Y$ be defined by

$$F(x) = \begin{cases} \{(u, -u^2) \in \mathbb{R}^2 : -1 < u \leq 0\}, & \text{if } x = -1, \\ [(x, 0), (x, -x^2)], & \text{otherwise.} \end{cases}$$

After a short calculation, we get $\mathrm{Min}(F, S) = \emptyset$ and $l - \mathrm{Min}(F, S) = \{-1\}$.

The following examples show that there is no relation between minimal and u-minimal solutions.

Example 6.15 [1] Let $X = \mathbb{R}$, $S = [0, 1]$, $Y = \mathbb{R}^2$, $C = \mathbb{R}^2_+$, and $F : S \rightrightarrows Y$ be defined by

$$F(x) = \begin{cases} \{(u, v) \in \mathbb{R}^2 : u^2 + v^2 = x^2, v > 0\}, & \text{if } x \neq -1, 0, \\ (-1/2, 1), & \text{if } x = -1, \\ (1/2, 1), & \text{if } x = 0. \end{cases}$$

We can easily check that $\mathrm{Min}(F, S) = \emptyset$ and $u - \mathrm{Min}(F, S) = \{-1\}$.

Example 6.16 Let $X=\mathbb{R}$, $S = [0, 1]$, $Y=\mathbb{R}^2$, $C = \mathbb{R}^2_+$, and $F : S \rightrightarrows Y$ be defined by

$$F(x) = \begin{cases} [[(2,2),(3,3)]], & \text{if } x = 0, \\ [[(0,0),(4,4)]], & \text{otherwise.} \end{cases}$$

where $[[(a,b),(c,d)]] = \{(y_1, y_2) : a \leq y_1 \leq c, b \leq y_2 \leq d\}$. After a short calculation, we get $\text{Min}(F, S) = (0, 1]$ and $u - \text{Min}(F, S) = \{0\}$.

We now recall the notions of optimal solutions of the problem (SOP) with respect to the relations $\preceq_C^*$ and $\prec_C^*$, where $* \in \{ml, mu\}$. For the set order relations $\preceq_C^*$ and $\prec_C^*$, we assume that Y is a normed space, $F(x) \in \mathcal{B}(Y)$ for all $x \in S$, $K := C$ is a closed convex and pointed cone with $\text{int}C \neq \emptyset$ and $F(x) \neq \emptyset$ for all $x \in X$.

Definition 6.14 Let $* \in \{ml, mu\}$. An element $\bar{x} \in S$ is called

(a) a *$*$-minimal solution* of the problem (SOP) with respect to $\preceq_C^*$ if there does not exist any $x \in S$ such that $F(x) \preceq_C^* F(\bar{x})$ and $F(x) \neq F(\bar{x})$, that is, either $F(x) \not\preceq_C^* F(\bar{x})$ or $F(x) = F(\bar{x})$ for any $x \in S$;

(b) a *$*$-weak minimal solution* of the problem (SOP) with respect to $\prec_C^*$ if

$$F(x) \prec_C^* F(\bar{x}) \text{ for some } x \in S \;\Rightarrow\; F(\bar{x}) \prec_C^* F(x).$$

We denote the set of $*$-minimal and $*$-weak minimal solutions of the problem (SOP) by $* - \text{Min}(F, S)$ and $* - \text{WMin}(F, S)$, respectively.

Remark 6.9 (a) Since the set order relations $\preceq_C^*$, $* \in \{ml, mu\}$, are partial order, Definition 6.14(a) can also be written as follows:
An element $\bar{x} \in S$ is said to be a *$*$-minimal solution* of the problem (SOP) if

$$F(x) \preceq_C^* F(\bar{x}) \text{ for some } x \in S \;\Rightarrow\; F(\bar{x}) = F(x).$$

Furthermore, if $\preceq_C^t$ is partial order for any $t \in \{l, u, s, c, m, mc, mn, \lambda\}$, then the above also holds true for Definition 6.12(a).

(b) Definition 6.12(c) and Definition 6.14(b) can also be written as follows:
An element $\bar{x} \in S$ is said to be a t-weak minimal solution ($*$-weak minimal solution) of the problem (SOP) if there does not exist any $x \in S$ such that $F(x) \prec_C^t F(\bar{x})$ $(F(x) \prec_C^* F(\bar{x}))$.

Clearly, $* - \text{Min}(F, S) \subseteq * - \text{WMin}(F, S)$. However, the reverse inclusion may not hold.

Example 6.17 Let $X = \mathbb{R}$, $S = [0, 1]$, $Y = \mathbb{R}^2$, $C = \mathbb{R}^2_+$, and $F : S \rightrightarrows Y$ be defined by

$$F(x) = \begin{cases} \{(u, v) \in \mathbb{R}^2 : u^2 + v^2 \leq 4, u > 0, v > 0\}, & \text{if } x = 0, \\ (0, 3) \times (0, 3), & \text{otherwise.} \end{cases}$$

Then, $0 \in ml - \text{WMin}(F, S)$ but $0 \notin ml - \text{Min}(F, S)$.

Example 6.18 Let $X = \mathbb{R}$, $S = [-1, 0]$, $Y = \mathbb{R}^2$, $C = \mathbb{R}^2_+$, and $F : S \rightrightarrows Y$ be defined by

$$F(x) = \begin{cases} \{(u, v) \in \mathbb{R}^2 : u^2 + v^2 \leq 9, u < 0, v < 0\}, & \text{if } x = 0, \\ (0, -1) \times (0, -1), & \text{otherwise.} \end{cases}$$

Then, $0 \in mu - \mathrm{WMin}(F, S)$ but $0 \notin mu - \mathrm{Min}(F, S)$.

Definition 6.15 Let S be a nonempty convex subset of X. A set-valued map $F : S \rightrightarrows Y$ is said to be

(a) *strictly natural ml-type C-quasi-convex* on S if for all $x_1, x_2 \in S$ with $x_1 \neq x_2$ and all $t \in (0, 1)$, there exists $\lambda \in [0, 1]$ such that

$$F(tx_1 + (1 - t)x_2) \prec_C^{ml} \lambda F(x_1) + (1 - \lambda)F(x_2).$$

(b) *strictly natural u-type C-quasi-convex* on S if for all $x_1, x_2 \in S$ with $x_1 \neq x_2$ and all $t \in (0, 1)$, there exists $\lambda \in [0, 1]$ such that

$$F(tx_1 + (1 - t)x_2) \prec_C^{mu} \lambda F(x_1) + (1 - \lambda)F(x_2).$$

Proposition 6.10 *Assume that S is a convex subset of X, $F : S \rightrightarrows Y$ is a strictly natural l-type C-quasi-convex map on S with nonempty compact values. Then,* $ml - \mathrm{Min}(F, S) = ml - \mathrm{WMin}(F, S)$.

Proposition 6.11 *Assume that S is a convex subset of X, $F : S \rightrightarrows Y$ is a strictly natural u-type C-quasi-convex map on S with nonempty compact values. Then,* $mu - \mathrm{Min}(F, S) = mu - \mathrm{WMin}(F, S)$.

We now give the following example to show that an ml-minimal solution may not be a minimal solution and vice-versa.

Example 6.19 Let $X = \mathbb{R}$, $S = [-1, 1]$, $Y = \mathbb{R}^2$, $C = \mathbb{R}^2_+$, and $F : S \rightrightarrows Y$ be defined by

$$F(x) = \begin{cases} (0, 4) \times (0, 4), & \text{if } x = -1, \\ [0, 2] \times [0, 2], & \text{otherwise.} \end{cases}$$

Then, $\mathrm{Min}(F, S) = (-1, 1]$ and $ml - \mathrm{Min}(F, S) = [-1, 1]$.

In the above example, if $F(-1) = [0, 4] \times [0, 4]$ is replaced by $F(-1) = (0, 4) \times (0, 4)$, then $\mathrm{Min}(F, S) = [-1, 1]$ and $ml - \mathrm{Min}(F, S) = \{-1\}$.

We now give the following examples to show that an mu-minimal solution may not be a minimal solution and vice-versa.

Example 6.20 Let $X = \mathbb{R}$, $S = \{0, 1\}$, $Y = \mathbb{R}^2$, $C = \mathbb{R}^2_+$, and $F : S \rightrightarrows Y$ be defined by

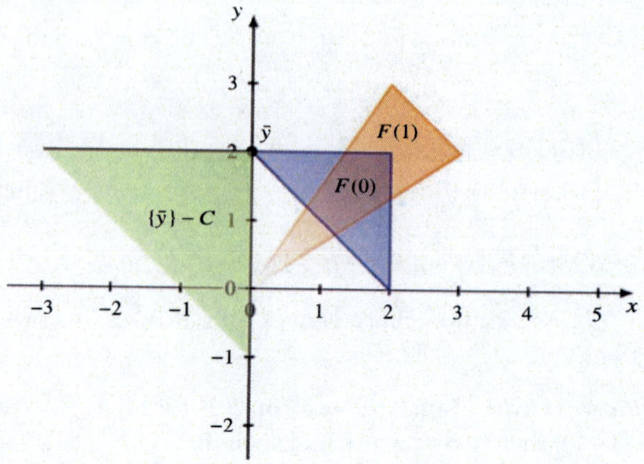

Fig. 6.12 Illustration of Example 6.20 with $C = \mathbb{R}^2_+$

$$F(x) = \begin{cases} \text{conv}\{(0,0),(2,3),(3,2)\}, & \text{if } x = 0, \\ \text{conv}\{(2,0),(0,2),(2,2)\}, & \text{if } x = 1. \end{cases}$$

From Fig. 6.12, we can see that there does not exist any $\bar{y} \in F(1)$ such that $(\{\bar{y}\} - C) \cap F(S) = \{\bar{y}\}$. Therefore, $1 \notin \text{Min}(F, S)$ but $1 \in mu - \text{Min}(F, S)$ because $F(0) \npreceq^{mu}_C F(1)$.

Example 6.21 Let $X = \mathbb{R}, S = \{0, 1\}, Y = \mathbb{R}^2, C = \mathbb{R}^2_+$, and $F : S \rightrightarrows Y$ be defined by

$$F(x) = \begin{cases} \{(x, y) \in \mathbb{R}^2 : x^2 + y^2 \leq 1\}, & \text{if } x = 0, \\ \{(\frac{-\sqrt{2}}{2}, \frac{-\sqrt{2}}{2})\}, & \text{if } x = 1. \end{cases}$$

Then we can easily obtain $\text{Min}(F, S) = \{0, 1\}$ but $ml - \text{Min}(F, S) = \{1\}$. Indeed, $F(1)\dot{-}F(0) = \emptyset$. Therefore, $F(1)\dot{-}F(0) \cap C = \emptyset$. Thus, $F(0) \npreceq^{mu}_C F(1)$ and therefore $1 \in mu - \text{Min}(F, S)$. Moreover, $F(0)\dot{-}F(1) \neq \emptyset$ (see Fig. 6.13). Therefore, $F(1)\dot{-}F(0) \cap C \neq \emptyset$. Thus, $F(1) \preceq^{mu}_C F(0)$ and therefore $0 \notin mu - \text{Min}(F, S)$.

The following example shows that an mu-minimal solution may not be a u-minimal solution and vice-versa.

Example 6.22 Let $X = \mathbb{R}$, $Y = \mathbb{R}^2$, $S = [0, 1]$, and $C = \mathbb{R}^2_+$. Let $F : S \rightrightarrows Y$ be defined by

$$F(x) = \begin{cases} [0, -1] \times [0, -1], & \text{if } x = 0, \\ (0, -2) \times (0, -2), & \text{otherwise.} \end{cases}$$

Then,

$$mu - \text{Min}F = [0, 1] \quad \text{and} \quad u - \text{Min}F = (0, 1].$$

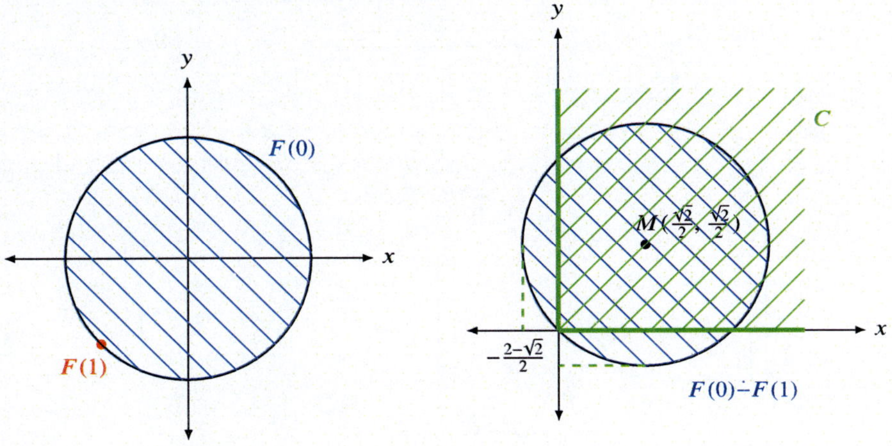

Fig. 6.13 Illustration of Example 6.21 with $C = \mathbb{R}^2_+$

Furthermore, if we replace the value of $F(x)$ for all $x \in (0, 1]$ by $[0, -2] \times [0, -2]$, then

$$mu - \text{Min}F = \{0\} \quad \text{and} \quad u - \text{Min}F = [0, 1].$$

Example 6.23 Let $X = \mathbb{R}$, $Y = \mathbb{R}^2$, $S = [-1, 1]$, and $C = \mathbb{R}^2_+$. Let $F : S \rightrightarrows Y$ be defined by

$$F(x) = \begin{cases} (0, 1) \times (0, 1), & \text{if } x = -1, \\ [0, 1/2] \times [0, 1/2], & \text{otherwise.} \end{cases}$$

We can easily see that

$$ml - \text{Min}F = [-1, 1] \quad \text{and} \quad l - \text{Min}F = (-1, 1].$$

Furthermore, if we replace the value of $F(-1)$ by $[0, 1] \times [0, 1]$, then

$$ml - \text{Min}F = \{-1\} \quad \text{and} \quad l - \text{Min}F = [-1, 1].$$

The following theorem shows that every weak minimal solution of (SOP) is a ml-weak minimal solution of the problem (SOP).

Theorem 6.5.1 [99] *If C is a closed convex pointed cone in Y with int$C \neq \emptyset$, then*

$$\text{WMin}(F, S) \subseteq ml - \text{WMin}(F, S).$$

From Example 6.17, it is clear that the reverse inclusion of the above theorem fails because $0 \in ml - \text{WMin}(F, S)$ but $0 \notin \text{WMin}(F, S)$.

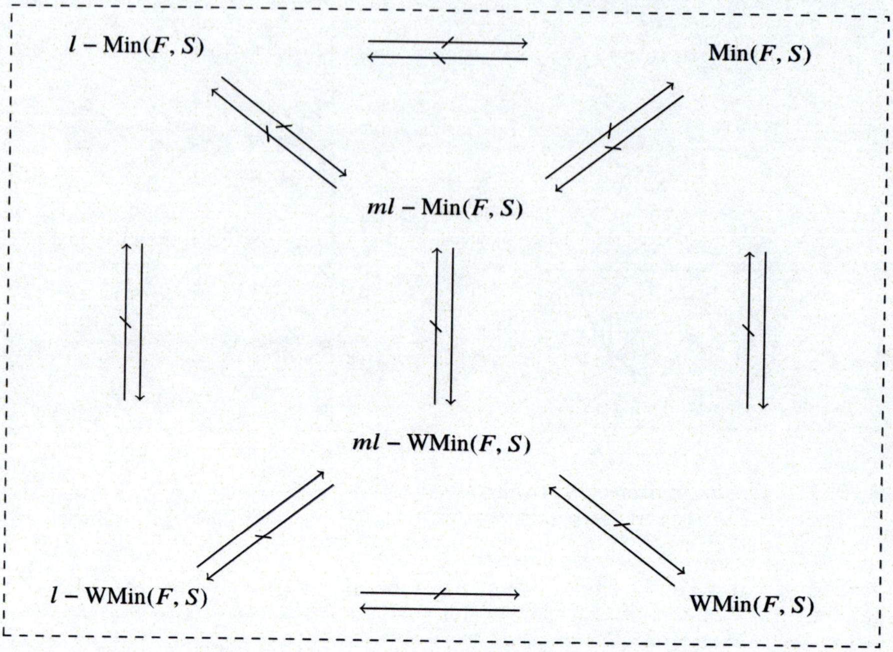

Fig. 6.14 Relationship between different kinds of solution concepts of the problem (SOP)

The following theorem shows that every l-weak (u-weak) minimal solution of (SOP) is a ml-weak (mu-weak) minimal solution of the problem (SOP).

Theorem 6.5.2 [99] *If C is a closed convex pointed cone in Y with $\text{int}C \neq \emptyset$, then*

$$u - \text{WMin}(F, S) \subseteq mu - \text{WMin}(F, S) \quad \textit{and} \quad l - \text{WMin}(F, S) \subseteq ml - \text{WMin}(F, S).$$

However, the converse of the above theorem may not hold. For instance, in Example 6.22, $0 \in mu - \text{WMin}(F, S)$ but $0 \notin u - \text{WMin}(F, S)$ and in Example 6.23, $-1 \in ml - \text{WMin}(F, S)$ but $-1 \notin l - \text{WMin}(F, S)$.

In the following diagram, we summarize the relations among various notions of minimal and weak minimal solutions involving the set order relations $\preceq_C^l$ and $\preceq_C^{ml}$. In a similar way, we can establish relations among various notions of minimal and weak minimal solutions using different set order relations (Fig. 6.14).

6.5.1 *Solution Concepts in Set Optimization with Respect to Variable Domination Structures*

This section introduces different concepts for minimal elements of a family of sets and solution concepts for the problem (SOP) with respect to variable ordering structures.

These concepts are defined based on set relations introduced in Definition 6.9. In addition, we present the relationship between the sets of different minimal elements.

Definition 6.16 [104] Let $\mathcal{A}$ be a family of nonempty sets in Y and $\mathcal{K} : Y \rightrightarrows Y$ be a set-valued map.

(a) A set $\bar{A} \in \mathcal{A}$ is called a *minimal element* of $\mathcal{A}$ with respect to $\preceq_t^{\mathcal{K}}$, $t \in \{l, u, cl, cu, pl, pu\}$, if

$$\forall A \in \mathcal{A}, \quad A \preceq_t^{\mathcal{K}} \bar{A} \quad \Rightarrow \quad \bar{A} \preceq_t^{\mathcal{K}} A.$$

(b) A set $\bar{A} \in \mathcal{A}$ is called a *strong minimal element* of $\mathcal{A}$ with respect to $\preceq_t^{\mathcal{K}}$, $t \in \{l, u, cl, cu, pl, pu\}$, if

$$\forall A \in \mathcal{A}, \quad \bar{A} \preceq_t^{\mathcal{K}} A.$$

(c) A set $\bar{A} \in \mathcal{A}$ is called a *strict minimal element* of $\mathcal{A}$ with respect to $\preceq_t^{\mathcal{K}}$, $t \in \{l, u, cl, cu, pl, pu\}$, if

$$\forall A \in \mathcal{A}, \quad A \preceq_t^{\mathcal{K}} \bar{A} \quad \Rightarrow \quad \bar{A} = A.$$

The sets of all minimal, strong minimal, and strict minimal elements of $\mathcal{A}$ with respect to $\preceq_t^{\mathcal{K}}$, $t \in \{l, u, cl, cu, pl, pu\}$, are denoted by $\text{Min}(\mathcal{A}, \preceq_t^{\mathcal{K}})$, $\text{SoMin}(\mathcal{A}, \preceq_t^{\mathcal{K}})$, and $\text{SiMin}(\mathcal{A}, \preceq_t^{\mathcal{K}})$, respectively.

Remark 6.10 (a) When $\mathcal{A}$ is a family of singleton sets and $\mathcal{K}(y)$ is a closed, convex and pointed cone for each $y \in Y$, then the definition of strictly minimal element of $\mathcal{A}$ with respect to $\preceq_t^{\mathcal{K}}$ reduces to the definition of nondominated element of $\mathcal{A}$ with respect to $\mathcal{K}$ (see [40, Definition 2.7]).

(b) If $\bar{A} \in \text{Min}(\mathcal{A}, \preceq_t^{\mathcal{K}})$, then for all $B \sim \bar{A}$, we have $B \in \text{Min}(\mathcal{A}, \preceq_t^{\mathcal{K}})$. From Definition 6.16, we obtain

$$\text{SiMin}(\mathcal{A}, \preceq_t^{\mathcal{K}}) \subseteq \text{Min}(\mathcal{A}, \preceq_t^{\mathcal{K}}) \quad \text{and} \quad \text{SoMin}(\mathcal{A}, \preceq_t^{\mathcal{K}}) \subseteq \text{Min}(\mathcal{A}, \preceq_t^{\mathcal{K}}).$$

However, neither $\text{SiMin}(\mathcal{A}, \preceq_t^{\mathcal{K}}) \subseteq \text{SoMin}(\mathcal{A}, \preceq_t^{\mathcal{K}})$ nor $\text{SoMin}(\mathcal{A}, \preceq_t^{\mathcal{K}}) \subseteq \text{SiMin}(\mathcal{A}, \preceq_t^{\mathcal{K}})$ always holds (see [104]).

(b)

$$\text{SoMin}(\mathcal{A}, \preceq_{cl}^{\mathcal{K}}) \subseteq \text{SoMin}(\mathcal{A}, \preceq_l^{\mathcal{K}}) \subseteq \text{SoMin}(\mathcal{A}, \preceq_{pl}^{\mathcal{K}})$$

$$\text{SiMin}(\mathcal{A}, \preceq_{pl}^{\mathcal{K}}) \subseteq \text{SiMin}(\mathcal{A}, \preceq_l^{\mathcal{K}}) \subseteq \text{SiMin}(\mathcal{A}, \preceq_{cl}^{\mathcal{K}})$$

$$\text{SoMin}(\mathcal{A}, \preceq_{cu}^{\mathcal{K}}) \subseteq \text{SoMin}(\mathcal{A}, \preceq_u^{\mathcal{K}}) \subseteq \text{SoMin}(\mathcal{A}, \preceq_{pu}^{\mathcal{K}})$$

$$\text{SiMin}(\mathcal{A}, \preceq_{pu}^{\mathcal{K}}) \subseteq \text{SiMin}(\mathcal{A}, \preceq_u^{\mathcal{K}}) \subseteq \text{SiMin}(\mathcal{A}, \preceq_{cu}^{\mathcal{K}}).$$

The following example illustrates that neither $\text{SiMin}(\mathcal{A}, \preceq_t^{\mathcal{K}}) \subseteq \text{SoMin}(\mathcal{A}, \preceq_t^{\mathcal{K}})$ nor $\text{SoMin}(\mathcal{A}, \preceq_t^{\mathcal{K}}) \subseteq \text{SiMin}(\mathcal{A}, \preceq_t^{\mathcal{K}})$ always holds.

Example 6.24 [104] Let

$$\begin{aligned} A_1 &= \{(y_1, y_2) \in \mathbb{R}^2 : 2 \leq y_1, y_2 \leq 3,\ y_1 + y_2 \leq 5\}, \\ A_2 &= \{(2, y_2) \in \mathbb{R}^2 : 2 \leq y_2 \leq 3\} \cup \{(y_1, 2) \in \mathbb{R}^2 : 2 \leq y_1 \leq 3\}, \\ A_3 &= \{(5, 5)\}, \\ A_4 &= \{(y_1, y_2) \in \mathbb{R}^2 : 3 \leq y_1 \leq 5,\ 0 \leq y_2 \leq 1\}, \end{aligned}$$

and the set-valued map $\mathcal{K} : \mathbb{R}^2 \rightrightarrows \mathbb{R}^2$ be defined as

$$\mathcal{K}(t) = \begin{cases} \{(d_1, d_2) : 0 \leq d_1 \leq 2d_2\}, & \text{if} t \in \mathbb{R}^2 \setminus \{(1, 3)\}, \\ \mathbb{R}^2_+, & \text{if} t = \{(1, 3)\}. \end{cases}$$

From Fig. 6.15, we can easily see that

$$A_1 \preceq_l^{\mathcal{K}} A_2, \quad A_1 \preceq_l^{\mathcal{K}} A_3, \quad A_1 \npreceq_l^{\mathcal{K}} A_4,$$

$$A_2 \preceq_l^{\mathcal{K}} A_1, \quad A_2 \preceq_l^{\mathcal{K}} A_3, \quad A_1 \npreceq_2^{\mathcal{K}} A_4,$$

$$A_3 \preceq_l^{\mathcal{K}} A_1, \quad A_3 \preceq_l^{\mathcal{K}} A_2, \quad A_3 \npreceq_l^{\mathcal{K}} A_4,$$

$$A_4 \npreceq_l^{\mathcal{K}} A_1, \quad A_4 \npreceq_l^{\mathcal{K}} A_2, \quad A_4 \preceq_l^{\mathcal{K}} A_1.$$

Let $\mathcal{A} := \{A_1, A_2, A_3\}$. Then we have

$$\text{Min}(\mathcal{A}, \preceq_l^{\mathcal{K}}) = \{A_1, A_2\}, \quad \text{SoMin}(\mathcal{A}, \preceq_l^{\mathcal{K}}) = \{A_1, A_2\}, \quad \text{SiMin}(\mathcal{A}, \preceq_l^{\mathcal{K}}) = \emptyset.$$

Let $\mathcal{A}' := \{A_1, A_2, A_3, A_4\}$. Then we have

$$\text{Min}(\mathcal{A}', \preceq_l^{\mathcal{K}}) = \{A_1, A_2, A_4\}, \quad \text{SoMin}(\mathcal{A}', \preceq_l^{\mathcal{K}}) = \{A_1, A_2\}, \quad \text{SiMin}(\mathcal{A}', \preceq_l^{\mathcal{K}}) = \{A_4\}.$$

Let $\mathcal{A}'' := \{A_3, A_4\}$. Then we have

$$\text{Min}(\mathcal{A}'', \preceq_l^{\mathcal{K}}) = \text{SoMin}(\mathcal{A}'', \preceq_l^{\mathcal{K}}) = \text{SiMin}(\mathcal{A}'', \preceq_l^{\mathcal{K}}) = \{A_4\}.$$

Proposition 6.12 *Let $\mathcal{A}$ be a family of sets in $P(Y)$, $S \in P(Y)$, and $|S|$ denote the number of elements in S. Then, for $t \in \{l, u, cl, cu, pl, pu\}$, the following statements hold.*

(a) *If* $|\text{SoMin}(\mathcal{A}, \preceq_t^{\mathcal{K}})| > 1$, *then* $\text{SiMin}(\mathcal{A}, \preceq_t^{\mathcal{K}}) = \emptyset$.
(b) *If* $|\text{SiMin}(\mathcal{A}, \preceq_t^{\mathcal{K}})| > 1$, *then* $\text{SoMin}(\mathcal{A}, \preceq_t^{\mathcal{K}}) = \emptyset$.
(c) *If* $\text{SoMin}(\mathcal{A}, \preceq_t^{\mathcal{K}}) \cap \text{SiMin}(\mathcal{A}, \preceq_t^{\mathcal{K}}) \neq \emptyset$, *then*

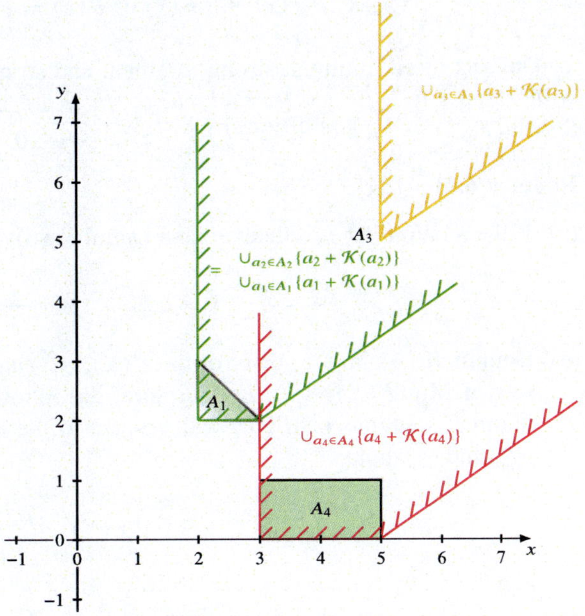

Fig. 6.15 Illustration of Example 6.24

$$\begin{cases} |\mathrm{SoMin}(\mathcal{A}, \preceq_t^{\mathcal{K}})| = |\mathrm{SiMin}(\mathcal{A}, \preceq_t^{\mathcal{K}})| = 1, \\ \mathrm{SoMin}(\mathcal{A}, \preceq_t^{\mathcal{K}}) = \mathrm{SiMin}(\mathcal{A}, \preceq_t^{\mathcal{K}}). \end{cases}$$

Now, we define the solution concepts of set optimization problem (SOP) with respect to the set order relations $\preceq_t^{\mathcal{K}}$, $t \in \{l, u, cl, cu, pl, pu\}$. Note that the solution concepts in the following definition are given in the preimage space X, whereas the solution concepts in Definition 6.16 are formulated in the image space Y.

Definition 6.17 [104] Let $F : X \rightrightarrows Y$ and $\mathcal{K} : Y \rightrightarrows Y$ be two set-valued maps such that $F(x)$ and $\mathcal{K}(y)$ are nonempty sets for all $x \in X$, $y \in Y$.

(a) An element $\bar{x} \in X$ is called a *minimal element* of (SOP) with respect to $\preceq_t^{\mathcal{K}}$, $t \in \{l, u, cl, cu, pl, pu\}$, if

$$x \in X, \quad F(x) \preceq_t^{\mathcal{K}} F(\bar{x}) \quad \Rightarrow \quad F(\bar{x}) \preceq_t^{\mathcal{K}} F(x).$$

(b) An element $\bar{x} \in X$ is called a *strong minimal element* of (SOP) with respect to $\preceq_t^{\mathcal{K}}$, $t \in \{l, u, cl, cu, pl, pu\}$, if

$$\forall\, x \in X \setminus \{\bar{x}\}, \quad F(\bar{x}) \preceq_t^{\mathcal{K}} F(x).$$

(c) An element $\bar{x} \in X$ is called a *strict minimal element* of (SOP) with respect to $\preceq_t^{\mathcal{K}}$, $t \in \{l, u, cl, cu, pl, pu\}$, if

$$x \in X, \quad F(x) \preceq_t^{\mathcal{K}} F(\bar{x}) \text{ or } F(x) = F(\bar{x}) \quad \Rightarrow \quad x = \bar{x}. \tag{6.8}$$

The sets of all minimal, strong minimal, and strict minimal elements of (SOP) with respect to $\preceq_t^{\mathcal{K}}$, $t \in \{l, u, cl, cu, pl, pu\}$, are denoted by $\text{Min}(F(X), \preceq_t^{\mathcal{K}})$, $\text{SoMin}(F(X), \preceq_t^{\mathcal{K}})$, and $\text{SiMin}(F(X), \preceq_t^{\mathcal{K}})$, respectively.

Remark 6.11 [104]

(a) If the relation $\preceq_t^{\mathcal{K}}$ is reflexive, then Definition 6.17(c) is equivalent to

$$x \in X, \quad F(x) \preceq_t^{\mathcal{K}} F(\bar{x}) \quad \Rightarrow \quad x = \bar{x}.$$

(b) Definition 6.17 implies that $\text{SoMin}(F(X), \preceq_t^{\mathcal{K}})$ and $\text{SiMin}(F(X), \preceq_t^{\mathcal{K}})$ are subsets of $\text{Min}(F(X), \preceq_t^{\mathcal{K}})$. Furthermore, the following relations for the sets of minimal solutions of (SOP) with respect to the lower set order relations $\preceq_l^{\mathcal{K}}$, $\preceq_{cl}^{\mathcal{K}}$, and $\preceq_{pl}^{\mathcal{K}}$ hold:

$$\text{SoMin}(F(X), \preceq_{cl}^{\mathcal{K}}) \subseteq \text{SoMin}(F(X), \preceq_l^{\mathcal{K}}) \subseteq \text{SoMin}(F(X), \preceq_{pl}^{\mathcal{K}})$$

and

$$\text{SiMin}(F(X), \preceq_{pl}^{\mathcal{K}}) \subseteq \text{SiMin}(F(X), \preceq_l^{\mathcal{K}}) \subseteq \text{SiMin}(F(X), \preceq_{cl}^{\mathcal{K}}).$$

Similarly, the following relations for the sets of minimal solutions of (SOP) with respect to the upper set order relations $\preceq_u^{\mathcal{K}}$, $\preceq_{cu}^{\mathcal{K}}$, and $\preceq_{pu}^{\mathcal{K}}$ hold:

$$\text{SoMin}(F(X), \preceq_{cu}^{\mathcal{K}}) \subseteq \text{SoMin}(F(X), \preceq_u^{\mathcal{K}}) \subseteq \text{SoMin}(F(X), \preceq_{pu}^{\mathcal{K}})$$

$$\text{SiMin}(F(X), \preceq_{pu}^{\mathcal{K}}) \subseteq \text{SiMin}(F(X), \preceq_u^{\mathcal{K}}) \subseteq \text{SiMin}(F(X), \preceq_{cu}^{\mathcal{K}}).$$

6.6 Existence of Solutions

It is well known that the semicontinuity for set-valued maps plays a significant role to study the set optimization problems. Kuroiwa [113] and Jahn and Ha [92] extended the concept of semicontinuities for set-valued maps by using the set order relations $\preceq_C^l$ and $\preceq_C^u$ and applied them to obtain the existence of solutions for set optimization problems. Hernández et al. [84] further used and investigated the semicontinuity for set-valued maps to study the existence of solutions of the problem (SOP) and the relation among solutions using vector approach and set approach. Very recently, Zhang and Huang [153] introduced the notion of lower semicontinuity from above and used it to obtain the existence of results and discussed the link between solutions of the problem (SOP) obtained by vector approach and set approach.

6.6.1 Generalized Semicontinuity for Set-Valued Maps

In this subsection, we introduce the notions of generalized semicontinuity for set-valued maps involving the partial set order relation $\preceq_C^{ml}$. Further, we study some properties of the generalized semicontinuity for set-valued maps, which are then applied to study the existence of solutions for set optimization problems.

Throughout this subsection, we assume that S is a nonempty subset of a Hausdorff topological vector space X and Y is a real normed space. Further, we assume that $F(x) \in \mathcal{B}(Y)$ for all $x \in S$, C is a closed convex and pointed cone with $\text{int}C \neq \emptyset$ and $F(x) \neq \emptyset$ for all $x \in X$.

Definition 6.18 The set-valued map $F : X \rightrightarrows Y$ is said to have

(a) $\preceq_C^{ml}$*-lower property* at $\bar{x} \in S$ if there exists a point $x \in S$ such that $F(x) \preceq_C^{ml} F(\bar{x})$;
(b) $\prec_C^{ml}$*-lower property* at $\bar{x} \in S$ if there exists a point $x \in S$ such that $F(x) \prec_C^{ml} F(\bar{x})$;
(c) *strictly* $\prec_C^{ml}$*-lower property* at $\bar{x} \in S$ if there exists a point $x \in S$ such that $F(x) \cap F(\bar{x}) = \emptyset$ and $F(x) \prec_C^{ml} F(\bar{x})$.

Definition 6.19 Let $\{A_\alpha\}_{\alpha\in I}$ be a net and $(I, <)$ be a directed set. The net $\{A_\alpha\}_{\alpha\in I}$ is said to be

(a) $\preceq_C^{ml}$*-increasing* if for $\alpha, \beta \in I$ with $\alpha < \beta$, we have $A_\alpha \preceq_C^{ml} A_\beta$;
(b) $\preceq_C^{ml}$*-decreasing* if for $\alpha, \beta \in I$ with $\alpha < \beta$, we have $A_\beta \preceq_C^{ml} A_\alpha$.

Definition 6.20 A set-valued map $F : S \rightrightarrows Y$ is said to be *ml-type Demi-lower semicontinuous* at $\bar{x} \in S$ if for any net $\{x_\alpha\}_{\alpha\in I}$ in S such that $x_\alpha \to \bar{x}$ and $\{F(x_\alpha)\}_{\alpha\in I}$ is a $\preceq_C^{ml}$-decreasing net, the following condition holds:

$$F(\bar{x}) \preceq_C^{ml} \text{Limsup}_\alpha (F(x_\alpha) + C),$$

where $\text{Limsup}_\alpha(F(x_\alpha) + C)$ denotes the set of all cluster points of $\{y_\alpha : y_\alpha \in (F(x_\alpha) + C)\}_{\alpha\in I}$.

We say that F is ml-type Demi-lower semicontinuous on S if it is ml-type Demi-lower semicontinuous at each point $\bar{x} \in S$.

Definition 6.21 Let X be a topological space. A set-valued map $F : X \rightrightarrows Y$ is said to be

(a) $\preceq_C^{ml}$*-lower semicontinuous from above* at $\bar{x} \in X$ if for any net $\{x_\alpha\}_{\alpha\in I}$ in X with $x_\alpha \to \bar{x}$ such that $\{F(x_\alpha)\}_{\alpha\in I}$ is a $\preceq_C^{ml}$-decreasing net, one has $F(\bar{x}) \preceq_C^{ml} F(x_\alpha)$ for all $\alpha \in I$;
(b) $\preceq_C^{ml}$*-upper semicontinuous from below* at $\bar{x} \in X$ if for any net $\{x_\alpha\}_{\alpha\in I}$ in X with $x_\alpha \to \bar{x}$ such that $\{F(x_\alpha)\}_{\alpha\in I}$ is a $\preceq_C^{ml}$-increasing net, one has $F(x_\alpha) \preceq_C^{ml} F(\bar{x})$ for all $\alpha \in I$.

We say that F is $\preceq_C^{ml}$-lower semicontinuous from above (respectively, $\preceq_C^{ml}$-upper semicontinuous from below) on X if it is $\preceq_C^{ml}$-lower semicontinuous from above (respectively, $\preceq_C^{ml}$-upper semicontinuous from below) at each point $\bar{x} \in X$.

Remark 6.12 The ml-type Demi-lower semicontinuity implies the $\preceq_C^{ml}$-lower semicontinuity from above, but the following example shows that the converse is not true.

Example 6.25 Let $S = \mathbb{R}^2$, $Y = \mathbb{R}^2$, and $C = \mathbb{R}^2_+$. Let $F : S \rightrightarrows Y$ be defined by

$$F(x) = \begin{cases} \{(0, 1)\}, & \text{if } x > 0, \\ \{(0, \varepsilon) : 0 < \varepsilon < 2\}, & \text{if } x = 0, \\ \{(0, -1)\}, & \text{if } x < 0. \end{cases}$$

At $\bar{x} = 0$, one can easily see that for any net $\{x_\alpha\}_{\alpha \in I}$ in S with $x_\alpha \to 0$, $\{F(x_\alpha)\}_{\alpha \in I}$ is a $\preceq_C^{ml}$-decreasing net and $F(0) \preceq_C^{ml} F(x_\alpha)$ for all $\alpha \in I$. Hence, F is $\preceq_C^{ml}$-lower semicontinuous from above at $\bar{x} = 0$. However, F is not ml-type Demi-lower semicontinuous at $\bar{x} = 0$. Indeed, taking a sequence $\{x_n\} = \{\frac{1}{n}\}_{n \in \mathbb{N}}$, we get $\text{Limsup}_{n \to +\infty}(F(x_n) + C) = C$. After a short calculation, we obtain $F(0) \not\preceq_C^{ml} \text{Limsup}_{n \to +\infty}(F(x_n) + C)$. Hence, F is not ml-type Demi-lower semicontinuous at $\bar{x} = 0$.

Definition 6.22 A set-valued map $F : X \rightrightarrows Y$ is said to be *$\preceq_C^{ml}$-lower semicontinuous* at $\bar{x} \in X$ if the set $\{x \in X : F(x) \preceq_C^{ml} F(\bar{x})\}$ is closed. We say that F is $\preceq_C^{ml}$-lower semicontinuous on X if it is $\preceq_C^{ml}$-lower semicontinuous at each point $\bar{x} \in X$.

Proposition 6.13 *If the set-valued map F is $\preceq_C^{ml}$-lower semicontinuous on X, then it is $\preceq_C^{ml}$-lower semicontinuous from above on X.*

Proof Let $\{x_\alpha\}_{\alpha \in I}$ be a net in X such that $x_\alpha \to \bar{x}$ and $F(x_\beta) \preceq_C^{ml} F(x_\alpha)$ for $\alpha < \beta$ with $\alpha, \beta \in I$. Then for each $\alpha \in I$, the net $\{x_\beta\}_{\alpha < \beta}$ satisfies $x_\beta \to \bar{x}$. By $\preceq_C^{ml}$-lower semicontinuity of F, one has $\bar{x} \in \{x \in X : F(x) \preceq_C^{ml} F(x_\alpha)\}$ for all $\alpha \in I$. This shows that F is $\preceq_C^{ml}$-lower semicontinuous from above on X. □

The following example shows that the reverse of the above proposition is not true.

Example 6.26 Let $X = \mathbb{R}$, $Y = \mathbb{R}$, and $C = \mathbb{R}_+$. Let $F : X \rightrightarrows Y$ be defined by

$$F(x) = \begin{cases} \{0\}, & \text{if } x > 0, \\ [0, 2), & \text{if } -1 < x \leq 0, \\ \{1\}, & \text{if } x \leq -1. \end{cases}$$

At $\bar{x} = 0$, one can easily see that for any net $\{x_\alpha\}_{\alpha \in I}$ in X with $x_\alpha \to 0$, $\{F(x_\alpha)\}_{\alpha \in I}$ is $\preceq_C^{ml}$-decreasing net and $F(0) \preceq_C^{ml} F(x_\alpha)$ for all $\alpha \in I$. Hence, F is $\preceq_C^{ml}$-lower semicontinuous from above at $\bar{x} = 0$. However, the set $\{x \in X : F(x) \preceq_C^{ml} F(0)\} = \{x \in X : -1 < x \leq 0\}$ is not closed. Hence, F is not $\preceq_C^{ml}$-lower semicontinuous at $\bar{x} = 0$.

Remark 6.13 In a similar way, we can introduce the notions of generalized semi-continuity with respect to other different kinds of set order relations.

6.6.2 *Existence of Solutions in Set Optimization Problems*

In this subsection, we study the existence of results for solutions of set optimization problems with respect to the partial set order relation $\preceq_C^{ml}$ by using generalized semicontinuity. Since existence results for other set order relations can be obtained in a similar way, we skip such a study.

Throughout this subsection, unless otherwise specified, we assume that S is a nonempty subset of a Hausdorff topological vector space X and Y is a real normed space. Further, we assume that $F(x) \in \mathcal{B}(Y)$ for all $x \in S$, C is a closed convex and pointed cone with $\text{int}C \neq \emptyset$ and $F(x) \neq \emptyset$ for all $x \in X$.

Let $A, B \in P(Y)$ and $\bar{x} \in S$, we write

$$A \sim B \quad \Leftrightarrow \quad A \preceq_C^{ml} B \text{ and } B \preceq_C^{ml} A,$$

$$E(\bar{x}, \preceq_C^{ml}) = \{x \in S : F(\bar{x}) \sim F(x)\},$$

and the level set of F at $\bar{x} \in S$ is given by

$$L(\bar{x}, \preceq_C^{ml}) = \{x \in S : F(x) \preceq_C^{ml} F(\bar{x})\}.$$

It is simple to verify that $E(\bar{x}, \preceq_C^{ml}) \subseteq L(\bar{x}, \preceq_C^{ml})$. The converse holds for a ml-minimal solution of the problem (SOP).

Proposition 6.14 *$\bar{x} \in ml - Min(F, S)$ if and only if $E(\bar{x}, \preceq_C^{ml}) = L(\bar{x}, \preceq_C^{ml})$.*

The following result is obvious and so we skip its proof.

Proposition 6.15 *If $\bar{x} \in ml - Min(F, S)$, then $E(\bar{x}, \preceq_C^{ml}) \subseteq ml - Min(F, S)$.*

Theorem 6.6.1 *Let S be a nonempty compact subset of a Hausdorff topological vector space X. If the set-valued map $F : S \rightrightarrows Y$ is $\preceq_C^{ml}$-lower semicontinuous from above on S, then the problem (SOP) has ml-minimal solution.*

Proof We define a relation $\preceq$ on the quotient set $P(Y)/\sim$ as follows: For any $[A]$ and $[B]$ in $P(Y)/\sim$, $[A] \preceq [B] \Leftrightarrow A \preceq_C^{ml} B$. Let $\{[F(x_\alpha)]\}_{\alpha \in I}$ be a totally ordered set in the quotient set $P(Y)/\sim$. Without loss of generality, let $\alpha, \beta \in I$ with $\alpha < \beta$ such that $[F(x_\beta)] \preceq [F(x_\alpha)]$. Then the compactness of S implies that there exist $\tilde{I} \subseteq I$ and a subnet $\{x_{\tilde{\alpha}}\}_{\tilde{\alpha} \in \tilde{I}}$ of $\{x_\alpha\}_{\alpha \in I}$ such that $x_{\tilde{\alpha}} \to \bar{x}$. Thus, by the $\preceq_C^{ml}$-lower semicontinuous from above of F, we know that $F(\bar{x}) \preceq_C^{ml} F(x_{\tilde{\alpha}})$ for all $\tilde{\alpha} \in \tilde{I}$. Hence, $[F(\bar{x})] \preceq [F(x_{\tilde{\alpha}})]$ for all $\tilde{\alpha} \in \tilde{I}$.

Next we prove that $[F(\bar{x})] \preceq [F(x_\alpha)]$ for all $\alpha \in I$. If it is not true, then there exists $\bar{\alpha} \in I$ such that $[F(\bar{x})] \npreceq [F(x_{\bar{\alpha}})]$. For each $\alpha' \in \tilde{I}$ with $\bar{\alpha} < \alpha'$, we have $[F(\bar{x}_{\alpha'})] \preceq$

$[F(x_{\tilde{\alpha}})]$. Since $[F(\bar{x})] \preceq [F(x_{\tilde{\alpha}})]$ for all $\tilde{\alpha} \in \tilde{I}$, one has $[F(\bar{x})] \preceq [F(x_{\alpha'})]$ and so $[F(\bar{x})] \preceq [F(x_{\tilde{\alpha}})]$, which is a contradiction. Therefore, $[F(\bar{x})] \preceq [F(x_\alpha)]$ for all $\alpha \in I$. Now by Zorn's lemma, we know that $\{[F(\bar{x})]\}_{x\in S}$ has a minimal element. That is, the problem (SOP) has a ml-minimal set.

□

Definition 6.23 Let S be a nonempty subset of a Hausdorff topological vector space and $F : S \rightrightarrows Y$ be a set-valued map. We say that S satisfies the condition (A) if for each net $\{x_\alpha\}_{\alpha\in I}$ in S such that $\{F(x_\alpha)\}_{\alpha\in I}$ is a $\preceq_C^{ml}$-decreasing net, there exist $\bar{I} \subseteq I$ and a subnet $\{x_{\tilde{\alpha}}\}_{\tilde{\alpha}\in\bar{I}}$ of $\{x_\alpha\}_{\alpha\in I}$ such that $\{x_{\tilde{\alpha}}\} \to \bar{x} \in S$.

Similar to the proof of Theorem 6.6.1, we can obtain the following theorem.

Theorem 6.6.2 *Let S be a nonempty subset of a Hausdorff topological vector space and $F : S \rightrightarrows Y$ be a set-valued map. If S satisfies the condition (A) and F is $\preceq_C^{ml}$-lower semicontinuous from above on S, then the problem (SOP) has a ml-minimal solution.*

6.6.3 Relation Between Minimal Solutions with Respect to Vector and Set Approach

In this subsection, we study the relations between minimal solutions for set optimization problems with respect to vector approach and set approach involving the partial set order relation $\preceq_C^{ml}$.

Throughout this subsection, unless otherwise specified, we assume that S is a nonempty subset of a Hausdorff topological vector space X and Y is a real normed space. Further, we assume that $F(x) \in \mathcal{B}(Y)$ for all $x \in S$, C is a closed convex and pointed cone with $\text{int}C \neq \emptyset$ and $F(x) \neq \emptyset$ for all $x \in X$.

Now, we show how the set relation $\preceq_C^{ml}$ can help to find the minimal solutions by vector approach.

Lemma 6.6.1 *If $\bar{x} \in Min(F, S)$ with $F(x) \preceq_C^{ml} F(\bar{x})$ for each $x \in S$, then $x \in Min(F, S)$.*

Proof Assume that $\bar{x} \in \text{Min}(F, S)$. Let $\bar{y} \in F(\bar{x})$ be such that $\bar{y} \in \min F(S)$. We only need to show that $\bar{y} \in F(x)$. Assume that $\bar{y} \notin F(x)$. Then by the hypothesis, we have $F(x) \preceq_C^{ml} F(\bar{x})$, that is, $F(x) \dot{-} F(\bar{x}) \cap (-C) \neq \emptyset$. Hence, there exists a $c \in C$ such that $-c \in F(x) \dot{-} F(\bar{x})$, equivalently $-c + F(\bar{x}) \subseteq F(x)$. Therefore, there exists $\bar{y} \in F(\bar{x})$ such that $-c + \bar{y} = y$ for some $y \in F(x)$. Then we have $\bar{y} = \tilde{y} + c$ for some $\tilde{y} \in F(x)$. This implies that $\bar{y} \in \tilde{y} + C$, that is, $\tilde{y} \preceq_C \bar{y}$ for some $\tilde{y} \in F(x)$ which contradicts to $\bar{y} \in \min F(S)$. Thus $\bar{y} \in F(x)$ and hence $x \in \text{Min}(F, S)$.

□

Proposition 6.16 *Let $\bar{x} \in Min(F, S)$. Then, only one of the following two assertions holds.*

(a) *$\bar{x}$ is a ml-minimal solution of the problem (SOP).*
(b) *There exists a minimal solution $\hat{x} \in Min(F, S)$ of the problem (SOP) such that $F(\hat{x}) \preceq_C^{ml} F(\bar{x})$ and $F(\hat{x}) \nsim F(\bar{x})$.*

Proof By the definition of ml-minimal solution, (b) is false if (a) holds. Assume that (a) does not hold. Then there exists $\hat{x} \in S$ such that $F(\hat{x}) \preceq_C^{ml} F(\bar{x})$ and $F(\hat{x}) \nsim F(\bar{x})$. Since $\bar{x} \in \text{Min}(F, S)$, by Lemma 6.6.1, we have that $\hat{x} \in \text{Min}(F, S)$ and (b) holds.

□

Definition 6.24 A set-valued map $F : S \rightrightarrows Y$ is said to be *strongly injective* on S if for any $x_1, x_2 \in S$, $F(x_2) \preceq_C^{ml} F(x_1)$ and $F(x_1) \npreceq_C^{ml} F(x_2)$ imply that $F(x_1) \cap F(x_2) \neq \emptyset$.

Lemma 6.6.2 *If $\bar{x} \in Min(F, S)$ and F is strongly injective on S, then $\bar{x} \in ml - Min(F, S)$.*

Proof Let $\bar{x} \in \text{Min}(F, S)$. Assume that $\bar{x} \notin ml - \text{Min}(F, S)$. Then there exists $\tilde{x} \in S$ such that $F(\tilde{x}) \preceq_C^{ml} F(\bar{x})$ and $F(\bar{x}) \neq F(\tilde{x})$. Since $\bar{x} \in \text{Min}(F, S)$, we can choose $\bar{y} \in F(\bar{x})$ such that $\bar{y} \in \min F(S)$. By Lemma 6.6.1, we have $\bar{y} \in F(\tilde{x})$, which contradicts the fact that F is strongly injective. Therefore, $\bar{x} \in ml - \text{Min}(F, S)$.

□

Theorem 6.6.3 *Let S be a nonempty subset of a Hausdorff topological vector space and $F : S \rightrightarrows Y$ satisfy $\preceq_C^{ml}$-lower property at $\bar{x} \in S$ with $F(\bar{x}) \cap F(\tilde{x}) = \emptyset$, $\tilde{x} \in S$. If $\min F(\bar{x}) \neq \emptyset$ and $\bar{x} \in ml - Min(F, S)$, then $\bar{x} \in Min(F, S)$.*

Proof Let $\bar{x} \in ml - \text{Min}(F, S)$ with $\min F(\bar{x}) \neq \emptyset$. Assume to the contrary that $\bar{x} \notin \text{Min}(F, S)$. Then $F(\bar{x}) \cap \min F(S) = \emptyset$. Since $\min F(\bar{x}) \subseteq F(\bar{x}) \subseteq F(S)$, we have $\min F(\bar{x}) \cap \min F(S) = \emptyset$. Since F has the $\preceq_C^{ml}$-lower property at $\bar{x} \in S$, there exists $\tilde{x} \in S$ such that $F(\tilde{x}) \prec_C^{ml} F(\bar{x})$. Then, there exists a $c \in C$ such that $-c + F(\bar{x}) \subseteq F(\tilde{x})$. Therefore, there exists $\bar{y} \in F(\bar{x})$ such that $-c + \bar{y} = \tilde{y}$ for some $\tilde{y} \in F(\tilde{x})$. This implies that $\bar{y} \in \tilde{y} + C$, that is, $\tilde{y} \preceq_C \bar{y}$ for some $\tilde{y} \in F(\tilde{x})$. Since $\bar{x} \in ml - \text{Min}(F, S)$, we have $F(\bar{x}) \prec_C^{ml} F(\tilde{x})$ and $F(\bar{x}) \neq F(\tilde{x})$. By $F(\bar{x}) \cap F(\tilde{x}) = \emptyset$ and $\tilde{y} \in F(\tilde{x})$, there exists a $\hat{y} \in F(\bar{x})$ such that $\hat{y} \preceq_C \tilde{y}$. Using the transitivity of the order relation $\preceq_C$, we get $\hat{y} \preceq_C \bar{y}$. Thus $\bar{y} \notin \min F(\bar{x})$, which contradicts the fact that $\min F(\bar{x}) \neq \emptyset$. Thus, we have $F(\bar{x}) \cap \min F(S) \neq \emptyset$ and hence $\bar{x} \in \text{Min}(F, S)$. □

6.7 Ekeland's Variational Principle for Set-Valued Maps

Ekeland's variational principle (in short, EVP) is one of the fundamental results from nonlinear analysis which was developed in the pioneer papers [45–47] by

I. Ekeland. One of the most important ideas of EVP is that in the absence of a known minimum, one can use EVP to reach close to a minimum. It is found that several other fundamental results from nonlinear analysis, namely, Caristi's fixed point theorem [23, 24], Takahashi's minimization theorem [147], Phelps's minimal element theorem [137, 138], etc., are equivalent to EVP in the sense that they can be achieved by using EVP and vice-versa. The EVP is one of the most powerful tools to deal with many applications in optimization, optimal control, global analysis, mathematical economy, partial differential equations, etc., see [3, 39, 45–47]. During the last three decades, EVP has been extended for vector-valued/set-valued maps and also under different space settings, see, for example, [3–6, 10, 12, 15, 17, 18, 20, 26–28, 50, 56, 58, 61, 63, 65, 71, 72, 76, 87, 89, 97, 131, 148, 149] and the references therein.

Ekeland's variational principle for vector-valued maps was explored by Németh [131], Tammer [148], and Isac [89]. However, each of these vector-valued versions have different conditions on the involved function. In 1998, Chen and Huang [27] unified these results. In [58, 59], a variational principle for a vector-valued map was presented as a consequence of the minimal point theorem on the product space. It is worth to mention that the minimal element theorems were established by Göpfert and Tammer [57] and further generalized by Göpfert, et al. [58, 59], Hamel and Löhne [71], Hamel [65], and Hamel and Tammer [74] on the product space $X \times Y$ in different settings. Such theorems played an important role to derive Ekeland's variational principle for vector-valued maps. These minimal element theorems are the extension of Phelps's minimal element theorem [137, 138].

Hamel and Löhne [72] considered a subset $\mathcal{A} \subseteq X \times P(Y)$, where X is a separated uniform space and Y is a topological vector space and introduced the following notation:

$$\mathcal{V}(\mathcal{A}) := \{V \in P(Y) : \exists x \in X : (x, V) \in \mathcal{A}\}.$$

Let Λ be the directed set and $\{q_\lambda\}_{\lambda \in \Lambda}$ be the family of quasi-metrics which generates the topology of the uniform space X. We write q_Λ if and only if an assertion holds for all $\lambda \in \Lambda$. Using the relation $\preceq^l_C$ and $\preceq^u_C$, Hamel and Löhne [72] introduced the following ordering relations on $X \times P(Y)$: For all $x_1, x_2 \in X$, $V_1, V_2 \in P(Y)$, and $k \in C \setminus -\mathrm{cl}C$,

$$(x_1, V_1) \preceq^k_l (x_2, V_2) \quad \Leftrightarrow \quad V_1 + q_\Lambda(x_1, x_2)k \preceq^l_C V_2,$$

and

$$(x_1, V_1) \preceq^k_u (x_2, V_2) \quad \Leftrightarrow \quad V_1 + q_\Lambda(x_1, x_2)k \preceq^u_C V_2.$$

Note that the previous relations can be read as

$$\text{for all } \lambda \in \Lambda, \quad V_1 + q_\lambda(x_1, x_2)k \preceq^l_C V_2,$$

and

$$\text{for all } \lambda \in \Lambda, \quad V_1 + q_\lambda(x_1, x_2)k \preceq^u_C V_2.$$

The relations $\preceq_l^k$ and $\preceq_u^k$ are reflexive and transitive on $X \times P(Y)$.

Hamel and Löhne [72] introduced the minimal element theorems for set-valued maps in the separated uniform spaces involving the set order relations $\preceq_C^l$ and $\preceq_C^u$. Such minimal element theorems are the extensions of minimal element theorems presented in [58, 59].

Moreover, they [72] introduced the concept of the domain of a set-valued map F for the set order relations $\preceq_C^l$ and $\preceq_C^u$ in the following way:

$$\preceq_C^l - \text{dom } F := \{x \in X : F(x) \preceq_C^l V \text{ for some nonempty } V \subseteq Y\},$$

and

$$\preceq_C^u - \text{dom } F := \{x \in X : F(x) \preceq_{C_1}^l V \text{ for some topologically bounded } V \subseteq Y\}.$$

They derived variational principle for set-valued maps involving the set order relation $\preceq_C^l / \preceq_C^u$.

In [12], we studied minimal element theorem, Ekeland's variational principle, Caristi's fixed point theorem, and Takahashi's minimization theorem involving set order relations $\preceq_k^{ml}$ and $\preceq_k^{mu}$ defined on $X \times P(Y)$ by using $\preceq_C^{ml}$ and $\preceq_C^{mu}$ as follows:

Let C be a solid convex cone in a normed space Y and (X, d) be a metric space. For all $x_1, x_2 \in X$, $V_1, V_2 \in P(Y)$, and $k \in \text{int}C$, define

$$(x_1, V_1) \preceq_k^{ml} (x_2, V_2) \quad \Leftrightarrow \quad V_1 + d(x_1, x_2)k \preceq_C^{ml} V_2,$$

and

$$(x_1, V_1) \preceq_k^{mu} (x_2, V_2) \quad \Leftrightarrow \quad V_1 + d(x_1, x_2)k \preceq_C^{mu} V_2.$$

It can be easily seen that the relations $\preceq_k^{ml}$ and $\preceq_k^{mu}$ are reflexive and transitive on $X \times P(Y)$.

We now consider the following assumptions.

Assumption 2 Let (X, d) be a complete metric space, Y be a real normed vector space, C be a solid closed convex pointed cone in Y, and $k \in \text{int}C$. Let $F : X \to P(Y)$ be a closed-valued map such that

(i) F is ml-bounded below (that is, there exists $V \in P(Y)$ such that $V \preceq_C^{ml} F(x)$ for all $x \in X$),
(ii) $\widetilde{S}(x) = \{\tilde{x} \in X : (\tilde{x}, F(\tilde{x})) \preceq_k^{ml} (x, F(x))\}$ is closed for all $x \in X$.

Assumption 3 Let (X, d) be a complete metric space, Y be a real normed vector space, C be a solid closed convex pointed cone in Y, and $k \in \text{int}C$. Let $F : X \to P(Y)$ be a closed-valued map such that

(i) F is mu-bounded below (that is, there exists $V \in P(Y)$ such that $V \preceq_C^{mu} F(x)$ for all $x \in X$),
(ii) $\widehat{S}(x) = \{\hat{x} \in X : (\hat{x}, F(\hat{x})) \preceq_k^{mu} (x, F(x))\}$ is closed for all $x \in X$.

The minimal element theorems involving the set order relations $\preceq_k^{ml}$ and $\preceq_k^{mu}$ on $X \times P(Y)$ are presented in [12]. Here, we mention such result only for the set order relation $\preceq_C^{ml}$.

Theorem 6.7.1 [12] *Let (X, d) be a complete metric space, Y be a real normed space, C be a solid closed convex pointed cone in Y, $k \in \text{int}C$, and $\mathcal{A} \subset X \times P(Y)$ be a nonempty set. Assume that the following conditions hold:*

(i) *$\mathcal{A}$ is ml-bounded below (that is, there exists $V \in P(Y)$ such that $V \preceq_C^{ml} P_{P(Y)}(\mathcal{A})$ for all $x \in X$);*
(ii) *For all $\preceq_k^{ml}$ −decreasing sequence $\{(x_n, V_n)\}_{n\in\mathbb{N}} \subset \mathcal{A}$ (that is, $(x_{n+1}, V_{n+1}) \preceq_k^{ml} (x_n, V_n)$ for all $n \in \mathbb{N}$), there exists $(x, V) \in \mathcal{A}$ such that $(x, V) \preceq_k^{ml} (x_n, V_n)$ for all $n \in \mathbb{N}$.*

Then for every $(x_0, V_0) \in \mathcal{A}$, there exists $(\bar{x}, \bar{V}) \in \mathcal{A}$ such that

(a) $(\bar{x}, \bar{V}) \preceq_k^{ml} (x_0, V_0)$,
(b) *for any $(\tilde{x}, \tilde{V}) \in \mathcal{A}$ such that $(\tilde{x}, \tilde{V}) \preceq_k^{ml} (\bar{x}, \bar{V})$, then $\tilde{x} = \bar{x}$.*

In [12], we established Ekeland's variational principle for set-valued maps involving the set order relations $\preceq_C^{ml}$ and $\preceq_C^{mu}$. Here we mention such result only for the set order relation $\preceq_C^{ml}$.

Theorem 6.7.2 [12] *Assume that the Assumption 2 holds. If for $k \in \text{int}C$ and $x_0 \in X$, $F(x_0) \not\subset F(X) + k + \text{int}C$, then there exists $\bar{x} \in X$ such that*

(a) $F(\bar{x}) + d(\bar{x}, x_0)k \preceq_C^{ml} F(x_0)$,
(b) $F(x) + d(\bar{x}, x)k \not\preceq_C^{ml} F(\bar{x})$ *for all* $x \neq \bar{x}$,
(c) $d(\bar{x}, x_0) \leq 1$.

In [12], we further obtained Caristi's fixed point theorems for set-valued maps under the set order relations $\preceq_C^{ml}$ and $\preceq_C^{mu}$. Here we mention such result only for the set order relation $\preceq_C^{ml}$.

Theorem 6.7.3 [12] *Suppose that the Assumption 2 and the following condition hold.*

(Caristi-$\preceq_C^{ml}$) Condition. *Let $T : X \to 2^X$ be a set-valued map such that for every $x \in X$, there exists $y \in T(x)$ such that*

$$F(y) + d(x, y)k \preceq_C^{ml} F(x).$$

Then T has a fixed point in X, that is, there exists $\bar{x} \in X$ with $\bar{x} \in T(\bar{x})$.

In [12], we also obtained Takahashi's minimization theorems for set-valued maps under the set order relations $\preceq_C^{ml}$ and $\preceq_C^{mu}$. Here we mention such result only for the set order relation $\preceq_C^{ml}$.

Theorem 6.7.4 [12] *Suppose that the Assumption 2 and the following condition hold.*

(Takahashi-$\preceq_C^{ml}$ Condition). *For every $y \in X$ with $F(y) \notin ml - \mathrm{WMin}(F, X)$, there exists $z \in X \setminus \{y\}$ such that*

$$F(z) + d(y, z)k \preceq_C^{ml} F(y).$$

Then there exists $\bar{x} \in X$ such that $F(\bar{x}) \in ml - \mathrm{WMin}(F, X)$.

We remark that the following implications hold

$$\text{Theorem 6.7.2} \quad \Leftrightarrow \quad \text{Theorem 6.7.3} \quad \Leftrightarrow \quad \text{Theorem 6.7.4}.$$

6.7.1 A Minimal Element Theorem and Ekeland's Principle with Mixed Set Order Relations

Throughout this subsection, unless otherwise specified, we assume that Y is a Hausdorff topological vector space and C is a nontrivial, solid convex cone. Let W be a nonempty set with a transitive relation $\preceq$ on W. We say that the sequence $\{w_n\}_{n\in\mathbb{N}} \subset W$ is $\preceq$-decreasing [76] if $w_{n+1} \preceq w_n$ for all $n \in \mathbb{N}$. We set $S_{\preceq}(w_0) := \{w \in W : w \preceq w_0\}$ for each $w_0 \in W$. Of course, $S_{\preceq} : W \rightrightarrows W$ is a set-valued map whose domain is dom $S_{\preceq} := \{w_0 \in W : S_{\preceq}(w_0) \neq \emptyset\}$. Clearly, $w \in S_{\preceq}(w_0) \Rightarrow S_{\preceq}(w) \subset S_{\preceq}(w_0)$ and dom $S_{\preceq} = W$ when $\preceq$ is a pre-order, that is, a reflexive and transitive order relation on W.

The following variational principle for minimal points on a pre-ordered set played a key role to establish the main results of this subsection.

Theorem 6.7.5 (Extended Brézis–Browder Principle) [21, 76] *Let $\preceq$ be a transitive relation and $\phi : W \to \overline{\mathbb{R}} := \mathbb{R} \cup \{\pm\infty\}$ be a function such that the following conditions hold.*

(i) *ϕ is $\preceq$-increasing (that is, $w_1 \preceq w_2$ implies $\phi(w_1) \leq \phi(w_2)$);*
(ii) *For every $\preceq$-decreasing sequence $\{w_n\}_{n\in\mathbb{N}} \subseteq W$, there exists $w \in W$ such that $w \preceq w_n$ for all $n \in \mathbb{N}$.*

Then for every $w_0 \in$ dom $S_{\preceq}$, there exists $\bar{w} \in S_{\preceq}(w_0)$ such that $\phi(\hat{w}) = \phi(\bar{w})$ for all $\hat{w} \in S_{\preceq}(\bar{w})$.

Let (X, d) be a metric space and $\lambda \in [0, 1]$. For $x_1, x_2 \in X$, $V_1, V_2 \in \Omega_C^{cb}$, and $k \in \mathrm{int}C$, in [6], we introduced the following set order relation $\preceq_k^{\lambda}$ on $X \times \Omega^{cb}$ as follows:

$$(x_1, V_1) \preceq_k^{\lambda} (x_2, V_2) \quad \Leftrightarrow \quad V_1 + d(x_1, x_2)k \preceq_C^{\lambda} V_2.$$

It can be easily seen that the set order relation $\preceq_k^{\lambda}$ is reflexive and transitive on $X \times \Omega^{cb}$.

By using the technique of [12, 72], but for the weighted set order relation $\preceq_k^{\lambda}$ on $X \times \Omega_C^{cb}$, we [6] established the following minimal element theorem. It is worth to mention that Hamel and Löhne [72] used the set relations $\preceq_k^{l}$ and $\preceq_k^{u}$, which are special cases of the set relation $\preceq_k^{\lambda}$, to obtain a minimal element theorem.

Let (X, d) be a complete metric space. For a set $\mathcal{A} \subseteq X \times \Omega^{cb}$, we denote by $P_{\Omega^{cb}}(\mathcal{A})$ the projection of $\mathcal{A}$ onto its second component, that is,

$$P_{\Omega^{cb}}(\mathcal{A}) = \{A \in \Omega^{cb} : \exists x \in X \text{ with } (x, A) \in \mathcal{A}\}.$$

Theorem 6.7.6 [6] *Let $\mathcal{A} \subset X \times \Omega^{cb}$ be a nonempty set. Assume that the following condition holds:*

(M) *For all $\preceq_k^\lambda$-decreasing sequence $\{(x_n, V_n)\}_{n\in\mathbb{N}} \subset \mathcal{A}$ (that is, $(x_{n+1}, V_{n+1}) \preceq_k^\lambda (x_n, V_n)$ for all $n \in \mathbb{N}$), there exists $(x, V) \in \mathcal{A}$ such that $(x, V) \preceq_k^\lambda (x_n, V_n)$ for all $n \in \mathbb{N}$.*

Then for every $(x_0, V_0) \in \mathcal{A}$, there exists $(\bar{x}, \bar{V}) \in \mathcal{A}$ such that

(a) $(\bar{x}, \bar{V}) \preceq_k^\lambda (x_0, V_0)$,
(b) *if $(\hat{x}, \hat{V}) \in \mathcal{A}$ such that $(\hat{x}, \hat{V}) \preceq_k^\lambda (\bar{x}, \bar{V})$, then $\hat{x} = \bar{x}$.*

Assumption 4 Let $F : X \to \Omega^{cb}$ be a C−closed-valued map such that

(i) F is bounded below (that is, there exists $V \in \Omega^{cb}$ such that $V \preceq_C^\lambda F(x)$ for all $x \in X$),
(ii) $\widehat{S}(x) = \{\hat{x} \in X : (\hat{x}, F(\hat{x})) \preceq_k^\lambda (x, F(x))\}$ is closed for all $x \in X$.

In [6], we established Ekeland's variational principle for set-valued maps involving set order relation $\preceq_C^\lambda$.

Theorem 6.7.7 *Assume that Assumption 4 is satisfied. If for $k \in \operatorname{int} C$ and $x_0 \in X$, $F(X) + k \not\preceq_C^\lambda F(x_0)$ holds, then there exists $\bar{x} \in X$ such that*

(a) $F(\bar{x}) + d(\bar{x}, x_0)k \preceq_C^\lambda F(x_0)$,
(b) $F(x) + d(\bar{x}, x)k \not\preceq_C^\lambda F(\bar{x})$ *for* $x \neq \bar{x}$,
(c) $d(\bar{x}, x_0) \leq 1$.

Remark 6.14 It is worth to mention that Theorems 6.7.6 and 6.7.7 are more general than [12, Theorems 4.4 and 4.6], [72, Theorems 5.1 and 6.1], and [76, Theorem 5.1] under certain assumptions. However, due to strong assumptions on the order relation $\preceq_C^\lambda$, Theorems 6.7.6 and 6.7.7 are not completely comparable with the results mentioned above.

We derived in [6] the following Caristi fixed point theorem for set-valued maps with set order relation $\preceq_C^\lambda$.

Theorem 6.7.8 *Suppose that Assumption 4 and the following condition hold.*

(Caristi-$\preceq_C^\lambda$) Condition. *Let $T : X \rightrightarrows X$ be a set-valued map with nonempty values such that for every $x \in X$, there exists $y \in T(x)$ satisfying*

$$F(y) + d(x, y)k \preceq_C^\lambda F(x).$$

Then T has a fixed point in X, that is, there exists $\bar{x} \in X$ such that $\bar{x} \in T(\bar{x})$.

We further obtained in [6] the following Takahashi minimization theorem for set-valued maps with mixed set order relation $\preceq_C^{\lambda}$.

Theorem 6.7.9 *Suppose that Assumption 4 and the following condition hold.*

(Takahashi-$\preceq_C^{\lambda}$ Condition). *For every $y \in X$ with $F(y) \notin \lambda - \mathrm{Min}(F, X)$, there exists $z \in X \setminus \{y\}$ such that*

$$F(z) + d(y, z)k \preceq_C^{\lambda} F(y).$$

Then there exists $\bar{x} \in X$ such that $F(\bar{x}) \in \lambda\text{-}\mathrm{Min}(F, X)$.

In [6], we verified that Theorems 6.7.7, 6.7.8, and 6.7.9 are equivalent to each other in the sense that each one can be derived by using the other.

6.7.2 Ekeland's Variational Principle for Set-Valued Maps in Quasi-Metric Spaces

We recall the definition of a quasi-metric space. For further details and definitions, we refer to [32].

Definition 6.25 Let X be a nonempty set. A *quasi-metric* on X is a function $q : X \times X \to \mathbb{R}_+ := [0, +\infty)$ that satisfies the following conditions:

(Q1) $q(x, y) \geq 0$ and $q(x, x) = 0$ for all $x \in X$;
(Q2) $q(x, y) \leq q(x, z) + q(z, y)$ for all $x, y, z \in X$;
(Q3) $q(x, y) = q(y, x) = 0 \Rightarrow x = y$ for all $x, y \in X$.

The set X equipped with a quasi-metric q is called a quasi-metric space and it is denoted by (X, q). If, in addition, the quasi-metric q satisfies the symmetry property, that is, $q(x, y) = q(y, x)$ for all $x, y \in X$, then q is called a metric. The topological space equipped with a quasi-metric is known as the Sorgenfrey line. Every quasi-metric space (X, q) can be viewed as a topological space on which the topology is induced by taking the collection of balls $\{\mathbb{B}_r(x) : r > 0\}$ as a base of the neighborhood filter for every $x \in X$, where the (left) ball $\mathbb{B}_r(x)$ is defined by

$$\mathbb{B}_r(x) := \{y \in X : q(x, y) < r\}.$$

We present some basic notions from quasi-metric spaces, which are needed in this subsection.

Definition 6.26 Let (X, q) be a quasi-metric space and Ω be a nonempty subset of X.

(a) We say that the sequence $\{x_n\} \subset X$ *(left-sequentially) converges* to $\bar{x} \in X$ if $\lim_{n\to\infty} q(x_k, x^*) = 0$, and it is denoted by $x_n \to x^* \in X$.

(b) We say that the set Ω is *left-sequentially closed* if for any sequence $x_n \to x^*$ with $\{x_n\} \subset \Omega$, $x^* \in \Omega$.

(c) We say that the sequence $\{x_n\} \subset X$ is *left-sequentially Cauchy* if for each $\beta \in \mathbb{N}$, there is a natural number N_β such that

$$q(x_n, x_m) < 1/\beta, \quad \text{for all } m \geq n \geq N_\beta.$$

(d) We say that the quasi-metric space (X, q) is *left-sequentially complete* if each left-sequentially Cauchy sequence is convergent and its limit belongs to X.

(e) A quasi-metric space is the *Hausdorff topological space* if

$$\left[\lim_{n\to\infty} q(x_n, \bar{x}) = 0 \text{ and } \lim_{n\to\infty} q(x_n, \bar{u}) = 0\right] \quad \Rightarrow \quad \bar{x} = \bar{u}. \tag{6.9}$$

(f) A quasi-metric space (X, q) ordered by a pre-order $\preceq$ (that is, a reflexive and transitive relation) is said to satisfy the *Hausdorff decreasing condition* if for every decreasing sequence $\{x_n\} \subset X$ and $\bar{x}, \bar{u} \in X$ with $\bar{x} \preceq \bar{u}$ the implication in (6.9) holds.

In 1983, Dancs, Hegedüs, and Medvegyev [37] (in short, DHM) established a fixed point theorem for set-valued maps on a complete metric space by using the generalized Picard iteration under some appropriate assumptions. Since then, many authors have generalized this fixed point theorem under different assumptions and in different settings. Recently, Bao et al. [17] extended DHM's fixed point theorem for parametric dynamic systems in quasi-metric spaces. Motivated by the result in [17], we, in [10], introduced the extended Picard iterative process for set-valued maps on the product spaces and obtained the extended version of DHM's fixed point theorem. We defined the extended Picard sequence in the following way:

Let X be a nonempty set, Y be a topological vector space, and $\Phi : X \times P(Y) \rightrightarrows X \times P(Y)$ be a set-valued map. We say that the sequence $\{(x_n, V_n)\}_{n\in\mathbb{N}}$ is an extended Picard sequence/iterative process if

$$(x_2, V_2) \in \Phi(x_1, V_1),\ (x_3, V_3) \in \Phi(x_2, V_2),\ \ldots,\ (x_n, V_n) \in \Phi(x_{n-1}, V_{n-1}),$$

for all $n \in \mathbb{N}$.

In [10], we established the following extended parametric fixed point theorem on the product space $X \times P(Y)$.

Theorem 6.7.10 *Let (X, q) be a complete Hausdorff quasi-metric space, Y be a topological vector space, and $\emptyset \neq \Gamma \subset X \times P(Y)$. Assume that the parametric dynamical system $\Phi : X \times P(Y) \rightrightarrows X \times P(Y)$ satisfies the following conditions:*

(F1) $(x, V) \in \Phi(x, V)$ *for all* $(x, V) \in \Gamma$.

(F2) *For all* $(x_1, V_1), (x_2, V_2) \in \Gamma$ *such that* $(x_2, V_2) \in \Phi(x_1, V_1)$, *we have* $\Phi(x_2, V_2) \subset \Phi(x_1, V_1)$.

(F3) *For each extended Picard sequence* $\{(x_n, V_n)\}_{n\in\mathbb{N}} \subset \Gamma$ *with* $x_n \to x^*$ *as* $n \to \infty$, *there exists* $V^* \in P(Y)$ *such that*

$$(x^*, V^*) \in \Gamma \textit{ and } (x^*, V^*) \in \Phi(x_n, V_n), \quad \textit{for all } n \in \mathbb{N}, \tag{6.10}$$

and

$$(x^*, V) \in \Gamma \cap \Phi(x^*, V^*) \textit{ implies } V = V^*. \tag{6.11}$$

(F4) *For each extended Picard sequence* $\{(x_n, V_n)\}_{n\in\mathbb{N}} \subset \Gamma$, $q(x_n, x_{n+1}) \to 0$ *as* $n \to \infty$.

Then for every $(x_0, V_0) \in \Gamma$, *there is an extended Picard sequence* $\{(x_n, V_n)\}_{n\in\mathbb{N}} \subset \Gamma$ *starting from* (x_0, V_0) *and ending at a fixed point* (x^*, V^*) *of* Φ *in the sense that* $\Phi(x^*, V^*) = \{(x^*, V^*)\}$.

From now onward, we assume that $\mathcal{K} : Y \rightrightarrows Y$ is a set-valued map and the following conditions hold.

- $\mathbf{0} \in \mathcal{K}(y)$.
- $\mathcal{K}(y) + \mathcal{K}(y) \subseteq \mathcal{K}(y)$ for all $y \in Y$.
- $[0, +\infty)k + \mathcal{K}(y) \subseteq \mathcal{K}(y)$ for all $y \in Y$ and all $k \in Y \setminus \{\mathbf{0}\}$.
- For all $y \in Y$ and all $v \in \mathcal{K}(y)$, we have

$$\mathcal{K}(y - v) \subseteq \mathcal{K}(y). \tag{6.12}$$

- For all $A, B, D, E \in P(Y)$, we have

$$\forall b \in B,\ \forall e \in E \text{ we have } \mathcal{K}(b) + \mathcal{K}(e) \subseteq \mathcal{K}(b + e). \tag{6.13}$$

- For all $y \in Y$ and all $v \in \mathcal{K}(y)$, we have

$$\mathcal{K}(y + v) \subseteq \mathcal{K}(y). \tag{6.14}$$

- For all $A, B, D, E \in P(Y)$, we have

$$\forall a \in A \text{ and } \forall d \in D \text{ we have } \mathcal{K}(a) + \mathcal{K}(d) \subseteq \mathcal{K}(a + d). \tag{6.15}$$

Let X be a quasi-metric space. For all $x_1, x_2 \in X$ and $V_1, V_2 \in P(Y)$, we [10] defined the set order relations $\preceq_k^u$ and $\preceq_k^l$ on $X \times P(Y)$ as follows:

$$(x_1, V_1) \preceq_k^u (x_2, V_2) \quad \Leftrightarrow \quad V_1 + q(x_2, x_1)k \preceq_u^{\mathcal{K}} V_2, \tag{6.16}$$

and

$$(x_1, V_1) \preceq_k^l (x_2, V_2) \quad \Leftrightarrow \quad V_1 + q(x_2, x_1)k \preceq_l^{\mathcal{K}} V_2. \tag{6.17}$$

We note that the above set order relations on $X \times P(Y)$ are pre-order. Note that the relation $\preceq_k^u$ is reflexive and transitive on $X \times P(Y)$ if (6.12) and (6.13) hold, and the relation $\preceq_k^l$ is reflexive and transitive on $X \times P(Y)$ if (6.14) and (6.15) hold.

Based on the idea of [49, 56, 59, 149], we, in [10], also defined the following order relations on $X \times P(Y)$, which are stronger than $\preceq_k^u$ and $\preceq_k^l$.

$$(x_1, V_1) \preceq_{k,h_k^u}^u (x_2, V_2) \quad \Leftrightarrow \quad \begin{cases} (x_1, V_1) = (x_2, V_2), \\ \text{or} \\ (x_1, V_1) \preceq_k^u (x_2, V_2) \text{ and } h_k^u(V_1) < h_k^u(V_2). \end{cases}$$

$$(x_1, V_1) \preceq_{k,h_k^l}^l (x_2, V_2) \quad \Leftrightarrow \quad \begin{cases} (x_1, V_1) = (x_2, V_2), \\ \text{or} \\ (x_1, V_1) \preceq_k^l (x_2, V_2) \text{ and } h_k^l(V_1) < h_k^l(V_2). \end{cases}$$

It can be easily seen that $\preceq_{k,h_k^u}^u$ and $\preceq_{k,h_k^l}^l$ are reflexive and transitive on $X \times P(Y)$.

Recently, Bao et al. [17, 19] have developed a constructive dynamical approach to prove the existence of a minimal element of a nonempty subset of the product space $X \times Y$ ordered by some preference. Such a result is called parametric minimal point theorem. They applied the parametric minimal point theorem to derive the Ekeland type variational principle for set-valued maps in the setting of quasi-metric spaces. They used the preferences given by a set-valued map $\mathcal{K} : Y \rightrightarrows Y$, but their approach depends on the vector approach. It is worth to mention that the set approach is used in [12, 72] to obtain minimal element theorems and the Ekeland type variational principle for set-valued maps in the setting of complete metric spaces with a constant ordering cone. Since the set-valued optimization problems with variable domination structures have their own importance not only in the theoretical areas but also in real-life applications (see, [40, 42, 44, 104]), we [10] extended the results of [17, 19] in the setting of product space $X \times P(Y)$ for the generalized variable upper/lower less set order relations $\preceq_u^{\mathcal{K}}/\preceq_l^{\mathcal{K}}$.

Definition 6.27 [10] Let X be a nonempty set, Y be a topological vector space, and $\Gamma \subset X \times P(Y)$ be a nonempty set pre-ordered by $\preceq_k^u$ [$\preceq_k^l$].

(a) A sequence $\{(x_n, V_n)\}_{n \in \mathbb{N}} \subset \Gamma$ is said to be *decreasing with respect to the pre-order* $\preceq_k^u$ [$\preceq_k^l$] if $(x_n, V_n) \preceq_k^u (x_{n-1}, V_{n-1})$ [$(x_n, V_n) \preceq_k^l (x_{n-1}, V_{n-1})$] for all $n \in \mathbb{N}$.

(b) An element $(\bar{x}, \bar{V})$ of Γ is said to be *partial minimal element with respect to the pre-order* $\preceq_k^u$ [$\preceq_k^l$] if $(x, V) \in \Gamma$ and $(x, V) \preceq_k^u (\bar{x}, \bar{V})$ [$(x, V) \preceq_k^l (\bar{x}, \bar{V})$], then $x = \bar{x}$.

(c) An element $(\bar{x}, \bar{V})$ of Γ is said to be *minimal element with respect to the pre-order* $\preceq_k^u$ [$\preceq_k^l$] if $(x, V) \in \Gamma$ and $(x, V) \preceq_k^u (\bar{x}, \bar{V})$ [$(x, V) \preceq_k^l (\bar{x}, \bar{V})$], then $(x, V) = (\bar{x}, \bar{V})$.

In [10], we derived the following minimal element theorem for the set order relation $\preceq_u^{\mathcal{K}}$.

Theorem 6.7.11 [10] *Let (X, q) be a Hausdorff quasi-metric space, Y be a topological vector space, $\mathcal{K} : Y \rightrightarrows Y$ be a set-valued map that satisfies (6.12) and (6.13), and $\Gamma \subset X \times P(Y)$ be a nonempty set. For a given $(x_0, V_0) \in \Gamma$, define the set*

$$\mathcal{A}_0 := \mathcal{A}(x_0, V_0) = \{(\tilde{x}, \tilde{V}) \in \Gamma : (\tilde{x}, \tilde{V}) \preceq_k^u (x_0, V_0)\}.$$

Let $\{(x_n, V_n)\}_{n\in\mathbb{N}} \subset \mathcal{A}_0$ be a $\preceq_k^u$-decreasing sequence such that the following conditions hold.

(M1) $q(x_n, x_{n+1}) \to 0$ *as* $n \to \infty$.
(M2) *If* $\{x_n\}$ *is a left-sequentially Cauchy sequence, then there exists* $(\bar{x}, \bar{V}) \in \mathcal{A}_0$ *such that* $(\bar{x}, \bar{V}) \preceq_k^u (x_n, V_n)$ *for all* $n \in \mathbb{N}$.

Then there is a decreasing sequence $\{(x_n, V_n)\}_{n\in\mathbb{N}} \subset \Gamma$ starting from (x_0, V_0) and ending at a partially minimal element $(\bar{x}, \bar{V})$ of Γ with respect to $\preceq_k^u$. If, furthermore, $(\bar{x}, \bar{V})$ satisfies the domination condition

$$(\bar{x}, V) \preceq_k^u (\bar{x}, \bar{V}) \quad \Rightarrow \quad V = \bar{V}, \quad \text{for all } (\bar{x}, V) \in \mathcal{A}_0, \tag{6.18}$$

then it can be chosen as a minimal element of the set Γ with respect to $\preceq_k^u$.

Moreover, if we replace $\preceq_k^u$ by $\preceq_{k,h_k^u}^u$, then, under the assumption (6.18), there is a decreasing sequence $\{(x_n, V_n)\}_{n\in\mathbb{N}} \subset \Gamma$ starting from (x_0, V_0) and ending at a minimal point $(\bar{x}, \bar{V})$ of Ω with respect to $\preceq_{k,h_k^u}^u$.

We also derived the following minimal element theorem for the set order relation $\preceq_l^{\mathcal{K}}$.

Theorem 6.7.12 [10] *Let (X, q) be a Hausdorff quasi-metric space, Y be a topological vector space, $\mathcal{K} : Y \rightrightarrows Y$ be a set-valued map that satisfy (6.14) and (6.15), and $\Gamma \subset X \times P(Y)$ be a nonempty set. For a given $(x_0, V_0) \in \Gamma$, define the set*

$$\mathcal{A}_0 := \mathcal{A}(x_0, V_0) = \{(\tilde{x}, \tilde{V}) \in \Gamma : (\tilde{x}, \tilde{V}) \preceq_k^l (x_0, V_0)\}.$$

Let $\{(x_n, V_n)\}_{n\in\mathbb{N}} \subset \mathcal{A}_0$ be a $\preceq_k^l$-decreasing sequence such that the following conditions hold.

(M1′) $q(x_n, x_{n+1}) \to 0$ *as* $n \to \infty$.
(M2′) *If* $\{x_n\}$ *is a left-sequentially Cauchy sequence, then there exists* $(\bar{x}, \bar{V}) \in \mathcal{A}_0$ *such that* $(\bar{x}, \bar{V}) \preceq_k^l (x_n, V_n)$ *for all* $n \in \mathbb{N}$.

Then, there is a decreasing sequence $\{(x_n, V_n)\}_{n\in\mathbb{N}} \subset \Gamma$ starting from (x_0, V_0) and ending at a partially minimal element $(\bar{x}, \bar{V})$ of Γ with respect to $\preceq_k^l$. If, furthermore, $(\bar{x}, \bar{V})$ satisfies the domination condition

$$(\bar{x}, V) \preceq_k^l (\bar{x}, \bar{V}) \quad \Rightarrow \quad V = \bar{V} \quad \text{for all } (\bar{x}, V) \in \mathcal{A}_0, \tag{6.19}$$

then it can be chosen as a minimal element of the set Γ with respect to $\preceq_k^l$.

Moreover, if we replace $\preceq_k^u$ *by* $\preceq_{k,h_k'}^l$*, then, under the assumption* (6.19)*, there is a decreasing sequence* $\{(x_n, V_n)\}_{n\in\mathbb{N}} \subset \Gamma$ *starting from* (x_0, V_0) *and ending at a minimal element* $(\bar{x}, \bar{V})$ *of* Γ *with respect to* $\preceq_{k,h_k'}^l$.

Recall that (X, q) be a quasi-metric space and Y be a topological vector space. A set-valued map $F : X \to P(Y)$ is said to be

(a) level-decreasingly-closed on dom$F$ with respect to $\preceq_u^{\mathcal{K}}$ $[\preceq_l^{\mathcal{K}}]$ if for any sequence $\{(x_n, V_n)\} \subset$ Graph$F$ such that $x_n \to \bar{x} \in X$ as $n \to \infty$ and $\{V_n\}$ is a sequence decreasing with respect to $\preceq_u^{\mathcal{K}}$ $[\preceq_l^{\mathcal{K}}]$, there exists $\bar{V} = F(\bar{x}) \in \mathrm{Min}(F(X); \preceq_u^{\mathcal{K}}$ $[\preceq_l^{\mathcal{K}}])$ such that $\bar{V} \preceq_u^{\mathcal{K}} [\preceq_l^{\mathcal{K}}] V_n$ for all $n \in \mathbb{N}$.
(b) quasi-bounded from below with respect to a closed convex cone C in Y if there is a bounded subset $M \subset Y$ such that $F(X) \subseteq M + C$.

In [10], we established the following Ekeland type variational principle for set-valued maps under variable ordering structures.

Theorem 6.7.13 [10] *Let* (X, q) *be a complete Hausdorff quasi-metric space,* Y *be a topological vector space, and* $\mathcal{K} : Y \rightrightarrows Y$ *be a set-valued map such that* (6.12) *and* (6.13) *hold. Let* C *be a convex cone in* Y *and* $F : X \to P(Y)$ *be a set-valued map which is quasi-bounded from below with respect to* C*. Assume that the following conditions are satisfied:*

(E1) *For every* $y \in Y$, $\mathcal{K}(y)$ *is a closed set in* Y.
(E2) *For any* $A, B \in P(Y)$*, if* $A \preceq_u^{\mathcal{K}} B$*, then* $\mathcal{K}(a) + \mathcal{K}(b) \subseteq \mathcal{K}(b)$ *for all* $a \in A$ *and* $b \in B$.
(E3) *For any sequence* $\{(x_n, V_n)\} \subset$ *Graph*F *such that* $x_n \to \bar{x} \in X$ *as* $n \to \infty$ *and* $\{V_n\}$ *is decreasing with respect to* $\preceq_u^{\mathcal{K}}$*, there exists a minimal element* $\bar{V} \in \mathrm{Min}(F(X); \preceq_u^{\mathcal{K}})$ *for which* $\bar{V} \preceq_u^{\mathcal{K}} V_n$ *for all* $n \in \mathbb{N}$.
(E4) $k \notin \mathrm{cl}(-C - \mathcal{K}(V_0)))$.

Then, for every $(x_0, F(x_0)) \in$ *Graph*F*, there exists* $(\bar{x}, F(\bar{x})) \in$ *Graph*F *with* $F(\bar{x}) \in \mathrm{Min}(F(X); \preceq_u^{\mathcal{K}})$ *such that*

(a) $F(\bar{x}) + q(x_0, \bar{x})k \preceq_u^{\mathcal{K}} F(x_0)$,
(b) $F(x) + q(\bar{x}, x)k \npreceq_u^{\mathcal{K}} F(\bar{x})$ *for all* $x \neq \bar{x}$.
Furthermore, assume that $F(\bar{x}) + k \npreceq_u^{\mathcal{K}} F(x_0)$ *for all* $k \in Y \setminus \{\mathbf{0}\}$ *and* $x_0 \in X$*, then*
(c) $q(x_0, \bar{x}) \leq 1$.

In [10], we also obtained the Ekeland type variational principle for set-valued maps involving variable order structure $\preceq_l^{\mathcal{K}}$.

Remark 6.15 Bao et al. [17] considered the main issues of Sen's capability theory [108, 142] and the variational rationality model of human behavior. They developed dynamical aspects of capability theory and discussed the major findings in this directions by applying the parametric fixed point theorem, parametric minimal element theorem, and Ekeland's variational principle. By using variational rationality [17, 18,

144, 145] technique, we can consider modeling the functionings/preferences dynamics in term of acceptable stays and changes, which mainly relates to the extended parametric fixed point theorem. Then we can find the functionings/preferences dynamics in term of worthwhile stays and changes, which relates to the obtained variational principle for maps with variable domination structures. Very recently, Bao et al. [20] also introduced a new version of Ekeland's variational principle in set optimization with domination structure and gave some applications to career development theories; in particular, changing the job process.

Acknowledgements In this research, the first author was supported by DST-SERB Project No. MTR/2017/000135 and the second author was supported by UGC-Dr. D.S. Kothari Post Doctoral Fellowship (DSKPDF) [F.4-2/2006 (BSR)/MA/19-20/0040]. All the authors acknowledge the constructive comments of the unknown referees which helped in bringing the chapter in the present form.

References

1. Alonso, M., Rodríguez-Marín, L.: Set-relations and optimality conditions in set-valued maps. Nonlinear Anal. **63**, 1167–1179 (2005)
2. Alonso, M., Rodríguez-Marín, L.: Optimality conditions for a nonconvex set-valued optimization problem. Comput. Math. Appl. **56**, 82–89 (2008)
3. Ansari, Q.H.: Metric Spaces—Including Fixed Point Theory and Set-Valued Maps. Narosa Publishing House, New Delhi (2010). Also Published by Alpha Science International Ltd. Oxford, U.K. (2010)
4. Ansari, Q.H.: Ekeland's variational principle and its extensions with applications. In: Almezel, S., Ansari, Q.H., Khamsi, M.A. (eds.) Topics in Fixed Point Theory, pp. 65–100. Springer, New York (2014)
5. Ansari, Q.H., Eshghinezhad, S., Fakhar, M.: Ekeland's variational principle for set-valued maps with applications to vector optimization in uniform spaces. Taiwan. J. Math. **18**(6), 1999–2020 (2014)
6. Ansari, Q.H., Hamel, A.H., Sharma, P.K.: Ekeland's variational principle with weighted set order relations. Math. Methods Oper. Res. **91**(1), 117–136 (2020)
7. Ansari, Q.H., Köbis, E., Sharma, P.K.: Characterizations of set relations with respect to variable domination structures via oriented distance function. Optimization **67**(9), 1389–1407 (2018)
8. Ansari, Q.H., Köbis, E., Sharma, P.K.: Characterizations of multiobjective robustness via oriented distance function and image space analysis. J. Optim. Theory Appl. **181**(3), 817–839 (2019)
9. Ansari, Q.H., Köbis, E., Yao, J.-C.: Vector Variational Inequalities and Vector Optimization - Theory and Applications. Springer, New York (2018)
10. Ansari, Q.H., Sharma, P.K.: Ekeland type variational principle for set-valued maps in quasimetric spaces with applications. J. Nonlinear Convex Anal. **20**(8), 1683–1700 (2019)
11. Ansari, Q.H., Sharma, P.K., Qin, X.: Characterizations of robust optimality conditions via image space analysis. Optimization **69**(9), 2063–2083 (2020)
12. Ansari, Q.H., Sharma, P.K., Yao, J.-C.: Minimal elements theorems and Ekeland's variational principle with new set order relations. J. Nonlinear Convex Anal. **19**(7), 1127–1139 (2018)
13. Araya, Y.: Four types of nonlinear scalarizations and some applications in set optimization. Nonlinear Anal. **75**, 3821–3835 (2012)

14. Araya, Y.: New types of nonlinear scalarizations in set optimization. In: Proceedings of the 3th Asian Conference on Nonlinear Analysis and Optimization, Matsue, Japan, pp. 1–15. Yakohama Publishers, Yakohama, Japan (2012)
15. Bao, T.Q., Mordukhovich, B.S.: Variational principles for set-valued maps with applications to multiobjective optimization. Control Cybern. **36**, 531–562 (2007)
16. Bao, T.Q., Mordukhovich, B.S.: Set-valued optimization in welfare economics. In: Kusuoka, S., Maruyama, T. (eds.) Advances in Mathematical Economics, vol. 13, pp. 113–153. Springer Japan, Tokyo (2010)
17. Bao, T.Q., Mordukhovich, B.S., Soubeyran, A.: Fixed points and variational principles with applications to capability theory of wellbeing via variational rationality. Set-Valued Var. Anal. **23**, 375–398 (2015)
18. Bao, T.Q., Mordukhovich, B.S., Soubeyran, A.: Variational analysis in psychological modeling. J. Optim. Theory Appl. **164**, 290–315 (2015)
19. Bao, T.Q., Mordukhovich, B.S., Soubeyran, A.: Minimal points, variational principles and variable preferences in set optimization. J. Nonlinear Convex Anal. **16**(8), 1511–1537 (2015)
20. Bao, T.Q., Soubeyran, A.: Variational principles in set optimization with domination structures and application to changing jobs. J. Appl. Numer. Optim. **1**(3), 217–241 (2019)
21. Brézis, B., Browder, F.E.: A general principle on ordered sets in nonlinear functional analysis. Adv. Math. **21**, 355–364 (1976)
22. Brink, C.: Power structures. Algebr. Univers. **30**, 177–216 (1993)
23. Caristi, J.: Fixed point theorems for mappings satisfying inwardness conditions. Trans. Am. Math. Soc. **215**, 241–251 (1976)
24. Caristi, J., Kirk, W.A.: Geometric fixed point theory and inwardness conditions. In: The Geometry of Metric and Linear Spaces. Lecture Notes in Mathematics, vol. 490, pp. 74–83. Springer, New York (1975)
25. Chen, J., Ansari, Q.H., Yao, J.-C.: Characterizations of set order relations and constrained set optimization problems via oriented distance function. Optimization **66**(11), 1741–1754 (2017)
26. Chen, G.Y., Huang, X.X.: Ekeland's ϵ-variational principle for set-valued mappings. Math. Methods Oper. Res. **48**(2), 181–186 (1998)
27. Chen, G.Y., Huang, X.X.: A unified approach to the existing three types of variational principles for vector-valued functions. Math. Methods Oper. Res. **48**(2), 349–357 (1998)
28. Chen, G.Y., Huang, X.X., Hou, S.H.: General Ekeland's variational principle for set-valued mappings. J. Optim. Theory Appl. **106**(1), 151–164 (2000)
29. Chen, G.Y., Jahn, J.: Optimality conditions for set-valued optimization problems. Math. Methods Oper. Res. **48**(2), 187–200 (1998)
30. Chen, J., Köbis, E., Köbis, M.A., Yao, J.-C.: A new set order relation in set-optimization. J. Nonlinear Convex Anal. **18**(4), 637–649 (2017)
31. Chiriaev, D., Walster, G.W.: Interval arithmetic specification. Available from: http://www.mscs.mu.edu/~globsol/walster-papers. html.[69] (1998)
32. Cobzaş, Ş.: Functional Analysis in Asymmetric Normed Spaces. Birkhäuser, Basel (2013)
33. Corley, H.W.: Existence and Lagrangian duality for maximization of set-valued functions. J. Optim. Theory Appl. **54**, 489–501 (1987)
34. Corley, H.W.: Optimality conditions for maximization of set-valued functions. J. Optim. Theory Appl. **58**(1), 1–10 (1988)
35. Crespi, G.P., Ginchev, I., Rocca, M.: First-order optimality conditions in set valued optimization. Math. Methods Oper. Res. **63**, 87–106 (2006)
36. Crespi, C.P., Mastrogiacomo, E.: Qualitative robustness of set-valued value-at-risk. Math. Methods Oper. Res. **91**, 25–54 (2020)
37. Dancs, S., Hegegedüs, M., Medvegyev, P.: A general ordering and fixed-point principle in complete metric space. Acta Sci. Math. **46**, 381–388 (1983)
38. Day, M.M.: Normed Linear Spaces. Springer, New York (1973)
39. De Figueiredo, D.G.: The Ekeland Variational Principle with Applications and Detours. Tata Institute of Fundamental Research, Bombay (1989)

40. Eichfelder, G.: Variable Ordering Structures in Vector Optimization. Springer, Heidelberg (2014)
41. Eichfelder, G., Jahn, J.: Vector and set optimization. In: Greco, S., Ehrgott, M., Figueira, J. (eds.) Multiple Criteria Decision Analysis: State of the Art Surveys, vol. 233, pp. 695–737. Springer, New York (2016)
42. Eichfelder, G., Pilecka, M.: Set approach for set optimization with variable ordering structures part I : set relations and relationship to vector approach. J. Optim. Theory Appl. **171**(3), 931–946 (2016)
43. Eichfelder, G., Pilecka, M.: Ordering structures and their applications. In: Rassias, T.M. (ed.) Applications of Nonlinear Analysis, vol. 134, pp. 265–304. Springer, Cham (2018)
44. Eichfelder, G., Pilecka, M.: Set approach for set optimization with variable ordering structures part II : scalarization approaches. J. Optim. Theory Appl. **171**(3), 947–963 (2016)
45. Ekeland, I.: Sur les problémes variationnels. C. R. Acad. Sci. Paris **275**, 1057–1059 (1972)
46. Ekeland, I.: On the variational principle. J. Math. Anal. Appl. **47**, 324–354 (1974)
47. Ekeland, I.: Nonconvex minimization problems. Bull. Am. Math. Soc. **1**(3), 443–474 (1979)
48. Fel'dman, M.M.: Sublinear operators defined on a cone. Sib. Math. J. **16**, 1005–1015 (1975)
49. Flores-Bazán, F., Gutiérrez, C., Novo, V.: A Brézis–Browder principle on partially ordered spaces and related ordering theorems. J. Math. Anal. Appl. **375**, 245–260 (2011)
50. Georgiev, P.G.: The strong Ekeland variational principle, the strong drop theorem and applications. J. Math. Anal. Appl. **131**, 1–21 (1988)
51. Georgiev, P.G., Tanaka, T.: Vector-valued set-valued variants of Ky Fan's inequality. J. Nonlinear Convex Anal. **1**, 245–254 (2000)
52. Gerth (Tammer), C.: Nichtkonvexe dualität in der vektoroptimierung (in German). Wiss. Z. TH Leuna-Merseburg **25**, 357–364 (1983)
53. Gerth (Tammer), C., Iwanow, I.: Dualität für nichtkonvexe vektoroptimierungs probleme (in German). Wiss. Z. Tech. Hochsch Ilmenau **2**, 61–81 (1985)
54. Gerth (Tammer), C., Weidner, P.: Nonconvex separation theorems and some applications in vector optimization. J. Optim. Theory Appl. **67**, 297–320 (1990)
55. Götz, A., Jahn, J.: The Lagrange multiplier rule in set-valued optimization. SIAM J. Optim. **10**(2), 331–344 (1999)
56. Göpfert, A., Riahi, A., Tammer, C., Zălinescu, C.: Variational Methods in Partially Ordered Spaces. Springer, New York (2003)
57. Göpfert, A., Tammer, C.: A new maximal point theorem. J. Anal. Appl. **14**(2), 379–390 (1995)
58. Göpfert, A., Tammer, C., Zălinescu, C.: A new minimal point theorem in product spaces. J. Anal. Appl. **18**(3), 767–770 (1999)
59. Göpfert, A., Tammer, C., Zălinescu, C.: On the vectorial Ekeland's Variational principle and minimal points in product spaces. Nonlinear Anal. **39**, 909–922 (2000)
60. Gutiérrez, C., Jiménez, B., Miglierina, E., Molho, E.: Scalarization in set optimization with solid and nonsolid ordering cones. J. Glob. Optim. **61**, 525–552 (2015)
61. Gutiérrez, C., Jiménez, B., Novo, V., Thibault, L.: Strict approximate solutions in set-valued optimization with applications to the approximate Ekeland variational principle. Nonlinear Anal. **73**, 3842–3855 (2010)
62. Gutiérrez, C., Miglierina, E., Molho, E., Novo, V.: Pointwise well-posedness in set optimization with cone proper sets. Nonlinear Anal. **75**, 1822–1833 (2012)
63. Ha, T.X.D.: Some variants of the Ekeland's variational principle for a set valued map. J. Optim. Theory Appl. **124**(1), 187–206 (2005)
64. Ha, T.X.D.: Optimality conditions for several types of efficient solutions of set-valued optimization problems. In: Pardalos, P., Rassias, T.M, Khan, A.A. (eds.) Nonlinear Analysis and Variational Problems, pp. 305–324. Springer, Heidelberg (2010)
65. Hamel, A.H.: Equivalents to Ekeland's variational principle in uniform spaces. Nonlinear Anal. **62**(5), 913–924 (2005)
66. Hamel, A.H.: Translative sets and functions and their applications to risk measure theory and nonlinear separation. IMPA preprint D 21/2006 (2006)

67. Hamel, A.H., Heyde, F., Löhne, A., Rudloff, B., Schrage, C.: Set optimization–a rather short introduction. In: Hamel, A.H., Heyde, F., Löhne, A., Rudloff, B., Schrage, C. (eds.) Set Optimization and Applications – The State of the Art, From Set Relations to Set-valued Risk Measures, pp. 65–141. Springer (2015)
68. Hamel, A.H., Heyde, F., Löhne, A., Rudloff, B., Schrage, C.: Set Optimization and Applications - The State of the Art: From Set Relations to Set-valued Risk Measures. Springer, Berlin (2015)
69. Hamel, A.H., Heyde, F., Rudloff, B.: Set-valued risk measures conical market models. Math. Financ. Econ. **5**(1), 1–28 (2010)
70. Hamel, A.H., Kostner, D.: Cone distribution functions and quantiles for multivariate random variable. J. Multivar. Anal. **167**, 97–113 (2018)
71. Hamel, A.H., Löhne, A.: A minimal point theorem in uniform spaces. In: Agarwal, R.P., O'Regan, D. (eds.) Nonlinear Analysis and Applications: To V. Lakshmikantham on His 80th Birthday, vol. 1, pp. 577–593. Kluwer Academic Publisher, Dordrecht (2003)
72. Hamel, A.H., Löhne, A.: Minimal elements theorems and Ekeland's variational principle with set relations. J. Nonlinear Convex Anal. **7**, 19–37 (2006)
73. Hamel, A.H., Löhne, A.: A set optimization approach to zero-sum matrix games with multi-dimensional payoffs. Math. Methods Oper. Res. **88**, 369–397 (2018)
74. Hamel, A.H., Tammer, C.: Minimal elements for product orders. Optimization **57**(2), 263–275 (2008)
75. Hamel, A.H., Visetti, D.: The value functions approach and Hopf-Lax formula for multiobjective costs via set optimization. J. Math. Anal. Appl. **483**(1), Article 123605 (2020)
76. Hamel, A.H., Zălinescu, C.: Minimal elements theorem revisited. J. Math. Anal. Appl. **486**(2), Article 123935?? (2020)
77. Han, Y., Huang, N.J.: Well-posedness and stability of solutions for set optimization problems. Optimization **66**(1), 17–33 (2017)
78. Han, Y., Wang, S.H., Huang, N.J.: Arcwise connectedness of the solution sets for set optimization problems. Oper. Res. Lett. **47**, 168–172 (2019)
79. Hernández, E.: A survey of set optimization problems with set solutions. In: Hamel, A.H., Heyde, F., Löhne, A., Rudloff, B., Schrage, C. (eds.) Set Optimization and Applications – The State of the Art. From Set Relations to Set-valued Risk Measures, pp. 142–158. Springer (2015)
80. Hernández, E., Rodríguez-Marín, L.: Nonconvex scalarization in set optimization with set-valued maps. J. Math. Anal. Appl. **325**, 1–18 (2007)
81. Hernández, E., Rodríguez-Marín, L.: Existence theorems for set optimization problems. Nonlinear Anal. **67**, 1726–1736 (2007)
82. Hernández, E., Rodríguez-Marín, L.: Lagrangian duality in set-valued optimization. J. Optim. Theory Appl. **134**, 119–134 (2007)
83. Hernández, E., Rodríguez-Marín, L.: Weak and strong subgradients of set-valued maps. J. Optim. Theory Appl. **149**, 352–365 (2011)
84. Hernández, E., Rodríguez-Marín, L., Sama, M.: On solutions of set-valued optimization problems. Comput. Math. Appl. **60**, 1401–1408 (2010)
85. Heyde, F.: Coherent risk measures and vector optimization. In: Küfer, K.-H. et al. (eds.) Multicriteria Decision Making and Fuzzy Systems. Theory, Methods and Applications. Shaker Verlag, Aachen (2006)
86. Hiriart-Urruty, J.-B.: Tangent cones, generalized gradients and mathematical programming in Banach spaces. Math. Oper. Res. **4**(1), 79–97 (1979)
87. Huang, X.X.: A new variant of Ekeland's variational principle for set-valued maps. Optimization **52**(1), 53–63 (2003)
88. Ide, J., Köbis, E.: Concepts of efficiency for uncertain multi-objective optimization problems based on set order relations. Math. Methods Oper. Res. **80**(1), 99–127 (2014)
89. Isac, G.: The Ekeland's principle and the Pareto ϵ-efficiency. In: Tamiz, M. (ed.) Multi-Objective Programming and Goal Programming, Theory and Applications, vol. 432. Springer (1996)

90. Jahn, J.: Vector Optimization: Theory Applications and Extensions. Springer, Berlin (2004)
91. Jahn, J.: Vectorization in set optimization. J. Optim. Theory Appl. **167**, 783–795 (2015)
92. Jahn, J., Ha, T.X.D.: New order relations in set optimization. J. Optim. Theory Appl. **148**, 209–236 (2011)
93. Jahn, J., Rauh, R.: Contingent epiderivatives and set-valued optimization. Math. Methods Oper. Res. **46**(2), 193–211 (1997)
94. Karaman, E., Güvenç, İ.A., Soyertem, M.: Optimality conditions in set-valued optimization problems with respect to a partial order relation by using subdifferentials. Optimization (2020). https://doi.org/10.1080/02331934.2020.1728270
95. Karaman, E., Soyertem, M., Güvenç, İ.A.: Optimality conditions in set-valued optimization problem with respect to a partial order relation via directional derivative. Taiwan. J. Math. **24**(3), 709–722 (2020)
96. Karaman, E., Soyertem, M., Güvenç, İ.A., Tozkan, D., Küçük, M., Küçük, Y.: Partial order relations on family of sets and scalarizations for set optimization. Positivity **22**(3), 783–802 (2018)
97. Khan, A.A., Tammer, C., Zălinescu, C.: Set-Valued Optimization - An Introduction with Applications. Springer, Berlin (2015)
98. Khoshkhabar-Amiranloo, S., Khorram, E., Soleimani-Damaneh, M.: Nonlinear scalarization functions and polar cone in set optimization. Optim. Lett. **11**, 521–535 (2017)
99. Khushboo, Lalitha, C.: Scalarizations for a set optimization problem using generalized oriented distance function. Positivity **23**(5), 1195–1213 (2019)
100. Klamroth, K., Köbis, E., Schöbel, A., Tammer, C.: A unified approach for different concepts of robustness and stochastic programming via nonlinear scalarizing functionals. Optimization **62**(5), 649–671 (2013)
101. Köbis, E.: Variable ordering structures in set optimization. J. Nonlinear Convex Anal. **18**, 1571–1589 (2017)
102. Köbis, E.: Set optimization by means of variable order relations. Optimization **66**, 1991–2005 (2017)
103. Köbis, E., Köbis, M.A.: Treatment of set order relations by means of a nonlinear scalarization functional: a full characterization. Optimization **65**(10), 1805–1827 (2016)
104. Köbis, E., Le, T.T., Tammer, C.: A generalized scalarization method in set optimization with respect to variable domination structure. Vietnam J. Math. **46**, 95–125 (2018)
105. Köbis, E., Le, T.T., Tammer, C., Yao, J.-C.: A new scalarizing functional in set optimization with respect to variable domination structures. Appl. Anal. Optim. **1**(2), 301–326 (2017)
106. Kostner, D.: Multi-criteria decision making via multivariate quantiles. Math. Methods Oper. Res. **91**, 73–88 (2020)
107. Krasnosel'skij, M.A.: Positive Solutions of Operator Equations. P. Noordhoff Ltd., Groningen (1964)
108. Kuklys, W.: Amartya Sen's Capability Approach: Theoretical Insights and Empirical Applications. Springer, Berlin (2005)
109. Kuroiwa, D.: Some criteria in set-valued optimization. Sūrikaisekikenkyūsho Kōkyūroku **985**, 171–176 (1997)
110. Kuroiwa, D.: Lagrange duality of set-valued optimization with natural criteria. Sūrikaisekikenkyūsho Kōkyūroku 1068 (1998)
111. Kuroiwa, D.: The natural criteria in set-valued optimization. Sūrikaisekikenkyūsho Kōkyūroku **1031**, 85–90 (1998)
112. Kuroiwa, D.: On set-valued optimization. Nonlinear Anal. **47**(2), 1395–1400 (2001)
113. Kuroiwa, D.: Existence theorems of set optimization with set-valued maps. J. Inf. Optim. Sci. **24**(1), 73–84 (2003)
114. Kuroiwa, D., Nuriya, T.: A generalized embedding vector space in set optimization. In: Proceedings of the Fourth International Conference on Nonlinear Analysis and Convex Analysis, pp. 297–304. Yakohama Publishers, Yakohama (2006)
115. Kuroiwa, D., Tanaka, T., Ha, T.X.D.: On cone convexity of set-valued maps. Nonlinear Anal. **30**, 1487–1496 (1997)

116. Kuwano, I.: Some minimax theorems for set-valued maps and their applications. Nonlinear Anal. **109**, 85–102 (2014)
117. Kuwano, I., Tanaka, T.: Continuity of cone-convex functions. Optim. Lett. **6**, 1847–1853 (2012)
118. Kuwano, I., Tanaka, T., Yamada, S.: Characterization of nonlinear scalarizing functions for set valued maps. In: Takahashi, W., Tanaka, T. (eds.) Nonlinear Analysis and Convex Analysis, pp. 193–204. Yokohama Publishers, Yokohama (2009)
119. Le, T.T.: Set optimization with respect to variable domination structure. Ph.D. thesis, Martin-Luther-University Halle-Wittenberg (2018)
120. Le, T.T.: Multiobjective approaches based on variable ordering structures for intensity problems in radiotherapy treatment. Rev. Investig. Oper. **39**(3), 426–448 (2018)
121. Lee, I.-K., Kim, M.-S., Elber, E.: Polynomial/rational approximation of Minkowski sum boundary curves. Graph. Model. Image Process. **60**(2), 136–165 (1998)
122. Li, J.: The optimality conditions for vector optimization of set-valued maps. J. Math. Anal. Appl. **237**, 413–424 (1999)
123. Lin, L.-J.: Optimization of set-valued functions. J. Math. Anal. Appl. **186**, 30–51 (1994)
124. Lozano-Pérez, T.: Spatial planning: a configuration space approach. IEEE Trans. Comput. **32**(2), 108–120 (1983)
125. Luc, D.T.: On scalarizing method in vector optimization. In: Fandel, G., Grauer, M., Kurzhanski, A., Wierzbicki, A.P. (eds.) Large-Scale Modelling and Interactive Decision Analysis. Lecture Notes in Economics and Mathematical Systems, vol 273. Springer, Berlin (1986)
126. Luc, D.T.: Scalarization of vector optimization problems. J. Optim. Theory Appl. **55**, 85–102 (1987)
127. Luc, D.T.: Theory of Vector Optimization. Springer, Berlin (1989)
128. Luc, D.T.: Contingent derivative of set-valued maps and applications to vector optimization. Math. Program. Ser. A **50**(1), 99–111 (1991)
129. Maeda, T.: On optimization problems with set-valued objective maps. Appl. Math. Comput. **217**, 1150–1157 (2010)
130. Maeda, T.: On optimization problems with set-valued objective maps: existence and optimality. J. Optim. Theory Appl. **153**, 263–279 (2012)
131. Németh, A.B.: A nonconvex vector minimization problem. Nonlinear Anal. **10**(7), 669–678 (1986)
132. Neukel, N.: Order relations of sets and its application in socio-economics. Appl. Math. Sci. **7**, 5711–5739 (2013)
133. Neukel, N.: Order relations for the cryptanalysis of substitution ciphers on the basis of linguistic data structures as an optimal strategy (2019). http://www.m-hikari.com/fbooks.html
134. Nishnianidze, M.N.: Fixed points of monotone multivalued operators. Bull. Georg. Acad. Sci. **114**(3), 489–491 (1984)
135. Nishizawa, S., Onodsuka, M., Tanaka, T.: Alternative theorems for set-valued maps based on a nonlinear scalarization. Pac. J. Optim. **1**(1), 147–159 (2005)
136. Pallaschke, D., Urbañski, R.: Pairs of Compact Convex Sets. Kluwer Academic Publishers, Dordrecht (2002)
137. Phelps, R.R.: Convex Functions, Monotone Operators and Differentiability. Springer, Berlin (1989)
138. Phelps, R.R.: Convex Functions, Monotone Operators and Differentiability, 2nd edn. Springer, Berlin (1993)
139. Preechasilp, P., Wangkeeree, R.: A note on semicontinuity of the solution mapping for parametric set optimization problems. Optim. Lett. **13**, 1085–1094 (2019)
140. Rubinov, A.M.: Sublinear operators and their applications. Russ. Math. Surv. **32**(4), 115–175 (1977)
141. Sach, P.H.: New nonlinear scalarization functions and applications. Nonlinear Anal. **75**(4), 2281–2292 (2012)
142. Sen, A.: Commodities and Capabilities. Oxford University Press, Oxford (1985)

143. Serra, J. (ed.): Image Analysis and Mathematical Morphology. Academic Press, London (1982)
144. Soubeyran, A.: Variational rationality, a theory of individual stability and change, worthwhile and ambidextry behaviors. Preprint, GREQAM, Aix-Marseille University (2009)
145. Soubeyran, A.: Variational rationality and the unsatisfied man: routines and the course pursuit between aspirations, capabilities and beliefs. Preprint, GREQAM, Aix-Marseille University (2010)
146. Sun Microsystems: Interval Arithmetic Programming Reference. Palo Alto, USA (2000)
147. Takahashi, W.: Existence theorems generalizing fixed point theorems for multivalued mappings. In: Baillon, J.-B., Théra, M. (eds.) Fixed Point Theory and Applications. Pitman Research Notes in Mathematics, vol. 252, pp. 397–406. Longman, Harlow (1991)
148. Tammer, C.: A generalization of Ekeland's variational principle. Optimization **25**(2), 129–141 (1992)
149. Tammer, C., Zălinescu, C.: Vector variational principles for set-valued functions. Optimization **60**, 839–857 (2011)
150. Young, R.C.: The algebra of many-valued quantities. Math. Ann. **104**(1), 260–290 (1931)
151. Xu, Y.D., Li, S.J.: A new nonlinear scalarization function and applications. Optimization **65**(1), 207–231 (2016)
152. Zaffaroni, A.: Degrees of efficiency and degrees of minimality. SIAM J. Control Optim. **42**(3), 1071–1086 (2003)
153. Zhang, C., Huang, N.: Set optimization problems of generalized semi-continuous set-valued maps with applications. Positivity **25**, 353–367 (2021)

Chapter 7
Characterizations and Generating Efficient Solutions to Interval Optimization Problems

Amit Kumar Debnath and Debdas Ghosh

Abstract In this study, we propose a method to obtain complete efficient solution sets of the optimization problems with interval-valued functions. The proposed method is based on the cone method for multiobjective optimization problems. Toward developing the method, a bi-objective characterization of efficient solutions to the problem under consideration is reported. In addition, we provide a saddle point characterization of efficient solutions to the problem with the help of a newly defined Lagrangian function. Finally, we provide an algorithmic implementation of the proposed method and support it with two numerical examples.

Keywords Interval optimization · Efficient solutions · Saddle point · Cone method

Abbreviations

IVF	Interval-valued function
IOP	Interval optimization problem
ES	Efficient solution
NS	Nondominated solution
POS	Pareto optimal solution
SP	Saddle point

A. K. Debnath · D. Ghosh (✉)
Department of Mathematical Sciences, Indian Institute of Technology (BHU), Varanasi 221005, India
e-mail: debdas.mat@iitbhu.ac.in

A. K. Debnath
e-mail: amitkdebnath.rs.mat18@itbhu.ac.in

V. Laha et al. (eds.), *Optimization, Variational Analysis and Applications*,
Springer Proceedings in Mathematics & Statistics 355,
https://doi.org/10.1007/978-981-16-1819-2_7

7.1 Introduction

In the present realistic decision-making field, optimization problems with vague parameters, such as fuzzy-valued and interval-valued parameters, play vital roles due to the inevitable presence of uncertainty and imprecision in real-life phenomena. The optimization problems with interval-valued functions (IVFs) are known as Interval Optimization Problems (IOPs). Generally, the coefficients of IVFs of IOPs are considered as compact intervals. Due to the practical importance of IOP, it has been a substantial topic of research for the past two decades.

7.1.1 Literature Survey

In [18], Moore proposed interval arithmetic to deal with intervals and IVFs. After that, Ishibuchi and Tanaka [13] developed a few ordering relations for intervals. It is to mention that the interval arithmetic, proposed in [18], has a few limitations, such as it cannot provide the additive inverse of a nondegenerate interval, i.e., an interval whose boundary points are different. Initially, this difficulty was fixed by applying a rule of the difference of intervals, known as H-difference [12]. However, it is found that the H-difference is quite restrictive [9]. Finally, the difficulty of getting additive inverse for any interval has been resolved by the development of a new rule of the difference of intervals, known as gH-difference of intervals [19]. Very recently, Ghosh et al. [8] investigated variable ordering relations for intervals.

Based on the aforesaid interval arithmetic and ordering relations, many researchers, see [1–3, 5, 9–11, 15, 16, 20, 21], to mention a few have proposed various results and methods for characterizing and obtaining the solutions to IOPs. In [3], the concepts of continuity and differentiability have been discussed for the one variable IVFs. Chalco-Cano et al. [2] and Ghosh [5] illustrated calculus and algebras of gH-differentiable IVFs. The concepts of fractional derivatives for one-dimensional IVFs have been developed by Lupulescu [16]. In [9], the concepts of Gâteaux and Fréchet derivatives for IVFs have been developed; in addition, the optimality conditions for an IOP whose objective function is Gâteaux differentiable has also been found in [9]. Ghosh et al. [10], further, characterized the solutions to IOPs by parametric representations of IVFs. For general nonlinear IOPs, Wu [20] and Chalco-Cano et al. [1] presented the Karush–Kuhn–Tucker (KKT) optimality conditions. A generalized KKT condition to obtain the solution to IOPs has been reported in [11]. In [21], some duality theorems for IOPs in weak and strong senses have been provided. In [15], a numerical technique to solve a quadratic IOP has been presented. In order to obtain a solution to an IOP, Ghosh [5] introduced a Newton method and a quasi-Newton method [6] for IOPs.

7.1.2 Motivation and Contribution

From the literature on IOPs, it is observed that an IOP generally has an infinite number of solutions; for instance, see Example 7.1 of this chapter. Although there are many theories and techniques, please see [1, 4, 5, 9, 10, 14–16, 21] for details to obtain the solutions to IOPs, but it is found that none of the techniques endeavor to generate the complete solution set of an IOP.

In this chapter, we develop a technique to generate the complete solution set of an IOP and illustrate its algorithmic implementation. In order to develop the technique, we characterize the solutions to IOPs in the light of conventional bi-objective optimization problems. In addition, we study a saddle point criterion to find a condition by which the saddle point of an IOP will be its efficient solution and vice versa.

7.1.3 Organization

The proposed work is delineated in the following way. The next section briefly presents interval arithmetic, dominance relations of intervals, and the concept of IVFs. The concepts of IOPs and their efficient solutions are also included in Sect. 7.2. Further, two types of characterizations—bi-objective characterization and saddle point characterization—are illustrated in Sect. 7.3. In Sect. 7.4, a technique to capture the entire efficient solution set is illustrated, and its algorithmic implementation is provided. Finally, in Sect. 7.5, a brief conclusion and a few future scopes of this study are provided.

7.2 Preliminaries

In this section, at first, we illustrate the arithmetic of intervals and the concept of dominance relation of intervals that are used throughout the chapter. Thereafter, we study IVFs and the concept of optimal solutions for IOPs.

7.2.1 Interval Arithmetic

Let $I(\mathbb{R})$ be the set of all compact intervals. Bold capital letters are used to represent the elements of $I(\mathbb{R})$. Also, an element $\mathbf{A}$ of $I(\mathbb{R})$ is represented by its corresponding small letter in the way $\mathbf{A} = \left[\underline{a}, \overline{a}\right]$.

For any two elements $\mathbf{A} = [\underline{a}, \overline{a}]$ and $\mathbf{B} = \left[\underline{b}, \overline{b}\right]$ in $I(\mathbb{R})$, the *addition* of $\mathbf{A}$ and $\mathbf{B}$ is defined by

$$\mathbf{A} \oplus \mathbf{B} = \left[\underline{a} + \underline{b}, \overline{a} + \overline{b}\right],$$

and *substraction* of **B** from **A** is defined by

$$\mathbf{A} \ominus \mathbf{B} = \left[\underline{a} - \overline{b}, \overline{a} - \underline{b}\right].$$

The *multiplication* of **A** by a real number γ is defined by

$$\gamma \odot \mathbf{A} = \begin{cases} [\gamma\underline{a}, \gamma\overline{a}] & \text{if } \gamma \geq 0 \\ [\gamma\overline{a}, \gamma\underline{a}] & \text{if } \gamma < 0. \end{cases}$$

Definition 7.1 (*gH-difference of intervals* [19]). The *gH-difference* of an interval **B** from an interval **A** is defined by

$$\mathbf{A} \ominus_{gH} \mathbf{B} = \left[\min\left\{\underline{a} - \underline{b}, \overline{a} - \overline{b}\right\}, \max\left\{\underline{a} - \underline{b}, \overline{a} - \overline{b}\right\}\right].$$

Definition 7.2 (*Dominance relations of intervals* [20]). Let **A**, $\mathbf{B} \in I(\mathbb{R})$. If $\mathbf{A} \preceq \mathbf{B}$, i.e., $\underline{a} \leq \underline{b}$ and $\overline{a} \leq \overline{b}$, then **B** is called *dominated* by **A**.

If $\mathbf{A} \prec \mathbf{B}$, i.e., either '$\underline{a} \leq \underline{b}$ and $\overline{a} < \overline{b}$' or '$\underline{a} < \underline{b}$ and $\overline{a} \leq \overline{b}$', then **B** is called strictly dominated by **A**.

7.2.2 Interval-Valued Functions and Interval Optimization Problems

Let $\mathcal{X}$ be a nonempty subset of $\mathbb{R}^n$. For each $x \in \mathcal{X}$, an IVF $\mathbf{F} : \mathcal{X} \to I(\mathbb{R})$ is presented by

$$\mathbf{F}(x) = \left[\underline{f}(x), \overline{f}(x)\right],$$

where $\underline{f}$ and $\overline{f}$ are real-valued functions on $\mathcal{X}$ such that

$$\underline{f}(x) \leq \overline{f}(x) \;\forall\, x \in \mathcal{X}.$$

Unless mentioned otherwise, throughout the chapter, we consider $\mathcal{X}$ as a nonempty subset of $\mathbb{R}^n$.

Definition 7.3 (*Convex IVF* [20]). Let $\mathcal{X}$ be convex. An IVF $\mathbf{F} : \mathcal{X} \to I(\mathbb{R})$ is called a *convex IVF* on $\mathcal{X}$ if for any two elements x_1 and x_2 in $\mathcal{X}$,

$$\mathbf{F}(\gamma x_1 + (1-\gamma)x_2) \preceq \gamma \odot \mathbf{F}(x_1) \oplus (1-\gamma) \odot \mathbf{F}(x_2) \;\forall\, \gamma \in [0, 1].$$

Remark 7.1 (*See* [20]). Let $\mathcal{X}$ be convex. An IVF $\mathbf{F}$ is convex on $\mathcal{X}$ if and only if $\underline{f}$ and $\overline{f}$ are convex on $\mathcal{X}$.

Remark 7.2 Let an IVF $\mathbf{F}$ be convex on a convex set $\mathcal{X}$. Then, for any nonzero $\mu = (\mu_1, \mu_2) \in [0, \infty)^2$, the function $\mu_1 \underline{f} + \mu_2 \overline{f}$ is convex on $\mathcal{X}$.

In the literature of interval optimization, an IOP is defined as follows:

$$\text{(IOP)} \qquad \begin{cases} \min & \mathbf{F}(x) \\ \text{subject to} & \mathbf{G}_i(x) \preceq \mathbf{0}, \ i \in \mathcal{I} = \{1, 2, \ldots, p\} \\ & x \in \mathcal{X}, \end{cases} \tag{7.1}$$

where $\mathbf{F} : \mathcal{X} \to I(\mathbb{R})$ and $\mathbf{G}_i : \mathcal{X} \to I(\mathbb{R})$ are IVFs for each $i \in \mathcal{I}$, and $\mathbf{0} = [0, 0]$. Throughout the chapter, for each $x \in \mathcal{X}$, we present $\mathbf{F}(x)$ and $\mathbf{G}_i(x)$, $i \in \mathcal{I}$, by

$$\mathbf{F}(x) = \left[\underline{f}(x), \overline{f}(x)\right] \text{ and } \mathbf{G}_i(x) = \left[\underline{g}_i(x), \overline{g}_i(x)\right], \text{ respectively.}$$

Denoting $\mathcal{S} = \left\{x \in \mathcal{X} \mid \mathbf{G}_i(x) \preceq \mathbf{0} \ \forall \, i \in \mathcal{I}\right\}$, we can precisely present the IOP (7.1) by

$$\min_{x \in \mathcal{S}} \mathbf{F}(x).$$

Since for each $i \in \mathcal{I}$,

$$\overline{g}_i(x) \leq 0 \Longleftrightarrow \mathbf{G}_i(x) = \left[\underline{g}_i(x), \overline{g}_i(x)\right] \preceq \mathbf{0} \ \forall \, x \in \mathcal{X},$$

the constraint set $\mathcal{S}$ of (7.1) can be presented by

$$\mathcal{S} = \left\{x \in \mathcal{X} \mid \overline{g}_i(x) \leq 0 \ \forall \, i \in \mathcal{I}\right\}.$$

Definition 7.4 (*Efficient solution (ES)* [21]). An $\bar{x} \in \mathcal{S}$ is known as an ES to the IOP (7.1) if $\nexists$ any $x \neq \bar{x} \in \mathcal{S}$ such that $\mathbf{F}(x) \prec \mathbf{F}(\bar{x})$.

Remark 7.3 A point $\bar{x} \in \mathcal{S}$ is an ES to the IOP (7.1) if and only if $\nexists$ any $x \neq \bar{x} \in \mathcal{S}$ such that $\left[\underline{f}(x), \overline{f}(x)\right] \prec \left[\underline{f}(\bar{x}), \overline{f}(\bar{x})\right]$, i.e., $\nexists$ any $x \neq \bar{x} \in \mathcal{S}$ such that the following relation holds:

$$\begin{aligned} \rho \Longleftrightarrow & \text{ either `}\underline{f}(x) \leq \underline{f}(\bar{x}) \text{ and } \overline{f}(x) < \overline{f}(\bar{x})\text{'} \\ & \text{ or `}\underline{f}(x) < \underline{f}(\bar{x}) \text{ and } \overline{f}(x) \leq \overline{f}(\bar{x})\text{'}. \end{aligned} \tag{7.2}$$

Definition 7.5 (*Nondominated solutions (NS) to IOP* [21]). If $\bar{x}$ in $\mathcal{S}$ is an ES to the IOP (7.1), then $\mathbf{F}(\bar{x})$ is known as a NS to the IOP (7.1).

Note 7.1 It is to note here that Wu [21] named the ES of this article as NS and NS of this article as nondominated objective value. However, throughout this article, we follow Definition 7.4 for ES and Definition 7.5 for NS to the IOP (7.1).

7.3 Characterizations of Efficient Solutions

This section provides two characterizations—bi-objective and saddle point—of the ESs to the IOP (7.1).

7.3.1 Bi-objective Characterization

In this subsection, we characterize the ESs to the IOP (7.1) with the help of the following conventional bi-objective optimization problem:

$$\text{(BOP)}\ \min_{x \in \mathcal{S}} f(x),\ \text{where}\ f(x) = \left(\underline{f}(x), \overline{f}(x)\right). \tag{7.3}$$

Definition 7.6 (*Pareto optimal solution (POS)* [7]). A point $\bar{x} \in \mathcal{S}$ is called a POS to the BOP (7.3) if $\nexists$ any $x \neq \bar{x} \in \mathcal{S}$ such that the relation ρ, defined in (7.2), holds.

Definition 7.7 (*NS to BOP* [7]). If an $\bar{x} \in \mathcal{S}$ is a POS to the BOP (7.3), then the corresponding vector $\left(\underline{f}(\bar{x}), \overline{f}(\bar{x})\right)$ is called a NS to the BOP (7.3).

Theorem 7.3.1 (Bi-objective characterization of ESs). *A point $\bar{x} \in \mathcal{S}$ is an ES to the IOP (7.1) if and only if $\bar{x}$ is a POS to the BOP (7.3).*

Proof The proof is obvious in the light of Remark 7.3 and Definition 7.6. □

Corollary 7.1 *Consider the IOP (7.1). Let $\mathcal{X}$ be convex, and $\mathbf{F}$ and $\mathbf{G}_i$ for all $i \in \mathcal{I}$ be convex IVFs on $\mathcal{X}$. Then, a point $\bar{x} \in \mathcal{S}$ is an ES to the IOP (7.1) if and only if $\exists$ a nonzero $\mu = (\mu_1, \mu_2) \in [0, \infty)^2$ such that $\bar{x}$ is an optimal solution to the following weighted optimization problem:*

$$\min_{x \in \mathcal{S}} \left(\mu_1 \underline{f}(x) + \mu_2 \overline{f}(x)\right).$$

Corollary 7.2 *If $\mathbf{F}(\bar{x}) = \left[\underline{f}(\bar{x}), \overline{f}(\bar{x})\right]$ is a NS to the IOP (7.1), then $\left(\underline{f}(\bar{x}), \overline{f}(\bar{x})\right)$ of $\mathbf{F}(\bar{x})$ is a NS to the BOP (7.3) and vice versa.*

Example 7.1 Consider the following IVF

$$\mathbf{F}(x) = [1, 1]x^5 \ominus [8, 8]x^4 \oplus [21, 22]x^3 \ominus [16, 18]x^2 \oplus [4, 5]$$

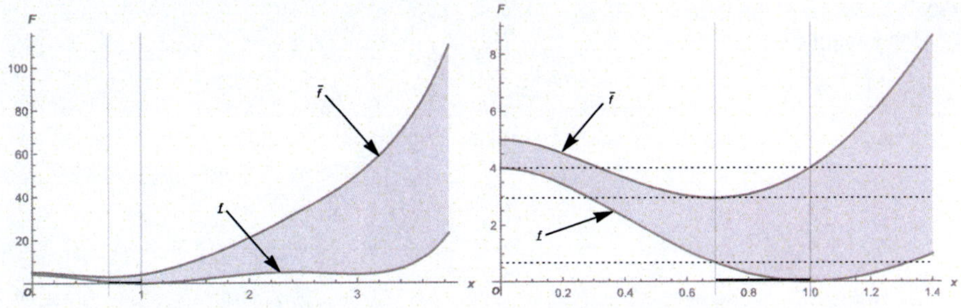

(a) The ES set of the IOP (7.4) (b) A closer view of the ES set of the IOP (7.4)

Fig. 7.1 The ES set with its closer view of the IOP (7.4) in Example 7.1

and the IOP

$$\min_{x\in[0,4]} \mathbf{F}(x). \tag{7.4}$$

The graph of the objective function

$$\mathbf{F}(x) = \left[\underline{f}(x), \overline{f}(x)\right] = \left[x^5 - 8x^4 + 21x^3 - 18x^2 + 4, x^5 - 8x^4 + 22x^3 - 16x^2 + 5\right]$$

of the IOP (7.4) is presented in Fig. 7.1 by the gray region. From Fig. 7.1a, b, it is evident that the ES set of the IOP (7.4) is the interval [0.692, 1], i.e, for each $\bar{x} \in [0.692, 1]$, the interval $\mathbf{F}(\bar{x})$ is a NS to the IOP (7.4). The ES set of the IOP (7.4) is presented in each of Fig. 7.1a, b by the bold black line segment on the x-axis.

Corresponding to the IOP (7.4), consider the following BOP:

$$\min_{x\in[0,4]} \left(\underline{f}(x), \overline{f}(x)\right), \tag{7.5}$$

i.e.,

$$\min_{x\in[0,4]} \left(x^5 - 8x^4 + 21x^3 - 18x^2 + 4, x^5 - 8x^4 + 22x^3 - 16x^2 + 5\right).$$

The objective space and the set of NSs to the BOP (7.5) are illustrated in Fig. 7.2 by the gray curve and the bold black arc, respectively. From Fig. 7.1b and Fig. 7.2, it is evident that, for each $\bar{x} \in [0.692, 1]$, the point $\left(\underline{f}(\bar{x}), \overline{f}(\bar{x})\right)$ is a NS to the BOP (7.5).

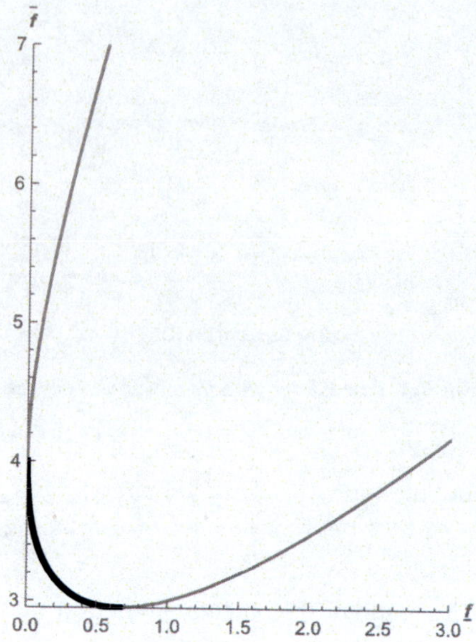

Fig. 7.2 NS set of the BOP (7.5) in Example 7.1

7.3.2 *Saddle Point Characterization*

In this subsection, we define a *saddle point* of the IOP (7.1) and characterize its ESs with the help of the following *Lagrangian function*, which is real-valued. For a given $\mu = (\mu_1, \mu_2) \in [0, \infty)^2$, a *Lagrangian function* corresponding to the IOP (7.1) is defined by

$$L_\mu(x, \vartheta) = \mu_1 \underline{f}(x) + \mu_2 \overline{f}(x) + \sum_{i=1}^{p} \vartheta_i \overline{g}_i(x), \tag{7.6}$$

where $\vartheta = \left(\vartheta_1, \vartheta_2, \ldots \vartheta_p\right)$ is a nonzero element of $[0, \infty)^p$.

Definition 7.8 (*Saddle point (SP)*). A point $\bar{x} \in \mathcal{S}$ is said to be a SP of the IOP (7.1) if for a given $\mu = (\mu_1, \mu_2) \in [0, \infty)^2$, $\exists$ a $\bar{\vartheta} = \left(\bar{\vartheta}_1, \bar{\vartheta}_2, \ldots \bar{\vartheta}_p\right)$, such that $\left(\bar{x}, \bar{\vartheta}\right)$ satisfies

$$L_\mu(\bar{x}, \vartheta) \leq L_\mu(\bar{x}, \bar{\vartheta}) \leq L_\mu(x, \bar{\vartheta}) \; \forall \, x \in \mathcal{S} \text{ and } \vartheta \in [0, \infty)^p. \tag{7.7}$$

We say that the IOP (7.1) satisfies the *Slater constraint qualification* if $\exists$ an x' in $\mathcal{X}$ such that

$$\mathbf{G}_i(x') \prec \mathbf{0} \; \forall \, i \in \mathcal{I}. \tag{7.8}$$

Theorem 7.3.2 (SP characterization of ESs). *Consider the IOP (7.1). Let $\mathcal{X}$ be convex and the IVFs $\mathbf{F}$ and $\mathbf{G}_i$ for all $i \in \mathcal{I}$ be convex on $\mathcal{X}$. Further, suppose that the IOP (7.1) satisfies the Slater constraint qualification (7.8). Then, $\bar{x} \in \mathcal{X}$ is an ES* *to the IOP (7.1) if and only if* $\exists$ *a* $\bar{\vartheta} = \left(\bar{\vartheta}_1, \bar{\vartheta}_2, \ldots \bar{\vartheta}_p\right) \in [0, \infty)^p$, *such that*

$$\sum_{i=1}^{p} \bar{\vartheta}_i \overline{g}_i(\bar{x}) = 0 \tag{7.9}$$

and $(\bar{x}, \bar{\vartheta})$ *satisfies the SP criterion (7.7) for a nonzero* $\mu = (\mu_1, \mu_2) \in [0, \infty)^2$.

Proof We claim that the constraint set $\mathcal{S}$ of the IOP (7.1) is convex. As the IVFs $\mathbf{G}_i$ for all $i \in \mathcal{I}$ are convex on $\mathcal{X}$, for $x_1, x_2 \in \mathcal{X}$, we have

$$\mathbf{G}_i\left(\gamma x_1 + (1-\gamma)x_2\right) \preceq \gamma \odot \mathbf{G}_i(x_1) \oplus (1-\gamma) \odot \mathbf{G}_i(x_2) \preceq \mathbf{0},$$

where $\gamma \in [0, 1]$. Thus, $\gamma x_1 + (1-\gamma)x_2 \in \mathcal{S}$ for all $\gamma \in [0, 1]$. Hence, our claim is true.

Let $\bar{x}$ be an ES to the IOP (7.1). Since $\mathbf{F}$ and $\mathbf{G}_i$ for all $j \in \mathcal{J}$ are convex, in view of Corollary 7.1, $\exists$ a nonzero $\mu = (\mu_1, \mu_2) \in [0, \infty)^2$ so that $\bar{x}$ is a solution to

$$\min_{x \in \mathcal{S}} \left(\mu_1 \underline{f}(x) + \mu_2 \overline{f}(x)\right).$$

Therefore,

$$\left(\mu_1 \underline{f}(x) + \mu_2 \overline{f}(x)\right) \geq \left(\mu_1 \underline{f}(\bar{x}) + \mu_2 \overline{f}(\bar{x})\right) \ \forall\, x \in \mathcal{S}. \tag{7.10}$$

Consider the following function on $\mathcal{X}$:

$$\psi(x) = \mu_1 \left(\underline{f}(x) - \underline{f}(\bar{x})\right) + \mu_2 \left(\overline{f}(x) - \overline{f}(\bar{x})\right).$$

Since $\mathbf{F}(\bar{x})$ is a fixed interval and $\mathbf{F}$ is a convex IVF on $\mathcal{X}$, the function $\mathbf{F}(x) \ominus_{gH} \mathbf{F}(\bar{x})$ is a convex IVF on $\mathcal{X}$. Therefore, by Remark 7.1, ψ is convex on $\mathcal{X}$. From Eq. (7.10), we then observe that the following system is inconsistent:

$$\begin{cases} \psi(x) < 0, \\ \overline{g}_i(x) < 0 \ \forall\, i \in \mathcal{I}, \\ x \in \mathcal{X}. \end{cases}$$

Hence, by the generalized Gordan theorem of alternatives on convex functions [17], we obtain $\xi \geq 0$ and $\beta_i \geq 0$ for all $i \in \mathcal{I}$ such that

$$\xi\, \psi(x) + \sum_{i=1}^{p} \beta_i\, \overline{g}_i(x) \geq 0 \ \forall\, x \in \mathcal{X}. \tag{7.11}$$

Further, it can be claimed that $\xi > 0$. On the contrary, let $\xi = 0$. Then, the inequality (7.11) yields

$$\sum_{i=1}^{p} \beta_i \, \overline{g}_i(x) \geq 0 \; \forall \, x \in \mathcal{X}. \tag{7.12}$$

As the IOP (7.1) is assumed to satisfy the Slater constraint qualification, $\exists$ an $x' \in \mathcal{X}$ such that

$$\mathbf{G}_i(x') = \left[\underline{g}_i(x'), \overline{g}_i(x')\right] \prec \mathbf{0} = [0, 0]$$

for all $i \in \mathcal{I}$.

Therefore, $\sum_{i=1}^{p} \beta_i \, \overline{g}_i(x') < 0$, which is clearly contradictory to Eq. (7.12). Therefore, ξ must be greater than 0. So, Eq. (7.11) yields

$$\psi(x) + \sum_{i=1}^{p} \bar{\vartheta}_i \, \overline{g}_i(x) \geq 0 \; \forall \, x \in \mathcal{X}, \tag{7.13}$$

where

$$\bar{\vartheta}_i = \frac{\beta_i}{\xi} \geq 0 \; \forall \, i \in \mathcal{I}.$$

As $\psi(\bar{x}) = 0$, from (7.13), we obtain

$$\sum_{i=1}^{p} \bar{\vartheta}_i \, \overline{g}_i(\bar{x}) \geq 0. \tag{7.14}$$

Next, since $\bar{x} \in \mathcal{S}$, we get

$$\mathbf{G}_i(\bar{x}) = \left[\underline{g}_i(\bar{x}), \overline{g}_i(\bar{x})\right] \preceq \mathbf{0} = [0, 0] \; \forall \, i \in \mathcal{I}. \tag{7.15}$$

Thus,

$$\sum_{i=1}^{p} \bar{\vartheta}_i \, \overline{g}_i(\bar{x}) \leq 0. \tag{7.16}$$

Hence, by the inequalities (7.14) and (7.16), we have

$$\sum_{i=1}^{p} \bar{\vartheta}_i \, \overline{g}_i(\bar{x}) = 0. \tag{7.17}$$

Therefore, by (7.6), we obtain

$$
\begin{aligned}
L_\mu(x,\bar{\vartheta}) - L_\mu(\bar{x},\bar{\vartheta}) &= \mu_1 \underline{f}(x) + \mu_2 \overline{f}(x) + \sum_{i=1}^{p} \bar{\vartheta}_i \overline{g}_i(x) - \mu_1 \underline{f}(\bar{x}) - \mu_2 \overline{f}(\bar{x}) \\
&= \mu_1 \left(\underline{f}(x) - \underline{f}(\bar{x})\right) + \mu_2 \left(\overline{f}(x) - \overline{f}(\bar{x})\right) + \sum_{i=1}^{p} \bar{\vartheta}_i \overline{g}_i(x) \\
&= \psi(x) + \sum_{i=1}^{p} \bar{\vartheta}_i\, \overline{g}_i(x) \\
&\geq 0 \text{ by the inequality (7.13).}
\end{aligned} \tag{7.18}
$$

Further, in view of (7.15), we have

$$
\sum_{i=1}^{p} \vartheta_i \overline{g}_i(\bar{x}) \leq 0 \; \forall \; \vartheta = \left(\vartheta_1, \vartheta_2, \ldots \vartheta_p\right) \in [0, \infty)^p. \tag{7.19}
$$

Thus,

$$
\begin{aligned}
L_\mu(\bar{x},\vartheta) &= \mu_1 \underline{f}(\bar{x}) + \mu_2 \overline{f}(\bar{x}) + \sum_{i=1}^{p} \vartheta_i \overline{g}_i(\bar{x}) \\
&\geq \mu_1 \underline{f}(\bar{x}) + \mu_2 \overline{f}(\bar{x}) \;\text{ by the inequality (7.19)} \\
&= \mu_1 \underline{f}(\bar{x}) + \mu_2 \overline{f}(\bar{x}) + \sum_{i=1}^{p} \bar{\vartheta}_i \overline{g}_i(\bar{x}) \;\text{ by Eq. (7.17)} \\
&= L_\mu(\bar{x},\bar{\vartheta}).
\end{aligned} \tag{7.20}
$$

Hence, for all $x \in \mathcal{S}$ and $\vartheta \in [0, \infty)^p$, in view of the inequalities (7.18) and (7.20), we have

$$
L_\mu(\bar{x},\vartheta) \leq L_\mu(\bar{x},\bar{\vartheta}) \leq L_\mu(x,\bar{\vartheta}).
$$

Therefore, $(\bar{x}, \bar{\vartheta})$ satisfies SP criteria (7.7) for the IOP (7.1).

For the converse part, we use the method of contradiction. Let $\exists$ a nonzero $\mu = (\mu_1, \mu_2) \in [0, \infty)^2$ such that the corresponding L_μ of IOP (7.1) satisfies the SP criterion (7.7) at $(\bar{x}, \bar{\vartheta})$ for some $\bar{\vartheta}$ with the property (7.9). Hence, for all $\vartheta = \left(\vartheta_1, \vartheta_2, \ldots \vartheta_p\right) \in [0, \infty)^p$, we get

$$
\begin{aligned}
& L_\mu(\bar{x},\vartheta) \leq L_\mu(\bar{x},\bar{\vartheta}) \leq L_\mu(x,\bar{\vartheta}) \;\forall\; x \in \mathcal{S} \\
\Longrightarrow\; & L_\mu(\bar{x},\bar{\vartheta}) \leq L_\mu(x,\bar{\vartheta}) \;\forall\; x \in \mathcal{S}.
\end{aligned}
$$

Further, if possible, let $\bar{x}$ be not the ES to the IOP (7.1). Hence, $\exists$ an $x' \in \mathcal{S}$ such that

$$
\begin{aligned}
& \mathbf{F}(x') \prec \mathbf{F}(\bar{x}) \\
\Longrightarrow \quad & \left[\underline{f}(x'), \overline{f}(x')\right] \prec \left[\underline{f}(\bar{x}), \overline{f}(\bar{x})\right] \\
\Longrightarrow \quad & \mu_1 \underline{f}(x') + \mu_2 \overline{f}(x') < \mu_1 \underline{f}(\bar{x}) + \mu_2 \overline{f}(\bar{x}) \\
\Longrightarrow \quad & \mu_1 \underline{f}(x') + \mu_2 \overline{f}(x') + \sum_{i=1}^{p} \bar{\vartheta}_i \overline{g}_i(x') < \mu_1 \underline{f}(\bar{x}) + \mu_2 \overline{f}(\bar{x}) \\
& \text{since } \mathbf{G}_i(x') \preceq \mathbf{0} \text{ for each } i \in \mathcal{I} \\
\Longrightarrow \quad & \mu_1 \underline{f}(x') + \mu_2 \overline{f}(x') + \sum_{i=1}^{p} \bar{\vartheta}_i \overline{g}_i(x') < \mu_1 \underline{f}(\bar{x}) + \mu_2 \overline{f}(\bar{x}) + \sum_{i=1}^{p} \bar{\vartheta}_i \overline{g}_i(\bar{x}) \\
& \text{by Eq. (7.9)} \\
\Longrightarrow \quad & L_\mu(x', \bar{\vartheta}) < L_\mu(\bar{x}, \bar{\vartheta}),
\end{aligned}
$$

which contradicts the SP criterion at $(\bar{x}, \bar{\vartheta})$ of L_μ. Therefore, $\bar{x}$ is an ES to the IOP (7.1). □

In the following example, we verify Theorem 7.3.2.

Example 7.2 Consider the following IOP:

$$
\left.
\begin{array}{ll}
\min & \mathbf{F}(x_1, x_2) = [4, 6] \odot x_1^2 \oplus [7, 10] \odot x_2^2 \oplus [-5, 2], \\
\text{subject to} & \mathbf{G}_1(x_1, x_2) = [1, 2] \odot x_1^2 \ominus [-3, 4] \odot x_2 \oplus [-9, -7] \preceq \mathbf{0}, \\
& \mathbf{G}_2(x_1, x_2) = [2, 3] \odot x_1 \oplus [1, 3] \odot x_2^2 \ominus [6, 10] \preceq \mathbf{0}, \\
& -2 \leq x_1 \leq 2,\ 1 \leq x_2 \leq 4.
\end{array}
\right\} \quad (7.21)
$$

Here the set $\mathcal{X} = \left\{(x_1, x_2) \in \mathbb{R}^2 \mid -2 \leq x_1 \leq 2,\ 1 \leq x_2 \leq 4\right\}$ is convex. On $\mathcal{X}$, the functions $\mathbf{F}$, $\mathbf{G}_1$, and $\mathbf{G}_2$ can be explicitly expressed as

$$
\begin{aligned}
\mathbf{F}(x_1, x_2) &= \left[\underline{f}(x), \overline{f}(x)\right] = \left[4x_1^2 + 7x_2^2 - 5,\ 6x_1^2 + 10x_2^2 + 2\right], \\
\mathbf{G}_1(x_1, x_2) &= \left[\underline{g}_1(x), \overline{g}_1(x)\right] = \left[x_1^2 - 4x_2 - 9,\ 2x_1^2 + 3x_2 - 7\right], \\
\text{and } \mathbf{G}_2(x_1, x_2) &= \left[\underline{g}_2(x), \overline{g}_2(x)\right] = \left[2x_1 + x_2^2 - 10,\ 3x_1 + 3x_2^2 - 6\right].
\end{aligned}
$$

Since the functions $\underline{f}$, $\overline{f}$, $\underline{g}_1$, $\overline{g}_1$, $\underline{g}_2$, and $\overline{g}_2$ are convex on $\mathcal{X}$, it can be said by Remark 7.1 that the functions $\mathbf{F}$, $\mathbf{G}_1$, and $\mathbf{G}_2$ are convex on $\mathcal{X}$.

Since at the point $\bar{x} = (0, 1) \in \mathcal{X}$,

$$
\mathbf{G}_1(\bar{x}) = \left[\underline{g}_1(\bar{x}), \overline{g}_1(\bar{x})\right] = [-11, -4] \prec \mathbf{0} \text{ and } \mathbf{G}_2(\bar{x}) = \left[\underline{g}_2(\bar{x}), \overline{g}_2(\bar{x})\right] = [-9, -3] \prec \mathbf{0}.
$$

Therefore, the point $\bar{x}$ satisfies the Slater constraint qualification of the IOP (7.21).

It can be claimed that $\bar{x}$ is an ES of IOP (7.21). On contrary, let there exist two real numbers δ_1 and δ_2 with $(\delta_1, \delta_2 + 1) \in \mathcal{S}$ such that

$$\begin{aligned}
&\mathbf{F}(\delta_1, \delta_2 + 1) \prec \mathbf{F}(0, 1),\\
\text{or } &[4, 6] \odot \delta_1^2 \oplus [7, 10] \odot (\delta_2 + 1)^2 \oplus [-5, 2] \prec [2, 12],\\
\text{or } &[4\delta_1^2 + 7\delta_2^2 + 14\delta_2 + 2, 6\delta_1^2 + 10\delta_2^2 + 20\delta_2 + 12] \prec [2, 12],
\end{aligned}$$

which implies

$$\text{either } 4\delta_1^2 + 7\delta_2^2 + 14\delta_2 < 0 \text{ or } 6\delta_1^2 + 10\delta_2^2 + 20\delta_2 < 0.$$

Thus, $\delta_2 < 0$, which is not possible, as $1 \leq \delta_2 + 1 \leq 4$. Therefore, $\nexists$ any $x \neq \overline{x} \in \mathcal{S}$ such that $\mathbf{F}(x) \prec \mathbf{F}(\overline{x})$. Hence, $\bar{x} = (0, 1)$ is an ES of the IOP (7.21).

We show that $\exists$ a $\bar{\vartheta} \in [0, \infty)^2$ such that $(\bar{x}, \bar{\vartheta})$ satisfies SP criteria (7.7) for the IOP (7.21).

Let us choose $\bar{\vartheta} = \left(\bar{\vartheta}_1, \bar{\vartheta}_2\right) = (2, 0)$ for which we get

$$\bar{\vartheta}_1 \overline{g}_1(\bar{x}) + \bar{\vartheta}_2 \overline{g}_2(\bar{x}) = 0.$$

Therefore, for any $\mu = (\mu_1, \mu_2) \in [0, \infty)^2$ and for all $\vartheta = (\vartheta_1, \vartheta_2) \in [0, \infty)^2$, we have

$$\begin{aligned}
L_\mu(\bar{x}, \vartheta) &= \mu_1 \underline{f}(\bar{x}) + \mu_2 \overline{f}(\bar{x}) + \vartheta_1 \overline{g}_1(\bar{x}) + \vartheta_2 \overline{g}_2(\bar{x})\\
&= -5\mu_1 + 2\mu_2 - 3\vartheta_2\\
&\leq -5\mu_1 + 2\mu_2\\
&= \mu_1 \underline{f}(\bar{x}) + \mu_2 \overline{f}(\bar{x})\\
&= \mu_1 \underline{f}(\bar{x}) + \mu_2 \overline{f}(\bar{x}) + \bar{\vartheta}_1 \overline{g}_1(\bar{x}) + \bar{\vartheta}_2 \overline{g}_2(\bar{x})\\
&= L_\mu(\bar{x}, \bar{\vartheta}),
\end{aligned}$$

and for all $x \in \mathcal{S}$,

$$\begin{aligned}
L_\mu(x, \bar{\vartheta}) - L_\mu(\bar{x}, \bar{\vartheta}) &= \mu_1 \left(\underline{f}(x) - \underline{f}(\bar{x})\right) + \mu_2 \left(\overline{f}(x) - \overline{f}(\bar{x})\right) + \bar{\vartheta}_1 \left(\overline{g}_1(x) - \overline{g}_1(\bar{x})\right)\\
&= (4\mu_1 + 6\mu_2 + 4)\, x_1^2 + (7\mu_1 + 10\mu_2) \left(x_2^2 - 1\right) + 6(x_2 - 1)\\
&\geq 0,
\end{aligned}$$

i.e.,

$$L_\mu(\bar{x}, \bar{\vartheta}) \leq L_\mu(x, \bar{\vartheta}) \text{ for all } x \in \mathcal{S}.$$

Hence, $(\bar{x}, \bar{\vartheta})$ satisfies SP criteria (7.7) for the IOP (7.21).

7.4 Generating the Complete Efficient Solution Set of IOPs

This section deals with a method to generate the complete ES set of the IOP (7.1). Due to Theorem 7.3.1, each POS to the BOP (7.3) is an ES to the IOP (7.1). Hence, we attempt to generate the complete POS set to the BOP (7.3). To do so, we apply the *cone method* [7] on the BOP (7.3). The cone method provides all the POSs and weak POSs of any multiobjective optimization problem [7]. In the following, we briefly sketch the cone method for the BOP (7.3).

In order to apply the cone method on the BOP (7.3), we assume that

(i) each of $\underline{f}$ and $\overline{f}$ of the BOP (7.3) has nonnegative minimum value on $\mathcal{S}$,
(ii) the sets $\mathcal{S}$ and $\mathcal{Y} = f(\mathcal{S})$ are compact.

This method relies on the fact that an $\bar{x} \in \mathcal{S}$ is a POS if and only if $f(\mathcal{S}) \cap \left(f(\bar{x}) - \mathbb{R}^2_{\geqq}\right) = \{f(\bar{x})\}$, where

$$\mathbb{R}^2_{\geqq} = \left\{x = (x_1, x_2) \in \mathbb{R}^2 \middle| \, x \geqq 0, \text{ i.e., } x_1 \geq 0, x_2 \geq 0\right\}.$$

The geometrical representation of the cone method for the BOP (7.3) is as follows. To capture a POS, we translate the nonpositive quadrant of $\mathbb{R}^2$, i.e., $-\mathbb{R}^2_{\geqq}$ along the direction of an unit vector $\hat{\alpha} = \left(\hat{\alpha}_1, \hat{\alpha}_2\right) \in \mathbb{R}^2_{\geqq}$ until it does not hit $\mathcal{Y}$. The translation process must be carried out in such a manner that the vertex of the cone is always kept up on the line of vector $z\hat{\alpha}$, where z is a real number. In that process, the cone $\mathcal{K} = \left(z\hat{\alpha} - \mathbb{R}^2_{\geqq}\right)$ with $z > 0$ can hit the boundary of $\mathcal{Y}$ in one of the following possible ways:

(i) The first contact portion of $\mathcal{K}$ with $\mathcal{Y}$ is the vertex of $\mathcal{K}$. Then, the contact point of $\mathcal{K}$ on $\mathcal{Y}$ is a NS to the BOP (7.3).
(ii) The first contact portion of $\mathcal{K}$ with $\mathcal{Y}$ is the one (or more) of the boundary line(s) of $\mathcal{K}$. In that case, if the contact portion of $\mathcal{Y}$ is a single point, then this point is also a NS to the BOP (7.3). Further, if the contact portion of $\mathcal{Y}$ is a set of points, then the extreme point of the contact portion is a NS to the BOP (7.3).

The pictorial representation of the cone method for the BOP (7.3) is depicted in Fig. 7.3, where it is assumed that each of $\underline{f}$ and $\overline{f}$ of the BOP (7.3) has zero minimum value on $\mathcal{S}$. The bold black arcs of the boundary of $\mathcal{Y}$ ($bd(\mathcal{Y})$) represent the region of NSs to the BOP (7.3).

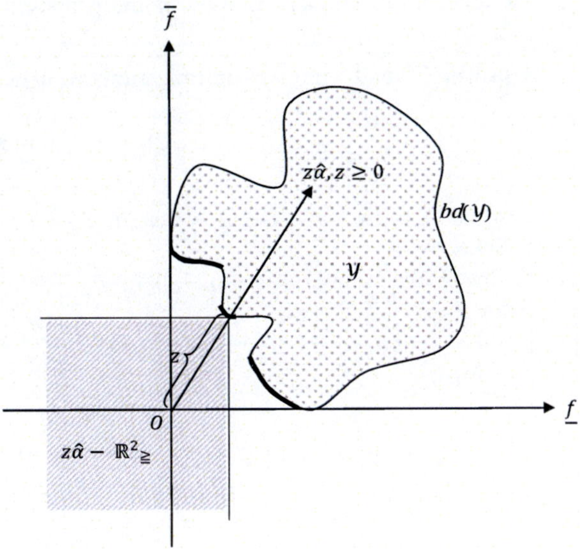

Fig. 7.3 Illustration of the cone method for the BOP (7.3)

Now, we consider the set $\mathcal{N} = \{y \mid z\hat{\alpha} \geqq f(x),\ y = f(x),\ x \in \mathcal{S},\ z \in \mathbb{R}\}$. For each particular value of z, $\mathcal{N}$ represents the region $\mathcal{K} \cap \mathcal{Y}$. If we try to minimize the region $\mathcal{K} \cap \mathcal{Y}$ by translating the cone $\mathcal{K}$ along the direction of $\hat{\alpha}$ such that $\mathcal{K} \cap \mathcal{Y} \neq \emptyset$, and in the optimum situation, if the intersecting region $\mathcal{K} \cap \mathcal{Y}$ contains only one point, then that point is certainly a NS to the BOP (7.3).

It is noteworthy that the intersecting region minimization process is the minimization of z-value satisfying the $z\hat{\alpha} \geqq f(x)$, $x \in \mathcal{S}$. Hence, to obtain a NS to the BOP (7.3), we need to solve the following optimization problem:

$$\text{CMIOP}(\hat{\alpha}) \begin{cases} \min & z \\ \text{subject to} & z\hat{\alpha}_1 \geqq \underline{f}(x), \\ & z\hat{\alpha}_2 \geqq \overline{f}(x), \\ & x \in \mathcal{S}. \end{cases} \tag{7.22}$$

Any solution $\bar{x} \in \mathcal{S}$ of the optimization problem (7.22) is a Pareto optimal (possibly weak) solution and $f(\bar{x})$ is a NS to the BOP (7.3). Solving the optimization problem (7.22) for various values of $\hat{\alpha}$, one can generate the complete POS set of the BOP (7.3) as well as the complete ES set of the IOP (7.1). Normally, we can consider the unit vector $\hat{\alpha} = (\cos\theta, \sin\theta)$, where $\theta \in [0, \frac{\pi}{2}]$, and to obtain various $\hat{\alpha}$, we have to consider different values of θ.

The algorithmic implementation of the proposed method is depicted in Algorithm 2.

Require: Given an interval optimization problem

$$\text{(IOP)} \qquad \min_{x \in \mathcal{S}} \mathbf{F}(x),$$

where $\mathcal{S} = \{x \in \mathcal{X} \mid \mathbf{G}_i(x) \preceq \mathbf{0} \; \forall \, i \in \mathcal{I}\} = \{x \in \mathcal{X} \mid \overline{g}_i(x) \leq 0 \; \forall \, i \in \mathcal{I}\}$.

Set $\mathcal{E} \leftarrow \emptyset$

Give n, the number of grid points for θ

for $\theta = 0 : \frac{\pi}{2n} : \frac{\pi}{2}$ **do**

  $\hat{\alpha} = (\cos\theta, \sin\theta)$

  Solve

$$\text{CMIOP}(\hat{\alpha}) \begin{cases} \min & z \\ \text{subject to} & z\hat{\alpha}_1 \geqq \underline{f}(x), \\ & z\hat{\alpha}_2 \geqq \overline{f}(x), \\ & x \in \mathcal{S}, \end{cases}$$

  $\mathcal{E} \leftarrow \mathcal{E} \bigcup \{\bar{x}\}$

end for

return The set $\mathcal{E}$ as the ES set to the problem (IOP)

Algorithm 2: To obtain complete ES set of an IOP

In MATLAB R2015a platform with Intel Core i5-2430M, 2:40 GHz CPU, 3 GB RAM, 32-bit Windows 7 environment, applying Algorithm 2 on the IOP of Example 7.1 for 10 and 25 values of $\hat{\alpha}$ (i.e., $n = 10, 25$ in Algorithm 2), we obtain 5 and 13 ESs, respectively, to the IOP (7.4). It is observed that all of the generated ESs lie in the interval [0.692, 1].

In Figs. 7.4a and 7.5a, the objective space of the IOP (7.4) is depicted by the shaded region. The ESs and corresponding NSs to the IOP (7.4), generated by the Algorithm 2 for $n = 10$ are, respectively, presented by the black dots on the x-axis and the black vertical line segments in Fig. 7.4a. Similarly, Fig. 7.5a presents for $n = 25$.

On the other hand, in Fig. 7.4b (and Fig. 7.5b), depicting the objective space of the BOP (7.5) by gray curve, the NSs to the BOP (7.5) corresponding to NSs to the IOP (7.4) of Fig. 7.4a (and Fig. 7.5a) are presented by the black dots on the objective space.

From Figs. 7.4a and 7.5a, it is noteworthy that if we increase the number of $\hat{\alpha}$-directions in Algorithm 2, we can generate more ESs to the IOP (7.4). Evidently, the more the $\hat{\alpha}$'s, the more the generated ESs to the IOP (7.4). Ideally, as $n \to \infty$, Algorithm 2 will generate the complete ES set of the IOP (7.4).

Similarly, by applying Algorithm 2 on the IOP of Example 7.2 for 25 values of $\hat{\alpha}$ (i.e., $n = 25$), we obtain the ES (0, 1) to the IOP (7.21). The objective space and the

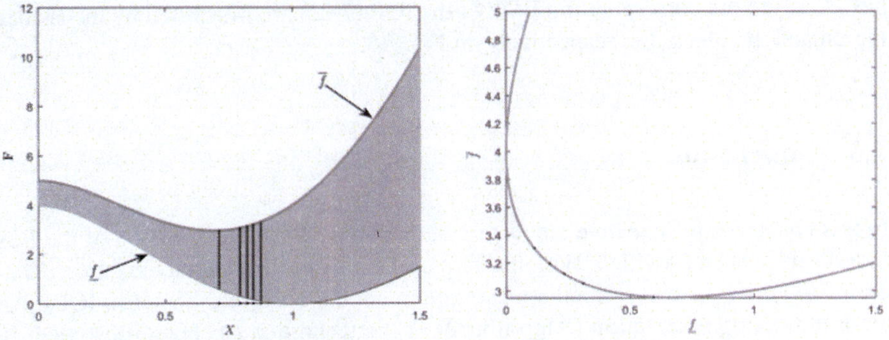

(a) ESs and NSs to the IOP (7.4) (b) NSs to the BOP (7.5)

Fig. 7.4 ESs and NSs to the IOP (7.4) and the corresponding BOP (7.5) obtained by Algorithm 2 for $n = 10$

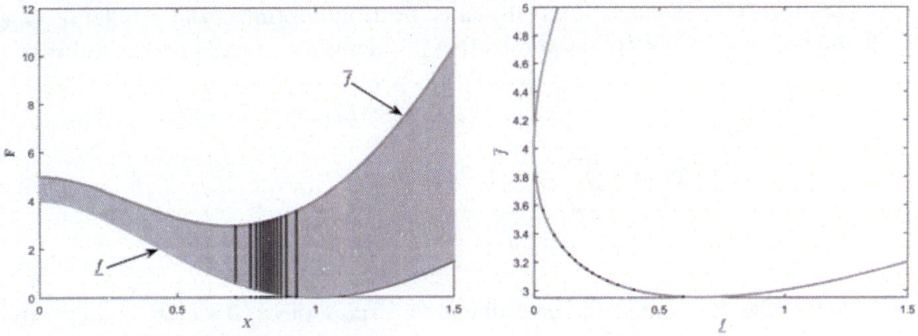

(a) ESs and NSs to the IOP (7.4) (b) NSs to the BOP (7.5)

Fig. 7.5 ESs and NSs to the IOP (7.4) and the corresponding BOP (7.5) obtained by Algorithm 2 for $n = 25$

Fig. 7.6 Generated NS to the BOP corresponding to the IOP (7.21) by the Algorithm 2

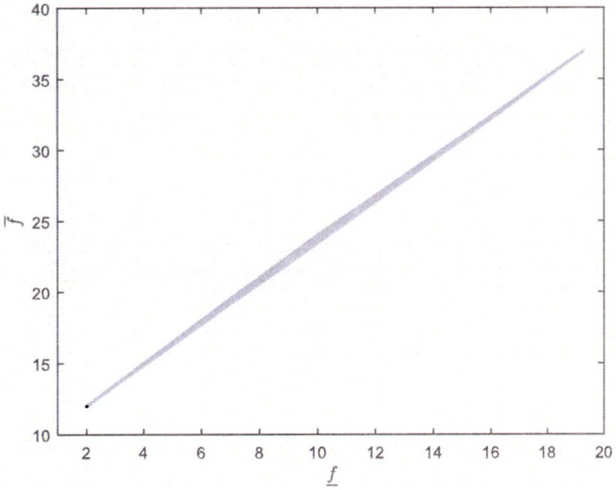

NS (2, 12) to the corresponding BOP of the IOP (7.21) are illustrated by the shaded region and the black dot, respectively, in Fig. 7.6.

7.5 Conclusion

In this chapter, a bi-objective characterization and a SP characterization of the ESs to IOPs have been provided. With the help of the bi-objective characterization (Theorem 7.3.1), we have described a technique to get the complete ES set to IOPs. An algorithmic implementation (Algorithm 2) of the technique has been illustrated. In order to find the SP characterization, we have studied a SP criterion for IOPs and proposed a condition for which the SP of an IOP will be its ES and vice versa (Theorem 7.3.2).

In the next step of this study, we will attempt to apply the proposed technique to solve practical IOPs, such as interval-valued portfolio optimization problems. One of the interval-valued portfolio optimization problems may be defined as follows:

$$\begin{aligned} \text{min} \quad & (x_1, x_2, \ldots x_n) \odot \mathbf{Q} \odot (x_1, x_2, \ldots, x_n)^t \\ \text{subject to} \quad & \sum_{i=1}^{n} x_i = 1, \\ & x_i \geq 0,\ i = 1, 2, \ldots, n, \end{aligned}$$

where x_i is the proportion of the investment corresponding to ith asset, and $\mathbf{Q}$ is the interval-valued risk, i.e., variance–covariance matrix of the interval-valued returns. Also, we shall endeavor to find a method to obtain the complete ES set of multiobjective IOPs.

Acknowledgements The authors extend a sincere thanks to the reviewers for their valuable comments. Financial support through Early Career Research Award (ECR/2015/000467), Science and Engineering Research Board, Government of India, is gratefully acknowledged.

References

1. Chalco-Cano, Y., Lodwick, W.A., Rufian-Lizana, A.: Optimality conditions of type KKT for optimization problem with interval-valued objective function via generalized derivative. Fuzzy Optim. Decis. Mak. **12**, 305–322 (2013)
2. Chalco-Cano, Y., Maqui-Huamán, G.G., Silva, G.N., Jiménez-Gamero, M.D.: Algebra of generalized Hukuhara differentiable interval-valued functions: review and new properties. Fuzzy Sets Syst. **375**, 53–69 (2019)
3. Chalco-Cano, Y., Rufian-Lizana, A., Roman-Flores, H., Jimenez-Gamero, M.D.: Calculus for interval-valued functions using generalized Hukuhara derivative and applications. Fuzzy Sets Syst. **219**, 49–67 (2013)

4. Chen, S.H., Wu, J., Chen, Y.D.: Interval optimization for uncertain structures. Finite Elem. Anal. Des. **40**, 1379–1398 (2004)
5. Ghosh, D.: Newton method to obtain efficient solutions of the optimization problems with interval-valued objective functions. J. Appl. Math. Comput. **53**, 709–731 (2017)
6. Ghosh, D.: A quasi-Newton method with rank-two update to solve interval optimization problems. Int. J. Appl. Comput. Math. **3**(3), 1719–1738 (2017)
7. Ghosh, D., Chakraborty, D.: A new Pareto set generating method for multi-criteria optimization problems. Oper. Res. Lett. **42**, 514–521 (2014)
8. Ghosh, D., Debnath, A.K., Pedrycz, W.: A variable and a fixed ordering of intervals and their application in optimization with interval-valued functions. Int. J. Approx. Reason. **121**, 187–205 (2020)
9. Ghosh, D., Chauhan, R.S., Mesiar, R., Debnath, A.K.: Generalized Hukuhara Gâteaux and Fréchet derivatives of interval-valued functions and their application in optimization with interval-valued functions. Inf. Sci. **510**, 317–340 (2020)
10. Ghosh, D., Ghosh, D., Bhuiya, S.K., Patra, L.K.: A saddle point characterization of efficient solutions for interval optimization problems. J. Appl. Math. Comput. **58**(1–2), 193–217 (2018)
11. Ghosh, D., Singh, A., Shukla, K.K., Manchanda, K.: Extended Karush-Kuhn-Tucker condition for constrained interval optimization problems and its application in support vector machines. Inf. Sci. **504**, 276–292 (2019)
12. Hukuhara, M.: Intégration des applications measurables dont la valeur est un compact convexe. Funkc. Ekvacioj **10**, 205–223 (1967)
13. Ishibuchi, H., Tanaka, H.: Multiobjective programming in optimization of the interval objective function. Eur. J. Oper. Res. **48**(2), 219–225 (1990)
14. Jianga, C., Xiea, H.C., Zhanga, Z.G., Hana, X.: A new interval optimization method considering tolerance design. Eng. Optim. **47**(12), 1637–1650 (2015)
15. Liu, S.T., Wang, R.T.: A numerical solution method to interval quadratic programming. Appl. Math. Comput. **189**(2), 1274–1281 (2007)
16. Lupulescu, V.: Fractional calculus for interval-valued functions. Fuzzy Sets Syst. **265**, 63–85 (2015)
17. Mangasarian, O.L.: Nonlinear Programming, Classics edn. Society for Industrial and Applied Mathematics (1994)
18. Moore, R.E.: Method and Applications of Interval Analysis, Classics edn. Society for Industrial and Applied Mathematics (1987)
19. Stefanini, L.: A generalization of Hukuhara difference and division for interval and fuzzy arithmetic. Fuzzy Sets Syst. **161**, 1564–1584 (2010)
20. Wu, H.C.: The Karush-Kuhn-Tucker optimality conditions in an optimization problem with interval-valued objective function. Eur. J. Oper. Res. **176**, 46–59 (2007)
21. Wu, H.C.: On interval-valued non-linear programming problems. J. Math. Anal. Appl. **338**(1), 299–316 (2008)

Chapter 8
Unconstrained Reformulation of Sequential Quadratic Programming and Its Application in Convex Optimization

R. Sadhu, C. Nahak, and S. P. Dash

Abstract A convex optimization problem with linear equality constraints is solved by the unconstrained minimization of a sequence of convex quadratic functions. The idea of sequential quadratic programming is combined with the concept of regularized gap function to construct an exact differentiable penalty function. A descent algorithm is proposed along with some numerical illustrations.

Keywords Convex optimization · Sequential quadratic programming · Regularized gap function · Exact penalty function · Unconstrained reformulation

8.1 Introduction

A large variety of practical problems involving decision-making (or system design, analysis, and operation) can be cast in the form of an optimization problem or some variation such as a multi-criterion optimization problem. Indeed, optimization has become an important tool in many areas. It is widely used in engineering, in electronic design automation, automatic control systems, and optimal design problems arising in civil, chemical, mechanical, and aerospace engineering. Optimization is used for problems arising in network design and operation, finance, supply chain management, scheduling, and many other areas. Recently in the year 2019 Xin-She Yang [1] introduced the essential ideas of algorithms and optimization techniques in the field of data mining and machine learning. The list of applications is still steadily expanding.

R. Sadhu · C. Nahak (✉)
Department of Mathematics, Indian Institute of Technology Kharagpur, Kharagpur 721302, India
e-mail: cnahak@maths.iitkgp.ernet.in

R. Sadhu
e-mail: rudrajitsadhu@gmail.com

S. P. Dash
NIC Office, Bhubaneswar 751001, India
e-mail: spdash@nic.in

V. Laha et al. (eds.), *Optimization, Variational Analysis and Applications*,
Springer Proceedings in Mathematics & Statistics 355,
https://doi.org/10.1007/978-981-16-1819-2_8

A convex optimization problem is one of the form

$$\text{minimize } f_0(x)$$

$$\text{subject to } f_i(x) \leq b_i,\ i = 1, \ldots, m,$$

where the functions $f_0, \ldots, f_m : \mathbb{R}^n \to \mathbb{R}$ are convex, i.e., satisfy $f_i(\alpha x + \beta y) \leq \alpha f_i(x) + \beta f_i(y)$ for all $x, y \in \mathbb{R}^n$ and all $\alpha, \beta \in \mathbb{R}$ with $\alpha + \beta = 1, \alpha \geq 0, \beta \geq 0$. The least-squares problem and linear programming problem are both special cases of the general convex optimization problem. A variety of applications of convex optimization are used in areas like probability and statistics, computational geometry, and data fitting.

There is, in general, no analytical formula for the solution of convex optimization problems, but there are very effective methods for solving them. Interior-point methods work very well in practice. Sequential quadratic programming (SQP) method is another most successful methods for solving constrained nonlinear optimization problems. This is an iterative procedure which generates a sequence of points (not necessarily feasible points), obtained by solving quadratic programming subproblems, and converges to the Karush–Kuhn–Tucker (KKT) point. This idea was first proposed by Wilson [2] in 1963. Since then the SQP method has been studied extensively by many researchers [3]. The readers may see Boggs [4], Gould et al. [5], Schittkowski et al. [6] for some good reviews on SQP algorithms.

To give an overview of the SQP method, we consider an equality constrained problem. At each iteration, the method solves a quadratic subproblem of the form

$$\begin{aligned} \text{(QP)} \quad & \min_x \ q(x) = \frac{1}{2} x^T H x + c^T x \\ & \text{subject to } Ax = b. \end{aligned} \tag{8.1}$$

Given a general nonlinear optimization problem, the quadratic subproblem (QP) is the quadratic approximation to the original problem at some current iterate. Thus the vector $c \in \mathbb{R}^n$ usually stands for the gradient vector of the objective function ∇f or the gradient of the Lagrangian; the $n \times n$ symmetric matrix H represents either the Hessian of the Lagrangian or an approximation to it and the solution x to the (QP) represents a search direction for the original problem. The linearization of the equality constraints at a current iterate of an optimization algorithm produces the system of linear equations $Ax = b$. We will assume here that A is an $m \times n$ matrix, with $m < n$, and that A has full row rank, i.e., the system constitutes m linearly independent equations. We also assume for convenience that H is positive semi-definite on the null space of the constraint matrix A, as this guarantees that (8.1) has a solution.

By first-order necessary optimality condition, the (QP) problem (8.1) associates with it a KKT system which is a system of $n + m$ linear equations in $n + m$ unknowns; this reduces the optimization problem to the problem of solving a system of linear equations. Several factorizing techniques are available for solving the KKT system directly, see [7–9] or else different iterative methods, see [10] can be used to solve the system up to a desired level of accuracy, these methods suit well for large systems.

In this chapter, we solve a convex optimization problem with equality constraints by using the SQP method. Instead of attempting to solve the KKT system associated with the SQP, here we propose an exact unconstrained reformulation of the problem (8.1) by using a projection map on the feasible set. The reformulation makes heavy use of *regularized gap function* for variational inequality problem, introduced by Fukushima [11]. Several variety of gap function for variational inequality are available in the literature [12–14]. A detailed survey on gap function is available in [15]. The regularized gap function was first used by Li and Peng [16] to propose an exact penalty function for the problem of minimizing a twice continuously differentiable function over a convex set. It was proved theoretically in [16] that under certain assumptions both the original and the reformulated problems have the same set of local and global solutions. Although from the theoretical point of view the reformulation is extremely sound, but its practical implementation to a general class of function encounters several difficulties. This motivates us to investigate the favorable cases, where the theory proposed in [16] is most applicable. The convex programming problem with equality constraints when solved via SQP suits extremely well to the proposed reformulation. Instead of solving the quadratic subproblem, we propose an unconstrained reformulation to it. This enables us to achieve the solution to the original convex programming problem by unconstrained minimization of a sequence of convex quadratic functions.

The chapter is structured as follows. In Sect. 8.2 we briefly discuss the mathematical backgrounds and introduce the exact penalty function for the general class of functions. Section 8.3 deals with the convex programming problem. The idea of sequential quadratic programming is used to implement an unconstrained reformulation to the convex problem. A geometrical illustration with a suitable example is provided in Sect. 8.4. Section 8.5 supports our work by providing four numerical examples. The conclusion is given in Sect. 8.6.

Notation: The following notation is used throughout the chapter, the vector norm $||x||$ is the Euclidean norm, i.e., $\|x\| = \sqrt{\sum_{i=1}^{n} x_i^2}$, where $x = (x_1, x_2, \ldots, x_n)^t$. x^k denotes the vector at kth iteration, $f_k = f(x^k)$, ∇f_k and $\nabla^2 f_k$, respectively, denotes the gradient and Hessaian of f at x^k. The matrix norm $||A||$ denotes the usual operator norm and $\|A\|_F$ denotes the Frobenius norm of the matrix A, $\|A\|_F = \sqrt{\sum_{i=1}^{m}\sum_{j=1}^{n} a_{ij}^2}$, where $A = (a_{ij})$ is a $m \times n$ matrix.

8.2 Mathematical Backgrounds

Let $f : \mathbb{R}^n \to \mathbb{R}$ be a twice continuously differentiable function and $S \subseteq \mathbb{R}^n$ is a closed convex set, a constrained optimization problem is

$$\min_{x \in S} f(x). \tag{8.2}$$

A point $\bar{x}$ is said to be a stationary point for the constrained minimization problem (8.2) if $\bar{x}$ satisfies the following variational inequality problem:

$$(y - x)^T \nabla f(x) \geqslant 0 \quad \text{for all} \quad y \in S. \tag{8.3}$$

In order to solve the variational inequality problem (8.3) [11] introduced the concept of regularized gap function and by virtue of which he reformulated the problem (8.3) as an equivalent optimization problem. Recently [16], by using the idea of regularized gap function, proposed an exact unconstrained reformulation of the problem (8.2). The gap function used by them in reformulation (8.2) is as follows:

$$\begin{aligned} G_\alpha(x) &= \max_{x \in S}\{(x - y)^T \nabla f(x) - \frac{1}{2\alpha}||x - y||^2\} \\ &= (x - H_\alpha(x))^T \nabla f(x) - \frac{1}{2\alpha}||x - H_\alpha(x)||^2 \\ &= \frac{\alpha}{2}||\nabla f(x)||^2 - \frac{1}{2\alpha}||(H_\alpha(x) - x) + \alpha \nabla f(x)||^2, \end{aligned} \tag{8.4}$$

where α is a positive penalty parameter dependent on the objective function f and

$$H_\alpha(x) = \text{Proj}_S(x - \alpha \nabla f(x)).$$

Here $\text{Proj}_S(x)$ denotes the unique orthogonal projection of the vector x onto the closed convex set S.

Thus the unconstrained reformulation of the problem (8.2) as proposed in [16] is

$$\min_{x \in \mathbb{R}^n} \; P_\alpha(x) = f(x) - G_\alpha(x). \tag{8.5}$$

Thus the explicit structure of $P_\alpha(x)$ is as follows:

$$P_\alpha(x) = f(x) + (H_\alpha(x) - x)^T \nabla f(x) + \frac{1}{2\alpha}||x - H_\alpha(x)||^2. \tag{8.6}$$

It is interesting to note that $P_\alpha(x)$ thus obtained is differentiable and its gradient is given by the following lemma:

Lemma 8.2.1 (see [16]) *Suppose that $P_\alpha(x)$ defined as in (8.6) and $f(x)$ is twice continuously differentiable then*

$$\nabla P_\alpha(x) = \frac{1}{\alpha}(I - \alpha \nabla^2 f(x))(x - H_\alpha(x)). \tag{8.7}$$

The expression for the gradient of $P_\alpha(x)$ shows that any fixed point of $H_\alpha(x)$ is always a stationary point of $P_\alpha(x)$. The following characterization of $H_\alpha(x)$ stands a basis for many iterative algorithm.

Lemma 8.2.2 (see [17]) *A vector $\bar{x} \in \mathbb{R}^n$ is a solution of* (8.3) *if and only if $\bar{x} = H_\alpha(\bar{x})$.*

Thus from the above two lemma it is clear that any stationary point of problem (8.2) is also a stationary point of $P_\alpha(x)$. The equivalence of the original constrained minimization problem (8.2) and the unconstrained minimization of $P_\alpha(x)$ on $\mathbb{R}^n$ is evident from the following theorem:

Theorem 8.2.1 (See [16]) *For any $\bar{x} \in \mathbb{R}^n$, the following statements hold.*

(i) Suppose $\alpha||\nabla^2 f(\bar{x})|| < 1$. Then $\bar{x}$ is a local minimizer of $f(x)$ in S if and only if $\bar{x}$ is a local minimizer of $P_\alpha(x)$ in $\mathbb{R}^n$.
(ii) Suppose $\alpha||\nabla^2 f(x)|| < 1$ for all $x \in \mathbb{R}^n$. Then $\bar{x}$ is a global minimizer of $f(x)$ in S if and only if $\bar{x}$ is a global minimizer of $P_\alpha(x)$ in $\mathbb{R}^n$.

The theorem says that by proper choice of the penalty parameter α, the unconstrained minimization of the penalized objective function over whole of $\mathbb{R}^n$ will solve the original problem. Thus under suitable assumption every local and global minimizer of $P_\alpha(x)$ are also the local and global minimizer of original objective function over the feasible set S. In spite of this strong theoretical results, the exact penalty function proposed above has two major difficulties.

- The penalty parameter α can be chosen explicitly, only when, the norm of the Hessian matrix is bounded on the whole of $\mathbb{R}^n$.
- The construction of the penalty function demands the formula for orthogonal projection of a vector on the set S, i.e., the explicit form of $H_\alpha(x)$ is needed.

In order to circumvent these drawback, in this chapter we choose a convex optimization problem and illustrate the above reformulation in that context.

8.3 Unconstrained Reformulation of the Convex Programming

Let us consider a convex programming of the form

$$\text{(CP)} \quad \min_x \; f(x) \quad \text{subject to } Ax = b, \tag{8.8}$$

where $f : \mathbb{R}^n \to \mathbb{R}$ is a twice continuously differentiable convex function, A is a $m \times n$ $(m < n)$ full rank matrix of rank m and $b \in \mathbb{R}^m$. The motivation behind the SQP approach is to model (CP) at the current iterate x^k by a quadratic programming subproblem, then use the minimizer of the subproblem to define a new iterate x^{k+1}. Consider the following quadratic programming problem as an approximate model of (CP) at x^k:

$$
\begin{aligned}
(\mathrm{QP}_k) \qquad \min_d \; q_k(d) = \; & \frac{1}{2} d^T \nabla^2 f_k d + \nabla f_k^T d + f_k \\
\text{subject to } & Ad = b - Ax^k.
\end{aligned} \tag{8.9}
$$

The first-order necessary KKT optimality conditions of (QP_k) are as follows:

$$\nabla^2 f_k d + \nabla f_k - A^T \lambda = 0 \tag{8.10a}$$

$$Ad + Ax^k - b = 0, \tag{8.10b}$$

where λ is the Lagrange multiplier corresponding to the equality constraints. The KKT system (8.10) is a system of $n + m$ linear equations in $n + m$ unknowns $(d; \lambda)$. In existing Newton-SQP method (8.10) is solved to obtain $(d_k; \lambda_k)$, which is used to generate the next iterate point x^{k+1}.

In this chapter, instead of solving the KKT system (8.10) associated with the (QP_k), we aimed at unconstrained reformulation of the quadratic subproblem (QP_k). The theory discussed in the previous section is used in the context of the problem (QP_k). Now we proceed to construct the penalized objective function of the form:

$$P_\alpha^k(d) = q_k(d) - G_\alpha^k(d). \tag{8.11}$$

For simplicity of notation, in our further discussion we write $Q_k = \nabla^2 f_k$, $C_k = \nabla f_k$ and $b_k = b - Ax^k$.

As discussed in the previous section the construction of the penalty function $P_\alpha(x)$ requires

- formula for orthogonal projection on the affine subset S.
- a proper estimate of the penalty parameter α.

The orthogonal projection of a point $z \in \mathbb{R}^n$ on an affine subset $S = \{x \in \mathbb{R}^n | Ax = b_k\}$ is given by

$$\mathrm{Proj}_S(z) = [I_n - A^T(AA^T)^{-1}A]z + A^T(AA^T)^{-1}b_k. \tag{8.12}$$

Thus to construct P_α^k (8.11) we use the projection formulae (8.12) for computing the gap function G_α^k. By substituting (8.12) in (8.4), the gap function $G_\alpha^k(d)$ takes the form:

$$G_\alpha^k(d) = \frac{\alpha}{2}||Q_k d + C_k||^2 - \frac{1}{2\alpha}||A^T(AA^T)^{-1}b_k - A^T(AA^T)^{-1}A[(I_n - \alpha Q_k)d - \alpha C_k]||^2. \quad (8.13)$$

Therefore, by using the expression for $G_\alpha^k(d)$ in (8.11) and simplifying the expression (8.11), we obtain $P_\alpha^k(d)$ in the form:

$$P_\alpha^k(d) = d^T \bar{Q}_k d + \bar{c_k}^T d + \bar{d}_k, \quad (8.14)$$

where

$$\bar{Q}_k = \frac{1}{2\alpha}[I_n - (I_n - \alpha Q_k)(I_n - \bar{A})](I_n - \alpha Q_k)$$

$$\bar{c_k}^T = [C_k^T(I_n - \bar{A}) - \frac{1}{\alpha}b_k^T(AA^T)^{-1}A](I_n - \alpha Q_k)$$

$$\bar{d}_k = \frac{1}{2\alpha}b_k^T(AA^T)^{-1}b_k - \frac{\alpha}{2}C_k^T(I_n - \bar{A})C_k + C_k^T A^T(AA^T)^{-1}b_k + f_k$$

$$\bar{A} = A^T(AA^T)^{-1}A.$$

Thus, our required reformulated problem is

$$(\text{UQP}_k) \qquad \min_{d\in\mathbb{R}^n} \; P_\alpha^k(d). \quad (8.15)$$

The new objective function $P_\alpha^k(d)$ thus obtained is a quadratic polynomial in d, and hence twice continuously differentiable for any value of the penalty parameter $\alpha > 0$. Also, it is interesting to note that since the Hessian of the quadratic objective function $q_k(d)$ is the matrix Q_k, its norm is bounded throughout $\mathbb{R}^n$. Thus the penalty parameter $\alpha > 0$ in the reformulated objective function $P_\alpha^k(d)$ can be chosen explicitly, i.e., any α satisfying $0 < \alpha||Q_k||_F < 1$ will ensure the exactness of the reformulated problem. Moreover the function f being convex its Hessian matrix $Q_k = \nabla^2 f_k$ is positive semi-definite at each iterate point x^k. Also if we assume that Q_k is positive definite on the affine search space S, then (UQP$_k$) possesses a unique solution say d_k. Hence we generate the next iteration point $x^{k+1} = x^k + d_k$. The iteration process continues unless a suitable stopping criterion is satisfied.

The following theorem establishes the equality of the solution set of the quadratic subproblem (QP$_k$) and the reformulated unconstrained problem (UQP$_k$).

Theorem 8.3.2 *If $0 < \alpha||Q_k||_F < 1$, then d_k is a minimizer of $q_k(d) = \frac{1}{2}d^T Q_k d + C^T d + f_k$ in S if and only if d_k is minimizer of $P_\alpha^k(d)$ in $\mathbb{R}^n$.*

Proof The proof of the theorem follows from Theorem 3.1 [18]. □

Remark 8.1 The KKT system (8.10) associated with (QP$_k$) involves a system of $n + m$ linear equations in $n + m$ unknowns, this increases the dimensionality of the search space from n to $n + m$. The reformulated problem (UQP$_k$), however, deals with an unrestricted minimization of the objective function in the n-dimensional Euclidian space, thus maintaining the same dimensionality. The major computational cost involved in the construction of the exact penalty function P_α^k is a $m \times m$ matrix inversion. Thus when m is relatively small than n, the reformulation is extremely useful.

We state the algorithm for solving the convex programming via unconstrained-SQP (USQP) in its simplest form. The algorithm runs as follows:

Select a starting point x^0
do { Evaluate f_k, ∇f_k, $\nabla^2 f_k$ and b_k;
  Evaluate $\bar{Q}_k$, $\bar{c}_k$ and $\bar{d}_k$;
  Solve (UQP_k) to obtain d_k;
  Set $x^{k+1} \leftarrow x^k + d_k$;
} Until (A Convergence test is not satisfied);

Algorithm 1: Algorithm for convex programming via USQP

8.4 Geometrical Illustration

We plot the contour of the objective function and the constraint set of the example below. We start the (USQP) iteration with the initial guess $x^0 = (-4, -5)$. The sequence of iterates generated is plotted and we observe that it converges to the solution x^* in 5 steps.

$$\min \quad f(x, y) = 2x^4 + 3y^4 + 3x^2y^2 - xy$$
$$\text{Subject to } \; 3x + 2y = 6.$$

Contour plot of convex program with x^0 as the initial guess

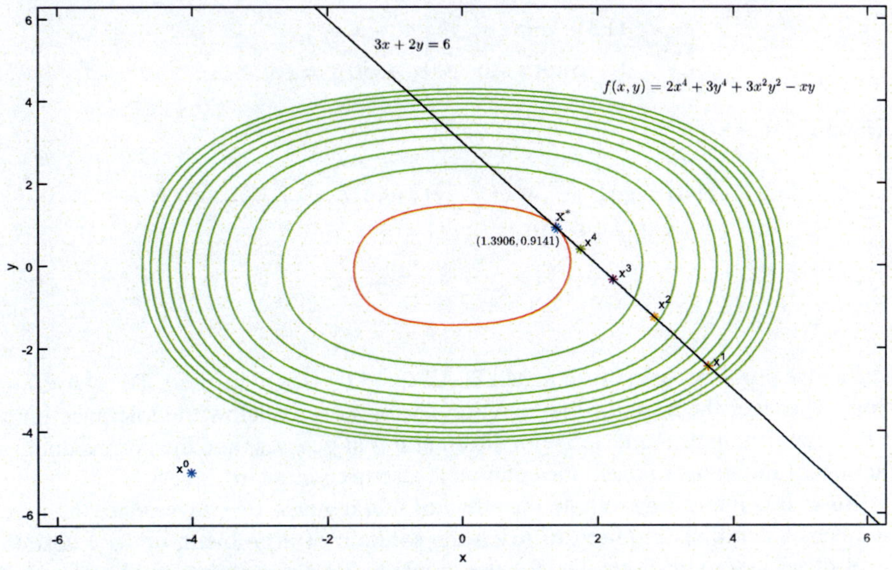

In this figure the contours of the objective function are plotted in green, the black line is the constraint set, and the $*$ denotes the iteration point. The contour of the objective corresponding to the optimal solution is plotted in red.

8.5 Numerical Examples

The following examples are examined with the proposed algorithm and the results are analyzed. The algorithm will terminate whenever the norm of the direction vector d_k comes close to zero. Each of the objective function below is chosen to be convex, along with a set of affine equality constraints.

Example 1:

$$\begin{aligned}\textbf{(P1)}: \min\ & x^4 + y^4 + z^4 - x - y - z + 1\\ \text{Subject to }\ & x + y - z = 1\\ & x - y + z = 1.\end{aligned}$$

Example 2:

$$\begin{aligned}\textbf{(P2)}: \min\ & 300x^2 + 100y^2 - 4300x - 2500y\\ \text{Subject to }\ & 300x + 200y = 1000.\end{aligned}$$

Example 3:

$$\begin{aligned}\textbf{(P3)} : \min \quad & e^{x^2} + e^{y^2} + x + y \\ \text{Subject to} \quad & 100x - 50y = 15.\end{aligned}$$

Example 4:

$$\begin{aligned}\textbf{(P4)} : \min \quad & (x^2 + y^2 + z^2)\log(1 + x^2 + y^2 + z^2) \\ \text{Subject to} \quad & 10x - 5y + z = 1 \\ & x - y + 3z = 2.\end{aligned}$$

The above problems are solved in MATLAB 2015 by USQP method. The algorithm stops whenever the norm of the direction vector d_k goes below the tolerance limit 10^{-8}. The examples are tested with different initial guesses, and the corresponding number of iterations to reach the solution is also accounted in Table 8.1.

Note: It is interesting to note the effect of initial guess, on convergence, and the number of iterations required to reach the solution of a problem, up to a desired level of accuracy. In particular, for the problem (P3), it is worth mentioning that whenever the initial guess is chosen to be (−10,10) the 'fmincon' (default program) of MATLAB fail to converge to the solution, whereas our method (USQP) converges to the solution in 25 steps.

Table 8.1 Numerical solution of the examples via USQP

	Minimizer x^*	Minimum value f^*	Initial guess x_0	No. of iteration
P1	(1,0.62996,0.62996)	0.05506	(1,1,1)	3
			(100,−300,500)	18
			(−10.1,−5.7,17)	10
P2	(1.9524,2.0714)	−12001	(1,1)	1
			(−100,50)	1
			(1000,5000)	1
P3	(−0.11280,−0.52562)	1.6926	(1,1)	6
			(−10,10)	25
			(5,−7)	57
P4	(−0.015066,−0.10264,0.63748)	0.14543	(1,1,1)	6
			(100,−500,77)	10
			(0,1000, −1500)	10
			(0,0,0)	8

8.6 Conclusion

In this chapter we have solved a convex programming problem with affine equality constraints via unconstrained sequential quadratic programming. By virtue of the exact penalty function proposed by [16], we provided an exact reformulation of the quadratic subproblem. A geometrical illustration of the scheme is given, and a suitable descent algorithm is proposed. The iteration scheme is tested and analyzed on a set of four test problems.

Acknowledgements The authors thank the anonymous reviewers very much for their constructive and detailed feedback.

References

1. Yang, X.-S.: Introduction to Algorihtms for Data Mining and Machine Learning. Elsevier Inc.(2019)
2. Wilson, R.B.: A simplicial algorithm for concave programming. Ph.D. Dissertation, Harvard University, Acta numerica (1963)
3. Chakraborty, S., Panda, G.: Two-phase-SQP method with higher-order convergence property. J. Oper. Res. Soc. China **4**(3), 385–396 (2016)
4. Boggs, P.T., Tolle, J.W.: Sequential quadratic programming. Acta Numer. **4**, 1–51 (1995)
5. Gould, N., Orban, D., Toint, P.: Numerical methods for large-scale nonlinear optimization. Acta Numer. **14**, 299–361 (2005)
6. Schittkowski, K., Yuan, Y.X.: Sequential quadratic programming methods. Wiley Encyclopedia of Operations Research and Management Science, pp. 147–224. New York (2011)
7. Gill, P.E., Murray, W., Saunders, M.A., Wright, M.H.: A Schur-complement method for sparse quadratic programming. DTIC Document (1987)
8. Nocedal, J., Wright, S.: Numerical Optimization. Springer Science (2006)
9. Schenk, O., Gärtner, K.: On fast factorization pivoting methods for sparse symmetric indefinite systems. Electron. Trans. Numer. Anal. **23**(1), 158–179 (2006)
10. Gould, N.I.M., Hribar, M.E., Nocedal, J.: On the solution of equality constrained quadratic programming problems arising in optimization. SIAM J. Sci. Comput. **23**(4), 1376–1395 (2001)
11. Fukushima, M.: Equivalent differentiable optimization problems and descent methods for asymmetric variational inequality problems. Math. Program. **53**(1–3), 99–110 (1992)
12. Larsson, T., Patriksson, M.: A class of gap functions for variational inequalities. Math. Program. **64**(1–3), 53–79 (1994)
13. Auchmuty, G.: Variational principles for variational inequalities. Numer. Funct. Anal. Optim. **10**(9–10), 863–874 (1989)
14. Auslender, A.: Optimisation Méthodes Numériques. Masson (1976)
15. Pappalardo, M., Mastroeni, G., Passacantando, M.: Merit functions: a bridge between optimization and equilibria. JOR **12**(1), 1–33 (2014)
16. Li, W., Peng, J.: Exact penalty functions for constrained minimization problems via regularized gap function for variational inequalities. J. Glob. Optim. **37**(1), 85–94 (2007)
17. Harker, P.T., Pang, J.-S.: Finite-dimensional variational inequality and nonlinear complementarity problems: a survey of theory, algorithms and applications. Math. Program. **48**(1–3), 161–220 (1990)
18. Sadhu, R., Nahak, C.: An exact convex reformulation of the quadratic programming. J. Adv. Math. Stud. **12**(2), 230–239 (2019)

Chapter 9
A Note on Quadratic Penalties for Linear Ill-Posed Problems: From Tikhonov Regularization to Mollification

Pierre Maréchal

Abstract The variational form of mollification fits in an extension of the generalized Tikhonov regularization. Using tools from variational analysis, we prove asymptotic consistency results for both this extended framework and the particular form of mollification that one obtains when building on the notion of target object.

Keywords Ill-posed problems · Regularization theory · Mollification

9.1 Introduction

Ill-posed inverse problems appear in many areas of applied mathematics, such as signal and image recovery, partial differential equations and statistics. Many of them take the form of a linear operator equation

$$Tf = g, \quad f \in F,$$

in which $T\colon F \to G$ is a bounded linear operator between the Hilbert spaces F and G and $g \in G$ is the data. Unfortunately, it frequently occurs that

$$\inf \left\{ \|Tf\| \;\middle|\; f \in (\ker T)^{\perp},\ \|f\| = 1 \right\} = 0,$$

a condition under which the pseudo-inverse $T^{\dagger}$ of T is unbounded. It results that the *natural* solution $T^{\dagger}g$ does not depend continuously on the data g and that the problem must be reformulated. Tikhonov regularization (see [15] and the references therein) initiated a vast theoretical corpus. It consists in approximating $T^{\dagger}$ by the bounded operator $R_{\alpha} = (T^*T + \alpha I)^{-1}T^*$, in which T^* denotes the adjoint of T and $\alpha > 0$ is a *regularization parameter*. The identity I may also be replaced by the more general selfadjoint operator Q^*Q, where Q is a bounded operator from F to some Hilbert

P. Maréchal (✉)
Institut de Mathématiques de Toulouse, Université Paul Sabatier, Toulouse, France
e-mail: pr.marechal@gmail.com

V. Laha et al. (eds.), *Optimization, Variational Analysis and Applications*,
Springer Proceedings in Mathematics & Statistics 355,
https://doi.org/10.1007/978-981-16-1819-2_9

space H. We then speak of *generalized Tikhonov regularization*. From the variational viewpoint, the generalized Tikhonov solution $f_\alpha = (T^*T + \alpha Q^*Q)^{-1}T^*g$ is well known to be the minimizer of the quadratic functional

$$\mathscr{F}_\alpha(f) := \| Tf - g \|^2 + \alpha \| Qf \|^2 . \tag{9.1}$$

In many cases, the solution space F is a functional space such as $L^2(\Re^d)$ or a subspace of it, and the quadratic penalty term $\alpha \| Qf \|^2$ may be used to enforce smoothness of the approximate solution. For example, Q may be a second-order differential operator (see [3, Chap. 8] for a detailed exposition).

Another way to promote smoothness is via the Fourier–Plancherel transform $\hat{f}$ of f: the variational counterpart of *mollification* [1, 6, 7, 9–11] essentially consists in penalizing $(1 - \hat{\varphi}_\alpha)\hat{f}$, in which φ_α is a convolution kernel indexed by $\alpha > 0$. The function φ_α is commonly defined, for $\alpha \in (0, 1]$, as

$$\varphi_\alpha(x) = \frac{1}{\alpha^d}\varphi\left(\frac{x}{\alpha}\right), \quad x \in \Re^d, \tag{9.2}$$

in which φ is a nonnegative integrable kernel function with unit integral, and the family $(\varphi_\alpha)_{\alpha\in(0,1]}$ is referred to as an *approximate unity*. The penalty term of mollification then takes the form $\| (I - C_\alpha) f \|^2$, in which

$$C_\alpha f = \varphi_\alpha * f.$$

Mollifiers were introduced in partial differential equations by Friedrichs [4, 16]. The term *mollification* has been used in regularization theory since the eighties. Mollification was developed in several directions. In the earlier works on the subject, mollifiers served the purpose of smoothing the data prior to inversion, whenever an explicit inversion formula was available (see [5, 12] and the references therein). In [8], an alternative approach was proposed, which gave rise to the so-called *method of approximate inverses*. In this approach, the operator under consideration is not assumed to have explicit inverse, but the adjoint equation has explicit solutions. This approach opens the way to application to a large class of inverse problems and can be extended to problems in Banach spaces [14]. A third approach appeared in the same period of time. In [7], a variational formulation of the idea of mollification was proposed, in the context of Fourier synthesis and deconvolution. This formulation was further studied and extended in [1, 6, 9, 11] and is the one we consider in this paper.

Unlike Tikhonov's regularization, mollification appeals to a parameter α which is not interpreted as a weighting of the penalty term, but rather as an *objective resolution*. Therefore, strictly speaking, mollification does not belong to the generalized Tikhonov family. However, obviously, letting α go to zero makes the penalization vanish in both cases. This suggests that Tikhonov and the mollification could be put in the same framework. To phrase it differently, we could widen the contours of the generalized Tikhonov regularization to the point of admitting mollification in its realm. This is what we propose to do here.

The paper is organized as follows. In Sect. 9.2 we consider the consistency issue in the aforementioned enlarged framework. In Sect. 9.3, we build on the notion of *target object*, absent from the original Tikhonov regularization, but present in the original works on mollification [1, 2, 7].

9.2 Generalizing Tikhonov Regularization

It is sometimes convenient to consider vector-valued regularization parameters. We may call *parameter choice rule* a function

$$\begin{aligned} \alpha : \Re_+ \times G &\longrightarrow \mathscr{P} \\ (\delta, g^\delta) &\longmapsto \alpha(\delta, g^\delta) \end{aligned}$$

in which $\mathscr{P}$ is a subset of $\Re^p_+ \setminus \{0\}$, and an *a priori parameter choice rule* the particular case for which α depends on its first argument only. Following [3, Definition 3.1], we now state:

Definition 9.1 A parametrized family (R_α) of bounded operators is a *regularization of* $T^\dagger$ if for every $g \in \mathscr{D}(T^\dagger)$, there exists a parameter choice rule α such that

(1) $\sup\big\{ \|\alpha(\delta, g^\delta)\| \mid g^\delta \in G,\ \|g^\delta - g\| \le \delta \big\} \to 0$ as $\delta \downarrow 0$;
(2) $\sup\big\{ \|R_{\alpha(\delta,g^\delta)} g^\delta - T^\dagger g\| \mid g^\delta \in G,\ \|g^\delta - g\| \le \delta \big\} \to 0$ as $\delta \downarrow 0$.

In this case, we say that the pair (R_α, α) is a convergent regularization method for solving $Tf = g$.

Recall that the domain of the operator $T^\dagger$ is the vector subspace $\mathscr{D}(T^\dagger) = \operatorname{ran} T + (\operatorname{ran} T)^\perp$, in which $E^\perp$ denotes the orthogonal complement of E. From [3, Proposition 3.4], we straightforwardly infer that:

Proposition 9.1 *If the family of bounded operators $(R_\alpha)_{\alpha\in\mathscr{P}}$ converges pointwise to $T^\dagger$ on $\mathscr{D}(T^\dagger)$ as $\alpha \to 0$ in $\mathscr{P}$, then $(R_\alpha)_{\alpha\in\mathscr{P}}$ is a regularization of $T^\dagger$ and, for every $g \in \mathscr{D}(T^\dagger)$, there exists an a priori parameter choice rule $\alpha(\delta)$ such that (R_α, α) is a convergent regularization method for solving $Tf = g$.*

The operators $T\colon F \to G$ and $Q\colon F \to H$ are said to satisfy Morozov's *completion condition* if there exists a constant $\gamma > 0$ such that

$$\forall f \in F, \quad \|Tf\|^2 + \|Qf\|^2 \ge \gamma \|f\|^2. \tag{9.3}$$

Under the completion condition, the operator $T^*T + Q^*Q$ admits a bounded inverse, as can be easily shown. In some cases of interest, it may happen that T^*T and Q^*Q can be *diagonalized* in the same Hilbert basis. In this case, it can be shown that

$$\forall f \in F, \quad \left\| (T^*T + Q^*Q)^{-1}T^*Tf \right\|_F \leq \| f \|_F. \tag{9.4}$$

The latter assumption is in force in the rest of this paper.

Theorem 9.2.1 *Let F, G be infinite dimensional Hilbert spaces and let $T : F \to G$ be injective. Let $Q_\alpha : F \to H$ be a family of operators such that*

- *for every fixed $\alpha \in \mathscr{P}$, T and Q_α satisfy the completion condition (9.3) and Condition (9.4);*
- *for every $f \in F$, $\| Q_\alpha f \| \to 0$ as $\alpha \to 0$ in $\Re^p$.*

Then, for every $g \in \mathscr{D}(T^\dagger) = \operatorname{ran} T + \operatorname{ran} T^\perp$, $f_\alpha := (T^*T + Q_\alpha^* Q_\alpha)^{-1}T^*g$ converges strongly to $f^\dagger$, the unique least square solution of the equation $Tf = g$.

Proof We shall prove that, for every $\mathscr{P}$-valued sequence (α_n) which converges to zero, the corresponding sequence (f_{α_n}) strongly converges to $f^\dagger$. By assumption, $g = Tf^\dagger + g^\perp$, in which $f^\dagger \in F$ and $g^\perp \in (\operatorname{ran} T)^\perp = \ker T^*$. We have

$$\begin{aligned} \| f_\alpha \|_F &= \left\| (T^*T + Q_\alpha^* Q_\alpha)^{-1}T^*(Tf^\dagger + g^\perp) \right\|_F \\ &= \left\| (T^*T + Q_\alpha^* Q_\alpha)^{-1}T^*Tf^\dagger \right\|_F \\ &\leq \| f^\dagger \|_F. \end{aligned}$$

In particular, the family f_α is bounded. Now, let (α_n) be a sequence in $\mathscr{P}$ which converges to 0. In order to simplify the notation, let $f_n := f_{\alpha_n}$ and $Q_n := Q_{\alpha_n}$. Since the sequence (f_n) is bounded, we can extract a weakly convergent subsequence (f_{n_k}). Let then $\tilde{f}$ be the weak limit of this subsequence. On the one hand,

$$T^*Tf_{n_k} \rightharpoonup T^*T\tilde{f} \quad \text{as} \quad k \to \infty \tag{9.5}$$

since T^*T is bounded. On the other hand,

$$Q_{n_k}^* Q_{n_k} f_{n_k} \rightharpoonup 0 \quad \text{as} \quad k \to \infty$$

since f_{n_k} is bounded and $Q_{n_k}^* Q_{n_k}$ converges pointwise to the null operator, so that

$$\begin{aligned} T^*Tf_{n_k} &= (T^*T + Q_{n_k}^* Q_{n_k}) f_{n_k} - Q_{n_k}^* Q_{n_k} f_{n_k} \\ &= T^*g - Q_{n_k}^* Q_{n_k} f_{n_k} \\ &= T^*Tf^\dagger - Q_{n_k}^* Q_{n_k} f_{n_k} \\ &\rightharpoonup T^*Tf^\dagger \end{aligned}$$

□

as $k \to \infty$. Together with (9.5), this shows that $T^*T\tilde{f} = T^*Tf^\dagger$, that is, by the injectivity of T, that $\tilde{f} = f^\dagger$. It follows that the whole sequence (f_n) converges weakly to $f^\dagger$. Finally, by the weak lower semicontinuity of the norm,

$$\| f^\dagger \| \le \liminf_{n\to\infty} \| f_n \| \le \limsup_{n\to\infty} \| f_n \| \le \| f^\dagger \|,$$

which establishes that $f_n \to f^\dagger$ as $n \to \infty$. ■

In the familiar case where $F = L^2(\mathfrak{R}^d)$ or subspaces of it and $H = L^2(\mathfrak{R}^d)$, the choice $Q_\alpha = I - C_\alpha$ corresponds to the mollification method described in the introduction. The previous theorem applies in this case since, as is well known, if (C_α) is as in (9.2), then

$$C_\alpha f \to f \quad \text{as} \quad \alpha \downarrow 0.$$

Notice that Morozov's completion condition is automatically satisfied in the important case where $F = L^2(V)$, the space of square integrable functions with essential support in V, whenever V is a compact domain. As a matter of fact, in this case, it follows from [1, Lemma 12 and Proposition 5] that there exists a positive constant ν_α such that

$$\forall f \in L^2(V), \quad \|(I - C_\alpha) f\|^2 \ge \nu_\alpha \| f \|^2.$$

9.3 Target Objects

In a number of cases, the operator T gives rise to an explicit *intertwining relationship*. By this, we mean the existence of a bounded operator $\Phi_\alpha : G \to G$ such that

$$TC_\alpha = \Phi_\alpha T. \tag{9.6}$$

Note that Eq. (9.6) constrains Φ_α only on the range of T. In order to extend its definition to the whole space G, we first use the unique bounded extension of Φ_α to the closure of ran T, and then extend it further by zero on $(\operatorname{ran} T)^\perp$. With this definition of Φ_α, it is easy to see that

$$\Phi_\alpha = \operatorname{cl}(TC_\alpha T^\dagger), \tag{9.7}$$

in which cl $(\cdot)$ denotes the extension by closure. More generally, it has been shown in [2] that whenever the operator $TC_\alpha T^\dagger$ is bounded, its closure to G minimizes $\Phi \mapsto \|\Phi T - TC_\alpha\|$ over all the bounded operators on G which vanish on $(\operatorname{ran} T)^\perp$.

At all events, we may consider the following variational form of mollification:

$$f_\alpha := \operatorname{argmin} \left\| Tf - \Phi_\alpha g \right\|^2 + \left\| (I - C_\alpha) f \right\|^2. \tag{9.8}$$

This form can be justified by the following heuristics. Since our *target object* is $C_\alpha f^\dagger$, the tautology $f^\dagger = C_\alpha f^\dagger + (I - C_\alpha) f^\dagger$ indicates that in addition to penalizing $(I - C_\alpha) f$ one should also aim at fitting the data corresponding to the mollified object. If $g \simeq Tf^\dagger$, then

$$\Phi_\alpha g \simeq T C_\alpha f^\dagger$$

by Eq. (9.6), whence the adequacy term in (9.8). The regularized solution is then given by

$$f_\alpha := \left(T^*T + (I - C_\alpha)^*(I - C_\alpha)\right)^{-1} T^* \Phi_\alpha g.$$

Important applications allow for the introduction of the *intertwining operator* Φ_β corresponding to approximate unities such as the above-defined families (C_α). We now review a few examples.

Example 9.1 In [7], the authors studied the problem of *spectral extrapolation*, which underlies *aperture synthesis* in astronomy and space imaging. This problem corresponds to the case where

$$T = frm[o]{-}{-}e_W U$$

with W a bounded domain containing an open set. Here, U denotes the Fourier–Plancherel operator. We refer to T_W as the Fourier truncation operator. Since $C_\alpha = U^{-1}[\hat{\varphi}_\alpha]U$, we see that

$$T C_\alpha = \mathbb{1}_W U U^{-1}[\hat{\varphi}_\alpha]U = [\hat{\varphi}_\alpha]\mathbb{1}_W U = [\hat{\varphi}_\alpha]T,$$

from which we infer that $\Phi_\alpha = [\hat{\varphi}_\alpha]$. ∎

Example 9.2 In the problem of deconvolution, as considered, e.g., in [6, 9], the situation is even simpler: since convolution operators commute, we readily see that $\Phi_\beta = C_\alpha$. ∎

Example 9.3 Finally, in computerized tomography [13], the underlying operator is the Radon transformation

$$(Tf)(\boldsymbol{\theta}, s) = \int f(\mathbf{x})\delta(s - \langle \boldsymbol{\theta}, \mathbf{x}\rangle)\, \mathbf{dx}, \quad \boldsymbol{\theta} \in \mathscr{S}^1, \quad s \in \mathfrak{R}.$$

A consequence of the so-called *Fourier slice theorem* is that, for any two functions f_1, f_2,

$$T(f_1 * f_2) = T f_1 \circledast T f_2,$$

in which $\circledast$ denotes the convolution with respect to the variable s. It follows that, in this case,

$$\Phi_\alpha = (g \mapsto T\varphi_\alpha \circledast g),$$

a relationship which was in force in [11]. ∎

We now establish a consistency theorem for the form of mollification given in (9.8).

Theorem 9.3.1 *Let $F = L^2(\Re^d)$ and let $T: F \to G$ be a bounded injective operator from F to the infinite dimensional Hilbert space G. Let $C_\alpha: F \to F$ be an approximate unity as in (9.2). Assume that, for every fixed $\alpha \in (0, 1]$, T and $I - C_\alpha$ satisfy the completion condition (9.3). Assume at last that, for every fixed $\alpha \in (0, 1]$, the intertwining operator Φ_α exists. Then, for every $g \in \mathscr{D}(T^\dagger) = \operatorname{ran} T + \operatorname{ran} T^\perp$, $f_\alpha := (T^*T + (I - C_\alpha)^*(I - C_\alpha))^{-1}T^*\Phi_\alpha g$ converges strongly to $f^\dagger$.*

Proof We shall prove that, for every positive sequence (α_n) converging to zero, $f_{\alpha_n} \to f^\dagger$ as $n \to \infty$. Let $g = Tf^\dagger + g^\perp$, with $f^\dagger \in F$ and $g^\perp \in (\operatorname{ran} T)^\perp$. Since $\Phi_\alpha g = TC_\alpha f^\dagger$, we have:

$$\begin{aligned} \| f_\alpha \|_F &= \| (T^*T + (I - C_\alpha)^*(I - C_\alpha))^{-1} T^*TC_\alpha f^\dagger \|_F \\ &\le \| C_\alpha f^\dagger \|_F \\ &\le \| \varphi_\alpha \|_1 \cdot \| f^\dagger \|_F = \| f^\dagger \|_F . \end{aligned}$$

The last equality stems from the fact that, in Eq. (9.2), φ is assumed to be positive and to have unit integral. Therefore, the family (f_α) is bounded. Let (α_n) be a sequence in $(0, 1]$ which converges to 0, and let $f_n := f_{\alpha_n}$, $C_n := C_{\alpha_n}$ and $\Phi_n := \Phi_{\alpha_n}$. Since the sequence (f_n) is bounded, we can extract a weakly convergent subsequence (f_{n_k}). Let then $\tilde{f}$ be the weak limit of this subsequence. On the one hand,

$$T^*Tf_{n_k} \rightharpoonup T^*T\tilde{f} \quad \text{as} \quad k \to \infty \tag{9.9}$$

since T^*T is bounded. On the other hand,

$$(I - C_{n_k})^*(I - C_{n_k})f_{n_k} \rightharpoonup 0 \quad \text{as} \quad k \to \infty$$

since f_{n_k} is bounded and $(I - C_{n_k})^*(I - C_{n_k})$ converges pointwise to the null operator, so that

$$\begin{aligned} T^*Tf_{n_k} &= (T^*T + (I - C_{n_k})^*(I - C_{n_k}))f_{n_k} - (I - C_{n_k})^*(I - C_{n_k})f_{n_k} \\ &= T^*\Phi_{n_k} g - (I - C_{n_k})^*(I - C_{n_k})f_{n_k} \\ &\rightharpoonup T^*Tf^\dagger \end{aligned}$$

□

as $k \to \infty$, since $T^*\Phi_{n_k} g = T^*TC_{n_k} f^\dagger$ goes to $T^*Tf^\dagger$. Together with (9.9), this shows that $T^*T\tilde{f} = T^*Tf^\dagger$, that is, by the injectivity of T, that $\tilde{f} = f^\dagger$. Therefore, the whole sequence (f_n) converges weakly to $f^\dagger$. Finally, by the weak lower semicontinuity of the norm,

$$\| f^\dagger \| \le \liminf_{n\to\infty} \| f_n \| \le \limsup_{n\to\infty} \| f_n \| \le \| f^\dagger \|,$$

which establishes that $f_n \to f^\dagger$ as $n \to \infty$. ■

9.4 Conclusion

We have shown that the variational form of mollification fits in an extension of the generalized Tikhonov regularization setting. Using tools from variational analysis, we have obtained asymptotic consistency results for both this extended framework and the particular form of mollification that one obtains when developing the notion of target object.

Acknowledgements The author wishes to thank Nathaël Alibaud for fruitful discussions on the subject, which led to significant improvements of this paper.

References

1. Alibaud, N., Maréchal, P., Saesor, Y.: A variational approach to the inversion of truncated Fourier operators. Inverse Probl. **25**(4), 045002 (2009)
2. Bonnefond, X., Maréchal, P.: A variational approach to the inversion of some compact operators. Pac. J. Optim. **5**(1), 97–110 (2009)
3. Engl, H.W., Hanke, M., Neubauer, A.: Regularization of Inverse Problems, vol. 375. Springer Science & Business Media (1996)
4. Friedrichs, K.O.: The identity of weak and strong extensions of differential operators. Trans. Am. Math. Soc. **55**(1), 132–151 (1944)
5. Hào, D.N.: A mollification method for ill-posed problems. Numer. Math. **68**, 469–506 (1994)
6. Hohage, T., Maréchal, P., Vanhems, A.: A mollifier approach to the deconvolution of probability densities. Part 2: Convergence rates (2020). Submitted
7. Lannes, A., Roques, S., Casanove, M.J.: Stabilized reconstruction in signal and image processing: I. partial deconvolution and spectral extrapolation with limited field. J. Mod. Opt. **34**(2), 161–226 (1987)
8. Louis, A.K., Maass, P.: A mollifier method for linear operator equations of the first kind. Inverse Probl. **6**(3), 427 (1990)
9. Maréchal, P., Simar, L., Vanhems, A.: A mollifier approach to the deconvolution of probability densities. Part 1: The methodology and its comparison to classical methods (2020). Submitted
10. Maréchal, P., Simo Tao Lee, W.C., Vanhems, A.: A mollifier approach to the nonparametric instrumental regression problem (2020). Submitted
11. Maréchal, P., Togane, D., Celler, A.: A new reconstruction methodology for computerized tomography: FRECT (fourier regularized computed tomography). IEEE Trans. Nucl. Sci. **47**(4), 1595–1601 (2000)
12. Murio, D.A.: The Mollification Method and the Numerical Solution of Ill-Posed Problems. Wiley (2011)
13. Natterer, F.: The Mathematics of Computerized Tomography. SIAM (2001)
14. Schuster, T.: The Method of Approximate Inverse: Theory and Applications, vol. 1906. Springer (2007)
15. Tikhonov, A.N., Arsenin, V.Y.: Methods for Solving Ill-Posed Problems. Wiley (1977)
16. Wikipedia contributors: Mollifier—Wikipedia, the free encyclopedia. https://en.wikipedia.org/w/index.php?title=Mollifier&oldid=950509587 (2020). Accessed 17 April 2020

Chapter 10
A New Regularization Method for Linear Exponentially Ill-Posed Problems

Walter Cedric Simo Tao Lee

Abstract This chapter provides a new regularization method which is particularly suitable for linear exponentially ill-posed problems. Under logarithmic source conditions (which have a natural interpretation in terms of Sobolev spaces in the aforementioned context), concepts of qualifications as well as order-optimal rates of convergence are presented. Optimality results under general source conditions expressed in terms of index functions are also studied. Finally, numerical experiments on three test problems attest the better performance of the new method compared to the well-known Tikhonov method in instances of exponentially ill-posed problems.

Keywords Exponentially ill-posed problems · Regularization method · Logarithmic source conditions · Order-optimality

10.1 Introduction

In this chapter, we are interested in the solution to the equation

$$Tx = y, \tag{10.1}$$

where $T : X \to Y$ is a linear bounded operator between two infinite dimensional Hilbert spaces X and Y with non-closed range. The data y belongs to the range of T and we assume that we only have approximated data y^δ satisfying

$$||y^\delta - y|| \leq \delta. \tag{10.2}$$

In such a setting, Eq. (10.1) is ill-posed in the sense that the Moore Penrose generalized inverse $T^\dagger$ of T which maps y to the best-approximate solution $x^\dagger$ of (10.1) is not continuous. Consequently a little perturbation on the data y may induce an

W. C. Simo Tao Lee (✉)
Institut de Mathématiques de Toulouse, Université Paul Sabatier, Toulouse, France
e-mail: wsimotao@math.univ-toulouse.fr

V. Laha et al. (eds.), *Optimization, Variational Analysis and Applications*,
Springer Proceedings in Mathematics & Statistics 355,
https://doi.org/10.1007/978-981-16-1819-2_10

arbitrarily large error in the solution $x^\dagger$. Instances of such ill-posed inverse problems are encountered in several fields in applied sciences among which: signal and image processing, computer tomography, immunology, satellite gradiometry, heat conduction problems, inverse scattering problems, statistics, and econometrics to name just a few, see, e.g., [11, 14, 20, 21, 31]. As a result of the ill-posedness of Eq. (10.1), a regularization method needs to be applied in order to recover from the noisy data y^δ a stable approximation x^δ of the solution $x^\dagger$. A regularization method can be regarded as a family of continuous operators $R_\alpha : Y \to X$ such that there exists a function $\Lambda : \mathbb{R}_+ \times Y \to \mathbb{R}_+$ satisfying the following: for every $y \in \mathcal{D}(T^\dagger) \subset Y$ and $y^\delta \in Y$ satisfying (10.2)

$$R_{\Lambda(\delta, y^\delta)} y^\delta \to x^\dagger \quad \text{as} \quad \delta \downarrow 0. \tag{10.3}$$

Some examples of regularizations methods are Tikhonov, Landweber, spectral cut-off, asymptotic regularization, approximate inverse, and mollification, see, e.g., [1, 7, 11, 21, 23, 24]. As a matter of fact, we would like to get estimates on the error committed while approximating $x^\dagger$ by $x^\delta = R_{\Lambda(\delta, y^\delta)} y^\delta$.

It is well known that for arbitrary $x^\dagger \in X$, the convergence of x^δ toward $x^\dagger$ is arbitrarily slow, see, e.g., [11, 35]. But still, by allowing smoothness of the solution $x^\dagger$, convergence rates could be established. Standard smoothness conditions known as Hölder type source condition take the form

$$x^\dagger \in X_\mu(\rho) = \left\{ (T^*T)^\mu w, \quad w \in X \quad \text{s.t.} \quad ||w|| \le \rho \right\}, \tag{10.4}$$

where μ and ρ are two positive constants. However such source conditions have shown their limitations as they are too restrictive in many problems and do not yield a natural interpretation. For this reason, general source conditions have been introduced in the following form:

$$x^\dagger \in X_\varphi(\rho) = \left\{ \varphi(T^*T) w, \quad w \in X \quad \text{s.t.} \quad ||w|| \le \rho \right\}, \tag{10.5}$$

where ρ is a positive constant and $\varphi : [0, ||T^*T||] \to \mathbb{R}_+$ is an index function, i.e., a non-negative monotonically increasing continuous function satisfying $\varphi(\lambda) \to 0$ as $\lambda \downarrow 0$. An interesting discussion on these source conditions can be found in [29] where the author explores how general source conditions of the form (10.5) are. Once the solution $x^\dagger$ satisfies a smoothness condition, i.e., $x^\dagger$ belongs to a proper subspace M of X, it is possible to derive convergence rates and the next challenge is about optimality. More precisely, for a regularization method $R : Y \to X$, we are interested in the worst case error:

$$\Delta(\delta, R, M) := \sup \left\{ ||Ry^\delta - x^\dagger||, \ x^\dagger \in M, \ y^\delta \in Y, \ \text{s.t.} \ ||y^\delta - Tx^\dagger|| \le \delta \right\}, \tag{10.6}$$

and we would like a regularization which minimizes this worst case error. In this respect, a regularization method $\bar{R} : Y \to X$ is said to be optimal if it achieves the minimum worst case error over all regularization methods, i.e., if

$$\Delta(\delta, \bar{R}, M) = \Delta(\delta, M) := \inf_R \Delta(\delta, R, M).$$

Similarly, a regularization is said to be order optimal if it achieves the minimum worst case error up to a constant greater than one, i.e., if

$$\Delta(\delta, \bar{R}, M) \leq C\Delta(\delta, M)$$

for some constant $C > 1$. When the subset M is convex and balanced, it is shown in [30] that

$$\omega(\delta, M) \leq \Delta(\delta, M) \leq 2\omega(\delta, M), \tag{10.7}$$

where $\omega(\delta, M)$ is the modulus of continuity of the operator T over M, i.e.,

$$\omega(\delta, M) = \sup\{||x||, \ x \in M, \ \text{s.t.} \ ||Tx|| \leq \delta\}. \tag{10.8}$$

In other words, we get the following:

$$\Delta(\delta, X_\varphi(\rho)) = O\left(\omega(\delta, X_\varphi(\rho))\right). \tag{10.9}$$

Recall that, under mild assumptions on the index function φ, the supremum defining the modulus of continuity is achieved and a simple expression of $\omega(\delta, X_\varphi(\rho))$ in term of function φ is available, see, e.g., [20, 28, 37]. Let us remind that a relevant notion in the study of optimality of a regularization method is qualification. In fact, the qualification of a regularization measures the capability of the method to take into account smoothness assumptions on the solution $x^\dagger$, i.e., the higher the qualification, the more the method is able to provide best rates for very smooth solutions.

Besides optimality, converse results and saturation results are also important aspects of regularization algorithms, see, [11, 27, 33, 34]. For converse results, we are interested in the following: given a particular convergence rate of $||x^\delta - x^\dagger||$ toward 0, which smoothness condition does the solution $x^\dagger$ needs to satisfy? Saturation results are about the maximal smoothness on the solution $x^\dagger$ for which a regularization method can still deliver the best rates of convergence. Finally, another significant aspect of regularization is the selection of the regularization parameter, i.e., finding a function $\Lambda(\delta, y^\delta)$ which guarantees convergence and possibly order-optimality.

Coming back to (10.5), notice that a very interesting subclass of general source conditions are logarithmic source conditions expressed as

$$x^\dagger \in X_{f_p}(\rho) = \left\{(-\ln(T^*T))^{-p}w, \quad w \in X \quad \text{s.t.} \quad ||w|| \leq \rho\right\}, \tag{10.10}$$

where p and ρ are positive constants and T satisfies $\|T^*T\| < 1$. Such smoothness conditions have clear interpretations in term of Sobolev spaces in exponentially ill-posed problems, see, e.g., [20, 37]. The latter class includes several problems of great importance such as backward heat equation, sideways heat equation, inverse problem in satellite gradiometry, control problem in heat equation, inverse scattering problems, and many others, see, [20]. Because of the importance of exponentially ill-posed problems, it is desirable to design regularization methods particularly suitable for this class of problems. It is precisely the aim of this chapter to provide such a regularization scheme.

In the next section, we define the new regularization method using both the variational formulation and the definition in terms of the so-called *generator* function g_α. A brief comparison with the Tikhonov method is done. Moreover basic estimates on the *generator* function g_α and its corresponding *residual* function r_α are also carried out.

Section 10.3 is devoted to optimality of the new method. Here we recall well-known optimality results under general source conditions of the form (10.5), see, [19, 20, 28, 32, 37]. For the specific case of logarithmic source conditions, qualification of the method is given and order-optimality is shown. Next we study optimality under general source conditions.

In Sect. 10.4, we present a comparative analysis of the new method with Tikhonov method, spectral cut-off, asymptotic regularization, and conjugate gradient.

Section 10.5 is about numerical illustrations. In this section, in order to confirm our prediction of better performance of the new method compared to Tikhonov and spectral cut-off in instance of exponentially ill-posed problems, we numerically compare the efficiency of the five regularization methods on three test problems coming from literature: A problem of image reconstruction taken from [36], a Fredholm integral equation of the first kind found in [2] and an inverse heat equation problem.

Finally in Sect. 10.6, for a fully applicability of the new method, we exhibit heuristic selection rules which fit with the new regularization technique. Moreover, we also compare the five regularization methods for each heuristic parameter choice rule under consideration.

10.2 The New Regularization Method

For the sake of simplicity, we assume henceforth that the operator T is injective. Hereafter, we set a positive number a such that the operator norm of T^*T is less than a, i.e., $||T^*T|| \leq a$. In the sequel, we assume that $a < 1$ which is always possible by scaling Eq. (10.1).

Let us consider the general variational formulation of a regularization method

$$x_\alpha = \arg\min_{x \in X} \mathscr{F}(Tx, y) + \mathscr{P}(x, \alpha), \tag{10.11}$$

where $\mathscr{F}(Tx, y)$ is the fit term, $\mathscr{P}(x, \alpha)$ is the penalty term, and $\alpha > 0$ is the regularization parameter. We recall that the fit term aims at fitting the model, the penalty term aims at introducing stability in the initial model $Tx = y$ and the regularization parameter α controls the level of regularization.

In most cases, the fit term $\mathscr{F}(Tx, y)$ is nothing but

$$\mathscr{F}(Tx, y) = ||Tx - y||^2 \tag{10.12}$$

and the penalty term depends on the regularization method. For instance, for Tikhonov regularization, $\mathscr{P}(x, \alpha)$ is given by

$$\mathscr{P}(x, \alpha) = \alpha||x||^2. \tag{10.13}$$

This penalization can sometimes compromises the quality of the resulting approximate solution x_α. Indeed, let $X = L^2(\mathbb{R}^n)$, then by Parseval identity, we see that

$$\mathscr{P}(x, \alpha) = \alpha||\hat{x}||^2_{L^2(\mathbb{R}^n)}, \tag{10.14}$$

where $\hat{x}$ is the Fourier transform of x. Equation (10.14) implies that the stability is introduced by uniformly penalizing all frequency components irrespective of the magnitude of frequencies. Yet, it is well known that the instability of the initial problem comes from high frequency components on the contrary to low frequency components.

Let us introduce the following penalty term where the regularization parameter α is no more defined as a weight but as an exponent:

$$\mathscr{P}(x, \alpha) = \|\left[I - (T^*T)^{\sqrt{\alpha}}\right]x\|^2. \tag{10.15}$$

In (10.15), $(T^*T)^{\sqrt{\alpha}}$ is defined via the spectral family $(E_\lambda)_\lambda$ associated with the self-adjoint operator T^*T, i.e.,

$$(T^*T)^{\sqrt{\alpha}}x = \int_{\lambda=0}^{||T^*T||_+} \lambda^{\sqrt{\alpha}} dE_\lambda x.$$

We keep the fit term defined in (10.12) and then the variational formulation of our new regularization method is given by

$$x_\alpha = \arg\min_{x\in X} ||Tx - y||^2 + \|\left[I - (T^*T)^{\sqrt{\alpha}}\right]x\|^2. \tag{10.16}$$

From the first-order optimality condition, we get that x_α is the solution to the linear equation:

$$\left[T^*T + \left(I - (T^*T)^{\sqrt{\alpha}}\right)^2\right]x = T^*y,$$

that is,

$$x_\alpha = \left[T^*T + \left(I - (T^*T)^{\sqrt{\alpha}}\right)^2\right]^{-1} T^*y. \tag{10.17}$$

From (10.17), we see that the new method can also be defined via the so-called *generator* function g_α, i.e.,

$$x_\alpha = g_\alpha(T^*T)T^*y, \tag{10.18}$$

with the function g_α defined by

$$g_\alpha(\lambda) = \frac{1}{\lambda + (1 - \lambda\sqrt{\alpha})^2}, \quad \lambda \in (0, ||T^*T||]. \tag{10.19}$$

Let us also define the *residual* function r_α corresponding to g_α as follows:

$$r_\alpha(\lambda) := 1 - \lambda g_\alpha(\lambda) = \frac{(1 - \lambda\sqrt{\alpha})^2}{\lambda + (1 - \lambda\sqrt{\alpha})^2}, \quad \lambda \in (0, ||T^*T||]. \tag{10.20}$$

The functions g_α and r_α defined in (10.19) and (10.20) are important since they will be repeatedly used in the convergence analysis of the regularization method. In fact, the regularization error $x^\dagger - x_\alpha$ and the propagated error $x_\alpha - x_\alpha^\delta$ are expressed via the functions r_α and g_α as follows:

$$x^\dagger - x_\alpha = r_\alpha(T^*T)x^\dagger, \quad x_\alpha - x_\alpha^\delta = g_\alpha(T^*T)T^*(y - y^\delta).$$

Finally, notice that the function g_α defined in (10.19) indeed satisfies the basic requirements for defining a regularization method, i.e.,

(a) g_α is continuous,
(b) $\forall \alpha > 0, \quad \sup_{\lambda \in (0, ||T^*T||]} \lambda g_\alpha(\lambda) \leq 1 < \infty$,
(c) $\lim_{\alpha \downarrow 0} g_\alpha(\lambda) = 1/\lambda$.

From (b) and (c), we deduce the convergence of the new regularization method by application of [11, Theorem 4.1]. Before going to optimality results, let us state some basic estimates (proven in the appendix) about the functions g_α and r_α.

Proposition 10.1 *Let the function g_α be defined by (10.19). Then for all $a < 1$ and $\alpha < 1$,*

$$\sup_{\lambda \in (0,a]} \sqrt{\lambda} g_\alpha(\lambda) = O\left(\frac{1}{\sqrt{\alpha}}\right). \tag{10.21}$$

Lemma 10.2.1 *For all α and λ satisfying $0 < \alpha \leq \lambda < 1$, the following estimates hold for the function r_α defined in (10.20):*

$$r_\alpha(\lambda) \leq \frac{9}{4}\left(\frac{\alpha|\ln(\lambda)|^2}{\lambda + \alpha|\ln(\lambda)|^2}\right). \tag{10.22}$$

10.3 Optimality Results

Before studying the optimality of the method presented in Sect. 10.2, we need first to recall general optimality results under source condition of the form (10.5). For doing so, let us specify assumptions on the function φ which defines the source set $X_\varphi(\rho)$.

Assumption 5 The function $\varphi : (0, a] \to \mathbb{R}_+$ is continuous, monotonically increasing and satisfies

(i) $\lim_{\lambda \downarrow 0} \varphi(\lambda) = 0$,
(ii) the function $\phi : (0, \varphi^2(a)] \to (0, a\varphi^2(a)]$ defined by

$$\phi(\lambda) = \lambda(\varphi^2)^{-1}(\lambda) \tag{10.23}$$

is convex.

□

Under Assumption 5 on the function φ, the following result from [37] holds and we can then define optimality under source condition (10.5).

Theorem 10.3.1 *Let $X_\varphi(\rho)$ be as in (10.5) and let Assumption 5 be fulfilled. Let the function ϕ be defined by (10.23). Then*

$$\omega(\delta, X_\varphi(\rho)) \leq \rho\sqrt{\phi^{-1}\left(\frac{\delta^2}{\rho^2}\right)}. \tag{10.24}$$

Moreover, if $\delta^2/\rho^2 \in \sigma\left(T^*T\varphi^2(T^*T)\right)$, then equality holds in (10.24).

A similar result to this theorem can be found in [20, Sect. 2], and [28, Sect. 3].

Remark 10.1 In [28], the results corresponding to Theorem 10.3.1 are given in term of the function $\Theta : (0, a] \to (0, a\varphi(a)]$ defined by

$$\Theta(\lambda) = \sqrt{\lambda}\varphi(\lambda). \tag{10.25}$$

Then, by simple computations, we can find that

$$\rho\sqrt{\phi^{-1}\left(\frac{\delta^2}{\rho^2}\right)} = \rho\,\varphi\left(\Theta^{-1}(\delta/\rho)\right). \tag{10.26}$$

In such a case, the convexity of the function ϕ defined in (10.23) is equivalent to the convexity of the function $\chi(\lambda) = \Theta^2\left((\varphi^2)^{-1}(\lambda)\right)$ and the condition $\delta^2/\rho^2 \in \sigma\left(T^*T\varphi^2(T^*T)\right)$ which allows to get the equality in (10.24) is equivalent to $\delta/\rho \in \sigma\left(\Theta(T^*T)\right)$.

From Theorem 10.3.1 and Remark 10.1, we can deduce that under the source condition (10.5) and Assumption 5, the best possible worst case error is $\rho\,\varphi\left(\Theta^{-1}(\delta/\rho)\right)$ whence the following definition.

Definition 10.1 (*Optimality under general source conditions*) Let Assumption 5 be satisfied and consider the source condition $x^\dagger \in X_\varphi(\rho)$. A regularization method $R(\delta) : Y \to X$ is said to be

1. *optimal* if $\Delta(\delta, R(\delta), X_\varphi(\rho)) \leq \rho\,\varphi\left(\Theta^{-1}(\delta/\rho)\right)$;
2. *order optimal* if $\Delta(\delta, R(\delta), X_\varphi(\rho)) \leq C\rho\,\varphi\left(\Theta^{-1}(\delta/\rho)\right)$ for some constant $C \geq 1$;
3. *quasi-order optimal* if for all $\epsilon > 0$, $\Delta(\delta, R(\delta), X_\varphi(\rho)) = O\left(f_\epsilon(\delta)\right)$, where the function $f_\epsilon : \mathbb{R}_+ \to \mathbb{R}_+$ converges to $\varphi\left(\Theta^{-1}(\delta/\rho)\right)$ as ϵ decreases to 0, i.e., for all $\delta > 0$, $f_\epsilon(\delta) \to \varphi\left(\Theta^{-1}(\delta/\rho)\right)$ as ϵ decreases to 0.

Having defined the optimality under general source conditions, let us now consider the particular case of logarithmic source conditions. For logarithmic source conditions, the function φ equals the function $f_p : (0, a] \to \mathbb{R}_+$ defined by

$$f_p(\lambda) = (-\ln(\lambda))^{-p}. \tag{10.27}$$

Next it is easy to see that the only point to check in Assumption 5 is the convexity of the function ϕ defined in (10.23). Precisely, for the index function f_p, this function is $\phi_p : (0, \ln(1/a)^{-2p}] \to (0, a\ln(1/a)^{-2p}]$ defined by

$$\phi_p(\lambda) = \lambda \exp(-\lambda^{-1/2p})$$

which was proven to be convex on the interval [0, 1] in [26]. In order to fulfill Assumption 5 and avoid the singularity of the function f_p at $\lambda = 1$, we assume that $a \leq \exp(-1) < 1$, i.e., $||T^*T|| \leq \exp(-1)$. Notice that this is not actually a restriction, since Eq. (10.1) can always be rescaled in order to meet this criterion.

Due to (10.24) it suffices to compute $\sqrt{\phi_p^{-1}\left(\delta^2/\rho^2\right)}$ in order to define the optimality in logarithmic source conditions. Thanks again to [26], we have that

$$\sqrt{\phi_p^{-1}(s)} = f_p(s)(1 + o(1)) \text{ as } s \to 0. \tag{10.28}$$

Hence, we deduce the following definition of optimality in case of logarithmic source condition.

Definition 10.2 (*Optimality under logarithmic source condition*) Consider logarithmic source condition (10.10), on defining f_p as in (10.27), a regularization method $R(\delta) : Y \to X$ is said to be

- *optimal* if $\Delta(\delta, R(\delta), X_{f_p}(\rho)) \leq \rho f_p(\delta^2/\rho^2)(1 + o(1))$ as $\delta \to 0$,
- *order optimal* if $\Delta(\delta, R(\delta), X_{f_p}(\rho)) \leq C\rho f_p(\delta^2/\rho^2)(1 + o(1))$ as $\delta \to 0$.

In the sequel, we are interested in optimality with respect to the noise level δ. In this respect, we can characterize the order-optimality under logarithmic source conditions as follows.

Remark 10.2 By definition of the function f_p, we get that $O(f_p(\delta^2/\rho^2)) = O(f_p(\delta))$ as $\delta \to 0$. Hence, equivalently to Definition 10.2, a regularization method $R(\delta) : Y \to X$ is said to be order optimal under logarithmic source condition if

$$\Delta(\delta, R(\delta), X_{f_p}(\rho)) = O(f_p(\delta)) \quad \text{as} \quad \delta \to 0.$$

10.3.1 Optimality Under Logarithmic Source Conditions

Having given all the necessary definitions, let us now study the optimality of the method proposed in Sect. 10.2.

Proposition 10.2 *The regularization g_α defined by (10.19) has qualification f_p. That is:*

$$\sup_{0<\lambda\leq a} |r_\alpha(\lambda)|\, f_p(\lambda) = O\left(f_p(\alpha)\right). \tag{10.29}$$

The proof the Proposition 10.2 heavily relies on the following lemma which is proven in the appendix.

Lemma 10.3.2 *Let p and α be two positive numbers with $\alpha \leq \bar{\alpha} < 1$, let $a \in (0, 1)$ and $\Psi_{p,\alpha} : (0, a] \to \mathbb{R}_+$ be the function defined by*

$$\Psi_{p,\alpha}(\lambda) = \frac{|\ln(\lambda)|^{2-p}}{\lambda + \alpha|\ln(\lambda)|^2}. \tag{10.30}$$

Then, the following hold:

(i) The function $\Psi_{p,\alpha}$ is well defined and differentiable on $(0, a]$, and its derivative is given by

$$\Psi'_{p,\alpha}(\lambda) = \frac{\lambda^{-1}|\ln(\lambda)|^{1-p}}{(\lambda + \alpha|\ln(\lambda)|^2)^2} h(\lambda), \tag{10.31}$$

where

$$h(\lambda) = \alpha p|\ln(\lambda)|^2 - \lambda\,(2 - p + |\ln(\lambda)|)\,. \tag{10.32}$$

(ii) If $p \leq 2$, there exists at least one $\lambda(\alpha, p)$ where h vanishes. Moreover for every such $\lambda(\alpha, p)$, the following holds

$$\lambda(\alpha, p) \simeq \alpha|\ln(\alpha)|, \tag{10.33}$$

that is, there exists two constants c_1 and c_2 depending on p only such that

$$c_1\alpha|\ln(\alpha)| \le \lambda(\alpha, p) \le c_2\alpha|\ln(\alpha)|.$$

Moreover, this result still holds if $p > 2$, $\lambda < c \le \exp(2-p)$ *and* α *is small.*

(iii) The supremum of the function $\Psi_{p,\alpha}$ *on* $(0, a]$ *satisfies*

$$\sup_{0<\lambda\le a} \Psi_{p,\alpha}(\lambda) = O\left(\alpha^{-1}|\ln(\alpha)|^{-p}\right). \tag{10.34}$$

Having stated the above lemma, the proof of Proposition 10.2 easily follows:

Proof If $\lambda \le \alpha$ then the monotonicity of the function f_p and the fact that the residual function r_α is bounded by 1 on $(0, a]$ yields (10.29). If $\lambda \ge \alpha$ then from Lemma 10.3.2, we deduce that

$$\sup_{0<\lambda\le a} \frac{\alpha|\ln(\lambda)|^2}{\lambda + \alpha|\ln(\lambda)|^2} f_p(\lambda) = O\left(f_p(\alpha)\right)$$

which together with Lemma 10.2.1 yields (10.29). □

From Proposition 10.2, we deduce the following optimality result.

Theorem 10.3.2 *Let* $p > 0$, $x^\dagger \in X_{f_p}(\rho)$, *and* $y^\delta \in Y$ *satisfying* (10.2) *with* $y = Tx^\dagger$. *Assume that* $||T^*T|| \le \exp(-1)$ *and let* $x(\delta) = g_{\alpha(\delta)}(T^*T)T^*y^\delta$ *with the function* g_α *being defined by* (10.19) *and let* $\alpha(\delta) = \Theta_p^{-1}(\delta)$ *with* Θ_p *defined by*

$$\Theta_p(\lambda) = \sqrt{\lambda}(\ln(1/\lambda))^{-p}. \tag{10.35}$$

Then the order-optimal estimate

$$||x^\dagger - x(\delta)|| = O\left(f_p(\delta)\right) \quad \text{as} \quad \delta \to 0 \tag{10.36}$$

holds. Thus the regularization g_α *defined by* (10.19) *is order optimal under logarithmic source conditions.*

Proof As usual, we start with the following splitting

$$||x^\dagger - x_\alpha^\delta|| \le ||x^\dagger - x_\alpha|| + ||x_\alpha - x_\alpha^\delta||. \tag{10.37}$$

Using that $x^\dagger - x_\alpha = r_\alpha(T^*T)x^\dagger$, $x_\alpha - x_\alpha^\delta = g_\alpha(T^*T)T^*(y - y^\delta)$ together with the source condition $x^\dagger \in X_{f_p(\rho)}$, we deduce that

$$||x^\dagger - x_\alpha|| \le C_1 \sup_{\lambda\in(0,a]} r_\alpha(\lambda)\, f_p(\lambda) \tag{10.38}$$

and

$$||x_\alpha - x_\alpha^\delta|| \le \delta\, C_2 \sup_{\lambda\in(0,a]} \sqrt{\lambda} g_\alpha(\lambda). \tag{10.39}$$

By applying the Propositions 10.1 and 10.2 to (10.38), (10.39) and using (10.37), we get that

$$\|x^\dagger - x_\alpha^\delta\| \le C_1' f_p(\alpha) + C_2' \frac{\delta}{\sqrt{\alpha}}, \tag{10.40}$$

where C_1' and C_2' are constants independent of α and λ. Hence, by taking $\alpha := \Theta_p^{-1}(\delta)$, the estimate in (10.36) follows from

$$\|x^\dagger - x(\delta)\| = O(f_p(\Theta_p^{-1}(\delta))) = O(f_p(\delta^2)) = O(f_p(\delta)).$$

Corollary 10.1 *Let $p > 0$, $x^\dagger \in X_{f_p}(\rho)$, and $y^\delta \in Y$ satisfying (10.2) with $y = Tx^\dagger$. Assume that $\|T^*T\| \le \exp(-1)$ and let $x(\delta) = g_{\alpha(\delta)}(T^*T)T^*y^\delta$ with the function g_α being defined by (10.19) and $\alpha(\delta) = \delta$. Then the order-optimal estimate*

$$\|x^\dagger - x(\delta)\| = O\left(f_p(\delta)\right) \quad as \quad \delta \to 0$$

holds. Thus the regularization g_α defined by (10.19) is order optimal under logarithmic source conditions with an a-priori parameter choice rule independent of the smoothness of the solution $x^\dagger$.

Proof By considering $\alpha(\delta) = \delta$ in (10.40), we get

$$\|x^\dagger - x_\alpha^\delta\| \le C_1' f_p(\delta) + C_2' \sqrt{\delta} = O(f_p(\delta)) \quad \text{as} \quad \delta \to 0,$$

since $\sqrt{\delta} = O(f_p(\delta))$ as $\delta \to 0$. □

The next proposition describes a Morozov-like discrepancy rule which leads to order-optimal convergence rates under logarithmic source conditions.

Proposition 10.3 *Let $p > 0$, $x^\dagger \in X_{f_p}(\rho)$, and $y^\delta \in Y$ satisfying (10.2) with $y = Tx^\dagger$. Assume that $\|T^*T\| \le \exp(-1)$ and consider the a-posteriori parameter choice rule*

$$\alpha(\delta, y^\delta) = \sup\left\{\alpha > 0, \quad \|Tx_\alpha^\delta - y^\delta\| \le \delta + \sqrt{\delta}\right\}. \tag{10.41}$$

Let $x(\delta) = g_{\alpha(\delta, y^\delta)}(T^*T)T^*y^\delta$ with the function g_α defined by (10.19), then the order-optimal estimate

$$\|x^\dagger - x(\delta)\| = O\left(f_p(\delta)\right) \quad as \quad \delta \to 0 \tag{10.42}$$

holds. Thus the regularization g_α defined by (10.19) is order optimal under logarithmic source conditions with the a-posteriori parameter choice rule defined by (10.41).

The proof of Proposition 10.3 is deferred to Appendix.

A Converse Result

Theorem 10.3.2 establishes that the logarithmic source condition (10.10) is sufficient to imply the rate $f_p(\delta)$ in (10.36). Now we are going to prove that the logarithmic source condition (10.10) is not only sufficient but also almost necessary. The following result based on [20, Theorem 8] establishes a converse result in the noise-free case for the new regularization method.

Theorem 10.3.3 *Let* $x_\alpha = g_\alpha(T^*T)Ty$ *with* $y = Tx^\dagger$ *and let the function* g_α *be defined in* (10.19)*. Then the estimate*

$$||x^\dagger - x_\alpha|| = O(f_p(\alpha)) \tag{10.43}$$

implies that $x^\dagger \in X_{f_q}(\rho)$ *for some* $\rho > 0$ *for all* $0 < q < p$.

The proof consists in checking that the function g_α defined in (10.19) satisfies all the conditions stated in Theorem 8 of [20]. More precisely, we just need to check that there exists a constant $C_g > 0$ such that

$$\sup_{\lambda \in (0, ||T^*T||]} g_\alpha(\lambda) \leq \frac{C_g}{\alpha}.$$

But, from (10.62), we see that the latter condition is obviously fulfilled.

10.3.2 Optimality Under General Source Conditions

Let us state the following quasi-optimal result under general source conditions.

Theorem 10.3.4 *Let* $p > 0$, $x^\dagger \in X_\varphi(\rho)$*, where* φ *is a concave index function satisfying Assumption* 5 *and* $y^\delta \in Y$ *satisfying* $||y - y^\delta|| \leq \delta$ *with* $y = Tx^\dagger$ *and* $\delta \leq \Theta(a)$*. Assume that* $||T^*T|| \leq a \leq \exp(-1)$ *and let* $x(\delta) = g_{\alpha(\delta)}(T^*T)T^*y^\delta$ *with the function* g_α *defined in* (10.19)*. For small positive* ϵ*, let* $\alpha(\delta) = \Theta_\epsilon^{-1}(\delta)$ *where the function* Θ_ϵ *is defined by* $\Theta_\epsilon(\lambda) = \lambda^{-\epsilon}\Theta(\lambda)$ *with* Θ *given in* (10.25)*.*

Then the estimate

$$||x^\dagger - x(\delta)|| = O\left((\Theta_\epsilon^{-1}(\delta))^{-\epsilon}\varphi(\Theta_\epsilon^{-1}(\delta))\right) \quad as \quad \delta \to 0$$

holds. Moreover, as $\epsilon \downarrow 0$, $(\Theta_\epsilon^{-1}(\delta))^{-\epsilon}\varphi(\Theta_\epsilon^{-1}(\delta)) \to \varphi(\Theta^{-1}(\delta))$*. Thus the regularization method defined via the function* g_α *given in* (10.19) *is quasi-order optimal under general source conditions.*

Proof We study two cases: $\alpha \geq \lambda$ and $\alpha < \lambda$. In the first case, $\sup_{(0, \exp(-1)]} r_\alpha(\lambda)\varphi(\lambda) \leq \varphi(\alpha)$ by monotonicity of the function φ and the order-optimality follows trivially. Let us study the main case when $\alpha < \lambda$. From Lemma 10.2.1, we get, for $\lambda \in (0, a]$,

$$
\begin{aligned}
r_\alpha(\lambda)\varphi(\lambda) &\le \frac{9}{4}|\ln(\lambda)|^2 \frac{\alpha}{\lambda+\alpha|\ln(\lambda)|^2}\varphi(\lambda)\\
&\le \frac{9}{4}|\ln(\alpha)|^2 \frac{\alpha}{\lambda+\alpha|\ln(a)|^2}\varphi(\lambda)\\
&\le \frac{9}{4}\alpha^{-\epsilon}(\alpha^{\epsilon/2}|\ln(\alpha)|)^2 \frac{\alpha}{\lambda+\alpha|\ln(a)|^2}\varphi(\lambda)\\
&\le \frac{9}{4}\frac{4}{\epsilon^2}\alpha^{-\epsilon} \frac{\alpha\lambda}{\lambda+\alpha|\ln(a)|^2}\frac{\varphi(\lambda)}{\lambda}\\
&\le \frac{9}{4}\frac{4}{\epsilon^2}\alpha^{-\epsilon} \frac{\alpha\lambda}{\lambda+\alpha|\ln(a)|^2}\frac{\varphi(\alpha)}{\alpha} \quad \text{by concavity of } \varphi\\
&\le C_\epsilon \alpha^{-\epsilon}\varphi(\alpha).
\end{aligned} \tag{10.44}
$$

□

Hence $\sup_{(0,a]} r_\alpha(\lambda)\varphi(\lambda) \le C_\epsilon \alpha^{-\epsilon}\varphi(\alpha)$. From (10.38) and (10.39), and (10.21) we get

$$||x^\dagger - x_\alpha^\delta|| \le C_\epsilon \alpha^{-\epsilon}\varphi(\alpha) + \frac{\delta}{\sqrt{\alpha}}.$$

By taking $\alpha(\delta) = \Theta_\epsilon^{-1}(\delta)$ with $\Theta_\epsilon(\lambda) = \lambda^{1/2-\epsilon}\varphi(\lambda)$, we get

$$||x^\dagger - x(\delta)|| = O\left((\Theta_\epsilon^{-1}(\delta))^{-\epsilon}\varphi(\Theta_\epsilon^{-1}(\delta))\right).$$

Now, it remains to show that $(\Theta_\epsilon^{-1}(\delta))^{-\epsilon}\varphi(\Theta_\epsilon^{-1}(\delta))$ converges to the optimal rate $\varphi(\Theta^{-1}(\delta))$ as ϵ goes to 0. Let $\alpha_* = \Theta^{-1}(\delta)$ and $\alpha_\epsilon = \Theta_\epsilon^{-1}(\delta)$, let us show that α_ϵ converges to α_* as ϵ goes to 0. By the monotonicity of Θ_ϵ for $\epsilon \in (0, 1/2)$ and the fact that $\delta \le \Theta(a)$ and $a < 1$, we get that, for all $\epsilon \in (0, 1/2)$,

$$\frac{\delta}{\Theta(a)} \le 1 < a^{-\epsilon} \quad \Rightarrow \quad \delta \le a^{-\epsilon}\Theta(a) = \Theta_\epsilon(a) \quad \Rightarrow \quad \alpha_\epsilon = \Theta_\epsilon^{-1}(\delta) \le a.$$

Hence $\alpha_\epsilon \in (0, a]$ and the sequence $(\alpha_\epsilon)_\epsilon$ is bounded and thus it admits a converging subsequence. Let $(\alpha_{\epsilon_n})_n$ a converging subsequence of $(\alpha_\epsilon)_\epsilon$, and let $\tilde{\alpha}$ be its limit. Let us show that $\tilde{\alpha} = \alpha_*$.

Since $\alpha_{\epsilon_n} \to \tilde{\alpha}$ and Θ is continuous, $\Theta(\alpha_{\epsilon_n}) \to \Theta(\tilde{\alpha})$. But $\Theta(\alpha_{\epsilon_n}) = \alpha_{\epsilon_n}^{\epsilon_n}\Theta(\alpha_*)$ since $\delta = \Theta(\alpha_*)$ and $\delta = \Theta_\epsilon(\alpha_\epsilon)$ for all small positive ϵ. So we get

$$\alpha_{\epsilon_n}^{\epsilon_n}\Theta(\alpha_*) \to \Theta(\tilde{\alpha}) \quad \text{i.e.,} \quad \alpha_{\epsilon_n}^{\epsilon_n} \to \frac{\Theta(\tilde{\alpha})}{\Theta(\alpha_*)}. \tag{10.45}$$

By the convergence of the sequence $(\alpha_{\epsilon_n})_n$, we get that $\alpha_{\epsilon_n}^{\epsilon_n} = \exp(\epsilon_n \ln(\alpha_{\epsilon_n}))$ converges to 1, (10.45) proves that $\Theta(\tilde{\alpha}) = \Theta(\alpha_*)$ and by bijectivity of the function Θ, we deduce that $\tilde{\alpha} = \alpha_*$. Since the sequence $(\epsilon_n)_n$ was arbitrarily chosen, we deduce that the whole sequence $(\alpha_\epsilon)_\epsilon$ converges to α_* as $\epsilon \downarrow 0$. Thus we deduce that $\alpha_\epsilon^{-\epsilon} \to 1$ and $\varphi(\alpha_\epsilon) \to \varphi(\alpha_*)$ which implies that

$$(\Theta_\epsilon^{-1}(\delta))^{-\epsilon}\varphi(\Theta_\epsilon^{-1}(\delta)) \to \varphi(\Theta^{-1}(\delta)).$$

For Holder type source conditions, Theorem 10.3.4 reduces to the following theorem.

Theorem 10.3.5 *Consider the setting of Theorem 10.3.4 with the function $\varphi(t) = t^\mu$, i.e., $x^\dagger \in (T^*T)^\mu$, then there exists an a-priori selection rule $\alpha(\delta)$ such that the following holds:*

$$\|x^\dagger - x^\delta_{\alpha(\delta)}\| = \begin{cases} O\left(\delta^{\frac{2\sigma}{2\sigma+1}}\right) & \forall \sigma < \mu, \text{ if } \mu \leq 1 \\ O\left(\delta^{\frac{2}{3}}\right) & , \text{ if } \mu > 1. \end{cases} \tag{10.46}$$

Remark 10.3 By defining a variant of the new regularization method where the approximate solution x^δ_α is defined as the solution of the optimization problem

$$x^\delta_\alpha = \arg\min_{x\in X} ||(T^*T)^{\sqrt{\alpha}}y^\delta - Tx||^2 + ||\left[I - (T^*T)^{\sqrt{\alpha}}\right]x||^2,$$

we can prove order-optimal rate under Holder type source condition but with a lower qualification index $\mu_0 = 1/2$. This variant is motivated by the mollification regularization method, where a target object defined as a smooth version of $x^\dagger$ is fixed prior to the regularization (see, e.g., [1, 7]). In this respect, the target object here is given as $(T^*T)^{\sqrt{\alpha}}x^\dagger$. This choice is legitimated by the smoothness property of the operator T and the fact that as α goes to 0, this target object converges to the solution $x^\dagger$. The study of this variant and the corresponding optimality results is beyond the scope of this chapter.

10.4 A Framework for Comparison

In the sequel, we are going to compare the new method with three continuous regularization methods: Tikhonov [38], spectral cut-off [11], Showalter [11] and one iterative regularization method: conjugate gradient [11, 21]. We recall that the first three methods (Tikhonov, spectral cut-off and Showalter) are linear methods on the contrary to conjugate gradient which is an iterative non-linear regularization method. Obviously the new method, Tikhonov, spectral cut-off and Showalter are members of the family of general regularization methods defined via a *generator* function. Roughly speaking, each regularization method is defined via a so-called *generator* function $g^{reg}_\alpha(\lambda)$ which converges pointwise to $1/\lambda$ as α goes to 0 and the regularized solution $x^\delta_{\alpha,reg}$ is defined by

$$x^\delta_{\alpha,reg} = g^{reg}_\alpha(T^*T)T^*y^\delta. \tag{10.47}$$

In this respect, the functions $g_\alpha^{reg}(\lambda)$ associated with Tikhonov, spectral cut-off, Showalter and the new method are defined as follows:

$$g_\alpha^{tik}(\lambda) = \frac{1}{\lambda+\alpha}, \quad g_\alpha^{sc}(\lambda) = \frac{1}{\lambda}1_{\{\lambda\geq\alpha\}}, \quad g_\alpha^{sw}(\lambda) = \frac{1-e^{-\lambda/\alpha}}{\lambda},$$
$$g_\alpha^{nrm}(\lambda) = \frac{1}{\lambda+(1-\lambda^{\sqrt{\alpha}})^2}, \tag{10.48}$$

where $\lambda \in (0, a]$ with $||T^*T|| \leq a < 1$.

Before getting into comparison of the new method to other regularization techniques, let us first point out a way of computing the regularized solution $x_{\alpha,nrm}^\delta$ of the new method.

10.4.1 Computation of the Regularized Solution $x_{\alpha,nrm}^\delta$

One way of computing the regularized solution $x_{\alpha,nrm}^\delta$ of the new method is by computing the singular value decomposition of operator T. That is to find a system (u_k, σ_k, v_k) such that

- the sequence $(u_k)_k$ forms a Hilbert basis of X,
- the sequence $(v_k)_k$ forms a Hilbert basis of the closure of the range of T,
- the sequence $(\sigma)_k$ is positive, decreasing and satisfies $Tu_k = \sigma_k v_k$ and $T^*v_k = \sigma_k u_k$.

Given that decomposition of T, it is trivial to see that the operator T^*T is diagonal in the Hilbert basis $(u_k)_k$. Therefore, given a function g defined on the interval $(0, \sigma_1^2)$, the operator $g(T^*T)$ is nothing but the diagonal operator defined on the Hilbert basis $(u_k)_k$ by $g(T^*T)u_k = g(\sigma_k^2)u_k$. Hence given the singular value decomposition (u_k, σ_k, v_k) of T, from (10.47) (with $reg = nrm$), the regularized solution $x_{\alpha,nrm}^\delta$ can be computed as

$$x_{\alpha,nrm}^\delta = \sum_k g_\alpha^{nrm}(\sigma_k^2)\,\langle T^*y^\delta, u_k\rangle\, u_k = \sum_k \frac{\sigma_k}{\sigma_k^2 + \left(1-\sigma_k^{2\sqrt{\alpha}}\right)^2}\,\langle y^\delta, v_k\rangle\, u_k. \tag{10.49}$$

Remark 10.4 The above singular value decomposition of operator T is only possible if T is a compact operator. However, it is important to notice that the new method does not apply only to compact operator. Indeed, the new method is based on the spectral family $(E_\lambda)_\lambda$ associated with the self-adjoint operator T^*T, and spectral family exists even for non-compact operator as pointed out in [11, Proposition 2.14]. This allows for the definition of a function applied to a self-adjoint non-compact operator. Of course, one might ask how we can compute the regularized solution $x_{\alpha,nrm}^\delta$ in such a case. By noticing that in practice, we always discretize Eq. (10.1) into matrix formulation, we can compute the singular value decomposition of the matrix representing the discretization of operator T and then apply (10.49) to compute $x^{\delta}_{\alpha,nrm}$.

It is important to notice that a crucial step in the computation of the regularized solution $x^{\delta}_{\alpha,nrm}$ is the singular value decomposition step which should be done rigorously especially for exponentially ill-posed problems. That is why we propose a state of the art algorithm as LAPACK's `dgesvd()` routine for SVD computation (see, e.g., [12, Sect. 8.6] for description of method). For an easy application, it is to be noted that this routine is implemented in the function `svd()` in `Matlab`. In Sect. 10.5, we will see that even for a very ill-conditioned matrix, we can still compute the regularized solution $x^{\delta}_{\alpha,nrm}$ very efficiently using the function `svd()` in `Matlab`.

Above, we saw that the new approximate solution $x^{\delta}_{\alpha,nrm}$ is computable using the singular value decomposition of operator T which might be delicate to compute. However, in some cases, there is an alternative for computing $x^{\delta}_{\alpha,nrm}$ when the operator $\log(T^*T)$ is explicitly known. Indeed, if the operator $\log(T^*T)$ is explicitly known, then the solution $u : \mathbb{R}_+ \to X$ to the initial value problem:

$$\begin{cases} u'(t) - \log(T^*T)u(t) = 0, \quad t \in \mathbb{R}_+ \\ u(0) = x, \end{cases} \tag{10.50}$$

evaluated at $t = \sqrt{\alpha}$ is nothing but $(T^*T)^{\sqrt{\alpha}}x$, i.e., $(T^*T)^{\sqrt{\alpha}}x = u(\sqrt{\alpha})$. Hence, through the resolution of the ordinary differential equation (10.50), the penalty term $\|\left[I - (T^*T)^{\sqrt{\alpha}}\right]x\|^2$ can be computed and this allows to compute the approximate solution $x^{\delta}_{\alpha,nrm}$.

An example of exponentially ill-posed problems for which the operator $\log(T^*T)$ is known is the backward heat equation. More precisely, let Ω be a smooth subset of $\mathbb{R}^n$ with $n \leq 3$ and $u : \Omega \times (0, \bar{t}] \to \mathbb{R}$ be the solution to the initial boundary value problem

$$\begin{cases} \frac{\partial u}{\partial t} = \Delta u, \quad \Omega \times (0, \bar{t}) \\ u(\cdot, 0) \;= f, \quad \Omega \\ u = 0 \;\text{ or }\; \frac{\partial u}{\partial \nu} = 0, \quad \text{on} \;\; \partial\Omega \times (0, \bar{t}]. \end{cases} \tag{10.51}$$

Assume we want to recover the initial temperature $f \in L^2(\Omega)$ given the final temperature $u(\cdot, \bar{t})$. By interpreting the heat Eq. (10.51) as an ordinary differential equation for the function $U : \left[0, \bar{t}\right] \to \mathcal{D}(\Delta) \subset L^2(\Omega)$, $t \to U(t) = u(\cdot, t)$, with the initial value $U(0) = f$, where

$$\mathcal{D}(\Delta) = H^2(\Omega) \cap H^1_0(\Omega) \;\text{ or }\; \mathcal{D}(\Delta) = \left\{ f \in H^2(\Omega), \;\; \frac{\partial f}{\partial \nu} = 0 \text{ on } \partial\Omega \right\},$$

we get that $U(t) = \exp(t\Delta) f$ for $t \in (0, \bar{t}]$, where $(\exp(t\Delta))_{t>0}$ is the strongly continuous semi-group generated by the unbounded self-adjoint linear operator Δ. This implies that the equation satisfied by the initial temperature f is nothing but

$$\exp(\bar{t}\Delta) f = u(\cdot, \bar{t}). \tag{10.52}$$

From (10.52), we deduce that $T^*T = \exp(2\bar{t}\Delta)$ and $\log(T^*T) = 2\bar{t}\Delta$ and thus operator $(T^*T)^{\sqrt{\alpha}}$ can be evaluated at a function $x \in L^2(\Omega)$ as the solution to the initial value problem

$$\begin{cases} u'(t) - 2\bar{t}\Delta u(t) = 0, & t \in \mathbb{R}_+ \\ u(0) = x, & \\ u(t) \in \mathcal{D}(\Delta), & \text{for } t \in \mathbb{R}_+ \end{cases} \tag{10.53}$$

evaluated at $t = \sqrt{\alpha}$.

In addition to the backward heat equation, there are other exponentially ill-posed problems for which $\log(T^*T)$ is known. This includes sideways heat equation (see [20, Sect. 8.3]) and more generally inverse heat conduction problems (see, e.g., [31, Sects. 3 & 4]).

10.4.2 Tikhonov Versus New Method

From the variational formulation of Tikhonov and the new method, we can see that both methods differ by the penalty term. For Tikhonov method, the penalty term is $\alpha\|x\|^2$ whereas for the new method, the penalty term is $\left\|\left[I - (T^*T)^{\sqrt{\alpha}}\right]x\right\|^2$. By considering $X = L^2(\mathbb{R}^n)$ for instance, by using the Parseval identity, we see that the penalty term is equal to $\alpha \left\|\hat{x}\right\|_{L^2(\mathbb{R}^n)}$. Therefore the weight α equally penalizes all frequency components irrespective of the magnitude of frequencies even though instability mainly comes from high frequency components. This is actually a drawback of the Tikhonov method which may induce an unfavorable trade-off between stability and fidelity to the model (see, e.g., [1], Fig. 10.4). On the contrary, for the new regularization method, high frequency components are much more regularized compared to low frequency components which are less and less regularized as the singular values increase to 1. In this way, we expect the new method to achieve a better trade-off between stability and fidelity to the model. Moreover, for exponentially ill-posed problems, the ill-posedness is accentuated due to the magnitude of singular values, the instability introduced by high frequency components are more pronounced and we expect the new regularization method to yield better approximations of $x^\dagger$.

10.4.3 Spectral Cut-Off Versus New Method

On the contrary to Tikhonov method, both spectral cut-off and the new method treat high frequency components and low frequency components differently. However, spectral cut-off regularized high frequency components by a mere cut-off and this may be too violent in several situations. Indeed even though high frequency components induce instability, they also carry some information which should not completely left out. For instance, for mildly ill-posed problems, this truncation will be very damaging on the quality of the approximation while for exponentially ill-posed problem, this truncation will be less damaging. A smooth transition (in term of regularization) from small singular values to other singular values would be more meaningful and desirable. This is actually what is done for the new method. Another advantage of the new method compared to spectral cut-off is the variational formulation of the new method which allows to add to the problem a-priori constraint on the solution (e.g., positivity, geometrical constraints, etc...).

10.4.4 Showalter Versus New Method

A major difference between Showalter method and the new method is that Showalter method does not have a variational formulation. Given that, for the Showalter method, it is not clear what is actually penalized in order to stabilize the problem. Moreover it would be difficult if not impossible to add a-priori constraints on the solution. Given a data y^δ, by inspecting the Showalter regularized solution which is given by $x_\alpha^\delta = \int_0^{1/\alpha} e^{-sT^*T} ds\, T^* y^\delta$, we see that the method introduces stability by truncating the integral $\int_0^{+\infty} e^{-sT^*T} ds\, T^* y^\delta = (T^*T)^{-1} T^* y^\delta$ on the interval $(0, 1/\alpha)$. On the other hand, we can see that, as the Tikhonov method, for all regularization parameter $\alpha > 0$, the *generator* function g_α^{sw} of Showalter method is strictly decreasing on the contrary to the generator function g_α^{nrm} of the new method which always exhibits a maximum close to $\lambda = 0$. This implies that the Showalter method cannot be seen as a smooth version of spectral cut-off which yields a smooth transition (in term of regularization) from high frequency components to low frequency components, on the contrary to the new method. Concerning the computation of the regularized solution $x_{\alpha,sw}^\delta$ for the Showalter method, it is important to notice that $x_{\alpha,sw}^\delta$ is the solution $u_\delta : \mathbb{R}_+ \to X$ of the initial value problem:

$$\begin{cases} u_\delta'(t) + T^*T u_\delta(t) = T^* y^\delta, & t \in \mathbb{R}_+ \\ u_\delta(0) = 0, \end{cases} \tag{10.54}$$

evaluated at $t = 1/\alpha$, i.e., $x_{\alpha,sw}^\delta = u_\delta(1/\alpha)$. By solving (10.54) using the forward finite difference of step size h, we get that u_δ can be approximated as

$$u_\delta(t+h) \approx u_\delta(t) + h\left[T^*y^\delta - T^*Tu_\delta(t)\right], \quad \text{with} \quad u_\delta(0) = 0. \tag{10.55}$$

10.4.5 Conjugate Gradient Versus New Method

Unlike all the other regularization methods under consideration (Tikhonov, spectral cut-off, Showalter, and the new method), the conjugate gradient method is an iterative non-linear regularization method. The conjugate gradient method regularizes Problem (10.1) by iteratively approximating $x^\dagger$ by the minimizer x_k of the functional $f(x) = ||Tx - y||^2$ on finite-dimensional Krylov subspaces

$$V_k = \text{span}\left\{T^*y, (T^*T)T^*y, ..., (T^*T)^{k-1}T^*y\right\},$$

where $k \geq 1$ and $k \in \mathbb{N}$. A major advantage of the conjugate gradient is the easy computation of regularized solution x_k (see, e.g., algorithm given in [21, Fig. 2.2]) and the fast convergence on the contrary to Landweber. However, as pointed out in [11, Theorem 7.6], the operator R_k which maps the data y to the regularized solution x_k is not always continuous contrarily to the new method. Moreover, compared to other regularization methods, there is no a-priori rules $k(\delta)$ such that $x^\delta_{k(\delta)}$ converges to $x^\dagger$ as $\delta \to 0$ [9].

A comparative plot of the *generator* functions g_α^{reg} associated with Tikhonov, spectral cut-off, Showalter, and the new method is given in Fig. 10.1.

Remark 10.5 On the contrary to generator functions of Tikhonov and Showalter, the generator function g_α^{nrm} associated with the new regularization always exhibits a maximum close to $\lambda = 0$ and the function always equals 1 at $\lambda = 0$. Indeed, it is trivial to check that both functions g_α^{tik} and g_α^{sw} are strictly decreasing for all $\alpha > 0$. Hence, the function g_α^{nrm} is the only one which can be seen as a smooth version of the function g_α^{sc} associated with spectral cut-off which has a very crude transition at $\lambda = \alpha$.

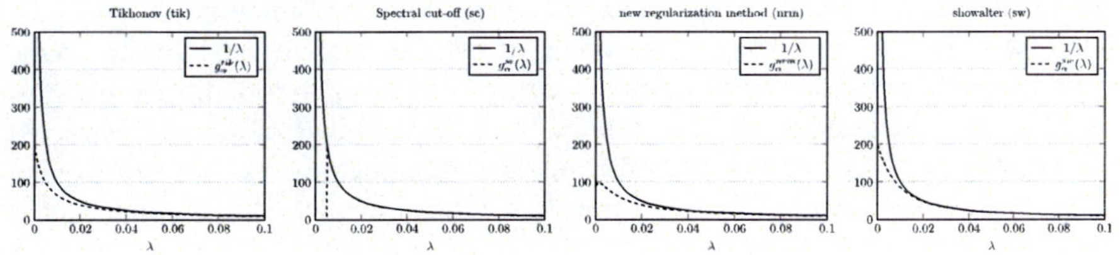

Fig. 10.1 Comparison generator function g_α^{reg} to function $\lambda \mapsto 1/\lambda$ for the four regularization methods (reg = tik,sc,nrm,sw)

10.5 Numerical Illustration

The aim here is to compare the performance of our new regularization method (`nrm`) to the classical Tikhonov method (`tik`), spectral cut-off (`tsvd`), Showalter (`sw`), and conjugate gradient (`cg`) for some (ill-posed) test problems. We consider three test problems. The first one is a problem of image reconstruction found in [36]. The second problem is a Fredholm integral equation of the first kind taken from [2] and the last one is an inverse heat problem. For the discretization of these problems, we use the functions `shaw()`, `baart()` and `heat()` of the `matlab` regularization tool package (see [18]). For the `heat()` and `shaw()` test problems, the discretization is done by collocation with approximation of integrals by quadrature rules. For the `baart()` test problem, the discretization is done by Galerkin methods with orthonormal box functions as basis functions. In the `matlab` regularization tool package, each of the functions `shaw()`, `baart()`, and `heat()` takes as input a discretization level n representing either the number of collocations points or the number of box functions considered on the interval [0, 1]. Given the input n, each function returns three outputs: a matrix A, a vector $x^\dagger$, and the vector y obtained by discretization without noise added. In this section, we considered the following discretization level for the `shaw()`, `baart()` and `heat()` test problem, respectively, $n_{shaw} = 160$, $n_{baart} = 150$ and $n_{heat} = 150$. For the simulations, we define noisy data $y_\xi = y + \xi$ where ξ is a random white noise vector. In order to compute the regularized solution $x^\delta_{\alpha,nrm}$ for the new method, we compute the SVD with the function `svd()` in `Matlab` and applied (10.49).

We consider a 4% noise level, the noise level being defined here by the ratio of the noise to the exact data. More precisely, given a noisy data $y_\xi = y + \xi$, the noise level is defined by $\sqrt{E(||\xi||^2)}/||y||$. In order to illustrate the ill-posedness of each test problem, we give on Fig. 10.2 the conditioning associated with each matrix A_{shaw}, A_{baart}, and A_{heat} obtained from the discretization of each problem.

We perform a Monte Carlo experiment of 3000 replications. In each replication, we compute the best relative error for each regularization method. Next we compute the minimum, maximum, average, and standard deviation errors (denoted by e_{min}, e_{max}, $\bar{e}$, $\sigma(e)$ over the 3000 replications for each schemes (`nrm` and `tik`, `tsvd`, `sw`, and `cg`). Figure 10.3 summarizes the results of the overall simulations.

In order to assess and compare the trade-off between stability and fidelity to the model for Tikhonov and the new method, we plot the curve of the conditioning versus relative error. The conditioning here is the condition number of the reconstructed operator $g^{reg}_\alpha(T^*T)$ associated with the regularization method. For instance, using the invariance of conditioning by inversion, for the new method, the conditioning

	shaw	baart	heat
cond(A)	2.3283×10^{19}	2.4561×10^{17}	1.2706×10^{49}

Fig. 10.2 Conditioning of the matrices A_{shaw}, A_{baart} and A_{heat} for $n_{shaw} = 160$, $n_{baart} = 150$, and $n_{heat} = 150$

baart

	nrm	tik	tsvd	sw	cg
e_{max}	0.34774	0.356887	0.346026	0.348538	0.346263
e_{min}	0.053773	0.055549	0.114955	0.051623	0.059753
$\bar{e}$	**0.16575**	**0.17078**	**0.19223**	**0.16581**	**0.19237**
$\sigma(e)$	**0.03993**	**0.04547**	**0.03466**	**0.04016**	**0.03488**

shaw

	nrm	tik	tsvd	sw	cg
e_{max}	0.186337	0.200609	0.181682	0.187038	0.186546
e_{min}	0.049132	0.048754	0.052527	0.047292	0.050999
$\bar{e}$	**0.13684**	**0.13857**	**0.15665**	**0.13746**	**0.15884**
$\sigma(e)$	**0.03307**	**0.03507**	**0.01955**	**0.03332**	**0.02269**

heat

	nrm	tik	tsvd	sw	cg
e_{max}	0.274714	0.283154	0.271837	0.270857	0.271285
e_{min}	0.100167	0.10736	0.115237	0.097847	0.106211
$\bar{e}$	**0.18857**	**0.19861**	**0.19007**	**0.18862**	**0.19365**
$\sigma(e)$	**0.02788**	**0.02756**	**0.02889**	**0.02725**	**0.02514**

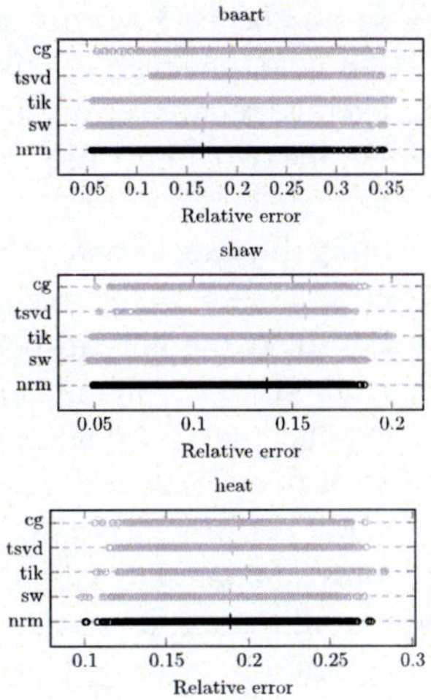

Fig. 10.3 Summary of the Monte Carlo experiment. On the right figure, the average relative error for each method is represented by the vertical stick

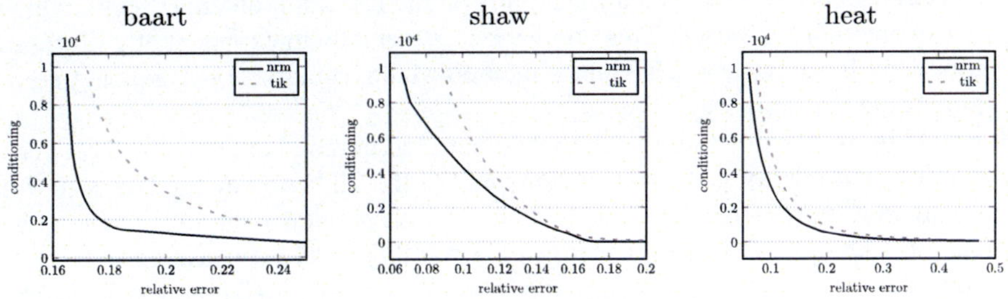

Fig. 10.4 Comparison of the trade-off between stability and accuracy of the new method (`nrm`) to Tikhonov (`tik`) for the three test problems: `shaw`, `baart`, and `heat`

corresponds to the condition number of the operator $T^*T + \left[I - (T^*T)^{\sqrt{\alpha}}\right]^2$ while for Tikhonov method, it corresponds to the condition number of $T^*T + \alpha I$. In this respect, for two regularization methods, the best one is the one whose curve is below the other one as it achieves the same relative errors with smaller conditioning. On Fig. 10.4, for each test problem, we compare the curve of conditioning versus relative error of the new method and Tikhonov method.

Notice that the first two problems (`shaw` and `baart`) are mildly ill-posed while the third problem (`heat`) is exponentially ill-posed.

Comments:

From Figs. 10.3 and 10.4, we can do the following comments:

- The new method always yields the smallest average relative errors among the five methods.

- From Fig. 10.3, we can see that both spectral cut-off and conjugate gradient yield the worst average relative errors except for the `heat` test problem where their average relative errors are smaller than the one of Tikhonov.
- For the two mildly ill-posed problems `shaw` and `baart`, Tikhonov method yields average relative errors close to the smallest one. On the contrary, for the exponentially ill-posed problem `heat`, Tikhonov method yields the worst average relative error among all the five methods.
- For the two mildly ill-posed problems (`shaw` and `baart`), the errors of the new method are not significantly smaller than those of Tikhonov on the contrary to the exponentially ill-posed problem (`heat`) where the new method produces smaller error than Tikhonov (about 5% smaller). This confirms our prediction about the better performance of the new method in instance of exponentially ill-posed problems compared to Tikhonov.
- For all three test problems, the new method performs better than spectral cut-off as could be expected. Moreover, the gap between the error is larger for the first two test problems which are mildly ill-posed. This also confirms the prediction about the poor performance of spectral cut-off for mildly ill-posed problems.
- On the contrary to the two mildly ill-posed problems (`shaw` and `baart`), spectral cut-off performs better than Tikhonov on the last test problem (`heat`), which is exponentially ill-posed. This emphasizes, especially in exponentially ill-posed problems, the drawback of Tikhonov method which regularizes all frequency components in the same way.
- From Fig. 10.4, we can see that the new method achieves a better trade-off between stability and fidelity to the model compared to the Tikhonov method. Indeed, for the three test problems the curve associated with the new method lies below the one of Tikhonov. This means that given a stability level κ (measured in term of conditioning), the new method provided a smaller error than Tikhonov. Conversely, for a given error level ϵ, the new method provides a lower conditioning of the reconstructed operator compared to Tikhonov. This also validates the prediction stated earlier.

10.6 Parameter Selection Rules

In this section, we are interested in the choice of the regularization parameter α. For practical purposes, we assume that we don't know the smoothness conditions satisfied by the unknown solution $x^\dagger$. Consequently, we are left with two types of parameter choice rules: A-posteriori rules which use information on the noise level δ and heuristic rules which depend only on the noisy data y^δ. However a huge default of a-posteriori parameter choice rules is their dependence on the noise level δ which, in practice, is hardly available or well estimated in most circumstances. In [8], it is shown how an underestimation or overestimation of the noise level δ may induce serious computation issues for the Morozov principle. Moreover, in [15], it is illustrated how heuristic rules may outperform sophisticated a-posteriori rules. Given those reasons, we turn to heuristic (or data driven) selection rules. We recall that, due to Bakushinskii véto [3], such rules are not convergent. But still, as mentioned earlier, heuristic rules may yield better approximations compared to sophisticated a-posteriori rules (see, e.g., [15]) and this is not surprising as the Bakushinskii result is based on worst case scenario.

We applied five noise-free parameter choice rules to the new method and the four regularization methods on the three test problems defined in Sect. 10.5: the generalized cross validation (GCV), the discrete quasi-optimality rule (DQO), two heuristic rules (H1 and H2), and a variant of the L-curve method (LCV) each described in [11, Sect. 4.5]. Roughly speaking, the parameter α chosen by each of those selection rules is as follows:

- The GCV rule consists in choosing $\hat{\alpha}$ as

$$\hat{\alpha} = \arg\min_{\alpha} \frac{||Tx_\alpha^\delta - y^\delta||}{\operatorname{tr}(r_\alpha(T^*T))},$$

  where r_α is the *residual* function associated with the regularization method under consideration. For the new method, r_α is defined in (10.20).
- The DQO method consists in discretizing the regularization parameter α as

$$\alpha_n = \alpha_0 q^n, \quad \alpha_0 \in (0, ||T^*T||], \quad \text{and} \quad 0 < q < 1.$$

  Next, the parameter $\hat{\alpha}$ is chosen as

$$\hat{\alpha} = \alpha_{\hat{n}} \quad \text{with} \quad \hat{n} = \arg\min_{n\in\mathbb{N}} ||x_{\alpha_{n+1}}^\delta - x_{\alpha_n}^\delta||. \tag{10.56}$$

  Recall that this rule defined by (10.56) is actually one of the variants of the continuous quasi-optimality rule defined by

$$\hat{\alpha} = \arg\min_{\alpha} ||\alpha \frac{\partial x_\alpha^\delta}{\partial \alpha}||.$$

- The third rule H1 taken in [11, Sect. 4.5] consists in choosing the parameter $\hat{\alpha}$ as

$$\hat{\alpha} = \arg\min_{\alpha} \frac{1}{\sqrt{\alpha}} ||Tx_\alpha^\delta - y^\delta||. \tag{10.57}$$

- The fourth rule H2 which is a variant of the third rule H1 consists in choosing the parameter $\hat{\alpha}$ as

$$\hat{\alpha} = \arg\min_{\alpha} \frac{1}{\alpha} ||T^*(Tx_\alpha^\delta - y^\delta)||. \tag{10.58}$$

- The variant of the L-curve (LCV) method considered here (see [11, Proposition 4.37]) consists in choosing the regularization parameter $\hat{\alpha}$ as

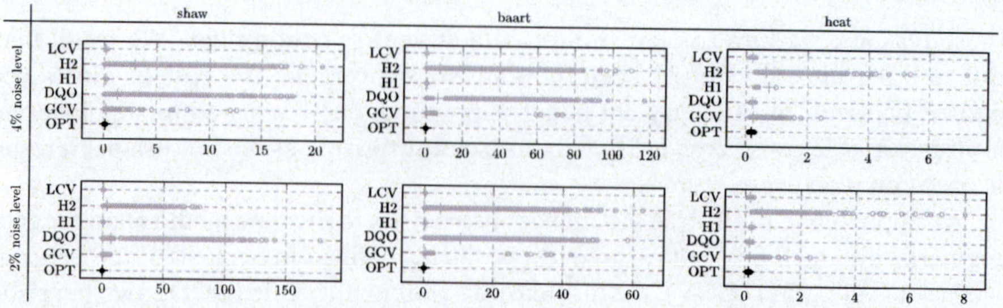

Fig. 10.5 Comparison of the relative error obtained by each selection rules (GCV, DQO, H1, H2, and LCV) for the two noise levels with the new method for the three tests problems `shaw`, `baart` and `heat`. On each plot, the x-axis corresponds to relative error and the vertical stick indicates the average relative error

$$\hat{\alpha} = \arg\min_{\alpha} ||x_\alpha^\delta||\,||Tx_\alpha^\delta - y^\delta||. \tag{10.59}$$

Recall that this rule actually tries to locate the parameter $\hat{\alpha}$ corresponding to the corner of the L-curve plot $||x_\alpha^\delta||$ versus $||Tx_\alpha^\delta - y^\delta||$ in a log-log scale. For more details about the L-curve method, see, e.g., [10, 16, 17].

For a comprehensive discussion of the above heuristic rules and conditions under which convergence is established, see [13, 25, 39] for GCV, [4–6, 22] for Quasi-optimality and [11, Sect. 4.5] for the rules H1, H2 and LCV.

For assessing the performance of each selection rule, we perform a Monte Carlo experiment of 3000 replications. For each replication, each test problem (`baart`, `shaw`, `heat`), and each regularization method (`nrm`, `tik`, `tsvd`, `sw` and `cg`), we compute the optimal regularization parameter α_{OPT}, the one chosen by each selection rule ($\alpha_{GCV}, \alpha_{DQO}, \alpha_{H1}, \alpha_{H2}, \alpha_{LCV}$). We also compute the corresponding relative errors:

$$\frac{||x^\dagger - x_{\alpha_{OPT}}^\delta||}{||x^\dagger||}, \quad \frac{||x^\dagger - x_{\alpha_{GCV}}^\delta||}{||x^\dagger||}, \quad \frac{||x^\dagger - x_{\alpha_{DQO}}^\delta||}{||x^\dagger||}, \quad \frac{||x^\dagger - x_{\alpha_{H1}}^\delta||}{||x^\dagger||}, \quad \frac{||x^\dagger - x_{\alpha_{H2}}^\delta||}{||x^\dagger||},$$

and $\frac{||x^\dagger - x_{\alpha_{LCV}}^\delta||}{||x^\dagger||}$. In order to analyze the convergence behavior of the selection rules, we consider two noise levels: 2% and 4%. The results are shown in Fig. 10.5 and Tables 10.1, 10.2, 10.3, 10.4, 10.5, 10.6, 10.7, 10.8, and 10.9.

From Tables 10.1, 10.2 and Fig. 10.5, we can see the following concerning the new regularization method:

- For the exponentially ill-posed problem `heat`, from Table 10.2 and the last column of Fig. 10.5, we can see that the discrete quasi-optimality rule and the variant of the L-curve are very efficient parameter choice rules for the new method. Indeed both the average relative errors and the average regularization parameters produced by the DQO and LCV rules are very near the optimal ones and decrease as the noise level decreases. Moreover, by looking at the standard deviation of the relative error

Table 10.1 Summary of the Monte carlo experiment with the five heuristic rules GCV, DQO, H1, H2, and LCV applied to the new method for the test problems `shaw` and `baart`. The x indicates columns where the average relative error is greater than 1

		shaw						baart					
		OPT	GCV	DQO	H1	H2	LCV	OPT	GCV	DQO	H1	H2	LCV
4%nl	e_{max}	0.18634	12.5834	x	0.20946	x	0.237316	0.347742	73.7316	x	0.349441	x	0.348385
	e_{min}	0.049132	0.055259	x	0.171478	x	0.096168	0.053773	0.143293	x	0.337422	x	0.181454
	$\bar{e}$	**0.13684**	**0.22309**	x	**0.18391**	x	**0.16075**	**0.16575**	**0.49844**	x	0.337422	x	0.181454
	$\sigma(e)$	**0.03307**	**0.44215**	x	0.00512	x	**0.01818**	**0.03993**	**3.10514**	x	**0.00165**	x	**0.03412**
	$\overline{reg.par.}$	**0.02095**	**0.02113**	x	**0.16317**	x	**0.03475**	**4.221e-3**	**6.81r-3**	x	**0.16309**	x	**0.02471**
2%nl	e_{max}	0.17458	6.32855	x	0.18501	x	0.245324	0.25839	42.4191	x	0.33493	x	0.273046
	e_{min}	0.03759	0.052238	x	0.16929	x	0.052099	0.05213	0.114199	x	0.307401	x	0.162947
	$\bar{e}$	**0.11391**	**0.17803**	x	**0.17507**	x	**0.12994**	**0.14712**	**0.42564**	x	**0.32104**	x	**0.19394**
	$\sigma(e)$	**0.03420**	**0.29453**	x	**0.00212**	x	**0.02828**	**0.03134**	**1.84369**	x	**0.00414**	x	**0.01828**
	$\overline{reg.par.}$	**7.58e-3**	**0.01144**	x	**0.11727**	x	**7.769e-3**	**2.627e-3**	**2.25e-3**	x	**0.0483**	x	**9.38e-3**

Table 10.2 Summary of the Monte carlo experiment with the five heuristic rules GCV, DQO, H1, H2, and LCV applied to the new method for the test problem `heat`

		heat					
		OPT	GCV	DQO	H1	H2	LCV
4%nl	e_{max}	0.274714	2.44329	0.279108	0.962294	7.22502	0.306221
	e_{min}	0.100167	0.109427	0.130933	0.267733	0.267816	0.101507
	$\bar{e}$	**0.18857**	**0.23329**	**0.205499**	**0.73711**	**0.647173**	**0.19349**
	$\sigma(e)$	**0.02788**	**0.13091**	**0.0209**	**0.30401**	**0.47841**	**0.02709**
	$\overline{reg.par.}$	**8.14e-4**	**6.235e-4**	**1.145e-3**	**0.64709**	**1.677e-4**	**8.842e-4**
2%nl	e_{max}	0.207777	2.34338	0.25523	0.261773	7.98943	0.289426
	e_{min}	0.073866	0.082679	0.081314	0.187784	0.228114	0.081314
	$\bar{e}$	**0.13947**	**0.17187**	**0.15295**	**0.2261**	**0.60093**	**0.16643**
	$\sigma(e)$	**0.01988**	**0.09237**	**0.02109**	**0.01094**	**0.51234**	**0.02929**
	$\overline{reg.par.}$	**5.204e-4**	**3.823e-4**	**6.909e-4**	**1.642e-3**	**8.736e-5**	**3.245e-4**

$\sigma(e)$, we see that those rules are very stable with respect to variations of the error term in y. Next, the GCV rule exhibit good average relative error, however, the GCV is not stable with respect to the noise in y, and this is shown by the spreading of dots along the x-axis or the corresponding large standard deviation $\sigma(e)$. Finally, the rule H2 is unstable and produces large relative errors norm whereas the rule H1 is more stable but does not yield satisfactory errors.

- For the mildly ill-posed test problems `shaw` and `baart`, the best heuristic rule for the new method is the variant of the L-curve method. Indeed, from Table 10.1 and two first columns of Fig. 10.5, we notice that the relative errors produced by the LCV rule are near the optimal ones. Moreover, the LCV rule is very stable with respect to the noise in y and both the relatives errors and the regularization parameters decrease as the noise level decreases. The second best rule is rule H1 which is also stable and convergent but produces relative errors larger than the one of LCV rule. Finally the rules DQO, GCV, and H2 are unstable and produce large relative error norm.

From Tables 10.3, 10.4, 10.5, 10.6, 10.7, 10.8, and 10.9, we apply the five selection rules GCV, DQO, H1, H2, and LCV to each regularization method. Obviously the GCV rule cannot be applied to conjugate gradient method due to its non-linear character. Although the DQO is originally designed for continuous regularization methods, notice that the rule defined in (10.56) can be applied to regularization methods with discrete regularization parameter such as truncated singular value decomposition and conjugate gradient. Indeed, we can applied the DQO rule to `tsvd` and `cg` by replacing $x_{\alpha_n}^{\delta}$ by x_k^{δ} in (10.56). Similarly the rules H1 and H2 originally designed for continuous regularization methods may be applicable to discrete regularization by defining the regularization parameter α as the inverse of the discrete parameter k. Following that idea, we applied the rules H1 and H2 to `tsvd` and `cg` by replacing α by $1/k$ in (10.57) and (10.58).

Table 10.3 Summary of the Monte Carlo experiment with GCV rule applied to nrm,tik,tsvd, and sw for the two noise levels on the three tests problems `shaw`, `baart` and `heat`. The x indicates columns where the average relative error is greater than 1

GCV		shaw				baart				heat			
		nrm	tik	tsvd	sw	nrm	tik	tsvd	sw	nrm	tik	tsvd	sw
4%nl	e_{max}	12.5834	5.5176	**x**	6.67616	73.7316	**x**	**x**	9.58465	2.44329	2.37791	**x**	2.69057
	e_{min}	0.055259	0.057251	**x**	0.052697	0.143293	**x**	**x**	0.152739	0.109427	0.111222	**x**	0.106857
	$\overline{e}$	**0.22309**	**0.38167**	**x**	**0.37981**	**0.49844**	**x**	**x**	**0.61592**	**0.23329**	**0.27623**	**x**	**0.23077**
	$\sigma(e)$	**0.44215**	**0.77228**	**x**	**0.81026**	**3.10514**	**x**	**x**	**1.08024**	**0.13091**	**0.18167**	**x**	**0.14511**
	$\overline{reg.par.}$	**0.02113**	**3.425e-3**	**x**	**0.02126**	**6.81e-3**	**x**	**x**	**4.376e-3**	**6.235e-4**	**2.816e-5**	**x**	**9.504e-5**
2%nl	e_{max}	6.32855	6.88993	**x**	8.85682	42.4191	**x**	**x**	4.83936	2.34338	3.20621	**x**	1.4366
	e_{min}	0.052238	0.047132	**x**	0.048763	0.114199	**x**	**x**	0.114434	0.082679	0.087569	**x**	0.080352
	$\overline{e}$	**0.17803**	**0.40427**	**x**	**0.42985**	**0.42564**	**x**	**x**	**0.37425**	**0.17187**	**0.21141**	**x**	**0.16403**
	$\sigma(e)$	**0.29453**	**0.94332**	**x**	**1.05855**	**1.84369**	**x**	**x**	**0.52374**	**0.09237**	**0.1747**	**x**	**0.09442**
	$\overline{reg.par.}$	**0.01144**	**1.487e-3**	**x**	**0.01023**	**2.25e-3**	**x**	**x**	**1.113e-3**	**3.823e-4**	**1.224e-5**	**x**	**3.608e-5**

Table 10.4 Summary of the Monte Carlo experiment with DQO rule applied to nrm,tik,tsvd,sw, and cg for the two noise levels on the three tests problems `shaw`, `baart` and `heat`. The x indicates columns where the average relative error is greater than 1

DQO		shaw					baart					heat				
		nrm	tik	tsvd	sw	cg	nrm	tik	tsvd	sw	cg	nrm	tik	tsvd	sw	cg
4%nl	e_{max}	x	1.0787	x	1.06752	x	x	5.18865	x	5.01141	x	0.279108	0.283843	x	0.330631	x
	e_{min}	x	0.06105	x	0.128984	x	x	0.128386	x	0.12797	x	0.130933	0.124984	x	0.14209	x
	$\bar{e}$	x	**0.24255**	x	**0.1676**	x	x	**0.30292**	x	**0.26028**	x	**0.2055**	**0.20955**	x	**0.19973**	x
	$\sigma(e)$	x	**0.12996**	x	**0.03473**	x	x	**0.56905**	x	**0.4043**	x	**0.02090**	**0.02349**	x	**0.02085**	x
	$reg.par.$	x	**8.16e-3**	x	**0.03038**	x	x	**8.146e-4**	x	**1.052e-3**	x	**1.145e-3**	**1.064e-4**	x	**1.486e-4**	x
2%nl	e_{max}	x	3.14131	x	2.83056	x	x	x	x	2.80503	x	0.25523	0.215776	x	0.312175	x
	e_{min}	x	0.043051	x	0.066213	x	x	x	x	0.081915	x	0.081314	0.085678	x	0.081303	x
	$\bar{e}$	x	**0.25222**	x	**0.23781**	x	x	x	x	**0.63955**	x	**0.15295**	**0.15585**	x	**0.16007**	x
	$\sigma(e)$	x	**0.31277**	x	**0.33941**	x	x	x	x	**0.47203**	x	**0.02109**	**0.01896**	x	**0.02309**	x
	$reg.par.$	x	**3.589e-3**	x	**0.02565**	x	x	x	x	**1.486e-4**	x	**6.909e-4**	**4.773e-5**	x	**7.363e-5**	x

Table 10.5 Summary of the Monte Carlo experiment with rule H1 applied to nrm,tik,tsvd,sw, and cg for the two noise levels on the two tests problems `shaw` and `baart`

H1		shaw					baart				
		nrm	tik	tsvd	sw	cg	nrm	tik	tsvd	sw	cg
4%nl	e_{max}	0.20946	0.256012	0.186581	0.248679	0.258152	0.349441	0.571788	0.348214	0.370115	0.346263
	e_{min}	0.171478	0.20863	0.169889	0.222574	0.159991	0.337422	0.372977	0.345054	0.338217	0.337497
	$\bar{e}$	**0.18391**	**0.23142**	**0.17112**	**0.23484**	**0.16948**	**0.34279**	**0.56375**	**0.34527**	**0.35003**	**0.34174**
	$\sigma(e)$	**5.12e-3**	**7.41e-3**	**1.44e-3**	**4.26e-3**	**5.95e-3**	**1.65e-3**	**4.39e-3**	**2.8e-4**	**4.19e-3**	**1.26e-3**
	$\overline{reg.par.}$	**0.16317**	**0.20483**	**0.25**	**0.42888**	**0.25025**	**0.16309**	**0.99972**	**0.5**	**0.20382**	**0.5**
2%nl	e_{max}	0.18501	0.201602	0.17421	0.226267	0.173244	0.334929	0.299022	0.345847	0.334085	0.34373
	e_{min}	0.16929	0.176076	0.169887	0.20998	0.163789	0.307401	0.187861	0.192709	0.30379	0.192243
	$\bar{e}$	**0.17507**	**0.1876**	**0.1702**	**0.21769**	**0.16824**	**0.32104**	**0.23703**	**0.34423**	**0.31878**	**0.341**
	$\sigma(e)$	**2.12e-3**	**3.77e-3**	**3.64e-4**	**2.61e-3**	**1.18e-3**	**4.14e-3**	**0.01658**	**0.01082**	**4.52e-3**	**8.65e-3**
	$\overline{reg.par.}$	**0.11728**	**0.07912**	**0.25**	**0.31221**	**0.25**	**0.0483**	**6.405e-3**	**0.49834**	**0.0483**	**0.49884**

Table 10.6 Summary of the Monte Carlo experiment with rule H2 applied to nrm,tik,tsvd,sw, and cg for the two noise levels on the three tests problems `shaw`, `baart` and `heat`. The x indicates columns where the average relative error is greater than 1

H2		shaw					baart					heat				
		nrm	tik	tsvd	sw	cg	nrm	tik	tsvd	sw	cg	nrm	tik	tsvd	sw	cg
4%nl	e_{max}	x	0.398139	x	1.06734	x	x	0.571788	x	3.89167	x	7.22502	0.968345	x	0.966995	x
	e_{min}	x	0.377665	x	0.128937	x	x	0.553372	x	0.151206	x	0.267816	0.967581	x	0.248926	x
	$\bar{e}$	x	**0.38798**	x	**0.16766**	x	x	**0.56381**	x	**0.28348**	x	**0.64717**	**0.968**	x	**0.96617**	x
	$\sigma(e)$	x	**2.96e-3**	x	**0.0347**	x	x	**2.67e-3**	x	**0.18975**	x	**0.47841**	**1.025e-4**	x	**0.01839**	x
	$\overline{reg.par.}$	x	**1**	x	**0.0306806**	x	x	**1**	x	**0.0356374**	x	**1.677e-4**	**1**	x	**0.999333**	x
2%nl	e_{max}	x	0.218418	x	1.2439	x	x	0.299022	x	2.05708	x	7.98943	0.413507	x	0.228358	x
	e_{min}	x	0.198588	x	0.066161	x	x	0.187861	x	0.118871	x	0.228114	0.389721	x	0.088958	x
	$\bar{e}$	x	**0.20743**	x	**0.16357**	x	x	**0.23703**	x	**0.21752**	x	**0.60093**	**0.40162**	x	**0.16603**	x
	$\sigma(e)$	x	**2.79e-3**	x	**0.02574**	x	x	**0.01658**	x	**0.17003**	x	**0.51234**	**3.941e-3**	x	**0.01742**	x
	$\overline{reg.par.}$	x	**0.13262**	x	**0.0273778**	x	x	**6.405e-3**	x	**6.403e-3**	x	**8.736e-5**	**1.04e-3**	x	**1.008e-4**	x

Table 10.7 Summary of the Monte Carlo experiment with rule H1 applied to nrm,tik,tsvd,sw, and cg for the two noise levels on the test problem `heat`. The x indicates columns where the average relative error is greater than 1

H1		heat				
		nrm	tik	tsvd	sw	cg
4%nl	e_{max}	0.962294	0.968345	x	0.966995	0.314004
	e_{min}	0.267733	0.967581	x	0.966207	0.196006
	$\bar{e}$	**0.73711**	**0.968**	**x**	**0.96665**	**0.2548**
	$\sigma(e)$	**0.3040**	**1.025e-4**	**x**	**1.065e-4**	**0.02256**
	$\overline{reg.par.}$	**0.64709**	**1**	**x**	**1**	**0.1522**
2%nl	e_{max}	0.261773	0.413507	x	0.576622	0.235608
	e_{min}	0.187784	0.25434	x	0.226524	0.118365
	$\bar{e}$	**0.2261**	**0.36496**	**x**	**0.52891**	**0.20326**
	$\sigma(e)$	**0.01094**	**0.04828**	**x**	**0.10989**	**0.01554**
	$\overline{reg.par.}$	**1.642e-3**	**8.135e-4**	**x**	**5.594e-3**	**0.1229**

From Tables 10.3, 10.4, 10.5, 10.6, 10.7, 10.8, and 10.9, we can do the following comments:

- The variant of the L-curve method defined through (10.59) is a very efficient heuristic parameter choice rule for each considered regularization method. Indeed, from Tables 10.8 and 10.9, by looking at the standard deviation $\sigma(e)$ of the relative error, we see that the LCV rule is stable for each regularization method, each test problem and each noise level. Next, the rule exhibits a convergent behavior for each test problem and each regularization method since the average relative error $\bar{e}$ and the average regularization parameter $\overline{reg.par.}$ decrease as the noise level decreases. Finally from Tables 10.3, 10.4, 10.5, 10.6, 10.7, 10.8, and 10.9, we find that the LCV rule always yields the smallest average relative error $\bar{e}$ among all the heuristic rules considered except in 4 cases (out of 30 cases in total) : `baart` test problem with 4% noise level for Showalter method and `heat` test problem with 2% noise level for the new method, Tikhonov and Showalter method. Notice that in each of those four cases, LCV rule yields the second best average relative error $\bar{e}$ after the DQO rule.
- For the exponentially ill-posed test problem `heat`, Table 10.10 summarizes the best heuristic rules for each regularization method:
- For the mildly ill-posed test problems `shaw` and `baart`, the best heuristic rule is always the LCV rule. For the new method, Tikhonov, truncated singular value decomposition and conjugate gradient, the LCV rule is followed by rule H1 whereas for the Showalter method, the LCV rule is followed by rule H2.
- For the exponentially ill-posed test problem `heat`, by comparing the five regularization methods combined each with its best heuristic selection rule among GCV, DQO, H1, H2, and LCV, we see that the new method equipped with the DQO rule (resp. the LCV rule) for 4% noise level (resp. for 2% noise level) yields the

Table 10.8 Summary of the Monte Carlo experiment with LCV rule applied to nrm,tik,tsvd,sw, and cg for the two noise levels on the two tests problems `shaw` and `baart`

LCV		shaw					baart				
		nrm	tik	tsvd	sw	cg	nrm	tik	tsvd	sw	cg
4%nl	e_{max}	0.237316	0.238876	0.186581	0.239964	0.186546	0.348385	0.362202	0.348214	0.348893	0.346263
	e_{min}	0.096168	0.082638	0.169889	0.087736	0.159991	0.181454	0.180919	0.345054	0.18049	0.337497
	$\bar{e}$	**0.16075**	**0.15554**	**0.17112**	**0.16065**	**0.16917**	**0.26562**	**0.26142**	**0.34527**	**0.27845**	**0.34174**
	$\sigma(e)$	**0.01818**	**0.02355**	**1.437e-3**	**0.01931**	**2.664e-3**	**0.03412**	**0.02758**	**2.804e-4**	**0.04829**	**1.26e-3**
	$\overline{reg.par.}$	**0.03472**	**8.899e-3**	**0.25**	**0.03802**	**0.25**	**0.02471**	**9.998e-3**	**0.5**	**0.04848**	**0.5**
2%nl	e_{max}	0.245324	0.243518	0.281998	0.245056	0.276412	0.273046	0.27934	0.240446	0.274307	0.243326
	e_{min}	0.052097	0.048994	0.146842	0.051787	0.060724	0.162974	0.14908	0.166265	0.15763	0.158944
	$\bar{e}$	**0.12994**	**0.12848**	**0.1598**	**0.12998**	**0.14911**	**0.19394**	**0.19351**	**0.17416**	**0.19351**	**0.17377**
	$\sigma(e)$	**0.02828**	**0.03026**	**0.01637**	**0.0282**	**0.0297**	**0.01828**	**0.01885**	**0.01036**	**0.01832**	**0.01068**
	$\overline{reg.par.}$	**7.769e-3**	**2.173e-3**	**0.2**	**3.491e-3**	**0.21322**	**9.38e-3**	**2.34e-3**	**0.33333**	**4.392e-3**	**0.33333**

Table 10.9 Summary of the Monte Carlo experiment with LCV rule applied to nrm,tik,tsvd,sw, and cg for the two noise levels on the test problem `heat`

LCV		heart				
		nrm	tik	tsvd	sw	cg
4%nl	e_{max}	0.306221	0.328995	0.364771	0.334528	0.345675
	e_{min}	0.101507	0.113791	0.134865	0.103206	0.120651
	$\overline{e}$	**0.19349**	**0.20276**	**0.2104**	**0.19778**	**0.19995**
	$\sigma(e)$	**0.02709**	**0.02928**	**0.03006**	**0.02904**	**0.02629**
	$\overline{reg.par.}$	**8.842e-4**	**5.708e-5**	**0.06756**	**1.253e-4**	**0.10016**
2%nl	e_{max}	0.289426	0.29641	0.345262	0.295178	0.30907
	e_{min}	0.081314	0.089006	0.100711	0.083142	0.086875
	$\overline{e}$	**0.16643**	**0.18146**	**0.17867**	**0.16141**	**0.16292**
	$\sigma(e)$	**0.02929**	**0.02828**	**0.03907**	**0.02992**	**0.03034**
	$\overline{reg.par.}$	**3.245e4**	**1.245e5**	**0.05105**	**2.233e-5**	**0.06903**

smallest average relative error $\bar{e}$ (about 2% smaller than the second best average relative error). For 4% noise level, the second smallest average relative error is achieved by Showalter method equipped with LCV rule whereas for 2% noise level, the second smallest average relative error is achieved by Tikhonov method equipped with DQO rule.

- For the two mildly ill-posed problems `shaw` and `baart`, by comparing the five regularization methods combined each with its best heuristic selection rule among GCV, DQO, H1, H2, and LCV, we notice there is no regularization method which always yields the smallest average relative error. For the `shaw` test problem, Tikhonov method with LCV rule yields the smallest average relative error $\bar{e}$. For the `baart` test problem, for 4% noise level, the smallest average relative error is obtained by the Showalter method equipped with the DQO rule. However, for this test problem, the DQO rule is not converging for the Showalter method as the average relative error $\bar{e}$ increases from 0.26028 to 0.63955 as the noise level decreases from 4% to 2%. If we discard Showalter with DQO rule, then for 4% noise level, the smallest average relative error is obtained by Tikhonov method equipped with LCV rule while for 2% noise level, the smallest average relative errors are obtained from conjugate gradient method equipped with LCV rule.

Remark 10.6 From Tables 10.1 , 10.2, 10.3, 10.4, 10.5, 10.6, 10.7, 10.8, and 10.9 we see that, the heuristic parameter choice rule LCV yields very satisfactory results for each considered regularization method. This reinforces the idea that the Bakushinskii véto [3] should not be seen as a limitation of heuristic parameter choice rule but rather as a safeguard to be taken into account.

In summary, we see that for the exponentially ill-posed test problem `heat`, the new regularization method always yields the smallest average relative error among the five considered regularization methods even when we consider heuristic parameter choice

Table 10.10 Summary best heuristic rules for each regularization method for the exponentially ill-posed test problem `heat`

	nrm	tik	tsvd	sw	cg
Best heuristic rules	DQO,LCV	DQO,LCV	LCV	DQO,LCV	LCV

rules. Hence in practical situation of exponentially ill-posed problems, we expect the new method to perform better than the other regularization methods (Tikhonov, truncated singular value decomposition, Showalter method, and conjugate gradient).

10.7 Conclusion

In this chapter, we presented a new regularization method which is particularly suitable for linear exponentially ill-posed problems. We study convergence analysis of the new method and we provided order-optimal convergence rates under logarithmic source conditions which has a natural interpretation in term of Sobolev spaces for exponentially ill-posed problems. For general source conditions expressed via index functions, we only provided quasi-order optimal rates. From the simulations performed, we saw that the new method performs better than Tikhonov method, spectral cut-off, Showalter, and conjugate gradient for the considered exponentially ill-posed problem, even with heuristic parameter choice rules. For the two mildly ill-posed problems treated, we saw that the new method actually yields results quite similar to those of Tikhonov and Showalter methods. The results of Sect. 10.6, where we applied five *error-free* selection rules to the five regularization methods suggest that the variant of the L-curve method defined in (10.59) and the discrete quasi-optimality rule defined in (10.56) are very efficient parameter choice rules for the new method in the context of exponentially ill-posed problem. In the context of mildly ill-posed problems, the results of experiments suggest that the LCV rule described in Sect. 10.6 is preferable.

Interesting perspectives would be a theoretical analysis of the LCV and DQO rules for the new regularization method in the framework of exponentially ill-posed problems in order to shed light on their good performances.

Acknowledgements The author would like to thank Pierre Maréchal and Anne Vanhems for their helpful comments, readings, and remarks.

10.8 Appendix

Proof of Proposition 10.1. Let us state the following standard inequality that we will use in the sequel:

$$\forall t \geq 0, \quad \exp(-t) \leq \frac{1}{1+t}. \tag{10.60}$$

Using (10.60) applied with $t = -\sqrt{\alpha}\ln(\lambda) \geq 0$, we get

$$1 - \exp(\sqrt{\alpha}\ln(\lambda)) \geq 1 - \frac{1}{1-\sqrt{\alpha}\ln(\lambda)} = \frac{-\sqrt{\alpha}\ln(\lambda)}{1-\sqrt{\alpha}\ln(\lambda)} = \sqrt{\alpha}\frac{|\ln(\lambda)|}{1+\sqrt{\alpha}|\ln(\lambda)|}. \tag{10.61}$$

But since $\alpha < 1$, $1+\sqrt{\alpha}|\ln(\lambda)| < 1+|\ln(\lambda)|$. Furthermore For all $\lambda \leq a < 1$, by the monotonicity of the function $t \longrightarrow |\ln(t)|/(1+|\ln(t)|) = -\ln(t)/(1-\ln(t))$ on $(0, 1)$, we get that

$$\frac{|\ln(t)|}{1+|\ln(t)|} \geq \frac{|\ln(a)|}{1+|\ln(a)|} \quad \forall t \in (0, a).$$

By applying the above inequality to (10.61) and taking the square, we get

$$\forall \lambda \in (0, a), \quad (1-\lambda^{\sqrt{\alpha}})^2 \geq M\alpha \quad \text{with} \quad M = \left(\frac{|\ln(a)|}{1+|\ln(a)|}\right)^2.$$

Whence the following inequality:

$$\frac{1}{\lambda + (1-\lambda^{\sqrt{\alpha}})^2} \leq \frac{1}{\lambda + M\alpha}, \tag{10.62}$$

which implies that

$$\sqrt{\lambda}g_\alpha(\lambda) \leq \frac{\lambda^{1/2}}{\lambda + M\alpha}. \tag{10.63}$$

It is rather straightforward to prove that the supremum over $\lambda \in (0, 1)$ of the right hand side of (10.63) is of order $\alpha^{-1/2}$ from which we deduce that

$$\sup_{\lambda\in(0,a]} \sqrt{\lambda}g_\alpha(\lambda) = O\left(\frac{1}{\sqrt{\alpha}}\right). \tag{10.64}$$

□

Proof of Lemma 10.2.1. Let $\lambda \in (0, 1)$. On the one hand, by applying the estimate $(1-\exp(t)) \geq -t/(1-t)$ which holds for all $t < 0$ to $t = \sqrt{\alpha}\ln(\lambda)$ and by taking squares, we have

$$(1-\lambda^{\sqrt{\alpha}})^2 \geq \frac{\alpha|\ln(\lambda)|^2}{(1+\sqrt{\alpha}|\ln(\lambda)|)^2}. \tag{10.65}$$

On the other hand, using the estimate $t^2 \geq (1-\exp(t))^2$ valid for all $t < 0$ to $t = \sqrt{\alpha}\ln(\lambda)$, we get

$$(1-\lambda^{\sqrt{\alpha}})^2 \leq \alpha|\ln(\lambda)|^2. \tag{10.66}$$

Now, for $\alpha \leq \lambda < 1$, $|\ln(\alpha)| \geq |\ln(\lambda)|$ which implies that $\sqrt{\alpha}|\ln(\lambda)| \leq \sqrt{\alpha}|\ln(\alpha)|$. Using the estimate $t^{\mu}\ln(1/t) \leq \mu$ which is true for all t in (0, 1) and every positive μ to $t = \lambda$ and $\mu = 1/2$, we deduce that

$$1 + \sqrt{\alpha}|\ln(\alpha)| \leq 3/2. \tag{10.67}$$

So, from (10.65) and (10.67), we deduce that

$$(1 - \lambda^{\sqrt{\alpha}})^2 \geq \frac{4}{9}\alpha|\ln(\lambda)|^2, \tag{10.68}$$

which implies that

$$r_{\alpha}(\lambda) \leq \frac{(1 - \lambda^{\sqrt{\alpha}})^2}{\lambda + (4/9)\alpha|\ln(\lambda)|^2}. \tag{10.69}$$

Finally, applying (10.66) and the fact that $\lambda \geq (4/9)\lambda$ to (10.69) yields (10.22). □

Proof of Lemma 10.3.2. (i) It is straightforward to check that (10.31) is indeed the derivative of the function $\Psi_{p,\alpha}$.

(ii) First notice that $\lim_{\lambda \to 0} h(\lambda) = +\infty$, hence, it suffices to find a $\bar{\lambda}$ such that $h(\bar{\lambda}) < 0$ to deduce the existence of a root of the function h on $(0, \bar{\lambda}]$. If $p < 2$, then $h(1) < 0$. If $p = 2$, then $h(\lambda) = |\ln(\lambda)|(2\alpha|\ln(\lambda)| - \lambda)$. Thus, $h(\lambda) < 0$ for λ close to 1 but smaller than 1. If $p > 2$, then $\lim_{\alpha \to 0} h(\lambda) = \lambda(p - 2 + \ln(\lambda)) < 0$ for all $\lambda < \exp(2 - p)$.

Now let us show that for every $\lambda(p, \alpha)$ which vanishes h, (10.33) holds.

$$h(\lambda) = 0 \implies \alpha = \lambda|\ln(\lambda)|^{-1}\left(\frac{2 - p + |\ln(\lambda)|}{p|\ln(\lambda)|}\right) \tag{10.70}$$

by monotonicity of the function $t \to (2 - p + t)/(pt)$ (irrespective of the sign of $2 - p$) and $t \to |\ln(\lambda)|$, we get that the function $l(\lambda) = \frac{2-p+|\ln(\lambda)|}{p|\ln(\lambda)|}$ is monotonic. If $p < 2$, the function l is increasing and we then get that, for all $\lambda \in (0, c]$ with $c < 1$,

$$\frac{1}{p} \leq l(\lambda) \leq l(c). \tag{10.71}$$

On the other hand, if $p \geq 2$, the function l is decreasing and for $\lambda \in (0, c]$ with $c < \exp(2 - p)$, we get

$$l(c) \leq l(\lambda) \leq 1/p. \tag{10.72}$$

From (10.70), (10.71) and (10.72), we deduce that

$$h(\lambda) = 0 \implies \alpha \sim \lambda|\ln(\lambda)|^{-1}. \tag{10.73}$$

From [37, Lemma 3.3], we get that

$$\alpha \sim \lambda |\ln(\lambda)|^{-1} \Rightarrow \lambda \sim \alpha |\ln(\alpha)|(1+o(1)) \quad \text{for} \quad \alpha \to 0.$$

This shows that the maximizers $\lambda(p, \alpha)$ of the function $\Psi_{p,\alpha}$ satisfies (10.33). Now let us deduce (10.34). We have

$$\alpha |\ln(\alpha)|^p \Psi_{p,\alpha}(\alpha |\ln(\alpha)|) = \frac{|\ln(\alpha)|^p \times |\ln(\alpha |\ln(\alpha)|)|^{2-p}}{|\ln(\alpha)| + |\ln(\alpha |\ln(\alpha)|)|^2} < |\ln(\alpha)|^p \times |\ln(\alpha |\ln(\alpha)|)|^{-p}.$$

With the change of variable $\varrho = |\ln(\alpha)|$ (i.e., $\alpha = \exp(-\varrho)$), we have

$$\begin{aligned} |\ln(\alpha)|^p \times |\ln(\alpha |\ln(\alpha)|)|^{-p} &= \frac{\varrho^p}{|\ln(\varrho \exp(-\varrho))|^p} \\ &= \frac{\varrho^p}{|-\varrho + \ln(\varrho)|^p} \\ &= \frac{\varrho^p}{(\varrho - \ln(\varrho))^p} \to 1 \quad \text{as} \quad \varrho \to \infty. \end{aligned}$$

This proves that

$$\alpha |\ln(\alpha)|^p \Psi_{p,\alpha}(\alpha |\ln(\alpha)|) = O(1)$$

and thus from (10.33), we deduce that (10.34) holds.

□

Proof of Proposition 10.3. For simplicity of notation, let $\alpha := \alpha(\delta, y^\delta)$. In order to establish (10.42), we are going to bound the terms $||x^\dagger - x_\alpha||$ and $||x_\alpha - x_\alpha^\delta||$ separately. Let us start with the regularization error term. Given that $x^\dagger \in X_{f_p}(\rho)$, we have $x^\dagger = f_p(T^*T)w$ and thus $x^\dagger - x_\alpha = r_\alpha(T^*T)x^\dagger = f_p(T^*T)r_\alpha(T^*T)w$. Hence by applying [20, Proposition 1] to $x^\dagger - x_\alpha$, we get

$$||x^\dagger - x_\alpha|| \leq ||r_\alpha(T^*T)w|| \sqrt{\phi_p^{-1}\left(||y - Tx_\alpha||^2/\rho^2\right)} \leq \rho \sqrt{\phi_p^{-1}\left(||y - Tx_\alpha||^2/\rho^2\right)}. \tag{10.74}$$

From (10.28) and (10.74), we deduce that

$$||x^\dagger - x_\alpha|| \leq \rho f_p\left(||y - Tx_\alpha||^2/\rho^2\right)(1+o(1)). \tag{10.75}$$

But

$$\begin{aligned} ||y - Tx_\alpha|| &\leq ||y^\delta - Tx_\alpha^\delta|| + ||y - Tx_\alpha - (y^\delta - Tx_\alpha^\delta)|| \\ &\leq \delta + \sqrt{\delta} + ||r_\alpha(T^*T)(y - y^\delta)|| \\ &\leq 2\delta + \sqrt{\delta} \\ &= \sqrt{\delta}(2\sqrt{\delta} + 1). \end{aligned} \tag{10.76}$$

From (10.75) and (10.76), we deduce that

$$\|x^\dagger - x_\alpha\| \leq \rho f_p\left(\delta(2\sqrt{\delta}+1)^2/\rho^2\right)(1+o(1)). \tag{10.77}$$

Using (10.77) and the fact that

$$\frac{f_p\left(\delta(2\sqrt{\delta}+1)^2/\rho^2\right)}{f_p(\delta)} = \left(\frac{-\ln(\delta)}{-\ln(\delta)-2\ln(1+2\sqrt{\delta})+2\ln(\rho)}\right)^p \to 1 \text{ as } \delta \to 0 \tag{10.78}$$

yields

$$\|x^\dagger - x_\alpha\| = O(f_p(\delta)) \quad \text{as} \quad \delta \to 0. \tag{10.79}$$

Now let us estimate the propagated data noise term. Let $\bar{\alpha} = q\alpha$ with $q \in (1, 2)$. From (10.41), since $\bar{\alpha} > \alpha$, we get

$$\|Tx_{\bar{\alpha}}^\delta - y^\delta\| > \delta + \sqrt{\delta}. \tag{10.80}$$

Therefore,

$$\begin{aligned} \|Tx_{\bar{\alpha}} - y\| &\geq \|Tx_{\bar{\alpha}}^\delta - y^\delta\| - \|T(x_{\bar{\alpha}}^\delta - x_{\bar{\alpha}}) - (y^\delta - y)\| \\ &> \delta + \sqrt{\delta} - \|r_{\bar{\alpha}}(T^*T)(y^\delta - y)\| \\ &\geq \delta + \sqrt{\delta} - \delta \\ &= \sqrt{\delta}. \end{aligned} \tag{10.81}$$

On the other hand, $\|Tx_{\bar{\alpha}} - y\| = \|T(x_{\bar{\alpha}} - x^\dagger)\| = \|(T^*T)^{1/2}(x_{\bar{\alpha}} - x^\dagger)\| = \|(T^*T)^{1/2} r_{\bar{\alpha}}(T^*T)x^\dagger\|$. By applying (10.44) with $\varphi(t) = \sqrt{t}$ and $\epsilon = 1/8$, we get that there exists a constant C such that $\|(T^*T)^{1/2}r_{\bar{\alpha}}(T^*T)x^\dagger\| \leq C\bar{\alpha}^{3/8}$. This implies that

$$\|Tx_{\bar{\alpha}} - y\| \leq C\bar{\alpha}^{3/8}. \tag{10.82}$$

From (10.81) and (10.82), we deduce that $\bar{\alpha}^{3/8} \geq \sqrt{\delta}/C$ which implies that $\bar{\alpha} \geq \bar{C}\delta^{4/3}$ with $\bar{C} = C^{-8/3}$. From (10.21), (10.39), the above lower bound of $\bar{\alpha}$ and the fact that $\alpha > \bar{\alpha}/2$, we get that, there exists a positive constant C' such that

$$\|x_\alpha - x_\alpha^\delta\| \leq C'\frac{\delta}{\sqrt{\alpha}} \leq C'\sqrt{2}\frac{\delta}{\sqrt{\bar{\alpha}}} \leq C'\sqrt{2/\bar{C}}\frac{\delta}{\sqrt{\delta^{4/3}}} = \delta^{1/3}C'\sqrt{2/\bar{C}}. \tag{10.83}$$

Given that $\delta^{1/3} = O(f_p(\delta))$ as $\delta \to 0$, we deduce that $\|x_\alpha - x_\alpha^\delta\| = O(f_p(\delta))$ as $\delta \to 0$ which together with (10.79) implies (10.42). □

References

1. Alibaud, N., Maréchal, P., Saesor, Y.: A variational approach to the inversion of truncated Fourier operators. Inverse Probl. **25**(4) (2009)
2. Baart, M.L.: The use of auto-correlation for pseudorank determination in noisy ill-conditioned linear least-squares problems. IMA J. Numer. Anal. **2**(2), 241–247 (1982)
3. Bakušinskiĭ, A.B.: Remarks on the choice of regularization parameter from quasioptimality and relation tests. Zh. Vychisl. Mat. i Mat. Fiz. **24**(8), 1258–1259 (1984)
4. Bauer, F., Kindermann, S.: Recent results on the quasi-optimality principle. J. Inverse Ill-Posed Probl. **17**(1), 5–18 (2009)
5. Bauer, F., Kindermann, S.: The quasi-optimality criterion for classical inverse problems. Inverse Probl. **24**(3) (2008)
6. Bauer, F., Reiß, M.: Regularization independent of the noise level: an analysis of quasi-optimality. Inverse Probl. **24**(5) (2008)
7. Bonnefond, X., Maréchal, P.: A variational approach to the inversion of some compact operators. Pac. J. Optim. **5**(1), 97–110 (2009)
8. Bonnefond, X., Maréchal, P., Simo Tao Lee, W.C.: A note on the Morozov principle via Lagrange duality. Set-Valued Var. Anal. **26**(2), 265–275 (2018)
9. Eicke, B., Louis, A.K., Plato, R.: The instability of some gradient methods for ill-posed problems. Numer. Math. **58**, 129–134 (1990)
10. Engl, H.W., Grever, W.: Using the L-curve for determining optimal regularization parameters. Numer. Math. **69**(1), 25–31 (1994)
11. Engl, H.W., Hanke, M., Neubauer, A.: Regularization of Inverse Problems. Mathematics and its Applications, vol. 375. Kluwer Academic Publishers Group, Dordrecht (1996)
12. Golub, G.H., Van Loan, C.F.: Matrix Computations. Johns Hopkins Studies in the Mathematical Sciences, 3rd edn. Johns Hopkins University Press, Baltimore, MD (1996)
13. Golub, G.H., Heath, M., Wahba, G.: Generalized cross-validation as a method for choosing a good ridge parameter. Technometrics **21**(2), 215–223 (1979)
14. Groetsch, C.W.: Inverse problems in the mathematical sciences. Vieweg Mathematics for Scientists and Engineers. Friedrich Vieweg & Sohn, Braunschweig (1993)
15. Hanke, M., Hansen, P.C.: Regularization methods for large-scale problems. Surveys Math. Indust. **3**(4), 253–315 (1993)
16. Hansen, P.C.: Analysis of discrete ill-posed problems by means of the L-curve. SIAM Rev. **34**(4), 561–580 (1992)
17. Hansen, P.C., O'Leary, D.P.: The use of the L-curve in the regularization of discrete ill-posed problems. SIAM J. Sci. Comput. **14**(6), 1487–1503 (1993)
18. Hansen, P.C.: Regularization tools version 4.0 for Matlab 7.3. Numer. Algorithms **46**(2), 189–194 (2007)
19. Hofmann, B., Mathé, P.: Analysis of profile functions for general linear regularization methods. SIAM J. Numer. Anal. **45**(3), 1122–1141 (2007)
20. Hohage, T.: Regularization of exponentially ill-posed problems. Numer. Funct. Anal. Optim. **21**(3—-4), 439–464 (2000)
21. Kirsch, A.: An Introduction to the Mathematical Theory of Inverse Problems. Applied Mathematical Sciences, vol. 120. Springer, New York (1996)
22. Leonov, A.S.: On the choice of regularization parameters by means of quasi-optimality and ratio criteria. Soviet. Math. Dokl. **19**(3) (1978)
23. Louis, A.K.: A unified approach to regularization methods for linear ill-posed problems. Inverse Probl. **15**(2), 489–498 (1999)
24. Louis, A.K., Mass, P.: A mollifier method for linear operator equations of the first kind. Inverse Probl. **6**(3), 427–440 (1990)
25. Lukas, M.A.: Asymptotic optimality of generalized cross-validation for choosing the regularization parameter. Numer. Math. **66**(1), 41–66 (1993)
26. Mair, B.A.: Tikhonov regularization for finitely and infinitely smoothing operators. SIAM J. Math. Anal. **25**(1), 135–147 (1994)

27. Mathé, P.: Saturation of regularization methods for linear ill-posed problems in Hilbert spaces. SIAM J. Numer. Anal. **42**(3), 968–973 (2004)
28. Mathé, P., Pereverzev, S.V.: Geometry of linear ill-posed problems in variable Hilbert scales. Inverse Probl. **19**(3), 789–803 (2003)
29. Mathé, P., Hofmann, B.: How general are general source conditions? Inverse Probl. **24**(1) (2008)
30. Micchelli, C.A., Rivlin, T.J.: A survey of optimal recovery. Optimal estimation in approximation theory. Proc. Internat. Sympos., Freudenstadt, 1976. Plenum Press, pp. 1–54 (1977)
31. Murio, D.A.: The Mollification Method and the Numerical Solution of Ill-posed Problems. A Wiley-Interscience Publication. Wiley, New York (1993)
32. Nair, M.T., Schock, E., Tautenhahn, U.: Morozov's discrepancy principle under general source conditions. Z. Anal. Anwendungen **22**(1), 199–214 (2003)
33. Neubauer, A.: On converse and saturation results for regularization methods, Beitäge zur angewandten Analysis und Informatik, pp. 262–270. Aachen, Shaker (1994)
34. Neubauer, A.: On converse and saturation results for Tikhonov regularization of linear ill-posed problems. SIAM J. Numer. Anal. **34**(2), 517–527 (1997)
35. Schock, E.: Approximate solution of ill-posed equations: arbitrarily slow convergence vs. superconvergence. Constructive methods for the practical treatment of integral equations (1984), pp. 234–243
36. Shaw Jr., C.M.: Improvement of the resolution of an instrument by numerical solution of an integral equation. J. Math. Anal. Appl. **37**, 83–112 (1972)
37. Tautenhahn, U.: Optimality for ill-posed problems under general source conditions. Numer. Funct. Anal. Optim. **19**(3–4), 377–398 (1998)
38. Tikhonov, A.N., Arsenin, V.Y.: Solutions of Ill-posed Problems. Wiley (1977)
39. Wahba, G.: Practical approximate solutions to linear operator equations when the data are noisy. SIAM J. Numer. Anal. **14**(4), 651–667 (1977)

Chapter 11
On Minimax Programming with Vanishing Constraints

Vivek Laha, Rahul Kumar, Harsh Narayan Singh, and S. K. Mishra

Abstract In this chapter, we deal with a class of minimax programming problems with vanishing constraints. We establish necessary and sufficient optimality results for such a feasible point to be an optimal solution. Moreover, we formulate Mond–Weir type dual model for such a minimax programming problem with vanishing constraints and obtain various duality results. Also, we apply some results obtained for minimax programming problem with vanishing constraints to a multiobjective optimization problem with vanishing constraints.

Keywords Vanishing constraints · Minimax programming · Generalized convexity · Mond-Weir duality · Multiobjective optimization · Weak efficient solutions

11.1 Introduction

Achtziger and Kanzow [1] studied mathematical programs with vanishing constraints for the first time. Hoheisel and Kanzow [2–4], Izmailov and Solodov [5] and Khare and Nath [6] derived optimality conditions for mathematical programs with vanishing constraints under several weak modified constraint qualifications. Mishra et al. [7] and Hu et al. [8] gave several dual models for the mathematical programs with vanish-

V. Laha (✉) · H. N. Singh · S. K. Mishra
Department of Mathematics, Institute of Science, Banaras Hindu University, Varanasi 221005, India
e-mail: laha.vivek333@gmail.com

H. N. Singh
e-mail: harshksingh92@gmail.com

S. K. Mishra
e-mail: bhu.skmishra@gmail.com

R. Kumar
Department of Mathematics, Government Chandravijay College, Dindori 481880, India
e-mail: kumarahul1992bhu@gmail.com

V. Laha et al. (eds.), *Optimization, Variational Analysis and Applications*,
Springer Proceedings in Mathematics & Statistics 355,
https://doi.org/10.1007/978-981-16-1819-2_11

ing constraints. Mishra et al. [9], Guu et al. [10], and Jayswal and Singh [11] studied multiobjective optimization problems with vanishing constraints. Kazemi and Kanzi [12], Kazemi et al. [13], Kanzi et al. [14], and Mokhtavayi et al. [15] obtained results for mathematical programs with vanishing constraints with nonsmooth data.

Schmitendorf [16] derived necessary and sufficient conditions of optimality for minimax programming problems. Mishra [17], Mehra and Bhatia [18], Studniarski and Taha [19], Antczak [20], Mandal and Nahak [21], and Zemkoho [22] studied minimax programming problems in detail. Nonsmooth minimax problems were worked out by Mishra and Shukla [23], Antczak [24], and Jayswal et al. [25] whereas second-order duality results were obtained by Mishra and Rueda [26], Ahmad et al. [27], Husain et al. [28], and Jayswal and Stancu–Minasian [29]. Lai and Chen [30], Lai et al. [31], and Lai and Liu [32] derived results for minimax programming in complex spaces. Semi-infinite minimax programming were dealt by Stefanescu and Stefanescu [33], and Upadhyay and Mishra [34]. Das and Nahak [35] analyzed set valued minimax programming problems.

The aim of this chapter is to study and analyze minimax programming problems with vanishing constraints. In Sect. 11.2, we derive necessary and sufficient optimality conditions for a feasible point to be an optimal solution of the minimax programs with vanishing constraints. In Sect. 11.3, we give parametric Mond–Weir type dual model to deal with the problem under consideration. In Sect. 11.4, we apply the obtained results to derive necessary and sufficient conditions for a multiobjective optimization problem with vanishing constraints. Section 11.5 concludes the findings of the chapter.

11.2 Optimality Conditions

Consider a minimax program with vanishing constraints (MMPVC) as follows:

$$\begin{aligned}
&\min_{x\in\mathbb{R}^n}\max_{1\le i\le k}\theta_i(x)\\
&\text{subject to}\\
&\Phi_i(x)\ge 0,\ \forall i\in L:=\{1,2,\ldots,l\},\\
&\psi_i(x)\Phi_i(x)\le 0,\ \forall i\in L,
\end{aligned}$$

where we assume that the functions $\theta_i,\ \Phi_i,\ \psi_i:\mathbb{R}^n\to\mathbb{R}$ are continuously differentiable and the set of all feasible solutions $\Omega:=\{x\in\mathbb{R}^n:\Phi_i(x)\ge 0,\ \psi_i(x)\Phi_i(x)\le 0,\ \forall i\in L\}$ is non empty and compact. The (MMPVC) may be connected to the following parametric mathematical program with vanishing constraints (PMPVC):

$$
\begin{aligned}
&\min \ \nu \\
&\text{subject to} \\
&\theta_i(x) \le \nu, \ \forall i \in K := \{1, 2, \ldots, k\}, \\
&\Phi_i(x) \ge 0, \ \forall i \in L := \{1, 2, \ldots, l\}, \\
&\psi_i(x)\Phi_i(x) \le 0, \ \forall i \in L, \\
&(x, \nu) \in \mathbb{R}^n \times \mathbb{R},
\end{aligned}
$$

where $\nu \in \mathbb{R}$ is any parameter and the set of all feasible solutions of the (PMPVC) is given by

$$\Omega \times V := \{(x, \nu) \in \mathbb{R}^n \times \mathbb{R} : \theta_i(x) \le \nu, \ \forall i \in K, \ \Phi_i(x) \ge 0, \ \psi_i(x)\Phi_i(x) \le 0, \ \forall i \in L\}.$$

On the lines of Crouzeix et al. [37] (MMPVC) and (PMPVC) may be related as follows:

Lemma 11.2.1 *If a point (x, ν) is a feasible solution of the (PMPVC), then x is a feasible solution of the (MMPVC). Moreover, if x is feasible for problem (MMPVC), then there exists $\nu \in \mathbb{R}$ such that (x, ν) is a feasible solution of the problem (PMPVC).*

Lemma 11.2.2 *A point $\bar{x}$ is a local minimum of the (MMPVC) with minimum value $\bar{\nu}$ if and only if a point $(\bar{x}, \bar{\nu})$ is a local minimum of (PMPVC) with minimum value $\bar{\nu}$.*

The following indexing will be useful for further analysis.

$$
\begin{aligned}
&I_+(\bar{x}) := \{i \in L : \Phi_i(\bar{x}) > 0\}; \\
&I_0 := \{i \in L : \Phi_i(\bar{x}) = 0\}; \\
&I_{+0} := \{i \in L : \Phi_i(\bar{x}) > 0, \Psi_i(\bar{x}) = 0\}; \\
&I_{+-} := \{i \in L : \Phi_i(\bar{x}) > 0, \Psi_i(\bar{x}) < 0\}; \\
&I_{0-} := \{i \in L : \Phi_i(\bar{x}) = 0, \Psi_i(\bar{x}) < 0\}; \\
&I_{00} := \{i \in L : \Phi_i(\bar{x}) = 0, \Psi_i(\bar{x}) = 0\}; \\
&I_{0+} := \{i \in L : \Phi_i(\bar{x}) = 0, \Psi_i(\bar{x}) > 0\}.
\end{aligned}
$$

Now, we prove parametric necessary optimality conditions for the (MMPVC).

Theorem 11.2.1 (Parametric necessary optimality conditions) *If $\bar{x}$ is a local minimum of (MMPVC) with minimum value $\bar{\nu}$ such that the modified Abadie constraint qualification (ACQ) is satisfied at $(\bar{x}, \bar{\nu})$, that is, $L^{VC}(\bar{x}, \bar{\nu}) \subseteq T(\bar{x}, \bar{\nu})$, where $T(\bar{x}, \bar{\nu})$ is the standard tangent cone of the (PMPVC) at $(\bar{x}, \bar{\nu})$ given by*

$$T(\bar{x}, \bar{\nu}) := \Big\{d \in \mathbb{R}^{n+1} : \exists\{(x^k, \nu^k)\} \subseteq \Omega \times V \text{ and } \{t^k\} \downarrow 0 \text{ such that } (x^k, \nu^k) \to (\bar{x}, \bar{\nu}) \text{ and } \frac{(x^k, \nu^k) - (\bar{x}, \bar{\nu})}{t^k} \to d\Big\}$$

and $L^{VC}(\bar{x}, \bar{\nu})$ *is the VC-linearized cone of the (PMPVC) at* $(\bar{x}, \bar{\nu})$ *given by*

$$L^{VC}(\bar{x}, \bar{\nu}) := \Big\{ d \in \mathbb{R}^{n+1} : \sum_{j=1}^{n} \frac{\partial \theta_i(\bar{x})}{\partial x_j} d_j \leq d_{n+1}, \forall i \in \{i \in K : \theta_i(\bar{x}) = \bar{\nu}\},$$
$$\sum_{j=1}^{n} \frac{\partial \Phi_i(\bar{x})}{\partial x_j} d_j = 0, \ \forall i \in I_{0+}(\bar{x}),$$
$$\sum_{j=1}^{n} \frac{\partial \Phi_i(\bar{x})}{\partial x_j} d_j \geq 0, \ \forall i \in I_{00}(\bar{x}) \cup I_{0-}(\bar{x}),$$
$$\sum_{j=1}^{n} \frac{\partial \psi_i(\bar{x})}{\partial x_j} d_j \leq 0, \ \forall i \in I_{00}(\bar{x}) \cup I_{+0}(\bar{x}) \Big\},$$

then there exists $\alpha_i \in \mathbb{R}(i \in K),\ \beta_i,\ \gamma_i \in \mathbb{R}(i \in L)$ *such that*

$$\sum_{i \in K} \alpha_i \nabla \theta_i(\bar{x}) - \sum_{i \in L} \beta_i \nabla \Phi_i(\bar{x}) + \sum_{i \in L} \gamma_i \nabla \psi_i(\bar{x}) = 0, \tag{11.1}$$

$$\alpha_i \geq 0, \ \alpha_i(\theta_i(\bar{x}) - \bar{\nu}) = 0, \ \forall i \in K, \sum_{i \in K} \alpha_i = 1, \tag{11.2}$$

$$\beta_i = 0, \ \forall i \in I_{+}(\bar{x}), \ \beta_i \geq 0, \ \forall i \in I_{00}(\bar{x}) \cup I_{0-}(\bar{x}), \ \beta_i \in \mathbb{R}, \ \forall i \in I_{0+}(\bar{x}), \tag{11.3}$$

$$\gamma_i = 0, \ \forall i \in I_{0+}(\bar{x}) \cup I_{0-}(\bar{x}), \ \gamma_i \geq 0, \ \forall i \in I_{00}(\bar{x}) \cup I_{+0}(\bar{x}). \tag{11.4}$$

Proof Since $\bar{x}$ is a local minimum of the (MMPVC) with the minimum value $\bar{\nu}$. Therefore, by the Lemma 11.2.2, $(\bar{x}, \bar{\nu})$ is a local minimum of the (PMPVC) with the minimum value $\bar{\nu}$. Also, as modified ACQ is satisfied at $(\bar{x}, \bar{\nu})$, by Theorem 1 in [1], there exists $\alpha_i \in \mathbb{R}(i \in K),\ \beta_i,\ \gamma_i \in \mathbb{R}(i \in L)$ such that

$$\sum_{i \in K} \alpha_i \nabla \theta_i(\bar{x}) - \sum_{i \in L} \beta_i \nabla \Phi_i(\bar{x}) + \sum_{i \in L} \gamma_i \nabla \psi_i(\bar{x}) = 0,$$
$$\alpha_i \geq 0, \ \alpha_i(\theta_i(\bar{x}) - \bar{\nu}) = 0, \ \forall i \in K, \sum_{i \in K} \alpha_i = 1,$$
$$\beta_i = 0, \ \forall i \in I_{+-}(\bar{x}) \cup I_{+0}(\bar{x}), \ \beta_i \geq 0, \ \forall i \in I_{00}(\bar{x}) \cup I_{0-}(\bar{x}), \ \beta_i \in \mathbb{R}, \ \forall i \in I_{0+}(\bar{x}),$$
$$\gamma_i = 0, \ \forall i \in I_{0+}(\bar{x}) \cup I_{0-}(\bar{x}), \ \gamma_i \geq 0, \ \forall i \in I_{00}(\bar{x}) \cup I_{+0}(\bar{x}).$$

This completes the proof. □

Now, we recall the notion of invexity introduced by Hanson [36] and derive sufficient optimality conditions for a feasible point to be optimal.

Definition 11.1 Let $\Omega \subseteq \mathbb{R}^n$ be nonempty, let $\theta : \mathbb{R}^n \to \mathbb{R}$ be continuously differentiable on an open set containing Ω and let $\eta : \mathbb{R}^n \times \mathbb{R}^n \to \mathbb{R}^n$ be a vector valued function. Then,

(a) the function θ is said to be *η-invex* at $\bar{x} \in \Omega$ over Ω, iff for any $x \in \Omega$, one has

$$\theta(x) - \theta(\bar{x}) \geq \langle \nabla\theta(\bar{x}), \eta(x, \bar{x}) \rangle;$$

(b) the function θ is said to be *η-quasiinvex* at $\bar{x} \in \Omega$ over Ω, iff for any $x \in \Omega$, one has

$$\theta(x) \leq \theta(\bar{x}) \implies \langle \nabla\theta(\bar{x}), \eta(x, \bar{x}) \rangle \leq 0;$$

(c) the function θ is said to be *η-pseudoinvex* at $\bar{x} \in \Omega$ over Ω, iff for any $x \in \Omega$, one has

$$\theta(x) < \theta(\bar{x}) \implies \langle \nabla\theta(\bar{x}), \eta(x, \bar{x}) \rangle < 0;$$

(d) the function θ is said to be *strictly η-pseudoinvex* at $\bar{x} \in \Omega$ over Ω, iff for any $x \in \Omega$, one has

$$\theta(x) \leq \theta(\bar{x}) \implies \langle \nabla\theta(\bar{x}), \eta(x, \bar{x}) \rangle < 0.$$

The following theorem gives sufficient conditions for the optimality of a feasible point of the (MMPVC).

Theorem 11.2.2 (Parametric sufficient optimality conditions) *Let $(\bar{x}, \bar{\nu})$ be a feasible solution of the (PMPVC) for which there exists $\alpha_i \in \mathbb{R}(i \in K)$, β_i, $\gamma_i \in \mathbb{R}(i \in L)$ such that conditions (11.1)–(11.4) are satisfied.*

Let us define the following index sets at $\bar{x} \in \Omega$

$$I_{00}^{++} := \{i \in I_{00} : \beta_i > 0,\ \gamma_i > 0\};$$
$$I_{00}^{+0} := \{i \in I_{00} : \beta_i > 0,\ \gamma_i = 0\};$$
$$I_{00}^{0+} := \{i \in I_{00} : \beta_i = 0,\ \gamma_i > 0\};$$
$$I_{0-}^{+0} := \{i \in I_{0-} : \beta_i > 0,\ \gamma_i = 0\};$$
$$I_{0+}^{+0} := \{i \in I_{0+} : \beta_i > 0,\ \gamma_i = 0\};$$
$$I_{0+}^{-0} := \{i \in I_{0+} : \beta_i < 0,\ \gamma_i = 0\};$$
$$I_{+0}^{0+} := \{i \in I_{+0} : \beta_i = 0,\ \gamma_i > 0\}.$$

Now, assume that $\sum_{i \in K} \alpha_i \theta_i$ is η-pseudoinvex and $-\Phi_i(i \in I_{00}^{++} \cup I_{00}^{+0} \cup I_{0-}^{+0} \cup I_{0+}^{+0})$, $\Phi_i(i \in I_{0+}^{-0})$, and $\psi_i(i \in I_{00}^{++} \cup I_{00}^{0+} \cup I_{+0}^{0+})$ are η-quasiinvex at $\bar{x}$ over Ω. Then,

(a) if $I_{00}^{++} \cup I_{00}^{0+} \cup I_{0+}^{-0} \cup I_{+0}^{0+} = \phi$, then $(\bar{x}, \bar{\nu})$ is a global minimum of the (PMPVC).

(b) *if* $I_{00}^{++} \cup I_{00}^{0+} = \phi$, *then* $(\bar{x}, \bar{\nu})$ *is a local minimum of the (PMPVC).*
(c) *if* $\bar{x}$ *is an interior point with respect to the set* $\Omega \cap \{x \in \mathbb{R}^n : \Phi_i(x) = 0, \psi_i(x) = 0,\ i \in I_{00}^{++} \cup I_{00}^{0+}\}$, *then* $(\bar{x}, \bar{\nu})$ *is a local minimum of the (PMPVC).*

Proof (a) Since $I_{00}^{++} \cup I_{00}^{0+} \cup I_{0+}^{-0} \cup I_{+0}^{0+} = \phi$. Therefore, for any $i \in I_{00}^{+0} \cup I_{0-}^{+0} \cup I_{0+}^{+0}$, $\Phi_i(\bar{x}) = 0$ and $\beta_i > 0$, by the η-quasiinvexity of $-\Phi_i (i \in I_{00}^{+0} \cup I_{0-}^{+0} \cup I_{0+}^{+0})$ at $\bar{x}$ over Ω, one has

$$\left\langle - \sum_{i \in I_{00}^{+0} \cup I_{0-}^{+0} \cup I_{0+}^{+0}} \beta_i \nabla \Phi_i(\bar{x}), \eta(x, \bar{x}) \right\rangle \leq 0,\ \forall x \in \Omega.$$

By the condition (11.1) in Theorem 11.2.1, it follows that

$$\left\langle \sum_{i \in K} \alpha_i \nabla \theta_i(\bar{x}), \eta(x, \bar{x}) \right\rangle \geq 0,\ \forall x \in \Omega.$$

Since $\sum_{i \in K} \alpha_i \theta_i$ is η-pseudoinvex at $\bar{x}$ over Ω, one has

$$\sum_{i \in K} \alpha_i \nu \geq \sum_{i \in K} \alpha_i \theta_i(x) \geq \sum_{i \in K} \alpha_i \theta_i(\bar{x}) = \sum_{i \in K} \alpha_i \bar{\nu},$$

which implies from (11.2) that $\bar{\nu} \leq \nu$ for any $(x, \nu) \in \Omega \times V$. Hence, $(\bar{x}, \bar{\nu})$ is a global minima of the (PMPVC).
(b) Here $I_{00}^{++} \cup I_{00}^{0+} = \phi$. Now, for any $i \in I_{0+}^{-0}$, $\beta_i < 0$, $\Phi_i(\bar{x}) = 0$, $\psi_i(\bar{x}) > 0$, which implies that $\Psi_i(x) > 0$ for any x sufficiently close to $\bar{x}$ and hence $\Phi_i(x) \leq 0$ for any x sufficiently close to $\bar{x}$. Now, since $\Phi_i(x) \geq 0$ for any $x \in \Omega$, therefore $\Phi_i(x) = 0$ for any $x \in \Omega$ sufficiently close to $\bar{x}$. By the η-quasiinvexity of $\Phi_i (i \in I_{0+}^{-0})$ at $\bar{x}$ over Ω, for any $x \in \Omega$ sufficiently close to $\bar{x}$, one has

$$\left\langle - \sum_{i \in I_{0+}^{-0}} \beta_i \nabla \Phi_i(\bar{x}), \eta(x, \bar{x}) \right\rangle \leq 0. \tag{11.5}$$

Similarly, for any $i \in I_{+0}^{0+}$, $\gamma_i > 0$, $\Phi_i(\bar{x}) > 0$, $\psi_i(\bar{x}) = 0$, which implies that $\Phi_i(x) > 0$ for any x sufficiently close to $\bar{x}$ and hence $\psi_i(x) \leq 0 = \psi_i(\bar{x})$ for any x sufficiently close to $\bar{x}$. By the η-quasiinvexity of $\psi_i (i \in I_{+0}^{0+})$ at $\bar{x}$ over Ω, for any $x \in \Omega$ sufficiently close to $\bar{x}$, one has

$$\left\langle \sum_{i \in I_{+0}^{0+}} \gamma_i \nabla \psi_i(\bar{x}), \eta(x, \bar{x}) \right\rangle \leq 0. \tag{11.6}$$

Adding inequalities (11.5) and (11.6), for any $x \in \Omega$ sufficiently close to $\bar{x}$, one has

$$\left\langle -\sum_{i\in I_{0+}^{-0}} \beta_i \nabla\Phi_i(\bar{x}) + \sum_{i\in I_{+0}^{0+}} \gamma_i \nabla\psi_i(\bar{x}), \eta(x,\bar{x}) \right\rangle \le 0. \tag{11.7}$$

By the condition (11.1) in Theorem 11.2.1, for any $x \in \Omega$ sufficiently close to $\bar{x}$, it follows that

$$\left\langle \sum_{i\in K} \alpha_i \nabla\theta_i(\bar{x}), \eta(x,\bar{x}) \right\rangle \ge 0.$$

Since $\sum_{i\in K} \alpha_i\theta_i$ is η-pseudoinvex at $\bar{x}$ over Ω, for any $x \in \Omega$ sufficiently close to $\bar{x}$, one has

$$\sum_{i\in K} \alpha_i \nu \ge \sum_{i\in K} \alpha_i\theta_i(x) \ge \sum_{i\in K} \alpha_i\theta_i(\bar{x}) = \sum_{i\in K} \alpha_i\bar{\nu},$$

which implies by (11.2) that $\bar{\nu} \le \nu$ for any $(x, \nu) \in \Omega \times V$ sufficiently close to $(\bar{x}, \bar{\nu})$. Hence, $(\bar{x}, \bar{\nu})$ is a local minima of the (PMPVC).

(c) Since $\bar{x}$ is an interior point with respect to the set $\Omega \cap \{x \in \mathbb{R}^n : \Phi_i(x) = 0,\ \psi_i(x) = 0,\ i \in I_{00}^{++} \cup I_{00}^{0+}\}$, therefore for any $x \in \Omega$ sufficiently close to $\bar{x}$, one has

$$\Phi_i(x) = 0,\ \psi_i(x) = 0,\ i \in I_{00}^{++} \cup I_{00}^{0+}.$$

Since $-\Phi_i (i \in I_{00}^{++})$ and $\psi_i (i \in I_{00}^{++} \cup I_{00}^{0+})$ are η-quasiinvex at $\bar{x}$ over Ω. Therefore, for any $x \in \Omega$ sufficiently close to $\bar{x}$, it follows that

$$\left\langle -\sum_{i\in I_{00}^{++}} \beta_i \nabla\Phi_i(\bar{x}) + \sum_{i\in I_{00}^{++}\cup I_{00}^{0+}} \gamma_i \nabla\psi_i(\bar{x}), \eta(x,\bar{x}) \right\rangle \le 0.$$

Using the conditions of cases (a) and (b) above, for any $x \in \Omega$ sufficiently close to $\bar{x}$, one has

$$\left\langle - \sum_{i\in I_{00}^{++}\cup I_{00}^{+0}\cup I_{0-}^{+0}\cup I_{0+}^{+0}\cup I_{0+}^{-0}} \beta_i \nabla\Phi_i(\bar{x}) + \sum_{i\in I_{00}^{++}\cup I_{00}^{0+}\cup I_{+0}^{0+}} \gamma_i \nabla\psi_i(\bar{x}), \eta(x,\bar{x}) \right\rangle \le 0. \tag{11.8}$$

By the condition (11.1) in Theorem 11.2.1, for any $x \in \Omega$ sufficiently close to $\bar{x}$, it follows that

$$\left\langle \sum_{i\in K} \alpha_i \nabla\theta_i(\bar{x}), \eta(x,\bar{x}) \right\rangle \ge 0.$$

Since $\sum_{i\in K} \alpha_i\theta_i$ is η-pseudoinvex at $\bar{x}$ over Ω, for any $x \in \Omega$ sufficiently close to $\bar{x}$, one has

$$\sum_{i\in K}\alpha_i\nu \geq \sum_{i\in K}\alpha_i\theta_i(x) \geq \sum_{i\in K}\alpha_i\theta_i(\bar{x}) = \sum_{i\in K}\alpha_i\bar{\nu},$$

which implies by (11.2) that $\bar{\nu} \leq \nu$ for any $(x, \nu) \in \Omega \times V$ sufficiently close to $(\bar{x}, \bar{\nu})$. Hence, $(\bar{x}, \bar{\nu})$ is a local minimum of the (PMPVC). □

A parameter-free necessary optimality condition is obtained by replacing $\bar{\nu}$ by $\theta_i(\bar{x})$ as follows:

Theorem 11.2.3 (Parameter-free necessary optimality conditions) *Let $\bar{x}$ be a local minimum of the (MMPVC) such that modified ACQ is satisfied at $\bar{x}$, that is, $L^{VC}(\bar{x}) \subseteq T(\bar{x})$, where $T(\bar{x})$ is the tangent cone of the (MMPVC) at $\bar{x}$ given by*

$$T(\bar{x}) := \left\{d \in \mathbb{R}^n : \exists\{x^k\} \subseteq \Omega \text{ and } \{t^k\} \downarrow 0 \text{ such that } x^k \to \bar{x} \text{ and } \frac{x^k - \bar{x}}{t^k} \to d\right\}$$

and $L^{VC}(\bar{x})$ is the VC-linearized cone of the (MMPVC) at $\bar{x}$ is given by

$$\begin{aligned} L^{VC}(\bar{x}) := \Bigg\{d \in \mathbb{R}^n : &\sum_{j=1}^{n}\frac{\partial\Phi_i(\bar{x})}{\partial x_j}d_j = 0,\ \forall i \in I_{0+}(\bar{x}),\\ &\sum_{j=1}^{n}\frac{\partial\Phi_i(\bar{x})}{\partial x_j}d_j \geq 0,\ \forall i \in I_{00}(\bar{x}) \cup I_{0-}(\bar{x}),\\ &\sum_{j=1}^{n}\frac{\partial\psi_i(\bar{x})}{\partial x_j}d_j \leq 0,\ \forall i \in I_{00}(\bar{x}) \cup I_{+0}(\bar{x})\Bigg\}, \end{aligned}$$

then there exists $\alpha_i \in \mathbb{R}(i \in K)$, β_i, $\gamma_i \in \mathbb{R}(i \in L)$ such that

$$\begin{aligned} &\sum_{i\in K}\alpha_i\nabla\theta_i(\bar{x}) - \sum_{i\in L}\beta_i\nabla\Phi_i(\bar{x}) + \sum_{i\in L}\gamma_i\nabla\psi_i(\bar{x}) = 0,\\ &\alpha_i \geq 0,\ \forall i \in K,\ \sum_{i\in K}\alpha_i = 1,\\ &\beta_i = 0,\ \forall i \in I_{+}(\bar{x}),\ \beta_i \geq 0,\ \forall i \in I_{00}(\bar{x}) \cup I_{0-}(\bar{x}),\ \beta_i \in \mathbb{R},\ \forall i \in I_{0+}(\bar{x}),\\ &\gamma_i = 0,\ \forall i \in I_{0+}(\bar{x}) \cup I_{0-}(\bar{x}) \cup I_{+-}(\bar{x}),\ \gamma_i \geq 0,\ \forall i \in I_{00}(\bar{x}) \cup I_{+0}(\bar{x}). \end{aligned}$$

11.3 Duality Results

A parametric Mond–Weir dual model to the (PMPVC), denoted by (PMWD-VC(x)), depending upon $x \in \Omega$, is given by

$$\max\ q$$
subject to

$$\sum_{i\in K}\alpha_i\nabla\theta_i(y)-\sum_{i\in L}\beta_i\nabla\Phi_i(y)+\sum_{i\in L}\gamma_i\nabla\psi_i(y)=0, \tag{11.9}$$

$$\alpha_i\geq 0,\ \alpha_i(\theta_i(y)-q)=0,\ \forall i\in K,\ \sum_{i\in K}\alpha_i=1, \tag{11.10}$$

$$-\beta_i\Phi_i(y)\geq 0,\ \forall i\in L,\ \beta_i\geq 0,\ \forall i\in I_+(x),\ \beta_i\in\mathbb{R},\ \forall i\in I_0(x), \tag{11.11}$$

$$\gamma_i\psi_i(y)\geq 0,\ \forall i\in L,\ \gamma_i\geq 0,\ \forall i\in I_{0-}(x)\cup I_{+-}(x),\ \gamma_i\leq 0,\ \forall i\in I_{0+}(x),$$
$$\gamma_i\in\mathbb{R},\ \forall i\in I_{+0}(x)\cup I_{00}(x). \tag{11.12}$$

The set of all feasible solutions of the (PMWD-VC(x)) is given by $S_{MW}(x)$ and the projection of $S_{MW}(x)$ on $\mathbb{R}^n$ is given by $pr_{\mathbb{R}^n}S_{MW}(x)$.

The following theorem establishes weak duality result between (MMPVC) and (PMWD-VC(x)).

Theorem 11.3.4 (Weak duality) *Let* $(x,\nu)\in\Omega\times V$ *and* $(y,q,\alpha,\beta,\gamma)\in\mathbb{R}^{n+1+k+2l}$ *be feasible solutions for the (PMPVC) and (PMWD-VC(x)), respectively. If* $\sum_{i\in K}\alpha_i\theta_i$ *is* η*-pseudoinvex and* $-\sum_{i\in L}\beta_i\Phi_i+\sum_{i\in L}\gamma_i\psi_i$ *is* η*-quasiinvex at* y*, then* $\nu\geq q$.

Proof Since $(x,\nu)\in\Omega\times V$ and $(y,q,\alpha,\beta,\gamma)\in\mathbb{R}^{n+1+k+2l}$ are feasible solutions for the (PMPVC) and (PMWD-VC(x)), respectively, therefore

$$\begin{aligned}-\beta_i\Phi_i(x)\leq 0\leq-\beta_i\Phi_i(y),\ \forall i\in I_+(x);\\ -\beta_i\Phi_i(x)=0\leq-\beta_i\Phi_i(y),\ \forall i\in I_0(x);\\ \gamma_i\psi_i(x)\leq 0\leq\gamma_i\psi_i(y),\ \forall i\in I_{+-}(x)\cup I_{0-}(x)\cup I_{0+}(x);\\ \gamma_i\psi_i(x)=0\leq\gamma_i\psi_i(y),\ \forall i\in I_{+0}(x)\cup I_{00}(x),\end{aligned}$$

which implies that

$$-\sum_{i\in L}\beta_i\Phi_i(x)+\sum_{i\in L}\gamma_i\psi_i(x)\leq-\sum_{i\in L}\beta_i\Phi_i(y)+\sum_{i\in L}\gamma_i\psi_i(y). \tag{11.13}$$

By η-quasiinvexity of $-\sum_{i\in L}\beta_i\Phi_i+\sum_{i\in L}\gamma_i\psi_i$ at y and inequality (11.13), it follows that

$$\left\langle-\sum_{i\in L}\beta_i\nabla\Phi_i(y)+\sum_{i\in L}\gamma_i\nabla\psi_i(y),\eta(x,y)\right\rangle\leq 0. \tag{11.14}$$

By dual feasibility condition (11.9) and inequality (11.14), one has

$$\left\langle \sum_{i\in K} \alpha_i \nabla\theta_i(y), \eta(x, y) \right\rangle \geq 0.$$

By the η-pseudoinvexity of $\sum_{i\in K} \alpha_i\theta_i$ at y, the above inequality gives

$$\sum_{i\in K} \alpha_i \nu \geq \sum_{i\in K} \alpha_i\theta_i(x) \geq \sum_{i\in K} \alpha_i\theta_i(y) \geq \sum_{i\in K} \alpha_i q,$$

which implies by (11.10) that $\nu \geq q$ and this completes the proof. □

We have the following weak duality result under stronger assumption of strictly η-pseudoinvexity.

Theorem 11.3.5 (Weak duality) *Let* $(x, \nu) \in \Omega \times V$ *and* $(y, q, \alpha, \beta, \gamma) \in \mathbb{R}^{n+1+k+2l}$ *be feasible solutions for the (PMPVC) and (PMWD-VC(x)), respectively. If* $\sum_{i\in K} \alpha_i\theta_i$ *is strictly* η*-pseudoinvex and* $-\sum_{i\in L} \beta_i\Phi_i + \sum_{i\in L} \gamma_i\psi_i$ *is* η*-quasiinvex at* y*, then* $\nu > q$.

The following theorem establishes strong duality result between (MMPVC) and (PMWD-VC(x)).

Theorem 11.3.6 (Strong duality) *If* $\bar{x}$ *is a local minimum of (MMPVC) with minimum value* $\bar{\nu}$ *such that modified ACQ is satisfied at* $(\bar{x}, \bar{\nu})$*, then there exist* $\bar{\alpha}_i \in \mathbb{R}(i \in K)$, $\bar{\beta}_i, \bar{\gamma}_i \in \mathbb{R}(i \in L)$ *such that* $(\bar{x}, \bar{\nu}, \bar{\alpha}, \bar{\beta}, \bar{\gamma})$ *is a feasible solution of the (PMWD-VC($\bar{x}$)). Moreover, if* $\sum_{i\in K} \bar{\alpha}\theta_i$ *is* η*-pseudoinvex and* $-\sum_{i\in L} \bar{\beta}\Phi_i + \sum_{i\in L} \bar{\gamma}\psi_i$ *is* η*-quasiinvex on* $pr_{\mathbb{R}^n} S_{MW}(\bar{x})$*, then* $(\bar{x}, \bar{\nu}, \bar{\alpha}, \bar{\beta}, \bar{\gamma})$ *is a global maximizer of the (PMWD-VC($\bar{x}$)).*

Proof Since $\bar{x}$ is a local minimum of (MMPVC) with minimum value $\bar{\nu}$ such that modified ACQ is satisfied at $(\bar{x}, \bar{\nu})$. Therefore, by Theorem 11.2.1, there exist $\bar{\alpha}_i \in \mathbb{R}(i \in K)$, $\bar{\beta}_i, \bar{\gamma}_i \in \mathbb{R}(i \in L)$ such that

$$\begin{aligned}
&\sum_{i\in K} \bar{\alpha}_i \nabla\theta_i(\bar{x}) - \sum_{i\in L} \bar{\beta}_i \nabla\Phi_i(\bar{x}) + \sum_{i\in L} \bar{\gamma}_i \nabla\psi_i(\bar{x}) = 0, \\
&\bar{\alpha}_i \geq 0,\ \bar{\alpha}_i(\theta_i(\bar{x}) - \bar{\nu}) = 0,\ \forall i \in K,\ \sum_{i\in K} \bar{\alpha}_i = 1, \\
&\bar{\beta}_i = 0,\ \forall i \in I_+(\bar{x}),\ \bar{\beta}_i \geq 0,\ \forall i \in I_{00}(\bar{x}) \cup I_{0-}(\bar{x}),\ \bar{\beta}_i \in \mathbb{R},\ \forall i \in I_{0+}(\bar{x}), \\
&\bar{\gamma}_i = 0,\ \forall i \in I_{0+}(\bar{x}) \cup I_{0-}(\bar{x}),\ \bar{\gamma}_i \geq 0,\ \forall i \in I_{00}(\bar{x}) \cup I_{+0}(\bar{x}),
\end{aligned}$$

which implies that $(\bar{x}, \bar{\nu}, \bar{\alpha}, \bar{\beta}, \bar{\gamma})$ is a feasible solution of the (PMWD-VC($\bar{x}$)). Since $\sum_{i\in K} \bar{\alpha}\theta_i$ is η-pseudoinvex and $-\sum_{i\in L} \bar{\beta}\Phi_i + \sum_{i\in L} \bar{\gamma}\psi_i$ is η-quasiinvex on $pr_{\mathbb{R}^n} S_{MW}(\bar{x})$, by the weak duality Theorem 11.3.4, for any feasible point $(y, q, \alpha, \beta, \gamma)$ of the (PMWD-VC($\bar{x}$)), one has

$$\bar{\nu} \geq q,$$

which implies that $(\bar{x}, \bar{\nu}, \bar{\alpha}, \bar{\beta}, \bar{\gamma})$ is an optimal solution of the (PMWD-VC($\bar{x}$)). □

Similarly, we have the following strong duality result.

Theorem 11.3.7 (Strong duality) *If $\bar{x}$ is a local minimum of (MMPVC) with minimum value $\bar{\nu}$ such that modified ACQ is satisfied at $(\bar{x}, \bar{\nu})$, then there exist $\bar{\alpha}_i \in \mathbb{R}(i \in K)$, $\bar{\beta}_i, \bar{\gamma}_i \in \mathbb{R}(i \in L)$ such that $(\bar{x}, \bar{\nu}, \bar{\alpha}, \bar{\beta}, \bar{\gamma})$ is a feasible solution of the (PMWD-VC($\bar{x}$)). Moreover, if $\sum_{i \in K} \bar{\alpha}\theta_i$ is strictly η-pseudoinvex and $-\sum_{i \in L} \bar{\beta}\Phi_i + \sum_{i \in L} \bar{\gamma}\psi_i$ is η-quasiinvex on $pr_{\mathbb{R}^n} S_{MW}(\bar{x})$, then $(\bar{x}, \bar{\nu}, \bar{\alpha}, \bar{\beta}, \bar{\gamma})$ is a strict global maximizer of the (PMWD-VC($\bar{x}$)).*

We have the following converse duality theorem between (MMPVC) and (PMWD-VC(x)).

Theorem 11.3.8 (Converse duality) *If $(\bar{y}, \bar{q}, \bar{\alpha}, \bar{\beta}, \bar{\gamma})$ is a feasible solution of the PMWD-VC(x) for every $x \in \Omega$ such that $\sum_{i \in K} \bar{\alpha}_i\theta_i$ is η-pseudoinvex and $-\sum_{i \in L} \bar{\beta}_i\Phi_i + \sum_{i \in L} \bar{\gamma}_i\psi_i$ is η-quasiinvex at $\bar{y}$, then $(\bar{y}, \bar{q})$ is a global minimizer of the (PMPVC).*

Proof Suppose to the contrary that $(\bar{y}, \bar{q})$ is not a global minimizer of the (PMPVC). Then, there exists $(\tilde{x}, \tilde{\nu}) \in \Omega \times V$ such that

$$\bar{q} > \tilde{\nu},$$

which implies that

$$\sum_{i \in K} \bar{\alpha}_i\theta_i(\bar{y}) = \sum_{i \in K} \bar{\alpha}_i\bar{q} > \sum_{i \in K} \bar{\alpha}_i\tilde{\nu} \geq \sum_{i \in K} \bar{\alpha}_i\theta_i(\tilde{x}).$$

By the η-pseudoinvexity of $\sum_{i \in K} \bar{\alpha}_i\theta_i$ at $\bar{y}$, one has

$$\left\langle \sum_{i \in K} \bar{\alpha}_i \nabla\theta_i(\bar{y}), \eta(\tilde{x}, \bar{y}) \right\rangle < 0. \tag{11.15}$$

By the feasibility of $(\tilde{x}, \tilde{\nu})$ for the (PMPVC) and the feasibility of $(\bar{y}, \bar{q}, \bar{\alpha}, \bar{\beta}, \bar{\gamma}) \in \mathbb{R}^{n+1+k+2l}$ for the (PMWD-VC(x)), one has

$$\begin{aligned}
-\bar{\beta}_i\Phi_i(\tilde{x}) \leq 0 \leq -\bar{\beta}_i\Phi_i(\bar{y}), &\quad \forall i \in I_+(\tilde{x});\\
-\bar{\beta}_i\Phi_i(\tilde{x}) = 0 \leq -\bar{\beta}_i\Phi_i(\bar{y}), &\quad \forall i \in I_0(\tilde{x});\\
\bar{\gamma}_i\psi_i(\tilde{x}) \leq 0 \leq \bar{\gamma}_i\psi_i(\bar{y}), &\quad \forall i \in I_{+-}(\tilde{x}) \cup I_{0-}(\tilde{x}) \cup I_{0+}(\tilde{x});\\
\bar{\gamma}_i\psi_i(\tilde{x}) = 0 \leq \bar{\gamma}_i\psi_i(\bar{y}), &\quad \forall i \in I_{+0}(\tilde{x}) \cup I_{00}(\tilde{x}),
\end{aligned}$$

which implies that

$$-\sum_{i\in L}\bar{\beta}_i\Phi_i(\tilde{x})+\sum_{i\in L}\bar{\gamma}_i\psi_i(\tilde{x})\leq -\sum_{i\in L}\bar{\beta}_i\Phi_i(\bar{y})+\sum_{i\in L}\bar{\gamma}_i\psi_i(\bar{y}). \tag{11.16}$$

By η-quasiinvexity of $-\sum_{i\in L}\bar{\beta}_i\Phi_i+\sum_{i\in L}\bar{\gamma}_i\psi_i$ at $\bar{y}$, the inequalities (11.16) implies that

$$\left\langle -\sum_{i\in L}\bar{\beta}_i\nabla\Phi_i(\bar{y})+\sum_{i\in L}\bar{\gamma}_i\nabla\psi_i(\bar{y}),\eta(\tilde{x},\bar{y})\right\rangle\leq 0. \tag{11.17}$$

Adding inequalities (11.15) and (11.17), one has

$$\left\langle\sum_{i\in K}\bar{\alpha}\nabla\theta_i(\bar{y})-\sum_{i\in L}\bar{\beta}_i\nabla\Phi_i(\bar{y})+\sum_{i\in L}\bar{\gamma}_i\nabla\psi_i(\bar{y}),\eta(\tilde{x},\bar{y})\right\rangle<0,$$

which is a contradiction to the feasibility of $(\bar{y},\bar{q},\bar{\alpha},\bar{\beta},\bar{\gamma})\in\mathbb{R}^{n+1+k+2l}$ for the (PMWD-VC(x)) for every $x\in\Omega$. Hence, $(\bar{y},\bar{q})$ is a global minimizer of the (PMPVC). □

Similarly, we can prove the following result.

Theorem 11.3.9 (Converse duality) *If $(\bar{y},\bar{q},\bar{\alpha},\bar{\beta},\bar{\gamma})$ is a feasible solution of the (PMWD-VC(x)) for every $x\in\Omega$ such that $\sum_{i\in K}\bar{\alpha}_i\theta_i$ is strictly η-pseudoinvex and $-\sum_{i\in L}\bar{\beta}_i\Phi_i+\sum_{i\in L}\bar{\gamma}_i\psi_i$ is η-quasiinvex at $\bar{y}$, then $(\bar{y},\bar{q})$ is a strict global minimizer of the (PMPVC).*

The following theorem establishes restricted converse duality result between the (PMPVC) and the (PMWD-VC(x)).

Theorem 11.3.10 (Restricted converse duality) *Let $(\bar{y},\bar{q},\bar{\alpha},\bar{\beta},\bar{\gamma})$ be a feasible solution of the (PMWD-VC(x)) for every $x\in\Omega$ and let $(\bar{x},\bar{\nu})$ be a feasible solution of the (PMPVC) such that $\bar{q}=\bar{\nu}$. If $\sum_{i\in K}\bar{\alpha}_i\theta_i$ is η-pseudoinvex and $-\sum_{i\in L}\bar{\beta}_i\Phi_i+\sum_{i\in L}\bar{\gamma}_i\psi_i$ is η-quasiinvex at $\bar{y}$, then $(\bar{x},\bar{\nu})$ is a global minimizer of the (PMPVC).*

Proof Suppose to contrary that $(\bar{x},\bar{\nu})$ is not a global minimizer of the (PMPVC). Then, there exists $(\tilde{x},\tilde{\nu})\in\Omega\times V$ such that

$$\bar{\nu}>\tilde{\nu},$$

which implies that

$$\bar{q}>\tilde{\nu},$$

which contradicts the weak duality Theorem 11.3.4 and hence the proof. □

Similarly, we have the following restricted converse duality theorem.

Theorem 11.3.11 (Restricted converse duality) *Let $(\bar{y}, \bar{q}, \bar{\alpha}, \bar{\beta}, \bar{\gamma})$ be a feasible solution of the (PMWD-VC(x)) for every $x \in \Omega$ and let $(\bar{x}, \bar{\nu})$ be a feasible solution of the (PMPVC) such that $\bar{q} = \bar{\nu}$. If $\sum_{i \in K} \bar{\alpha}_i \theta_i$ is strictly η-pseudoinvex and $-\sum_{i \in L} \bar{\beta}_i \Phi_i + \sum_{i \in L} \bar{\gamma}_i \psi_i$ is η-quasiinvex at $\bar{y}$, then $(\bar{x}, \bar{\nu})$ is a strict global minimizer of the (PMPVC).*

11.4 Applications to Multiobjective Optimization

In this section, we apply the optimality results obtained for minimax programs with vanishing constraints in the previous section to multiobjective optimization problems with vanishing constraints and obtain the corresponding optimality results.

Consider a multiobjective optimization problem with vanishing constraints (MOPVC) as follows:

$$
\begin{aligned}
&\min\ \theta(x) := (\theta_1(x), \ldots, \theta_k(x)) \\
&\text{subject to} \\
&\Phi_i(x) \geq 0,\ \ \forall i \in L := \{1, 2, \ldots, l\}, \\
&\psi_i(x)\Phi_i(x) \leq 0,\ \ \forall i \in L,
\end{aligned}
$$

where we assume that the functions $\theta_i,\ \Phi_i,\ \psi_i : \mathbb{R}^n \to \mathbb{R}$ are continuously differentiable and the set of all feasible solutions $\Omega := \{x \in \mathbb{R}^n : \Phi_i(x) \geq 0,\ \psi_i(x)\Phi_i(x) \leq 0,\ \forall i \in L\}$ is non empty and compact.

A vector $\bar{x} \in \Omega \subseteq \mathbb{R}^n$ is said to be a weak efficient solution of the (MOPVC), iff for all $x \in \Omega$, one has

$$
\theta(x) - \theta(\bar{x}) := (\theta_1(x) - \theta_1(\bar{x}), \ldots, \theta_k(x) - \theta_k(\bar{x})) \notin -int\mathbb{R}^k_+.
$$

The following theorem gives the Karush–Kuhn–Tucker necessary optimality condition for weak efficient solutions of the (MOPVC).

Theorem 11.4.12 (Karush–Kuhn–Tucker necessary optimality conditions) *Let $\bar{x} \in \Omega$ be a weak efficient solution of the (MOPVC). Further, assume that modified ACQ from Theorem 11.2.1 is satisfied at $(\bar{x}, 0)$. Then, there exist $\bar{\alpha}_i \in \mathbb{R} (i \in K),\ \bar{\beta}_i, \bar{\gamma}_i \in \mathbb{R} (i \in L)$ such that the following conditions hold:*

$$\sum_{i\in K}\bar{\alpha}_i\nabla\theta_i(\bar{x})-\sum_{i\in L}\bar{\beta}_i\nabla\Phi_i(\bar{x})+\sum_{i\in L}\bar{\gamma}_i\nabla\psi_i(\bar{x})=0,$$
$$\bar{\beta}_i=0,\ \forall i\in I_+(\bar{x}),\ \bar{\beta}_i\geq 0,\ \forall i\in I_{00}(\bar{x})\cup I_{0-}(\bar{x}),\ \bar{\beta}_i\in\mathbb{R},\ \forall i\in I_{0+}(\bar{x}),$$
$$\bar{\gamma}_i=0,\ \forall i\in I_{0+}(\bar{x})\cup I_{0-}(\bar{x}),\ \bar{\gamma}_i\geq 0,\ \forall i\in I_{00}(\bar{x})\cup I_{+0}(\bar{x}). \tag{11.18}$$

Proof Let $\bar{x}\in\Omega$ be a weak efficient solution of the problem (MOPVC) and define

$$\hat{\theta}_i(x):=\theta_i(x)-\theta_i(\bar{x}),\ i\in K,\ x\in\Omega.$$

Then, it can be easily verified that $\bar{x}$ is a global optimal solution with optimal value zero of the following minimax mathematical program with vanishing constraints:

$$\min_{x\in\Omega}\max_{i\in K}\hat{\theta}_i(x). \tag{11.19}$$

Indeed, let us write $\hat{\xi}(x):=\max_{i\in K}\hat{\theta}_i(x)$ and prove that

$$\hat{\xi}(\bar{x})\leq\hat{\xi}(x),\ \forall x\in\Omega. \tag{11.20}$$

Suppose to contrary that (11.20) does not hold, then there exists $\tilde{x}\in\Omega$ such that

$$\hat{\xi}(\bar{x})>\hat{\xi}(\tilde{x}).$$

Since $\hat{\xi}(\bar{x})=0$, so the above inequality implies that

$$\max_{i\in K}\{\theta_i(\tilde{x})-\theta_i(\bar{x})\}<0.$$

Thus,

$$\theta_i(\tilde{x})-\theta_i(\bar{x})<0,\ \forall i\in K,$$

which contradicts that $\bar{x}$ is a weak efficient solution of the (MOPVC). So, we can employ the parametric necessary optimality condition in Theorem 11.2.1, but applied to minimax problem (11.19). Thus, we find $\bar{\alpha}_i\in\mathbb{R}\ (i\in K),\ \bar{\beta}_i,\bar{\gamma}_i\in\mathbb{R}\ (i\in L)$ such that

$$\sum_{i\in K}\bar{\alpha}_i\nabla\hat{\theta}_i(\bar{x})-\sum_{i\in L}\bar{\beta}_i\nabla\Phi_i(\bar{x})+\sum_{i\in L}\bar{\gamma}_i\nabla\psi_i(\bar{x})=0,$$
$$\bar{\alpha}_i\geq 0,\ \bar{\alpha}_i(\hat{\theta}_i(\bar{x})-\hat{\nu})=0,\ \forall i\in K,\ \sum_{i\in K}\bar{\alpha}_i=1,$$
$$\bar{\beta}_i=0,\ \forall i\in I_+(\bar{x}),\ \bar{\beta}_i\geq 0,\ \forall i\in I_{00}(\bar{x})\cup I_{0-}(\bar{x}),\ \bar{\beta}_i\in\mathbb{R},\ \forall i\in I_{0+}(\bar{x}),$$
$$\bar{\gamma}_i=0,\ \forall i\in I_{0+}(\bar{x})\cup I_{0-}(\bar{x}),\ \bar{\gamma}_i\geq 0,\ \forall i\in I_{00}(\bar{x})\cup I_{+0}(\bar{x}), \tag{11.21}$$

where $\hat{\nu}$ is the corresponding optimal value of the objective function in problem (11.19) which is nothing else but zero. Then, it is clear that (11.21) implies (11.18). Hence, the proof is complete. □

In the following theorem we prove a sufficient condition for the existence of weak efficient solution of the (MOPVC).

Theorem 11.4.13 (Sufficient Optimality Conditions) *Let $\bar{x} \in \Omega$ satisfy (11.18). Suppose that $\sum_{i\in K} \bar{\alpha}_i\theta_i$ is η-pseudoinvex and $-\Phi_i (i \in I_{00}^{++} \cup I_{00}^{+0} \cup I_{0-}^{+0} \cup I_{0+}^{+0})$, $\Phi_i (i \in I_{0+}^{-0})$, and $\psi_i (i \in I_{00}^{++} \cup I_{00}^{0+} \cup I_{+0}^{0+})$ are η-quasiinvex at $\bar{x}$ over Ω. If $I_{00}^{++} \cup I_{00}^{0+} \cup I_{0+}^{-0} \cup I_{+0}^{0+} = \Phi$, then $\bar{x}$ is a weak efficient solution of the (MOPVC).*

Proof We proceed on similar lines of the proof of Theorem 11.4.12. Let $\bar{x} \in \Omega$ satisfy (11.18). Let us write

$$\hat{\theta}_i(x) := \theta_i(x) - \theta_i(\bar{x}), i \in K, x \in \Omega.$$

Then, it can be easily verified that $\bar{x}$ also satisfy the conditions (11.21) with $\hat{\alpha}_i := \frac{\bar{\alpha}_i}{\sum_{i=1}^{l} \bar{\alpha}_i}, i \in K,\ \hat{\beta}_i := \bar{\beta}_i,\ \hat{\gamma}_i := \bar{\gamma}_i \in \mathbb{R}\ (i \in L)$ and $\hat{\nu} = 0$.

Also, by η-pseudoinvexity of $\sum_{i\in K} \bar{\alpha}_i\theta_i$, it follows that $\sum_{i\in K} \hat{\alpha}_i\hat{\theta}_i$ is also η-pseudoinvex at $\bar{x} \in \Omega$ with respect to same kernel function η. Thus, applying the sufficient optimality criteria in Theorem 11.2.2, we conclude that $\bar{x}$ is a global optimal solution of the minimax programming problem with vanishing constraints

$$\min_{x\in\Omega} \max_{i\in K} \hat{\theta}_i(x).$$

This implies that,

$$\hat{\xi}(\bar{x}) \leq \hat{\xi}(x), \forall x \in \Omega, \quad where\ \hat{\xi}(x) := \max_{i\in K} \hat{\theta}_i(x).$$

This means that,

$$\max_{i\in K}\{\theta_i(x) - \theta_i(\bar{x})\} \geq 0, \quad \forall x \in \Omega,$$

which implies that

$$\theta(x) - \theta(\bar{x}) := (\theta_1(x) - \theta_1(\bar{x}), \ldots, \theta_k(x) - \theta_k(\bar{x})) \notin -int\mathbb{R}^k_+, \ \forall x \in \Omega.$$

Consequently, $\bar{x}$ is a weak efficient solution of the (MOPVC). □

11.5 Conclusions

In this chapter, we have analyzed a class of minimax program with vanishing constraints. We have obtained parametric necessary and parameter-free necessary optimality conditions for such a problem. Further, we have obtained sufficient optimality conditions under the assumptions of η-pseudoinvexity and η-quasiinvexity. We have formulated parametric Mond–Weir type dual model and derived several duality results. Also, we have utilized some results obtained for minimax program with vanishing constraints to derive necessary and sufficient optimality results for a multiobjective optimization problem with vanishing constraints.

Acknowledgements The authors are thankful to the anonymous referees who helped to improve the presentation of this chapter in its present form. The research of Dr. Vivek Laha is supported by UGC-BSR start up grant by University Grant Commission, New Delhi, India (Letter No. F.30-370/2017(BSR)) (Project No. M-14-40). The research of Prof. S.K. Mishra is supported by Department of Science and Technology, SERB, New Delhi, India through grant no.: MTR/2018/000121.

References

1. Achtziger, W., Kanzow, C.: Mathematical programs with vanishing constraints: optimality conditions and constraint qualifications. Math. Program. **114**(1), 69–99 (2008)
2. Hoheisel, T., Kanzow, C.: First-and second-order optimality conditions for mathematical programs with vanishing constraints. Appl. Math. **52**(6), 495–514 (2007)
3. Hoheisel, T., Kanzow, C.: Stationary conditions for mathematical programs with vanishing constraints using weak constraint qualifications. J. Math. Anal. Appl. **337**(1), 292–310 (2008)
4. Hoheisel, T., Kanzow, C.: On the Abadie and Guignard constraint qualifications for mathematical programmes with vanishing constraints. Optimization **58**(4), 431–448 (2009)
5. Izmailov, A.F., Solodov, M.V.: Mathematical programs with vanishing constraints: optimality conditions, sensitivity, and a relaxation method. J. Optim. Theory Appl. **142**(3), 501–532 (2009)
6. Khare, A., Nath, T.: Enhanced Fritz John stationarity, new constraint qualifications and local error bound for mathematical programs with vanishing constraints. J. Math. Anal. Appl. **472**(1), 1042–1077 (2019)
7. Mishra, S.K., Singh, V., Laha, V.: On duality for mathematical programs with vanishing constraints. Ann. Oper. Res. **243**(1–2), 249–272 (2016)
8. Hu, Q., Wang, J., Chen, Y.: New dualities for mathematical programs with vanishing constraints. Ann. Oper. Res. **287**(1), 233–255 (2020)
9. Mishra, S.K., Singh, V., Laha, V., Mohapatra, R.N.: On constraint qualifications for multiobjective optimization problems with vanishing constraints. Optimization Methods, Theory and Applications, pp. 95–135. Springer, Berlin, Heidelberg (2015)
10. Guu, S.M., Singh, Y., Mishra, S.K.: On strong KKT type sufficient optimality conditions for multiobjective semi-infinite programming problems with vanishing constraints. J. Inequal. Appl. **2017**(1), 1–9 (2017)
11. Jayswal, A., Singh, V.: The Characterization of Efficiency and Saddle Point Criteria for Multiobjective Optimization Problem with Vanishing Constraints. Acta Math. Sci. **39**(2), 382–394 (2019)
12. Kazemi, S., Kanzi, N.: Constraint qualifications and stationary conditions for mathematical programming with non-differentiable vanishing constraints. J. Optim. Theory Appl. **179**(3), 800–819 (2018)

13. Kazemi, S., Kanzi, N., Ebadian, A.: Estimating the Frechet normal cone in optimization problems with nonsmooth vanishing constraints. Iran J. Sci. Technol. A **43**(5), 2299–2306 (2019)
14. Kanzi, N., Barilla, D., Caristi, G.: Qualifications and stationary conditions for nonsmooth multiobjective mathematical programming problem with vanishing constraints. Numer. Comput. Theory Algorithms NUMTA **2019**, 138 (2019)
15. Mokhtavayi, H., Heidari, A., Kanzi, N.: Necessary and sufficient conditions for M-stationarity of nonsmooth optimization. Comp. Meth. Part. D. E. (2020). https://doi.org/10.22034/cmde.2020.30733.1459
16. Schmitendorf, W.E.: Necessary conditions and sufficient conditions for static minmax problems. J. Math. Anal. Appl. **57**(3), 683–693 (1977)
17. Mishra, S.K.: Generalized pseudo convex minmax programming. Opsearch **35**(1), 32–44 (1998)
18. Mehra, A., Bhatia, D.: Optimality and duality for minmax problems involving arcwise connected and generalized arcwise connected functions. J. Math. Anal. Appl. **231**(2), 425–445 (1999)
19. Studniarski, M., Taha, A.W.A.: A characterization of strict local minimizers of order one for nonsmooth static minmax problems. J. Math. Anal. Appl. **259**(2), 368–376 (2001)
20. Antczak, T.: Minimax programming under $(p, r)-$invexity. Eur. J. Oper. Res. **158**(1), 1–19 (2004)
21. Mandal, P., Nahak, C.: Minmax programming problems with $(p, r) - \rho - (\eta, \theta)$-invexity. Int. J. Math. Oper. Res. **5**(1), 121–143 (2013)
22. Zemkoho, A.B.: A simple approach to optimality conditions in minmax programming. Optimization **63**(3), 385–401 (2014)
23. Mishra, S.K., Shukla, K.: Nonsmooth minimax programming problems with $Vr-$invex functions. Optimization **59**(1), 95–103 (2010)
24. Antczak, T.: Nonsmooth minimax programming under locally Lipschitz $(\Phi, \rho)-$invexity. Appl. Math. Comp. **217**(23), 9606–9624 (2011)
25. Jayswal, A., Ahmad, I., Kummari, K., Al-Homidan, S.: On minimax programming problems involving right upper-Dini-derivative functions. J. Inequal. Appl. **2014**(1), 326 (2014)
26. Mishra, S.K., Rueda, N.G.: Second-order duality for nondifferentiable minimax programming involving generalized type I functions. J. Optim. Theory Appl. **130**(3), 479–488 (2006)
27. Ahmad, I., Husain, Z., Sharma, S.: Second-order duality in nondifferentiable minmax programming involving type-I functions. J. Comp. Appl. Math. **215**(1), 91–102 (2008)
28. Husain, Z., Jayswal, A., Ahmad, I.: Second order duality for nondifferentiable minimax programming problems with generalized convexity. J. Glob. Optim. **44**(4), 593 (2009)
29. Jayswal, A., Stancu-Minasian, I.: Higher-order duality for nondifferentiable minimax programming problem with generalized convexity. Nonlinear Anal-Theory Meth. Appl. **74**(2), 616–625 (2011)
30. Lai, H.L.H., Chen, J.C.J.: Optimality conditions for minimax programming of analytic functions. Taiwan. J. Math. **8**(4), 673–686 (2004)
31. Lai, H.C., Lee, J.C., Ho, S.C.: Parametric duality on minimax programming involving generalized convexity in complex space. J. Math. Anal. Appl. **323**(2), 1104–1115 (2006)
32. Lai, H.C., Liu, J.C.: Duality for nondifferentiable minimax programming in complex spaces. Nonlinear Anal-Theory Meth. Appl. **71**(12), 224–233 (2009)
33. Stefanescu, M.V., Stefanescu, A.: On semi-infinite minmax programming with generalized invexity. Optimization **61**(11), 1307–1319 (2012)
34. Upadhyay, B.B., Mishra, S.K.: Nonsmooth semi-infinite minmax programming involving generalized $(\Phi, \rho)-$invexity. J. Syst. Sci. Complex. **28**(4), 857–875 (2015)
35. Das, K., Nahak, C.: Set-valued minimax programming problems under generalized cone convexity. Rendiconti del Circolo Matematico di Palermo Series 2 **66**(3), 361–374 (2017)
36. Hanson, M.A.: On sufficiency of the Kuhn-Tucker conditions. J. Math. Anal. Appl. **80**, 545–550 (1981)
37. Crouzeix, J.P., Ferland, J.A., Schaible, S.: An algorithm for generalized fractional programs. J. Optim. Theory Appl. **47**, 35–49 (1985)

Chapter 12
On Minty Variational Principle for Nonsmooth Interval-Valued Multiobjective Programming Problems

Balendu Bhooshan Upadhyay and Priyanka Mishra

Abstract In this chapter, we consider a class of nonsmooth interval-valued multiobjective programming problems and a class of approximate Minty and Stampacchia vector variational inequalities. Under generalized approximate LU-convexity hypotheses, we establish the relations between the solutions of approximate Minty and Stampacchia vector variational inequalities and the approximate LU-efficient solutions of the nonsmooth interval-valued multiobjective programming problem. The results of this chapter extend and unify the corresponding results of [14, 22, 23, 30, 33] for nonsmooth interval-valued multiobjective programming problems.

Keywords Approximate LU-convexity · Approximate LU-efficient solutions · Interval-valued programming problems

12.1 Introduction

In multiobjective programming problems, two or more objective functions are minimized on some set of constraints. Usually, optimization problems are considered to deal with deterministic values, and therefore, we get precise solutions. However, in many real-life applications, optimization problems occur with uncertainty. Interval-valued optimization is one of the deterministic optimization models to deal with inexact, imprecise, or uncertain data. In interval-valued optimization, the coefficients of objective and constraint functions are compact intervals. To deal with the functions with interval coefficients, Moore [25, 26] introduced the concept of

B. B. Upadhyay (✉) · P. Mishra
Department of Mathematics, Indian Institute of Technology Patna, Patna 801103, India
e-mail: bhoodhan@iitp.ac.in

P. Mishra
e-mail: priyanka.iitp14@gmail.com

V. Laha et al. (eds.), *Optimization, Variational Analysis and Applications*, Springer Proceedings in Mathematics & Statistics 355,
https://doi.org/10.1007/978-981-16-1819-2_12

interval analysis. Wu [31] established the Karush–Kuhn–Tucker optimality conditions for interval-valued optimization problem. Antczak [1] established Fritz John and Karush–Kuhn–Tucker necessary and sufficient optimality conditions for nonsmooth interval-valued multiobjective programming problem. For more details about interval-valued optimization problems, we refer to [2, 8, 9, 16, 17] and the references cited therein.

The notion of efficiency or Pareto optimality is a widely used solution concept in multiobjective programming problems. Due to complexity of multiobjective programming problems, several variants of efficient solutions have been studied by many researchers, see [4, 5, 13, 15, 18] and the references cited therein. Loridan introduced the notion of ϵ-efficient solution for multiobjective programming problems. Recently, many authors have shown interest in the study of characterization and applications of approximate efficient solutions of multiobjective programming problems, see [12, 13, 21, 22] and the references cited therein.

In 1980, Giannessi [10] introduced the notion of vector variational inequality problems. Vector variational inequality problems have wider applications in optimization, optimal control, and economics equilibrium problems, see for example [7, 11, 19] and the references cited therein. The equivalence between the solutions of vector variational inequalities and solutions of multiobjective programming problems have been studied extensively by many authors, see [20, 24, 27–30, 32] and the references cited therein. Mishra and Laha [22] established the relations between the solutions of approximate vector variational inequalities and approximate efficient solution of the nonsmooth multiobjective programming problems. Further, Gupta and Mishra [14] extend the results of [22] for generalized approximate convex functions. Zhang et al. [33] established the relations between the solutions of interval-valued multiobjective programming problems and vector variational inequalities.

12.1.1 The Proposed Work

The novelty and contributions of our work are of three folds:

In the first fold, motivated by the work of Gupta and Mishra [14], we have introduced a new class of generalized approximate LU-convex functions, namely; approximate LU-pseudoconvex of type I, approximate LU-pseudoconvex of type II, approximate LU-quasiconvex of type I, and approximate LU-quasiconvex of type II functions. These classes of generalized approximate LU-convex functions are more general than the classes of generalized approximate convex functions used in Gupta and Mishra [14], Mishra and Laha [22] and Mishra and Upadhyay [23].

In the second fold, we extend the works of Lee and Lee [20], Mishra and Upadhyay [23] and Upadhyay et al. [30] for the class of interval-valued multiobjective programming problems.

In the third fold, we generalize the works of [14, 20, 23, 30] from Euclidean space to a more general space such as Banach space.

The rest of the chapter is organized as follows: In Sect. 12.2, some basic definitions and preliminaries are given which will be used throughout the sequel. In Sect. 12.3, we establish the relations between the solutions of approximate vector variational inequalities and approximate LU-efficient solutions of the nonsmooth interval-valued multiobjective programming problem by using generalized approximate LU-convex functions. The numerical example has also been given to justify the significance of these results.

12.2 Definition and Preliminaries

Let Ω be a Banach space and Ω^* be its dual space equipped with norms $\|.\|$ and $\|.\|_*$, respectively. Let $\langle ., . \rangle$ denotes the dual pair between Ω and Ω^* and Γ be a nonempty subset of Ω. Let $B(z; \delta)$ be an open ball centered at z and radius $\delta > 0$. Let $\mathbf{0}$ denotes the zero vector in $\mathbb{R}^n$.

For $z, y \in \mathbb{R}^n$, following notion for equality and inequalities will be used throughout the sequel:

(i) $z = y, \iff z_i = y_i, \forall i = 1, 2, \ldots, n;$
(ii) $z < y, \iff z_i < y_i, \forall i = 1, 2, \ldots, n;$
(iii) $z \leqq y, \iff z_i \leq y_i, \forall i = 1, 2, \ldots, n;$
(iv) $z \leq y, \iff z_i \leq y_i, \forall i = 1, 2, \ldots, n, i \neq j$ and $z_j < y_j$ for some j.

The following notions of interval analysis are from Moore [25].
Let $\mathcal{I}$ denotes the class of all closed intervals in $\mathbb{R}$. $A = [a^L, a^U] \in \mathcal{I}$ denotes a closed interval, where a^L and a^U denote the lower and upper bounds of A, respectively.

For $A = [a^L, a^U],\ B = [b^L, b^U] \in \mathcal{I}$, we have

(i) $A + B = \{a + b : a \in A \text{ and } b \in B\} = [a^L + b^L, a^U + b^U];$
(ii) $-A = \{-a : a \in A\} = [-a^U, -a^L];$
(iii) $A \times B = \{ab : a \in A \text{ and } b \in B\} = [\min_{ab}, \max_{ab}]$, where $\min_{ab} = \min\{a^L b^L, a^L b^U, a^U b^L, a^U b^U\}$ and $\max_{ab} = \max\{a^L b^L, a^L b^U, a^U b^L, a^U b^U\}$.

Then, we can show that

$$A - B = A + (-B) = [a^L - b^U, a^U - b^L],$$
$$kA = \{ka : a \in A\} = \begin{cases} [ka^L, ka^U], & k \geq 0, \\ |k|[-a^U, -a^L], & k < 0, \end{cases} \tag{12.1}$$

where $k \in \mathbb{R}$. The real number a can be considered as a closed interval $A_a = [a, a]$.

Let $A = [a^L, a^U],\ B = [b^L, b^U] \in \mathcal{I}$, then we define

1. $A \preceq_{LU} B \iff a^L \leq b^L$ and $a^U \leq b^U$,
2. $A \prec_{LU} B \iff A \preceq_{LU} B$ and $A \neq B$, that is, one of the following is satisfied:

a. $a^L < b^L$ and $a^U < b^U$; or

b. $a^L \leq b^L$ and $a^U < b^U$; or
c. $a^L < b^L$ and $a^U \leq b^U$.

Remark 12.1 $A = [a^L, a^U]$, $B = [b^L, b^U] \in \mathcal{I}$ are comparable if and only if $A \preceq_{LU} B$ or $A \succeq_{LU} B$. A and B are not comparable if one of the following holds:

$$a^L \leq b^L \text{ and } a^U > b^U; \; a^L < b^L \text{ and } a^U \geq b^U; \; a^L < b^L \text{ and } a^U > b^U;$$

$$a^L \geq b^L \text{ and } a^U < b^U; \; a^L > b^L \text{ and } a^U \leq b^U; \; a^L > b^L \text{ and } a^U < b^U.$$

Let $\mathbf{A} = (A_1, \ldots, A_n)$ be an interval-valued vector, where each component $A_k = [a_k^L, a_k^U]$, $k = 1, 2, \ldots, n$ is a closed interval. Let $\mathbf{A}$ and $\mathbf{B}$ be two interval-valued vectors, if A_k and B_k are comparable for each $k = 1, 2, \ldots, n$, then

1. $\mathbf{A} \preceq_{LU} \mathbf{B}$ if and only if $A_k \preceq_{LU} B_k$ for each $k = 1, 2, \ldots, n$;
2. $\mathbf{A} \prec_{LU} \mathbf{B}$ if and only if $A_k \preceq_{LU} B_k$ for each $k = 1, 2, \ldots, n$, and $A_r \prec_{LU} B_r$ for at least one index r.

The function $g : \mathbb{R}^n \to \mathcal{I}$ is called an interval-valued function, if $g(z) = [g^L(z), g^U(z)]$, where g^L and g^U are real-valued functions defined on $\mathbb{R}^n$ satisfying $g^L(z) \leq g^U(z)$, for all $z \in \mathbb{R}^n$.

Definition 12.1 ([24]) The set Γ is said to be a *convex set*, if for all $z, y \in \Gamma$, one has

$$z + \lambda(y - z) \in \Gamma, \;\; \forall \lambda \in [0, 1].$$

The following notions are from [6].

Definition 12.2 A function $g : \Gamma \to \mathbb{R}$ is said to be *Lipschitz* near $z_\circ \in \Gamma$, if there exist two positive constants $L, \delta > 0$, such that for all $y, z \in B(z_\circ; \delta) \cap \Gamma$, one has

$$|g(y) - g(z)| \leq L\|y - z\|.$$

The function g is *locally Lipschitz* on Γ, if it is Lipschitz near every $z \in \Gamma$.

Definition 12.3 Let $g : \Gamma \to \mathbb{R}$ be Lipschitz near $z \in \Gamma$. The *Clarke generalized directional derivative* of g at $z \in \Gamma$ in the direction $d \in \Omega$, is given as

$$g^\circ(z; d) := \limsup_{\substack{y \to z \\ t \downarrow 0}} \frac{g(y + td) - g(y)}{t}.$$

Definition 12.4 Let $g : \Gamma \to \mathbb{R}$ be *Lipschitz* near $z \in \Gamma$. The *Clarke generalized subdifferential* of g at $z \in \Gamma$ is given as

$$\partial^c g(z) := \{\xi \in \Omega^* : g^\circ(z; d) \geq \langle \xi, d \rangle, \;\; \forall d \in \Omega\}.$$

Definition 12.5 [13] A function $g : \Gamma \to \mathbb{R}$ is said to be *approximate convex* at $z_\circ \in \Gamma$, if for all $\varepsilon > 0$, there exists $\delta > 0$, such that for all $z, y \in B(z_\circ; \delta) \cap \Gamma$, one has

$$g(y) - g(z) \geq \langle \xi, y - z \rangle - \varepsilon \|y - z\|, \quad \forall \xi \in \partial^c g(z).$$

The following notions of generalized approximate convexity are from Bhatia et al. [3].

Definition 12.6 A function $g : \Gamma \to \mathbb{R}$ is said to be *approximate pseudoconvex of type I* at $z_\circ \in \Gamma$, if for all $\varepsilon > 0$, there exists $\delta > 0$, such that for all $z, y \in B(z_\circ; \delta) \cap \Gamma$, and if

$$\langle \xi, y - z \rangle \geq 0, \text{ for some } \xi \in \partial^c g(z),$$

then

$$g(y) - g(z) \geq -\varepsilon \|y - z\|.$$

Definition 12.7 A function $g : \Gamma \to \mathbb{R}$ is said to be *approximate pseudoconvex of type II (or strictly approximate pseudoconvex of type II)* at $z_\circ \in \Gamma$, if for all $\varepsilon > 0$, there exists $\delta > 0$, such that for all $z, y \in B(z_\circ; \delta) \cap \Gamma$, and if

$$\langle \xi, y - z \rangle + \varepsilon \|y - z\| \geq 0, \text{ for some } \xi \in \partial^c g(z),$$

then

$$g(y) \geq (>) g(z).$$

Definition 12.8 A function $g : \Gamma \to \mathbb{R}$ is said to be *approximate quasiconvex of type I* at $z_\circ \in \Gamma$, if for all $\varepsilon > 0$, there exists $\delta > 0$, such that for all $z, y \in B(z_\circ; \delta) \cap \Gamma$, and if

$$g(y) \leq g(z),$$

then

$$\langle \xi, y - z \rangle - \varepsilon \|y - z\| \leq 0, \ \forall \xi \in \partial^c g(z).$$

Definition 12.9 A function $g : \Gamma \to \mathbb{R}$ is said to be *approximate quasiconvex of type II (or strictly approximate quasiconvex of type II)* at $z_\circ \in \Gamma$, if for all $\varepsilon > 0$, there exists $\delta > 0$, such that for all $z, y \in B(z_\circ; \delta) \cap \Gamma$, and if

$$g(y) \leq (<) g(z) + \varepsilon \|y - z\|,$$

then

$$\langle \xi, y - z \rangle \leq 0, \ \forall \xi \in \partial^c g(z).$$

Definition 12.10 An interval-valued function $g : \Gamma \to I$ is said to be an *approximate LU-pseudoconvex function of type I (or approximate LU-pseudoconvex func-*

tion of type II) at $z_\circ \in \Gamma$, if and only if the real-valued functions $g^L(z)$ and $g^U(z)$ are approximate pseudoconvex functions of type I (or approximate pseudoconvex functions of type II) at $z_\circ \in \Gamma$.

Definition 12.11 An interval-valued function $g : \Gamma \to \mathcal{I}$ is said to be a *strictly approximate* LU*-pseudoconvex function of type II* at $z_\circ \in \Gamma$, if and only if the real-valued functions $g^L(z)$ and $g^U(z)$ are approximate pseudoconvex functions of type II and at least one of the $g^L(z)$ and $g^U(z)$ is strictly approximate pseudoconvex function of type II at $z_\circ \in \Gamma$.

Definition 12.12 An interval-valued function $g : \Gamma \to \mathcal{I}$ is said to be an *approximate* LU*-quasiconvex function of type I (approximate* LU*-quasiconvex function of type II)* at $z_\circ \in \Gamma$, if and only if the real-valued functions $g^L(z)$ and $g^U(z)$ are approximate quasiconvex functions of type I (or approximate quasiconvex function of type II) at $z_\circ \in \Gamma$.

We consider the following nonsmooth interval-valued multiobjective programming problem:

$$\text{(NIVMPP)} \quad \text{Minimize} \quad g(z) = (g_1(z), \ldots, g_p(z)),$$
$$\text{subject to} \quad z \in \Gamma,$$

where $g_i = [g_i^L, g_i^U] : \Gamma \to \mathcal{I}$, $i \in I := \{1, \ldots, p\}$ are locally Lipschitz interval-valued functions and Γ be a nonempty, closed, and convex subset of Ω.

The following notions of approximate LU-efficient solution are the adaptation of the notions of approximate efficient solution introduced by Mishra and Laha [22].

Let $\epsilon = (\varepsilon, \ldots, \varepsilon)$, a point $z_\circ \in \Gamma$ is said to be an approximate LU-efficient solution:

(ALUES)$_1$, if and only if for any sufficiently small $\varepsilon > 0$, there does not exist $\delta > 0$ such that, for all $z \in B(z_\circ; \delta) \cap \Gamma$, $z \neq z_\circ$, one has

$$g(z) \prec_{LU} g(z_\circ) + \epsilon \|z - z_\circ\|.$$

(ALUES)$_2$, if and only if for any sufficiently small $\varepsilon > 0$, there exists $\delta > 0$ such that, for all $z \in B(z_\circ; \delta) \cap \Gamma$, one has

$$g(z) \nprec_{LU} g(z_\circ) + \epsilon \|z - z_\circ\|.$$

(ALUES)$_3$, if and only if for any $\varepsilon > 0$, there exists $\delta > 0$ such that, for all $z \in B(z_\circ; \delta) \cap \Gamma$, one has

$$g(z) \nprec_{LU} g(z_\circ) - \epsilon \|z - z_\circ\|.$$

For more details about approximate efficient solution, we refer to [14, 22].
From now onward, $\epsilon := (\varepsilon, \ldots, \varepsilon)$, unless otherwise specified.

Now, for interval-valued functions, we formulate the following approximate Minty and Stampacchia vector variational inequalities in terms of Clarke subdifferential:

(AMVI)$_1$ To find $z_\circ \in \Gamma$ such that, for any sufficiently small $\varepsilon > 0$, there does not exist $\delta > 0$ such that, for all $z \in B(z_\circ;\delta) \cap \Gamma$, $z \neq z_\circ$ and $\xi_i^L \in \partial^c g_i^L(z)$ and $\xi_i^U \in \partial^c g_i^U(z)$, $i \in I$, one has

$$(\langle \xi_1^L, z - z_\circ\rangle, \ldots, \langle \xi_p^L, z - z_\circ\rangle) \leq \epsilon \|z - z_\circ\|,$$
$$(\langle \xi_1^U, z - z_\circ\rangle, \ldots, \langle \xi_p^U, z - z_\circ\rangle) \leq \epsilon \|z - z_\circ\|.$$

(AMVI)$_2$ To find $z_\circ \in \Gamma$ such that, for any sufficiently small $\varepsilon > 0$, there exists $\delta > 0$ such that, for all $z \in B(z_\circ;\delta) \cap \Gamma$ and $\xi_i^L \in \partial^c g_i^L(z)$ and $\xi_i^U \in \partial^c g_i^U(z)$, $i \in I$, one has

$$(\langle \xi_1^L, z - z_\circ\rangle, \ldots, \langle \xi_p^L, z - z_\circ\rangle) \nleq \epsilon \|z - z_\circ\|,$$
$$(\langle \xi_1^U, z - z_\circ\rangle, \ldots, \langle \xi_p^U, z - z_\circ\rangle) \nleq \epsilon \|z - z_\circ\|.$$

(AMVI)$_3$ To find $z_\circ \in \Gamma$ such that, for any $\varepsilon > 0$, there exists $\delta > 0$ such that, for all $z \in B(z_\circ;\delta) \cap \Gamma$ and $\xi_i^L \in \partial^c g_i^L(z)$ and $\xi_i^U \in \partial^c g_i^U(z)$, $i \in I$, one has

$$(\langle \xi_1^L, z - z_\circ\rangle, \ldots, \langle \xi_p^L, z - z_\circ\rangle) \nleq -\epsilon \|z - z_\circ\|,$$
$$(\langle \xi_1^U, z - z_\circ\rangle, \ldots, \langle \xi_p^U, z - z_\circ\rangle) \nleq -\epsilon \|z - z_\circ\|.$$

(ASVI)$_1$ To find $z_\circ \in \Gamma$ such that, for any $\varepsilon > 0$ sufficiently small, there exist $z \in \Gamma$, $z \neq z_\circ$, $\zeta_i^L \in \partial^c g_i^L(z_\circ)$ and $\zeta_i^U \in \partial^c g_i^U(z_\circ)$, $i \in I$, such that

$$(\langle \zeta_1^L, z - z_\circ\rangle, \ldots, \langle \zeta_p^L, z - z_\circ\rangle) \leq \epsilon \|z - z_\circ\|,$$
$$(\langle \zeta_1^U, z - z_\circ\rangle, \ldots, \langle \zeta_p^U, z - z_\circ\rangle) \leq \epsilon \|z - z_\circ\|.$$

(ASVI)$_2$ To find $z_\circ \in \Gamma$ such that, for any sufficiently small $\varepsilon > 0$, for all $z \in \Gamma$, $\zeta_i^L \in \partial^c g_i^L(z_\circ)$ and $\zeta_i^U \in \partial^c g_i^U(z_\circ)$, $i \in I$, one has

$$(\langle \zeta_1^L, z - z_\circ\rangle, \ldots, \langle \zeta_p^L, z - z_\circ\rangle) \nleq \epsilon \|z - z_\circ\|,$$
$$(\langle \zeta_1^U, z - z_\circ\rangle, \ldots, \langle \zeta_p^U, z - z_\circ\rangle) \nleq \epsilon \|z - z_\circ\|.$$

(ASVI)$_3$ To find $z_\circ \in \Gamma$ such that, for any $\varepsilon > 0$, there exists $\delta > 0$ such that, for all $z \in B(z_\circ;\delta) \cap \Gamma$, $\zeta_i^L \in \partial^c g_i^L(z_\circ)$ and $\zeta_i^U \in \partial^c g_i^U(z_\circ)$, $i \in I$, one has

$$(\langle \zeta_1^L, z - z_\circ\rangle, \ldots, \langle \zeta_p^L, z - z_\circ\rangle) \nleq -\epsilon \|z - z_\circ\|,$$
$$(\langle \zeta_1^U, z - z_\circ\rangle, \ldots, \langle \zeta_p^U, z - z_\circ\rangle) \nleq -\epsilon \|z - z_\circ\|.$$

Remark 12.2 If each g_i, $i \in I$ is real-valued function, then the above vector variational inequalties coincide with the vector variational inequalities given in [14, 22].

12.3 Relationship Among (NIVMPP), (ASVI) and (AMVI)

In this section, we establish some relationships between the nonsmooth interval-valued multiobjective programming problem (NIVMPP) and approximate vector variational inequalities (AMVI), (ASVI) under generalized approximate LU-convexity.

The following theorem states the condition under which an approximate LU-efficient solution becomes a solution of approximate Minty variational inequality.

Theorem 12.3.1 *Let each $g_i^L, g_i^U : \Gamma \to \mathbb{R}$, $i \in I$ be locally Lipschitz functions. Then,*

1. *if each g_i, $i \in I$ is approximate LU-pseudoconvex of type II at $z_\circ \in \Gamma$ and $z_\circ$ is an (ALUES)$_1$ of the (NIVMPP), then $z_\circ$ also solves (AMVI)$_1$;*
2. *if each g_i, $i \in I$ is approximate LU-pseudoconvex of type II at $z_\circ \in \Gamma$ and $z_\circ$ is an (ALUES)$_2$ of the (NIVMPP), then $z_\circ$ also solves (AMVI)$_2$;*
3. *if each g_i, $i \in I$ is strictly approximate LU-pseudoconvex of type II at $z_\circ \in \Gamma$ and $z_\circ$ is an (ALUES)$_3$ of the (NIVMPP), then $z_\circ$ also solves (AMVI)$_3$.*

Proof 1. On contrary assume that $z_\circ$ is an (ALUES)$_1$ of the (NIVMPP) but does not solves (AMVI)$_1$. Then, for some $\varepsilon > 0$ sufficiently small, there exists $\bar{\delta} > 0$, such that for all $z \in B(z_\circ; \bar{\delta}) \cap \Gamma$, $\xi_i^L \in \partial^c g_i^L(z)$ and $\xi_i^U \in \partial^c g_i^U(z)$, $i \in I$, we get

$$\begin{aligned}
&(\langle \xi_1^L, z - z_\circ\rangle, \ldots, \langle \xi_p^L, z - z_\circ\rangle) \leq \epsilon \|z - z_\circ\|,\\
&(\langle \xi_1^U, z - z_\circ\rangle, \ldots, \langle \xi_p^U, z - z_\circ\rangle) \leq \epsilon \|z - z_\circ\|,
\end{aligned}$$

that is,

$$\begin{aligned}
&\langle \xi_i^L, z_\circ - z\rangle + \varepsilon\|z - z_\circ\| \geq 0\\
&\langle \xi_i^U, z_\circ - z\rangle + \varepsilon\|z - z_\circ\| \geq 0, \ \forall i \in I, i \neq j,\\
&\text{and}\\
&\langle \xi_j^L, z_\circ - z\rangle + \varepsilon\|z - z_\circ\| > 0\\
&\langle \xi_j^U, z_\circ - z\rangle + \varepsilon\|z - z_\circ\| > 0, \ \text{for some } j \in I.
\end{aligned} \tag{12.2}$$

Since, each g_i, $i \in I$ is approximate LU-pseudoconvex of type II at $z_\circ \in \Gamma$, it follows that each g_i^L and g_i^U, $i \in I$ are approximate pseudoconvex of type II. Hence, for all $\varepsilon > 0$, there exists $\hat{\delta} > 0$, such that, for all $z \in B(z_\circ, \hat{\delta}) \cap \Gamma$, if

$$\langle \xi_i^L, z_\circ - z\rangle + \varepsilon\|z - z_\circ\| \geq 0, \ \text{for some } \xi_i^L \in \partial^c g_i^L(z), \ i \in I,$$

then

$$g_i^L(z) - g_i^L(z_\circ) \leq 0, \ \forall i \in I.$$

Similarly, if

$$\langle \xi_i^U, z_\circ - z\rangle + \varepsilon\|z - z_\circ\| \geq 0, \ \text{for some } \xi_i^U \in \partial^c g_i^U(z), \ i \in I,$$

then

$$g_i^U(z) - g_i^U(z_\circ) \leq 0, \ \forall i \in I.$$

Let $\delta := \min\{\hat{\delta}, \bar{\delta}\}$, from (12.2) and the definition of approximate LU-pseudoconvexity of type II, we have

$$\boldsymbol{g}(z) - \boldsymbol{g}(z_\circ) \preceq_{LU} \boldsymbol{0} \prec_{LU} \epsilon\|z - z_\circ\|,$$

for all $z \in B(z_\circ; \delta) \cap \Gamma$, which contradicts our assumption.

2. On contrary assume that $z_\circ$ is an $(\text{ALUES})_2$ of the (NIVMPP) but does not solves $(\text{AMVI})_2$. Then, for some $\varepsilon > 0$, sufficiently small and for all $\bar{\delta} > 0$, there exists $z \in B(z_\circ; \bar{\delta}) \cap \Gamma$, $\xi_i^L \in \partial^c g_i^L(z)$ and $\xi_i^U \in \partial^c g_i^U(z)$, $i \in I$, we get

$$(\langle \xi_1^L, z - z_\circ\rangle, \ldots, \langle \xi_p^L, z - z_\circ\rangle) \leq \epsilon\|z - z_\circ\|,$$
$$(\langle \xi_1^U, z - z_\circ\rangle, \ldots, \langle \xi_p^U, z - z_\circ\rangle) \leq \epsilon\|z - z_\circ\|,$$

that is

$$\begin{aligned} &\langle \xi_i^L, z_\circ - z\rangle + \varepsilon\|z - z_\circ\| \geq 0, \\ &\langle \xi_i^U, z_\circ - z\rangle + \varepsilon\|z - z_\circ\| \geq 0, \ \forall i \in I, i \neq j, \\ &\text{and} \\ &\langle \xi_j^L, z_\circ - z\rangle + \varepsilon\|z - z_\circ\| > 0, \\ &\langle \xi_j^U, z_\circ - z\rangle + \varepsilon\|z - z_\circ\| > 0, \ \text{for some } j \in I. \end{aligned} \tag{12.3}$$

Since, each g_i, $i \in I$ is approximate LU-pseudoconvex of type II at $z_\circ \in \Gamma$, it follows that each g_i^L and g_i^U, $i \in I$ are approximate pseudoconvex of type II. Hence, for all $\varepsilon > 0$, there exists $\hat{\delta} > 0$, such that whenever $z \in B(z_\circ; \hat{\delta}) \cap \Gamma$ and if

$$\langle \xi_i^L, z_\circ - z\rangle + \varepsilon\|z - z_\circ\| \geq 0, \ \text{for some } \xi_i^L \in \partial^c g_i^L(z), \ i \in I,$$

then

$$g_i^L(z) - g_i^L(z_\circ) \leq 0, \ \forall i \in I.$$

Similarly, if

$$\langle \xi_i^U, z_\circ - z\rangle + \varepsilon\|z - z_\circ\| \geq 0, \ \text{for some } \xi_i^U \in \partial^c g_i^U(z), \ i \in I,$$

then

$$g_i^U(z) - g_i^U(z_\circ) \leq 0, \ \forall i \in I.$$

Let $\delta := \min\{\hat{\delta}, \bar{\delta}\}$, then from (12.3) and the definition of approximate LU-convexity of type II, one has

$$\boldsymbol{g}(z) - \boldsymbol{g}(z_\circ) \preceq_{LU} \boldsymbol{0} \prec_{LU} \epsilon\|z - z_\circ\|,$$

for some $z \in B(z_\circ; \delta) \cap \Gamma$, which contradicts our assumption.

3. On contrary assume that $z_\circ$ is an $(\text{ALUES})_3$ of the (NIVMPP) but does not solves $(\text{AMVI})_3$. Then, for some $\varepsilon > 0$ and for all $\bar{\delta} > 0$, one has

$$(\langle \xi_1^L, z - z_\circ \rangle, \ldots, \langle \xi_p^L, z - z_\circ \rangle) \leq -\epsilon \|z - z_\circ\| < \epsilon \|z - z_\circ\|,$$
$$(\langle \xi_1^U, z - z_\circ \rangle, \ldots, \langle \xi_p^U, z - z_\circ \rangle) \leq -\epsilon \|z - z_\circ\| < \epsilon \|z - z_\circ\|,$$

for all $z \in B(z_\circ; \bar{\delta}) \cap \Gamma$, $\xi_i^L \in \partial^c g_i^L(z)$ and $\xi_i^U \in \partial^c g_i^U(z)$,
that is,

$$\begin{aligned} &\langle \xi_i^L, z_\circ - z \rangle + \varepsilon \|z - z_\circ\| > 0, \\ &\langle \xi_i^U, z_\circ - z \rangle + \varepsilon \|z - z_\circ\| > 0, \ \forall i \in I. \end{aligned} \tag{12.4}$$

Since, each g_i, $i \in I$ is strictly approximate LU-pseudoconvex of type II at $z_\circ \in \Gamma$, it follows that each g_i^L and g_i^U, $i \in I$ are approximate pseudoconvex of type II and atleast one of the g_i^L and g_i^U, $i \in I$ is strictly approximate pseudoconvex of type II at $z_\circ \in \Gamma$. Without loss of generality, assume that each g_i^L, $i \in I$ is strictly approximate pseudoconvex of type II. Hence, for all $\varepsilon > 0$, there exists $\hat{\delta} > 0$, such that whenever $z \in B(z_\circ; \hat{\delta}) \cap \Gamma$ and if

$$\langle \xi_i^L, z_\circ - z \rangle + \varepsilon \|z - z_\circ\| \geq 0, \ \text{for some } \xi_i^L \in \partial^c g_i^L(z), \ i \in I,$$

then

$$g_i^L(z) - g_i^L(z_\circ) < 0, \ \forall i \in I.$$

Similarly, if

$$\langle \xi_i^U, z_\circ - z \rangle + \varepsilon \|z - z_\circ\| \geq 0, \ \text{for some } \xi_i^U \in \partial^c g_i^U(z), \ i \in I,$$

then

$$g_i^U(z) - g_i^U(z_\circ) \leq 0, \ \forall i \in I.$$

Let $\delta := \min\{\bar{\delta}, \hat{\delta}\}$, from (12.4) and the definition of strictly approximate LU-pseudo convexity of type II, we have

$$g_i(z) - g_i(z_\circ) \prec_{LU} 0, \ \forall i \in I, \tag{12.5}$$

for all $z \in B(z_\circ; \delta) \cap \Gamma$.
From (12.5), we can get an $\varepsilon > 0$ sufficiently small, such that

$$\boldsymbol{g}(z) - \boldsymbol{g}(z_\circ) \prec_{LU} -\epsilon \|z - z_\circ\|,$$

which contradicts our assumption. □

Theorem 12.3.2 *Let each* $g_i^L, g_i^U : \Gamma \to \mathbb{R}$, $i \in I$ *be locally Lipschitz functions. Then*

1. *if each g_i, $i \in I$ is approximate LU-quasiconvex of type II at $z_\circ \in \Gamma$ and $z_\circ$ solves (ASVI)$_1$, then $z_\circ$ is also an (ALUES)$_1$ of the (NIVMPP);*
2. *if each g_i, $i \in I$ is approximate LU-quasiconvex of type II at $z_\circ \in \Gamma$ and $z_\circ$ solves (ASVI)$_2$, then $z_\circ$ is also an (ALUES)$_2$ of the (NIVMPP);*
3. *if each g_i, $i \in I$ is approximate LU-pseudoconvex of type II at $z_\circ \in \Gamma$ and $z_\circ$ solves (ASVI)$_3$, then $z_\circ$ is also an (ALUES)$_3$ of the (NIVMPP).*

Proof 1. On contrary assume that $z_\circ$ is a solution of (ASVI)$_1$ but not an (ALUES)$_1$ of the (NIVMPP). Then, for some $\varepsilon > 0$, sufficiently small, there exists $\bar{\delta} > 0$, such that

$$\boldsymbol{g}(z) - \boldsymbol{g}(z_\circ) \prec_{LU} \epsilon \|z - z_\circ\|, \tag{12.6}$$

for all $z \in B(z_\circ; \bar{\delta}) \cap \Gamma$, $z \neq z_\circ$. Since, each g_i, $i \in I$ is approximate LU-quasiconvex of type II at $z_\circ$, it follows that each g_i^L and g_i^U, $i \in I$ are approximate quasiconvex of type II at $z_\circ$. Hence, for all $\varepsilon > 0$, there exists $\hat{\delta} > 0$, such that for all $z \in B(z_\circ; \hat{\delta}) \cap \Gamma$, if

$$g_i^L(z) \leq g_i^L(z_\circ) + \varepsilon \|z - z_\circ\|, \ \forall i \in I,$$

then

$$\langle \zeta_i^L, z - z_\circ \rangle \leq 0, \ \forall \zeta_i^L \in \partial^c g_i^L(z_\circ), \ i \in I.$$

Similarly, if

$$g_i^U(z) \leq g_i^U(z_\circ) + \varepsilon \|z - z_\circ\|, \ \forall i \in I,$$

then

$$\langle \zeta_i^U, z - z_\circ \rangle \leq 0, \ \forall \zeta_i^U \in \partial^c g_i^U(z_\circ), \ i \in I.$$

Let $\delta := \min\{\bar{\delta}, \hat{\delta}\}$, from (12.6) and the definition of approximate LU-quasiconvexity of type II, one has

$$\begin{aligned} \langle \zeta_i^L, z - z_\circ \rangle &\leq 0 < \varepsilon \|z - z_\circ\|, \\ \langle \zeta_i^U, z - z_\circ \rangle &\leq 0 < \varepsilon \|z - z_\circ\|, \end{aligned}$$

for all $z \in B(z_\circ; \delta) \cap \Gamma$, $\zeta_i^L \in \partial^c g_i^L(z_\circ)$, $\zeta_i^U \in \partial^c g_i^U(z_\circ)$, $i \in I$, which contradicts our assumption.

2. Assume that $z_\circ$ is a solution of (ASVI)$_2$. Then, for any $\varepsilon > 0$ sufficiently small, for every $z \in \Gamma$, $\zeta_i^L \in \partial^c g_i^L(z_\circ)$ and $\zeta_i^U \in \partial^c g_i^U(z_\circ)$, $i \in I$, one has

$$\begin{aligned} (\langle \zeta_1^L, z - z_\circ \rangle, \ldots, \langle \zeta_p^L, z - z_\circ \rangle) &\nleq \epsilon \|z - z_\circ\|, \\ (\langle \zeta_1^U, z - z_\circ \rangle, \ldots, \langle \zeta_p^U, z - z_\circ \rangle) &\nleq \epsilon \|z - z_\circ\|, \end{aligned}$$

that is,

$$\begin{aligned} (\langle \zeta_1^L, z - z_\circ \rangle, \ldots, \langle \zeta_p^L, z - z_\circ \rangle) &\nleq 0, \\ (\langle \zeta_1^U, z - z_\circ \rangle, \ldots, \langle \zeta_p^U, z - z_\circ \rangle) &\nleq 0. \end{aligned} \tag{12.7}$$

Since, each g_i, $i \in I$ is approximate LU-quasiconvex of type II at $z_\circ$, it follows that each g_i^L and g_i^U, $i \in I$ are approximate quasiconvex of type II at $z_\circ$. Hence, for all $\varepsilon > 0$, there exists $\hat{\delta} > 0$, such that for all $z \in B(z_\circ, \hat{\delta}) \cap \Gamma$, if

$$g_i^L(z) \leq g_i^L(z_\circ) + \varepsilon \|z - z_\circ\|, \ \forall i \in I,$$

then

$$\langle \zeta_i^L, z - z_\circ \rangle \leq 0, \ \forall \zeta_i^L \in \partial^c g_i^L(z_\circ), \ i \in I.$$

Similarly, if

$$g_i^U(z) \leq g_i^U(z_\circ) + \varepsilon \|z - z_\circ\|, \ \forall i \in I,$$

then

$$\langle \zeta_i^U, z - z_\circ \rangle \leq 0, \ \forall \zeta_i^U \in \partial^c g_i^U(z_\circ), \ i \in I.$$

From (12.7) and the definition of approximate LU-quasiconvexity of type II, it follows that

$$\boldsymbol{g}(z) - \boldsymbol{g}(z_\circ) \not\prec_{LU} \epsilon \|z - z_\circ\|,$$

for all $z \in B(z_\circ; \delta) \cap \Gamma, \ z \neq z_\circ$. Therefore, $z_\circ$ is an (ALUES)$_2$ of the (NIVMPP).

3. On contrary assume that $z_\circ$ solves (ASVI)$_3$ but not an (ALUES)$_3$. Then, for some $\varepsilon > 0$, and for all $\bar{\delta} > 0$, there exists $z \in B(z_\circ; \bar{\delta}) \cap \Gamma$, such that

$$\boldsymbol{g}(z) - \boldsymbol{g}(z_\circ) \prec_{LU} -\epsilon \|z - z_\circ\|,$$

that is

$$\begin{aligned} g_i^L(z) - g_i^L(z_\circ) &< 0, \\ g_i^U(z) - g_i^U(z_\circ) &< 0, \ \forall i \in I. \end{aligned} \tag{12.8}$$

Since, each g_i, $i \in I$ is approximate LU-pseudoconvex of type II at $z_\circ$, it follows that each g_i^L and g_i^U, $i \in I$ are approximate pseudoconvex of type II at $z_\circ$. Hence, for all $\varepsilon > 0$, there exists $\hat{\delta} > 0$, such that for all $z \in B(z_\circ; \hat{\delta}) \cap \Gamma$, if

$$\langle \zeta_i^L, z - z_\circ \rangle + \varepsilon \|z - z_\circ\| \geq 0, \ \text{for some } \zeta_i^L \in \partial^c g_i^L(z_\circ), \ i \in I,$$

then

$$g_i^L(z) - g_i^L(z_\circ) \geq 0, \ \forall i \in I.$$

Similarly, if

$$\langle \zeta_i^U, z - z_\circ \rangle + \varepsilon \|z - z_\circ\| \geq 0, \ \text{for some } \zeta_i^U \in \partial^c g_i^U(z_\circ), \ i \in I,$$

then

$$g_i^U(z) - g_i^U(z_\circ) \geq 0, \ \forall i \in I.$$

Let $\delta := \min\{\hat{\delta}, \bar{\delta}\}$, from (12.8) and the definition of approximate LU-pseudoconvexity of type II, one has

$$\begin{aligned}\langle \zeta_i^L, z - z_\circ \rangle &< -\varepsilon \|z - z_\circ\|, \\ \langle \zeta_i^U, z - z_\circ \rangle &< -\varepsilon \|z - z_\circ\|, \ \forall i \in I,\end{aligned} \tag{12.9}$$

for some $z \in B(z_\circ; \delta) \cap \Gamma$ and all $\zeta_i^L \in \partial^c g_i^L(z_\circ)$, $\zeta_i^U \in \partial^c g_i^U(z_\circ)$, $i \in I$, which contradicts our assumption. □

The following corollary is a direct consequence of Theorems 12.3.1 and 12.3.2.

Corollary 12.1 *Let each $g_i^L, g_i^U : \Gamma \to \mathbb{R}$, $i \in I$ be locally Lipschitz functions. Then,*

1. *if each g_i, $i \in I$ is approximate LU-quasiconvex of type II and approximate LU-pseudoconvex of type II at $z_\circ \in \Gamma$. Let $z_\circ$ is a solution of $(ASVI)_1$, then $z_\circ$ is also a solution of $(AMVI)_1$.*
2. *if each g_i, $i \in I$ is approximate LU-quasiconvex of type II and approximate LU-pseudoconvex of type II at $z_\circ \in \Gamma$. Let $z_\circ$ is a solution of $(ASVI)_2$, then $z_\circ$ is also a solution of $(AMVI)_2$.*
3. *if each g_i, $i \in I$ is strictly approximate LU-pseudoconvex of type II at $z_\circ \in \Gamma$. Let $z_\circ$ is a solution of $(ASVI)_3$, then $z_\circ$ is also a solution of $(AMVI)_3$.*

Now, to illustrate the significance of Theorems 12.3.1, 12.3.2 and Corollary 12.1, we have the following example.

Example 12.1 Consider the following nonsmooth interval-valued multiobjective programming problem

$$\text{(P)} \qquad \begin{aligned} &\text{Minimize} \quad \mathbf{g}(z) = (g_1(z), g_2(z)) \\ &\text{subject to} \quad z \in \Gamma \subseteq \mathbb{R}, \end{aligned}$$

where $\Gamma = [-1, 1]$ and $g_1, g_2 : \Gamma \to \mathcal{I}$ are defined as

$$g_1^L(z) = \begin{cases} z^3 + z, & z \geq 0, \\ 2z, & z < 0, \end{cases} \quad g_1^U(z) = \begin{cases} z^3 + 2z, & z \geq 0, \\ z, & z < 0, \end{cases}$$

and

$$g_2^L(z) = \begin{cases} z - z^2, & z \geq 0, \\ 2z, & z < 0, \end{cases} \quad g_2^U(z) = \begin{cases} z + 1, & z \geq 0, \\ 2z + e^z, & z < 0. \end{cases}$$

The Clarke generalized subdifferentials of g_1 and g_2 are given by

$$\partial^c g_1^L(z) = \begin{cases} 3z^2 + 1, & z > 0, \\ [1, 2], & z = 0, \\ 2, & z < 0, \end{cases} \quad \partial^c g_1^U(z) = \begin{cases} 3z^2 + 2, & z > 0, \\ [1, 2], & z = 0, \\ 1, & z < 0, \end{cases}$$

and

$$\partial^c g_2^L(z) = \begin{cases} 1 - 2z, & z > 0, \\ [1, 2], & z = 0, \\ 2, & z < 0, \end{cases} \qquad \partial^c g_2^U(z) = \begin{cases} 1, & z > 0, \\ [1, 3], & z = 0, \\ 2 + e^z, & z < 0, \end{cases}$$

For any $0 < \varepsilon < 1$, let $\delta = \frac{1}{10}$, such that for all $z, y \in B(0; \delta) \cap \Gamma$, $\xi_1^L \in \partial^c g_1^L(z)$, $\xi_1^U \in \partial^c g_1^U(z)$, $\xi_2^L \in \partial^c g_2^L(z)$ and $\xi_2^U \in \partial^c g_2^U(z)$, one has

$$\langle \xi_1^L, y - z \rangle + \varepsilon \|y - z\| = \begin{cases} (3z^2 + 1)(y - z) + \varepsilon\|y - z\| > 0, & z > 0, y > 0, y - z > 0; \\ (3z^2 + 1)(y - z) + \varepsilon\|y - z\| < 0, & z > 0, y > 0, y - z < 0; \\ (3z^2 + 1)(y - z) + \varepsilon\|y - z\| < 0, & z > 0, y \le 0; \\ 2(y - z) + \varepsilon\|y - z\| > 0, & z < 0, y \ge 0; \\ 2(y - z) + \varepsilon\|y - z\| > 0, & z < 0, y < 0, y - z > 0; \\ 2(y - z) + \varepsilon\|y - z\| < 0, & z < 0, y < 0, y - z < 0; \\ k_1(y - z) + \varepsilon\|y - z\| > 0, & z = 0, y > 0, k_1 \in [1, 2]; \\ k_1(y - z) + \varepsilon\|y - z\| < 0, & z = 0, y < 0, k_1 \in [1, 2], \end{cases}$$

$$\langle \xi_1^U, y - z \rangle + \varepsilon \|y - z\| = \begin{cases} (3z^2 + 2)(y - z) + \varepsilon\|y - z\| > 0, & z > 0, y > 0, y - z > 0; \\ (3z^2 + 2)(y - z) + \varepsilon\|y - z\| < 0, & z > 0, y > 0, y - z < 0; \\ (3z^2 + 2)(y - z) + \varepsilon\|y - z\| < 0, & z > 0, y \le 0; \\ (y - z) + \varepsilon\|y - z\| > 0, & z < 0, y < 0, y - z > 0; \\ (y - z) + \varepsilon\|y - z\| < 0, & z < 0, y < 0, y - z < 0; \\ (y - z) + \varepsilon\|y - z\| > 0, & z < 0, y \ge 0; \\ k_2(y - z) + \varepsilon\|y - z\| > 0, & z = 0, y > 0, k_2 \in [1, 2]; \\ k_2(y - z) + \varepsilon\|y - z\| < 0, & z = 0, y < 0, k_2 \in [1, 2], \end{cases}$$

$$\langle \xi_2^L, y - z \rangle + \varepsilon \|y - z\| = \begin{cases} (1 - 2z)(y - z) + \varepsilon\|y - z\| > 0, & z > 0, y > 0, y - z > 0; \\ (1 - 2z)(y - z) + \varepsilon\|y - z\| < 0, & z > 0, y > 0, y - z < 0; \\ (1 - 2z)(y - z) + \varepsilon\|y - z\| < 0, & z > 0, y \le 0; \\ 2(y - z) + \varepsilon\|y - z\| > 0, & z < 0, y < 0, y - z > 0; \\ 2(y - z) + \varepsilon\|y - z\| < 0, & z < 0, y < 0, y - z < 0; \\ 2(y - z) + \varepsilon\|y - z\| > 0, & z < 0, y \ge 0; \\ t_1(y - z) + \varepsilon\|y - z\| > 0, & z = 0, y > 0, t_1 \in [1, 2]; \\ t_1(y - z) + \varepsilon\|y - z\| < 0, & z = 0, y < 0, t_2 \in [1, 2]; \end{cases}$$

and

$$\langle \xi_2^U, y - z \rangle + \varepsilon \|y - z\| = \begin{cases} (y - z) + \varepsilon\|y - z\| > 0, & z > 0, y > 0, y - z > 0; \\ (y - z) + \varepsilon\|y - z\| < 0, & z > 0, y > 0, y - z < 0; \\ (y - z) + \varepsilon\|y - z\| < 0, & z > 0, y \le 0; \\ (2 + e^z)(y - z) + \varepsilon\|y - z\| > 0, & z < 0, y < 0, y - z > 0; \\ (2 + e^z)(y - z) + \varepsilon\|y - z\| < 0, & z < 0, y < 0, y - z < 0; \\ (2 + e^z)(y - z) + \varepsilon\|y - z\| > 0, & z < 0, y \ge 0; \\ t_2(y - z) + \varepsilon\|y - z\| > 0, & z = 0, y > 0, t_2 \in [1, 3]; \\ t_2(y - z) + \varepsilon\|y - z\| < 0, & z = 0, y < 0, t_2 \in [1, 3]. \end{cases}$$

Also,

$$g_1^L(y) - g_1^L(z) = \begin{cases} (y-z)(y^2+zy+z^2+1), & z > 0, y > 0, y - z > 0; \\ y^3 + y - 2z, & z < 0, y > 0; \\ 2(y-z), & z < 0, y < 0, y - z > 0; \\ y^3 + y, & z = 0, y > 0, \end{cases}$$
$$> 0,$$

$$g_1^U(y) - g_1^U(z) = \begin{cases} (y-z)(y^2+zy+z^2+2), & z > 0, y > 0, y - z > 0; \\ y^3 + 2y - z, & z < 0, y > 0; \\ y-z, & z < 0, y < 0, y - z > 0; \\ y^3 + 2y, & z = 0, y > 0, \end{cases}$$
$$> 0,$$

$$g_2^L(y) - g_2^L(z) = \begin{cases} (y-z)(1-z-y), & z > 0, y > 0, y - z > 0; \\ y - y^2 - 2z, & z < 0, y > 0; \\ 2(y-z), & z < 0, y < 0, y - z > 0; \\ y - y^2, & z = 0, y > 0, \end{cases}$$
$$> 0,$$

and

$$g_2^U(y) - g_2^U(z) = \begin{cases} y-z, & z > 0, y > 0, y - z > 0; \\ y + 1 - 2z - e^z, & z < 0, y > 0; \\ 2(y-z) + e^y - e^z, & z < 0, y < 0, y - z > 0; \\ y, & z = 0, y > 0, \end{cases}$$
$$> 0.$$

Hence, $g_1 = [g_1^L, g_1^U]$ and $g_2 = [g_2^L, g_2^U]$ are approximate LU-pseudoconvex of type II at $z_\circ = 0$.
Evidently, $z_\circ = 0$, solves (ASVI)$_3$. Since, for any $z > 0$, $z \in B(z_\circ;\delta) \cap \Gamma$, $\zeta_1^L \in \partial^c g_1^L(z_\circ)$, $\zeta_1^U \in \partial^c g_1^U(z_\circ)$, $\zeta_2^L \in \partial^c g_2^L(z_\circ)$ and $\zeta_2^U \in \partial^c g_2^U(z_\circ)$, we have

$$\langle \zeta_1^L, z - z_\circ\rangle + \varepsilon\|z - z_\circ\| = k_1 z + \varepsilon\|z\| > 0,\ k_1 \in [1,2],$$
$$\langle \zeta_1^U, z - z_\circ\rangle + \varepsilon\|z - z_\circ\| = k_2 z + \varepsilon\|z\| > 0,\ k_2 \in [1,2],$$
$$\langle \zeta_2^L, z - z_\circ\rangle + \varepsilon\|z - z_\circ\| = t_1 z + \varepsilon\|z\| > 0,\ t_1 \in [1,2],$$
$$\text{and } \langle \zeta_2^U, z - z_\circ\rangle + \varepsilon\|z - z_\circ\| = t_2 z + \varepsilon\|z\| > 0,\ t_2 \in [1,3],$$

that is

$$((\zeta_1^L, z - z_\circ), \langle \zeta_2^L, z - z_\circ\rangle) \nleq -\epsilon\|z - z_\circ\|,$$
$$(\langle \zeta_1^U, z - z_\circ\rangle, \langle \zeta_2^U, z - z_\circ\rangle) \nleq -\epsilon\|z - z_\circ\|.$$

Moreover, $z_\circ = 0$ is an (ALUES)$_3$ of the problem (P). Since, for any $\varepsilon > 0$, let $\delta = \frac{1}{2}$, such that for all $z > 0$, $z \in B(z_\circ; \delta) \cap \Gamma$, we have

$$g_1^L(z) - g_1^L(z_\circ) + \varepsilon\|z - z_\circ\| = z^3 + z + \varepsilon\|z\| > 0,$$
$$g_1^U(z) - g_1^U(z_\circ) + \varepsilon\|z - z_\circ\| = z^3 + 2z + \varepsilon\|z\| > 0,$$
$$g_2^L(z) - g_2^L(z_\circ) + \varepsilon\|z - z_\circ\| = z - z^2 + \varepsilon\|z\| > 0,$$
$$g_2^U(z) - g_2^U(z_\circ) + \varepsilon\|z - z_\circ\| = z + \varepsilon\|z\| > 0,$$

that is

$$\mathbf{g}(z) - \mathbf{g}(z_\circ) + \epsilon\|z - z_\circ\| \nless_{LU} \mathbf{0}.$$

Furthermore, $z_\circ = 0$ solves (AMVI)$_3$. Since, for any $\varepsilon > 0$, sufficiently small, let $\delta = \frac{1}{2}$, such that for all $z > 0$, $z \in B(z_\circ; \delta) \cap \Gamma$, $\xi_1^L \in \partial^c g_1(z)$, $\xi_1^U \in \partial^c g_1^U(z)$, $\xi_2^L \in \partial^c g_2^L(z)$ and $\xi_2^U \in \partial^c g_2^U(z)$, we have

$$\langle \xi_1^L, z - z_\circ\rangle + \varepsilon\|z - z_\circ\| = 3z^3 + z + \varepsilon\|z\| > 0,$$
$$\langle \xi_1^U, z - z_\circ\rangle + \varepsilon\|z - z_\circ\| = 3z^3 + 2z + \varepsilon\|z\| > 0,$$
$$\langle \xi_2^L, z - z_\circ\rangle + \varepsilon\|z - z_\circ\| = 1 - 2z + \varepsilon\|z\| > 0,$$
$$\langle \xi_2^U, z - z_\circ\rangle + \varepsilon\|z - z_\circ\| = z + \varepsilon\|z\| > 0,$$

that is

$$(\langle \xi_1^L, z - z_\circ\rangle, \langle \xi_2^L, z - z_\circ\rangle) \nleq -\epsilon\|z - z_\circ\|,$$
$$(\langle \xi_1^U, z - z_\circ\rangle, \langle \xi_2^U, z - z_\circ\rangle) \nleq -\epsilon\|z - z_\circ\|.$$

12.4 Conclusions

In this chapter, we have considered a class of nonsmooth interval-valued multiobjective programming problems (NIVMPP) and certain classes of approximate Minty and Stampacchia vector variational inequalities; namely, (AMVI)$_1$, (AMVI)$_2$, (AMVI)$_3$, (ASVI)$_1$, (ASVI)$_2$, and (ASVI)$_3$. We have established the equivalence among the solutions of these vector variational inequalities and the approximate LU-efficient solutions; namely, (ALUES)$_1$, (ALUES)$_2$, (ALUES)$_3$ of the nonsmooth interval-valued multiobjective programming problem (NIVMPP). The numerical example has been given to justify the significance of these results. The results of the chapter extend and unify the corresponding results of [14, 22, 23, 30, 33] to a more general class of nonsmooth optimization problems, namely, nonsmooth interval-valued multiobjective programming problem (NIVMPP).

Acknowledgements The first author is supported by the Science and Engineering Research Board, Department of Science and Technology, Government of India, through grant number "ECR/2016/001961." The authors would like to express their thanks to the anonymous reviewers for their very valuable comments and suggestions to improve the quality of the chapter.

References

1. Antczak, T.: Optimality conditions and duality results for nonsmooth vector optimization problems with the multiple interval-valued objective function. Acta Math. Sci. Ser. B **37**, 1133–1150 (2017)
2. Antczak, T.: Exactness property of the exact absolute value penalty function method for solving convex nondifferentiable interval-valued optimization problems. J. Optim. Theory Appl. **176**, 205–224 (2018)
3. Bhatia, D., Gupta, A., Arora, P.: Optimality via generalized approximate convexity and quasi-efficiency. Optim. Lett. **7**, 127–135 (2013)
4. Chinchuluun, A., Pardalos, P.M.: A survey of recent developments in multiobjective optimization. Ann. Oper. Res. **154**, 29–50 (2007)
5. Chinchuluun, A., Pardalos, P.M., Migdalas, A., Pitsoulis, A.: Pareto Optimality. Game Theory and Equilibria. Springer, Berlin (2008)
6. Clarke, F.H.: Optimization and Nonsmooth Analysis. Wiley, New York (1983)
7. Dafermos, S.: Exchange price equilibria and variational inequalities. Math. Program. **46**, 391–402 (1990)
8. Ghosh, D.: Newton method to obtain efficient solutions of the optimization problems with interval-valued objective functions. J. Appl. Math. Comput. **53**, 709–731 (2017)
9. Ghosh, D., Chauhan, R.S., Mesiar, R., Debnath, A.K.: Generalized Hukuhara Gâteaux and Fréchet derivatives of interval-valued functions and their application in optimization with interval-valued functions. Inform. Sci. **510**, 317–340 (2020)
10. Giannessi, F.: Theorems of the alternative, quadratic programs and complementarity problems. In: Cottle, R.W., Giannessi, F., Lions, J.L. (eds.) Variational Inequalities and Complementarity Problems, pp. 151–186. Wiley, Chichester (1980)
11. Giannessi, F.: On Minty variational principle. In: Giannessi, F., Komlósi, S., Rapcśak, T. (eds.) New Trends in Mathematical Programming, pp. 93–99. Kluwer Academic Publishers, Dordrecht (1997)
12. Gupta, D., Mehra, A.: Two types of approximate saddle points. Numer. Funct. Anal. Optim. **29**, 532–550 (2008)
13. Gupta, A., Mehra, A., Bhatia, D.: Approximate convexity in vector optimisation. Bull. Aust. Math. Soc. **74**, 207–218 (2006)
14. Gupta, P., Mishra, S.K.: On Minty variational principle for nonsmooth vector optimization problems with generalized approximate convexity. Optimization **67**, 1157–1167 (2018)
15. Hanson, M.A., Mond, B.: Necessary and sufficient conditions in constrained optimization. Math. Program. **37**, 51–58 (1987)
16. Jayswal, A., Ahmad, I., Banerjee, J.: Nonsmooth interval-valued optimization and saddle-point optimality criteria. Bull. Malays. Math. Sci. Soc. **39**, 1391–1411 (2016)
17. Jayswal, A., Stancu-Minasian, I., Banerjee, J.: Optimality conditions and duality for interval-valued optimization problems using convexifactors. Rend. Circ. Mat. Palermo **65**, 17–32 (2016)
18. Jeyakumar, V., Mond, B.: On generalised convex mathematical programming. J. Austral. Math. Soc. Ser. B. **34**, 43–53 (1992)
19. Kinderlehrer, D., Stampacchia, G.: An Introduction to Variational Inequalities and Their Applications. Academic Press, London (1980)

20. Lee, G.M., Lee, K.B.: Vector variational inequalities for nondifferential convex vector optimization problems. J. Glob. Optim. **32**, 597–612 (2005)
21. Loridan, P.: ϵ-solutions in vector minimization problems. J. Optim. Theory Appl. **43**, 265–276 (1984)
22. Mishra, S.K., Laha, V.: On Minty variational principle for nonsmooth vector optimization problems with approximate convexity. Optim. Lett. **10**, 577–589 (2016)
23. Mishra, S.K., Upadhyay, B.B.: Some relations between vector variational inequality problems and nonsmooth vector optimization problems using quasi efficiency. Positivity **17**, 1071–1083 (2013)
24. Mishra, S.K., Upadhyay, B.B.: Pseudolinear Functions and Optimization. Chapman and Hall, CRC Press (2014)
25. Moore, R.E.: Interval Analysis. Prentice-Hall, Englewood Cliffs, New Jersey (1966)
26. Moore, R.E.: Methods and Applications of Interval Analysis. SIAM Studies in Applied Mathematics, Philadelphia (1979)
27. Upadhyay, B.B., Mishra, P.: On generalized Minty and Stampacchia vector variational-like inequalities and nonsmooth vector optimization problem involving higher order strong invexity. J. Sci. Res. **64**, 182–191 (2020)
28. Upadhyay, B.B., Mishra, P.: On vector variational inequalities and vector optimization problems. In: Soft Computing: Theories and Applications, pp. 257–267. Springer, Singapore (2020)
29. Upadhyay, B.B., Mishra, P., Mohapatra, R.N., Mishra, S.K.: On the applications of nonsmooth vector optimization problems to solve generalized vector variational inequalities using convexificators. Adv. Intell. Sys. Comput. https://doi.org/10.1007/978-3-030-21803-4_66
30. Upadhyay, B.B., Mohapatra, R.N., Mishra, S.K.: On relationships between vector variational inequality and nonsmooth vector optimization problems via strict minimizers. Adv. Nonlinear Var. Inequal. **20**, 1–12 (2017)
31. Wu, H.-C.: The Karush-Kuhn-Tucker optimality conditions in an optimization problem with interval-valued objective function. Eur. J. Oper. Res. **176**, 46–59 (2007)
32. Yang, X.Q.: Vector variational inequality and vector pseudolinear optimization. J. Optim. Theory Appl. **95**, 729–734 (1997)
33. Zhang, J., Zheng, Q., Ma, X., Li, L.: Relationships between interval-valued vector optimization problems and variational inequalities. Fuzzy Optim. Decis. Mak. **15**, 33–55 (2016)

Chapter 13
On Constraint Qualifications for Multiobjective Optimization Problems with Switching Constraints

Yogendra Pandey and Vinay Singh

Abstract In this chapter, we consider multiobjective optimization problems with switching constraint (MOPSC). We introduce linear independence constraint qualification (LICQ), Mangasarian–Fromovitz constraint qualification (MFCQ), Abadie constraint qualification (ACQ), and Guignard constraint qualification (GCQ) for multiobjective optimization problems with switching constraint (MOPSC). Further, we introduce the notion of Weak stationarity, Mordukhovich stationarity, and Strong stationarity, i.e., W-stationarity, M-stationarity, and S-stationarity, respectively, for the MOPSC. Also, we present a survey of the literature related to existing constraint qualifications and stationarity conditions for mathematical programs with equilibrium constraints (MPEC), mathematical programs with complementarity constraints (MPCC), mathematical programs with vanishing constraints (MPVC), and for mathematical programs with switching constraints (MPSC). We establish that the M-stationary conditions are sufficient optimality conditions for the MOPSC using generalized convexity. Further, we propose a Wolfe-type dual model for the MOPSC and establish weak duality and strong duality results under assumptions of generalized convexity.

Keywords Switching constraints · Constraint qualifications · Optimality conditions · Duality

13.1 Introduction

We consider the following multiobjective optimization problems with switching constraints (MOPSC):

Y. Pandey (✉)
Department of Mathematics, Satish Chandra College, Ballia 277001, India
e-mail: pandeyiitb@gmail.com

V. Singh
Department of Mathematics, National Institute of Technology, Aizawl 796012, Mizoram, India
e-mail: vinaybhu1981@gmail.com

V. Laha et al. (eds.), *Optimization, Variational Analysis and Applications*,
Springer Proceedings in Mathematics & Statistics 355,
https://doi.org/10.1007/978-981-16-1819-2_13

$$
\begin{aligned}
\text{(MOPSC)} \qquad & \min\ (f_1(x), \ldots, f_m(x)) \\
& \text{subject to} : g_i(x) \leq 0,\ i = 1, \ldots, p, \\
& h_i(x) = 0,\ i = 1, \ldots, q, \\
& G_i(x) H_i(x) = 0, \quad i = 1, \ldots, l.
\end{aligned}
$$

All the functions $f_1, \ldots, f_m, g_1, \ldots, g_p, h_1, \ldots, h_q, G_1, \ldots, G_l, H_1, \ldots, H_l : \mathbb{R}^n \to \mathbb{R}$ are assumed to be continuously differentiable.

In optimal control, the concept of control switching became very important, for details see, [13, 14, 23, 27, 35, 59, 61, 63, 67] and references therein. Mathematical programs with switching constraints (MPSC) are related to mathematical programs with vanishing constraints (MPVC) and mathematical programs with complementarity constraints (MPCC) (see [38, 51]). Similarly, multiobjective optimization problems with switching constraints (MOPSC) are also closely related to multiobjective optimization problems with vanishing constraints (MOPVC), see [44]. Mehlitz [42] introduced the notions of weak, Mordukhovich, and strong stationarities for mathematical programs with switching constraints (MPSC). Recently, Kanzow et al. [34] proposed several relaxation schemes for the MPSC.

Constraint qualifications are regularity conditions for Kuhn–Tucker necessary optimality in nonlinear programming problems. The Slater constraint qualification, the weak Arrow–Hurwicz–Uzawa constraint qualification, the weak reverse convex constraint qualification, the Kuhn–Tucker constraint qualification, the linear independence constraint qualification (LICQ), the Mangasarian–Fromovitz constraint qualification (MFCQ), the Abadie constraint qualification (ACQ), and the Guignard constraint qualification (GCQ) are some of the important constraint qualifications among several constraint qualifications in nonlinear programming problems (see, [1, 24, 41]). Many authors studied these constraint qualifications and found relations for different types of optimization problems under smooth and nonsmooth environments. We refer to [9, 12, 22, 36, 37, 39, 40, 56, 57] for more details about several constraint qualifications and relationships among them for nonlinear programming problems and multiobjective programming problems.

Motivated by the above-mentioned works our aim is to study several constraint qualifications and stationarity conditions of the MOPSC. The chapter is structured as follows: We begin with some preliminary results in Sect. 13.2. Section 13.3 is dedicated to the study of constraint qualifications like LICQ, MFCQ, generalized ACQ, and generalized GCQ for the MOPSC. In Sect. 13.4, we introduce weak stationarity (W-stationarity), Mordukhovich stationarity (M-stationarity), and strong stationarity (S-stationarity) for the MOPSC. In Sect. 13.5, we establish that the M-stationary conditions are sufficient optimality conditions for the MOPSC using generalized convexity. In Sect. 13.6, we propose a Wolfe type dual for the MOPSC and establish weak duality and strong duality results under assumptions of generalized convexity. In Sect. 13.7, we discuss some future research work.

13.2 Preliminaries

This section contains some preliminaries which will be used throughout the chapter. Consider the following multiobjective optimization problem (MOP):

$$\begin{aligned} \text{(MOP)} \qquad & \hat{f}(x) := (\hat{f}_1(x), \ldots, \hat{f}_{\hat{m}}(x)) \\ s.t. \quad & \hat{g}_i(x) \le 0, \forall i = 1, 2, \ldots, \hat{p}, \\ & \hat{h}_i(x) = 0, \forall i = 1, 2, \ldots, \hat{q}, \end{aligned} \tag{13.1}$$

where all the functions $\hat{f}_i, \hat{g}_i, \hat{h}_i : \mathbb{R}^n \to \mathbb{R}$ are continuously differentiable. Set F to be the feasible set of the MOP.

For each $k = \{1, \ldots, \hat{m}\} \subset \mathbb{N}$, the nonempty sets $\hat{S}^k$ and $\hat{S}$ are given as follows:

$$\begin{aligned} \hat{S}^k := \{x \in \mathbb{R}^n |\ & \hat{g}_i(x) \le 0, \forall i = 1, 2, \ldots, \hat{p}, \\ & \hat{h}_i(x) = 0, \forall i = 1, 2, \ldots, \hat{q}, \\ & \hat{f}_i(x) \le \hat{f}_i(\bar{x}), \forall 1, 2, \ldots, \hat{m}, \text{ and i} \neq \text{k}\}, \end{aligned} \tag{13.2}$$

and

$$\begin{aligned} \hat{S} := \{x \in \mathbb{R}^n |\ & \hat{g}_i(x) \le 0, \forall i = 1, 2, \ldots, \hat{p}, \\ & \hat{h}_i(x) = 0, \forall i = 1, 2, \ldots, \hat{q}, \\ & \hat{f}_i(x) \le \hat{f}_i(\bar{x}), \forall i = 1, 2, \ldots, \hat{m}\}. \end{aligned} \tag{13.3}$$

The following concept of the linearized cone to $\hat{S}$ at $\bar{x} \in \hat{S}$ was introduced in [39] for the MOP.

Definition 13.1 The *linearized cone* to $\hat{S}$ at $\bar{x} \in \hat{S}$ is the set $L\left(\hat{S}; \bar{x}\right)$ given by

$$\begin{aligned} L\left(\hat{S}; \bar{x}\right) := \{d \in \mathbb{R}^n |\ & \nabla \hat{g}_i(\bar{x})^T d \le 0, \forall i \in I_{\hat{g}}, \\ & \nabla \hat{h}_i(\bar{x})^T d = 0, \forall i \in I_{\hat{h}}, \\ & \nabla \hat{f}_i(\bar{x})^T d \le 0, \forall i \in I_{\hat{f}}\}. \end{aligned}$$

where

$$\begin{aligned} I_{\hat{g}} &:= \{i \in \{1, \ldots, \hat{p}\} \mid \hat{g}_i(\bar{x}) = 0\}, \\ I_{\hat{h}} &:= \{1, \ldots, \hat{q}\}, \\ I_{\hat{f}} &:= \{1, \ldots, \hat{m}\}. \end{aligned}$$

Some of the important convex cones that play a vital role in optimization are the polar cone, tangent cone, and normal cone. The notion of tangent cones may be considered a generalization of the tangent concept in a smooth case to that in a nonsmooth case.

For the sake of convenience, let us recall the definition of a well-known concept having a crucial role to define constraint qualifications.

Definition 13.2 ([8, 58]) Let $\hat{S}$ be a nonempty subset of $\mathbb{R}^n$. The *tangent cone* to $\hat{S}$ at $\bar{x} \in cl\hat{S}$ is the set $T\left(\hat{S}; \bar{x}\right)$ defined by

$$T(\hat{S}; \bar{x}) := \left\{ d \in \mathbb{R}^n | \exists \{x^n\} \subseteq \hat{S}, \{t_n\} \downarrow 0 : x^n \to \bar{x}, \frac{x^n - \bar{x}}{t_n} \to d \right\},$$

where $cl\hat{S}$ denotes the closure of $\hat{S}$.

The following definitions of constraint qualifications for the MOP are taken from [39].

Definition 13.3 Let $\bar{x} \in F$ be a feasible solution of the MOP. Then the *linear independence constraint qualification* (LICQ) holds at $\bar{x}$, if the gradients

$$\nabla \hat{f}_i(\bar{x}) \left(i \in I_{\hat{f}}\right),$$
$$\nabla \hat{g}_i(\bar{x}) \left(i \in I_{\hat{g}}\right),$$
$$\nabla \hat{h}_i(\bar{x}) \left(i \in I_{\hat{h}}\right),$$

are linearly independent.

Definition 13.4 Let $\bar{x} \in F$ be a feasible solution of the MOP. Then the *Mangasarian–Fromovitz constraint qualification* (MFCQ) holds at $\bar{x}$, if the gradients

$$\nabla \hat{f}_i(\bar{x}) \left(i \in I_{\hat{f}}\right),$$
$$\nabla \hat{h}_i(\bar{x}) \left(i \in I_{\hat{h}}\right),$$

are linearly independent, and the system

$$\nabla \hat{f}_i(\bar{x})^T d = 0 \forall i \in I_{\hat{f}},$$
$$\nabla \hat{g}_i(\bar{x})^T d < 0, \forall i \in I_{\hat{g}},$$
$$\nabla \hat{h}_i(\bar{x})^T d = 0, \forall i \in I_{\hat{h}},$$

has a solution $d \in \mathbb{R}^n$.

Definition 13.5 Let $\bar{x} \in F$ be a feasible solution of the MOP. Then the *Abadie constraint qualification* (ACQ) holds at $\bar{x}$ if

$$L\left(\hat{S}; \bar{x}\right) \subseteq T\left(\hat{S}; \bar{x}\right).$$

Definition 13.6 Let $\bar{x} \in F$ be a feasible solution of the MOP. Then the *generalized Abadie constraint qualification* (GACQ) holds at $\bar{x}$ if

$$L\left(\hat{S}; \bar{x}\right) \subseteq \bigcap_{k=1}^{\hat{m}} T\left(\hat{S}^k; \bar{x}\right).$$

The following concept of efficiency was introduced by Pareto [52].

Definition 13.7 Let $\bar{x} \in F$ be a feasible solution of the MOP . Then $\bar{x}$ is said to be a *local efficient solution* of the MOP , if there exists a number $\delta > 0$ such that, there is no $x \in F \bigcap B(\bar{x}; \delta)$ satisfying

$$\hat{f}_i(x) \leq \hat{f}_i(\bar{x}), \forall i = 1, \ldots, \hat{m},$$

$$\hat{f}_i(x) < \hat{f}_i(\bar{x}), \text{ at least one } i,$$

where $B(\bar{x}; \delta)$ denotes the open ball of radius δ and centre $\bar{x}$.

Definition 13.8 Let $\bar{x} \in F$ be a feasible solution of the MOP. Then $\bar{x}$ is said to be an *efficient solution* of the MOP, if there is no $x \in F$ satisfying

$$\hat{f}_i(x) \leq \hat{f}_i(\bar{x}), \forall i = 1, \ldots, \hat{m},$$

$$\hat{f}_i(x) < \hat{f}_i(\bar{x}), \text{ at least one } i.$$

The following definitions and results are taken from [41].

Definition 13.9 Let f be a differentiable real-valued function defined on a nonempty open convex set $X \subseteq \mathbb{R}^n$. Then the function f is said to be *pseudoconvex* at $\bar{x} \in X$ if the following implication holds:

$$x, \bar{x} \in X, \ \langle \nabla f(\bar{x}), x - \bar{x} \rangle \geq 0 \Rightarrow f(x) \geq f(\bar{x}).$$

Equivalently,

$$x, \bar{x} \in X, \ f(x) < f(\bar{x}) \Rightarrow \langle \nabla f(\bar{x}), x - \bar{x} \rangle < 0.$$

Definition 13.10 Let f be a differentiable real-valued function defined on a nonempty open convex set $X \subseteq \mathbb{R}^n$. Then the function f is said to be quasiconvex at $\bar{x} \in X$ iff the following implication holds:

$$x, \bar{x} \in X, \quad f(x) \leq f(\bar{x}) \Rightarrow \langle \nabla f(\bar{x}), x - \bar{x} \rangle \leq 0.$$

13.3 Constraint Qualifications for Multiobjective Optimization Problems with Switching Constraint

The standard constraint qualifications for nonlinear optimization problems (LICQ or MFCQ) are always violated at every feasible point for mathematical programs with equilibrium constraints (MPEC)(see, [65]), for mathematical programs with complementarity constraints (MPCC) (see, [60]), for mathematical programs with vanishing constraints (MPVC)(see, [29]) and for mathematical programs with switching constraints (MPSC)(see,[42]).

Ye [66] introduced several constraint qualifications for the KKT-type necessary optimality conditions involving Mordukhovich co-derivatives for mathematical problems with variational inequality constraints (MPVIC). The standard Abadie constraint qualification is unlikely to be satisfied by the MPEC, the MPVC, and MPSC. Flegel and Kanzow [16] introduced the modified Abadie constraint qualification for the MPEC. Ye [64] proposed new constraint qualifications namely MPEC weak reverse convex constraint qualification, MPEC Arrow–Hurwicz–Uzawa constraint qualification, MPEC Zangwill constraint qualification, MPEC Kuhn–Tucker constraint qualification, MPEC Abadie constraint qualification. He also proved the relationship among them. For more details about several new constraint qualifications for the MPEC, the MPCC and the MPVIC, (see, [10, 11, 18–21, 25, 26]).

Hoheisel and Kanzow [30] introduced the Abadie and Guignard constraint qualifications for mathematical programs with vanishing constraints. Mishra et al. [44] introduced suitable modifications in constraint qualifications like Cottle constraint qualification, Slater constraint qualification, Mangasarian–Fromovitz constraint qualification, linear independence constraint qualification, linear objective constraint qualification, generalized Guignard constraint qualification for multiobjective optimization problems with vanishing constraints and established relationships among them. We refer to [2, 28, 29, 31, 32] and references their in for more details about constraint qualifications for the MPVC.

Recently, Ardakani et al. [3] introduced two new Abadie-type constraint qualifications and presented some necessary conditions for properly efficient solutions of the problem, using convex subdifferential for multiobjective optimization problems with nondifferentiable convex vanishing constraints. Mehlitz [42] introduced MPSC-tailored versions of MFCQ and LICQ and studied MPSC-tailored versions of the Abadie and Guignard constraint qualification for the MPSC.

Given a feasible point $\bar{x} \in S$, we consider the following index sets:

$$\begin{aligned}
I_g(\bar{x}) &:= \{i = 1, 2, \ldots, p \, : \, g_i(\bar{x}) = 0\},\\
\alpha := \alpha(\bar{x}) &= \{i = 1, 2, \ldots, l \, : \, G_i(\bar{x}) = 0, \, H_i(\bar{x}) \neq 0\},\\
\beta := \beta(\bar{x}) &= \{i = 1, 2, \ldots, l \, : \, G_i(\bar{x}) = 0, \, H_i(\bar{x}) = 0\},\\
\gamma := \gamma(\bar{x}) &= \{i = 1, 2, \ldots, l \, : \, G_i(\bar{x}) \neq 0, \, H_i(\bar{x}) = 0\}.
\end{aligned}$$

Let us define a feasible set S of MOPSC by

$$S := \{x \in \mathbb{R}^n : \ g_i(x) \leq 0, \forall i = 1, 2, \ldots, p, \\ h_i(x) = 0, \forall i = 1, 2, \ldots, q, \\ G_i(x)H_i(x) = 0, \forall i = 1, 2, \ldots, l\}.$$

Consider the following function:

$$\eta_i(x) := G_i(x)H_i(x), \forall i = 1, 2, \ldots, l \tag{13.4}$$

its gradient is given by

$$\nabla \eta_i(x) = G_i(x)\nabla H_i(x) + H_i(x)\nabla G_i(x), \forall i = 1, 2, \ldots, l. \tag{13.5}$$

By the definition of the index sets, we get

$$\nabla \eta_i(\bar{x}) = \begin{cases} H_i(\bar{x})\nabla G_i(\bar{x}), & \text{if} \quad i \in \alpha, \\ 0, & \text{if} \quad i \in \beta, \\ G_i(\bar{x})\nabla H_i(\bar{x}), & \text{if} \quad i \in \gamma, \end{cases} \tag{13.6}$$

For each $k = 1, 2, \ldots, m$, the nonempty sets S^k and S are defined as follows:

$$S^k := \{x \in \mathbb{R}^n | \ g_i(x) \leq 0, \forall i = 1, 2, \ldots, p, \\ h_i(x) = 0, \forall i = 1, 2, \ldots, q, \\ G_i(x)H_i(x) = 0, \forall i = 1, 2, \ldots, r, \\ f_i(x) \leq f_i(\bar{x}), \forall i = 1, 2, \ldots, m, i \neq k\},$$

and

$$S := \{x \in \mathbb{R}^n | \ g_i(x) \leq 0, \forall i = 1, 2, \ldots, p, \\ h_i(x) = 0, \forall i = 1, 2, \ldots, q, \\ G_i(x)H_i(x) = 0, \forall i = 1, 2, \ldots, r, \\ f_i(x) \leq f_i(\bar{x}), \forall i = 1, 2, \ldots, m, \}.$$

The following result gives the standard linearized cone to S^k, $k = 1, 2, \ldots, m$, at an efficient solution $\bar{x} \in S$ of the MOPSC .

Lemma 13.3.1 *Let $\bar{x} \in S$ be an efficient solution of the MOPSC. Then, the linearized cone to S^k, $k = 1, 2, \ldots, m$, at $\bar{x}$ is given by*

$$L(S^k;\bar{x}) = \{d \in \mathbb{R}^n | \ \nabla f_i(\bar{x})^T d \leq 0, \forall i \in I_f, i \neq k,$$
$$\nabla g_i(\bar{x})^T d \leq 0, \forall i \in I_g,$$
$$\nabla h_i(\bar{x})^T d = 0, \forall i \in I_h, \tag{13.7}$$
$$\nabla H_i(\bar{x})^T d = 0, \forall i \in \gamma,$$
$$\nabla G_i(\bar{x})^T d = 0, \forall i \in \alpha\}.$$

Proof Let $\eta_i(x) = G_i(x)H_i(x), \forall i = 1, 2, \dots, r$. By the definitions of the index sets and in view of Definition of the linearized cone to S^k, $k = 1, \dots, m$ at $\bar{x} \in S^k$ is given by

$$L(S^k;\bar{x}) = \{d \in \mathbb{R}^n | \ \nabla f_i(\bar{x})^T d \leq 0, \forall i \in I_f, i \neq k,$$
$$\nabla g_i(\bar{x})^T d \leq 0, \forall i \in I_g,$$
$$\nabla h_i(\bar{x})^T d = 0, \forall i \in I_h,$$
$$\nabla \eta_i(\bar{x})^T d = 0, \forall i \in \alpha \cup \gamma\}.$$

We know that $\nabla \eta_i(\bar{x}) = G_i(\bar{x})\nabla H_i(\bar{x}) + H_i(\bar{x})\nabla G_i(\bar{x})$,

$$\nabla \eta_i(\bar{x})^T d = 0$$

implies

$$G_i(\bar{x})\nabla H_i(\bar{x})^T d + H_i(\bar{x})\nabla G_i(\bar{x})^T d = 0.$$

Since, $G_i(\bar{x}) = 0, \ \forall i \in \alpha$, and $H_i(\bar{x}) = 0, , \ \forall i \in \gamma$, we get

$$L(S^k;\bar{x}) = \{d \in \mathbb{R}^n | \ \nabla f_i(\bar{x})^T d \leq 0, \forall i \in I_f, i \neq k,$$
$$\nabla g_i(\bar{x})^T d \leq 0, \forall i \in I_g,$$
$$\nabla h_i(\bar{x})^T d = 0, \forall i \in I_h,$$
$$\nabla H_i(\bar{x})^T d = 0, \forall i \in \gamma, \tag{13.8}$$
$$\nabla G_i(\bar{x})^T d = 0, \forall i \in \alpha\}.$$

We introduce a tightened nonlinear multiobjective optimization problem (TNLMOP) derived from the MOPSC depending on an efficient solution $\bar{x} \in S$ as follows

$$\text{(TNLMOP)} \qquad f(x) := (f_1(x), \dots, f_m(x))$$
$$s.t. \quad g_i(x) \leq 0, \forall i = 1, 2, \dots, p,$$
$$h_i(x) = 0, \forall i = 1, 2, \dots, q, \tag{13.9}$$
$$G_i(x) = 0, \forall i \in \alpha \cup \beta,$$
$$H_i(x) = 0, \forall i \in \gamma \cup \beta.$$

The feasible set of the TNLMOP is a subset of the feasible set of MOPSC.

Definition 13.11 Let $\bar{x} \in S$ be a feasible point of the MOPSC. If LICQ holds for TNLMOP at $\bar{x}$. Then $\bar{x}$ is said to satisfy LICQ–MOPSC.

Definition 13.12 Let $\bar{x} \in S$ be a feasible point of the MOPSC. If MFCQ holds for TNLMOP at $\bar{x}$. Then $\bar{x}$ is said to satisfy MFCQ–MOPSC.

From the Definitions 13.3 and 13.4 [39] for TNLMOP, one has

$$\text{LICQ} \implies \text{MFCQ}.$$

Therefore,

$$\text{LICQ–MOPSC} \implies \text{MFCQ-MOPSC}.$$

13.3.1 A Generalized Guignard and Abadie CQ for MOPSC

For each $k = 1, 2, \ldots, m$, the nonempty sets $\overline{S}^k$ and $\overline{S}$ are defined as follows:

$$\begin{aligned}\overline{S}^k := \{x \in \mathbb{R}^n |\ & g_i(x) \le 0, \forall i = 1, 2, \ldots, p,\\ & h_i(x) = 0, \forall i = 1, 2, \ldots, q,\\ & G_i(x) = 0, \forall i \in \alpha \cup \beta,\\ & H_i(x) = 0, \forall i \in \gamma \cup \beta,\\ & f_i(x) \le f_i(\bar{x}), \forall i = 1, 2, \ldots, m, i \ne k\},\end{aligned}$$

and

$$\begin{aligned}\overline{S} := \{x \in \mathbb{R}^n |\ & g_i(x) \le 0, \forall i = 1, 2, \ldots, p,\\ & h_i(x) = 0, \forall i = 1, 2, \ldots, q,\\ & G_i(x) = 0, \forall i \in \alpha \cup \beta,\\ & H_i(x) = 0, \forall i \in \gamma \cup \beta,\\ & f_i(x) \le f_i(\bar{x}), \forall i = 1, 2, \ldots, m\}.\end{aligned}$$

The linearized cone to $\overline{S}^k$ at $\bar{x} \in \overline{S}^k$ is given by

$$\begin{aligned}L(\overline{S}^k; \bar{x}) = \{d \in \mathbb{R}^n |\ & \nabla f_i(\bar{x})^T d \le 0, \forall i = 1, \ldots, m, i \ne k,\\ & \nabla g_i(\bar{x})^T d \le 0, \forall i \in I_g,\\ & \nabla h_i(\bar{x})^T d = 0, \forall i \in I_h,\\ & \nabla G_i(\bar{x})^T d = 0, \forall i \in \alpha \cup \beta,\\ & \nabla H_i(\bar{x})^T d = 0, \forall i \in \gamma \cup \beta\}.\end{aligned} \tag{13.10}$$

$$L(\overline{S};\bar{x}) = \{d \in \mathbb{R}^n |\ \nabla f_i(\bar{x})^T d \leq 0, \forall i = 1, \ldots, m,$$
$$\nabla g_i(\bar{x})^T d \leq 0, \forall i \in I_g,$$
$$\nabla h_i(\bar{x})^T d = 0, \forall i \in I_h, \quad (13.11)$$
$$\nabla G_i(\bar{x})^T d = 0, \forall i \in \alpha \cup \beta,$$
$$\nabla H_i(\bar{x})^T d = 0, \forall i \in \gamma \cup \beta\}.$$

We have the following relation:

$$L\left(\overline{S};\bar{x}\right) = \bigcap_{k=1}^{m} L\left(\overline{S}^k;\bar{x}\right). \quad (13.12)$$

Definition 13.13 Let $\bar{x} \in X$ be any feasible solution to the MOPSC. Then, a *Generalized Abadie Constraint Qualification* (GACQ) for the MOPSC, denoted by GACQ–MOPSC, holds at $\bar{x}$, if

$$L\left(\overline{S};\bar{x}\right) \subseteq \bigcap_{k=1}^{m} T\left(S^k;\bar{x}\right).$$

The following constraint qualification gives a sufficient condition to the GACQ–MOPVC.

Definition 13.14 Let $\bar{x} \in X$ be any feasible solution to the TNLMOP. Then a *Generalized Abadie Constraint Qualification* (GACQ) for the TNLMOP, denoted by GACQ–TNLMOP, holds at $\bar{x}$, if

$$L\left(S;\bar{x}\right) \subseteq \bigcap_{k=1}^{m} T\left(\overline{S}^k;\bar{x}\right).$$

Note 13.1 The standard GACQ gives a sufficient condition for the GACQ–MOPVC to hold. Since $L\left(\overline{S};\bar{x}\right) \subseteq L\left(S;\bar{x}\right)$.

The following lemma is about relationships between GACQ–TNLMOP and GACQ–MOPSC.

Lemma 13.3.2 *If the GACQ–TNLMOP holds at $\bar{x}$ then the standard GACQ and the GACQ–MOPVC both are satisfied at $\bar{x}$.*

Proof We know that

$$\overline{S}^k \subset S^k \ \ \forall k = 1, 2 \ldots m$$

and

$$T(\overline{S}^k;\bar{x}) \subset T(S^k;\bar{x}) \;\; \forall k = 1, 2 \ldots m.$$

Hence,

$$\bigcap_{k=1}^{m} T\left(\overline{S}^k;\bar{x}\right) \subset \bigcap_{k=1}^{m} T\left(S^k;\bar{x}\right).$$

From Definition 13.14, we have

$$L\left(\overline{S};\bar{x}\right) \subseteq L\left(S;\bar{x}\right) \subseteq \bigcap_{k=1}^{m} T\left(\overline{S}^k;\bar{x}\right) \subset \bigcap_{k=1}^{m} T\left(S^k;\bar{x}\right).$$

Therefore, GACQ–MOPSC holds at $\bar{x}$.
By Definitions 13.13 and 13.14, we obtain

$$\text{GACQ–TNLMOP} \implies \text{GACQ–MOPSC}$$

Now, we discuss the relationship between tangent cone $T\left(\overline{S}^k;\bar{x}\right)$, k=1, 2, ..., m, and the linearized cone $L(S;\bar{x})$.

Lemma 13.3.3 *Let $\bar{x} \in$ be a feasible solution of the MOPSC. Then we have*

$$\bigcap_{k=1}^{m} clcoT\left(\overline{S}^k;\bar{x}\right) \subseteq L\left(S;\bar{x}\right).$$

Proof The proof follows on the lines of the proof of Lemma 3.1 [42]. □

Definition 13.15 Let $\bar{x} \in X$ be any feasible solution to the TNLMOP. Then a *Generalized Guignard Constraint Qualification* (GGCQ) for the TNLMOP, denoted by GGCQ–TNLMOP, holds at $\bar{x}$, if

$$L\left(S;\bar{x}\right) \subseteq \bigcap_{k=1}^{m} clcoT\left(\overline{S}^k;\bar{x}\right).$$

Definition 13.16 Let $\bar{x} \in X$ be any feasible solution to the MOPSC. Then, a *Generalized Guignard Constraint Qualification* (GGCQ) for the MOPSC, denoted by GGCQ–MOPSC, holds at $\bar{x}$, if

$$L\left(\overline{S};\bar{x}\right) \subseteq \bigcap_{k=1}^{m} clcoT\left(S^k;\bar{x}\right).$$

The following result gives the relationship between the GGCQ–TNLMOP and the GGCQ–MOPVC.

Lemma 13.3.4 *Let $\bar{x} \in X$ be any feasible solution of the MOPVC. If the GGCQ-TNLMOP holds at $\bar{x}$, then the GGCQ–MOPVC also holds at $\bar{x} \in X$.*

Proof Assume that $\bar{x} \in X$ is a feasible solution of the MOPSC and GGCQ–TNLMOP holds at $\bar{x}$, then

$$L(S; \bar{x}) \subseteq \bigcap_{k=1}^{m} clcoT\left(\overline{S}^k; \bar{x}\right). \tag{13.13}$$

Also,

$$\overline{S}^k \subset S^k \ \forall k = 1, 2 \ldots m$$

and

$$T(\overline{S}^k; \bar{x}) \subset T(S^k; \bar{x}) \ \forall k = 1, 2 \ldots m.$$

Hence

$$\bigcap_{k=1}^{m} clcoT\left(\overline{S}^k; \bar{x}\right) \subset \bigcap_{k=1}^{m} clcoT\left(S^k; \bar{x}\right). \tag{13.14}$$

We always have

$$L\left(\overline{S}; \bar{x}\right) \subseteq L(S; \bar{x}). \tag{13.15}$$

From Eqs. (13.13), (13.14) and (13.15), we get

$$L\left(\overline{S}; \bar{x}\right) \subseteq \bigcap_{k=1}^{m} clcoT\left(S^k; \bar{x}\right).$$

Therefore, GGCQ–MOPVC holds at $\bar{x} \in X$. This completes the proof.

In the following lemma, we derive a relationship between the GACQ–MOPSC and the GGCQ–MOPSC.

Lemma 13.3.5 *Let $\bar{x} \in X$ be a feasible solution of the MOPSC. If the GACQ–MOPSC holds at $\bar{x}$ then the GGCQ-MOPSC is satisfied.*

Proof Assume $\bar{x} \in X$ be a feasible solution of the MOPSC and that GACQ–MOPSC holds at $\bar{x}$. From Definition 13.13, we have

$$L\left(\overline{S}; \bar{x}\right) \subseteq \bigcap_{k=1}^{m} T\left(S^k; \bar{x}\right).$$

Since

$$T\left(S^k; \bar{x}\right) \subseteq clcoT\left(S^k; \bar{x}\right),$$

we have

$$\bigcap_{k=1}^{m} T\left(S^k; \bar{x}\right) \subseteq \bigcap_{k=1}^{m} clcoT\left(S^k; \bar{x}\right).$$

Which implies

$$L\left(\overline{S}; \bar{x}\right) \subseteq \bigcap_{k=1}^{m} clcoT\left(S^k; \bar{x}\right).$$

Therefore, the GGCQ–MOPSC is satisfied at $\bar{x}$. This completes the proof. □

By Lemma 13.3.5, we have

$$\text{GACQ–MOPSC} \implies \text{GGCQ–MOPSC.}$$

13.4 Stationary Conditions for MOPSC

The standard nonlinear programming has only one dual stationary condition, i.e., the Karush–Kuhn–Tucker condition, but we have various stationarity concepts for mathematical programs with equilibrium constraints (MPEC), mathematical program with complementarity constraints (MPCC), mathematical program with vanishing constraints (MPVC), and mathematical program with switching constraints (MPSC).

Outrata [50] introduced the notion of Mordukhovich stationary point (M-stationary) for mathematical programs with equilibrium constraints (MPEC). Scheel and Scholtes [60] introduced the concept of strong-stationary point (S-stationary) and Clarke-stationary (C-stationary) for the mathematical program with complementarity constraints (MPCC). Flegel and Kanzow [15] introduced the concept of Alternatively, stationary point (A-stationary) for the MPEC. Further, Flegel and Kanzow [17] proved that M-stationarity is the first-order optimality condition under a weak Abadie-type constraint qualification for the MPEC.

Ye [64] introduced various stationarity conditions and obtained new constraint qualifications for the considered MPEC. Hoheisel and Kanzow [29] introduced several stationarity conditions for mathematical programs with vanishing constraints (MPVC) using weak constraint qualifications. Ardali et al. [4] studied several new constraint qualifications, GS-stationarity concepts, and optimality conditions for a nonsmooth mathematical program with equilibrium constraints based on the convexificators. Mehlitz [42] introduced notions of weak stationary point (W-stationary), Mordukhovich stationary point (M-stationary), strong stationary point (S-stationary) for mathematical program with vanishing constraints (MPVC) and obtain that the S-stationarity conditions of the MPSC equal its KKT conditions in a certain sense.

In this section, we introduce the notion of weak stationarity, Mordukhovich stationarity, and strong stationarity, i.e., W-stationarity, M-stationarity, and S-stationarity, respectively for the MOPSC.

The following stationarity conditions can be treated as a multiobjective analog of the stationarity conditions for scalar optimization problem with switching constraint introduced in [42].

Definition 13.17 (*W-stationary point*) A feasible point $\bar{x}$ of MOPSC is called a *weak stationary point* (W-stationary point) if there exists $\lambda = (\lambda^g, \lambda^h, \lambda^G, \lambda^H) \in \mathbb{R}^{p+q+2l}$, and $\theta_i > 0, i \in \{1, \ldots, m\}$ such that following conditions hold:

$$0 = \sum_{i=1}^{m} \theta_i \nabla f_i(\bar{x}) + \sum_{i \in I_g} \lambda_i^g \nabla g_i(\bar{x}) + \sum_{i=1}^{q} \lambda_i^h \nabla h_i(\bar{x}) + \sum_{i=1}^{l} [\lambda_i^G \nabla G_i(\bar{x}) + \lambda_i^H \nabla H_i(\bar{x})],$$

$$\forall i \in I^g(\bar{x}) : \lambda_i^g \geq 0,$$

$$\forall i \in \alpha(\bar{x}) : \lambda_i^H = 0,$$

$$\forall i \in \gamma(\bar{x}) : \lambda_i^G = 0.$$

Definition 13.18 (*M-stationary point*) A feasible point $\bar{x}$ of MOPSC is called a *Mordukhovich stationary point* (M-stationary point) if there exists $\lambda = (\lambda^g, \lambda^h, \lambda^G, \lambda^H) \in \mathbb{R}^{p+q+2l}$, and $\theta_i > 0, i \in \{1, \ldots, m\}$ such that following conditions hold:

$$0 = \sum_{i=1}^{m} \theta_i \nabla f_i(\bar{x}) + \sum_{i \in I_g} \lambda_i^g \nabla g_i(\bar{x}) + \sum_{i=1}^{q} \lambda_i^h \nabla h_i(\bar{x}) + \sum_{i=1}^{l} [\lambda_i^G \nabla G_i(\bar{x}) + \lambda_i^H \nabla H_i(\bar{x})],$$

$$\forall i \in I^g(\bar{x}) : \lambda_i^g \geq 0,$$

$$\forall i \in \alpha(\bar{x}) : \lambda_i^H = 0,$$

$$\forall i \in \gamma(\bar{x}) : \lambda_i^G = 0,$$

$$\forall i \in \beta(\bar{x}) : \lambda_i^G \lambda_i^H = 0.$$

Definition 13.19 (*S-stationary point*) A feasible point $\bar{x}$ of MOPSC is called a *strong stationary point* (S-stationary point) if there exists $\lambda = (\lambda^g, \lambda^h, \lambda^G, \lambda^H) \in \mathbb{R}^{p+q+2l}$, and $\theta_i > 0, i \in \{1, \ldots, m\}$ such that following conditions hold:

$$0 = \sum_{i=1}^{m} \theta_i \nabla f_i(\bar{x}) + \sum_{i \in I_g} \lambda_i^g \nabla g_i(\bar{x}) + \sum_{i=1}^{q} \lambda_i^h \nabla h_i(\bar{x}) + \sum_{i=1}^{l} [\lambda_i^G \nabla G_i(\bar{x}) + \lambda_i^H \nabla H_i(\bar{x})],$$

$$\forall i \in I^g(\bar{x}) : \lambda_i^g \geq 0,$$

$$\forall i \in \alpha(\bar{x}) : \lambda_i^H = 0,$$

$$\forall i \in \gamma(\bar{x}) : \lambda_i^G = 0,$$

$$\forall i \in \beta(\bar{x}) : \lambda_i^G = 0 \text{ and } \lambda_i^H = 0.$$

By Definitions 13.17, 13.18 and 13.19, we have

$$S - stationarity \implies M - stationarity \implies W - stationarity.$$

13.5 Sufficient Optimality Conditions for the MOPSC

Mordukhovich [46] established necessary optimality conditions for multiobjective equilibrium programs with equilibrium constraints in finite-dimensional spaces based on advanced generalized differential tools of variational analysis. Bao et al. [5] studied multiobjective optimization problems with equilibrium constraints (MOPECs) described by generalized equations in the form

$$0 \in G(x, y) + Q(x, y),$$

where mappings G and Q are set-valued.

Bao et al. [5] established a necessary optimality conditions for the MOPEC using tools of variational analysis and generalized differentiation. Mordukhovich [48] derived new qualified necessary optimality conditions for the MOPEC in finite- and infinite-dimensional spaces. Movahedian and Nobakhtian [49] derived a necessary optimality result on any Asplund space and established sufficient optimality conditions for nonsmooth MPEC in Banach spaces. Recently, Pandey and Mishra [53] introduced the concept of Mordukhovich stationary point in terms of the Clarke subdifferentials and established that M-stationarity conditions are strong KKT-type sufficient optimality conditions for the multiobjective semi-infinite mathematical programming problem with equilibrium constraints.

We divide the index sets as follows. Let
$T^+ := \{i : \lambda_i^h > 0\}, \quad T^- := \{i : \lambda_i^h < 0\}$
$\beta^+ := \{i \in \beta : \lambda_i^G > 0, \lambda_i^H > 0\},$
$\beta_G^+ := \{i \in \beta : \lambda_i^G = 0, \lambda_i^H > 0\}, \quad \beta_G^- := \{i \in \beta : \lambda_i^G = 0, \lambda_i^H < 0\},$
$\beta_H^+ := \{i \in \beta : \lambda_i^H = 0, \lambda_i^G > 0\}, \quad \beta_H^- := \{i \in \beta : \lambda_i^H = 0, \lambda_i^G < 0\},$
$\alpha^+ := \{i \in \alpha : \lambda_i^G > 0\}, \quad \alpha^- := \{i \in \alpha : \lambda_i^G < 0\},$
$\gamma^+ := \{i \in \gamma : \lambda_i^H > 0\}, \quad \gamma^- := \{i \in \gamma : \lambda_i^H < 0\}.$

Definition 13.20 Let $\bar{x} \in X$ be a feasible point of the MOPSC. We say that the *No Nonzero Abnormal Multiplier Constraint Qualification* (NNAMCQ) is satisfied at

$\bar{x}$, if there is no nonzero vector $\lambda = (\lambda^g, \lambda^h, \lambda^G, \lambda^H) \in \mathbb{R}^{p+q+2l}$, such that

$$0 \in \sum_{i \in I_g} \lambda_i^g \nabla g_i(\bar{x}) + \sum_{i=1}^{q} \lambda_i^h \nabla h_i(\bar{x}) + \sum_{i=1}^{l} [\lambda_i^G \nabla G_i(\bar{x}) + \lambda_i^H \nabla H_i(\bar{x})],$$

$$\forall i \in I^g(\bar{x}) : \lambda_i^g \geq 0,$$

$$\forall i \in \alpha(\bar{x}) : \lambda_i^H = 0,$$

$$\forall i \in \gamma(\bar{x}) : \lambda_i^G = 0,$$

and

$$\forall i \in \beta(\bar{x}) : \lambda_i^G \lambda_i^H = 0.$$

The following theorem shows that the MOPSC M-stationary conditions are a KKT type sufficient optimality conditions for weakly efficient solution of the MOPSC.

Theorem 13.5.1 *Let $\bar{x} \in X$ be a feasible point of the MOPSC and the M-stationarity conditions hold at $\bar{x}$. Suppose that each $f_i (i = 1, \ldots, m)$ is pseudoconvex at $\bar{x}$, $g_j (j \in J(\bar{x}))$, $h_i (i \in T^+)$, $-h_i (i \in T^-)$, $G_i (i \in \alpha^+ \cup \beta_H^+ \cup \beta^+)$, $-G_i$ $(i \in \alpha^- \cup \beta_H^-)$, $H_i (i \in \gamma^+ \cup \beta_G^+ \cup \beta^+)$, $-H_i (i \in \gamma^- \cup \beta_G^-)$ are quasiconvex at $\bar{x}$. If $\alpha^- \cup \gamma^- \cup \beta_G^- \cup \beta_H^- = \phi$, then $\bar{x}$ is a weakly efficient solution for MOPSC.*

Proof Assume that $\bar{x}$ is not a weakly efficient solution for MOPSC. Then there exists a feasible point x for MOPSC such that such that

$$f_i(x) < f_i(\bar{x}) \quad \forall i = 1, \ldots, m.$$

Since each f_i is pseudoconvex, we have

$$\langle \nabla f_i(\bar{x}), x - \bar{x} \rangle < 0. \tag{13.16}$$

Also $\eta_i > 0$ for all $i \in \{1, \ldots, m\}$, we get

$$\left\langle \sum_{i=1}^{m} \eta_i \nabla f_i(\bar{x}), x - \bar{x} \right\rangle < 0. \tag{13.17}$$

Since $\bar{x}$ is MOPSC M-stationary point, we have

$$-\sum_{i \in I_g} \lambda_i^g \nabla g_i(\bar{x}) - \sum_{i=1}^{q} \lambda_i^h \nabla h_i(\bar{x}) - \sum_{\alpha \cup \beta} \lambda_i^G \nabla G_i(\bar{x}) - \sum_{\beta \cup \gamma} \lambda_i^H \nabla H_i(\bar{x}) = \sum_{i=1}^{m} \eta_i \nabla f_i(\bar{x}). \tag{13.18}$$

By Eq. (13.17), we get

$$\left\langle \left(\sum_{i \in I_g} \lambda_i^g \nabla g_i(\bar{x}) + \sum_{i=1}^{q} \lambda_i^h \nabla h_i(\bar{x}) + \sum_{\alpha \cup \beta} \lambda_i^G \nabla G_i(\bar{x}) + \sum_{\beta \cup \gamma} \lambda_i^H \nabla H_i(\bar{x}) \right), x - \bar{x} \right\rangle > 0. \tag{13.19}$$

For each $i \in I_g(\bar{x}),\ g_i(x) \leq 0 = g_i(\bar{x})$. Hence, by quasiconvexity of g_i, we have

$$\langle \nabla g_i(\bar{x}), x - \bar{x} \rangle \leq 0\,. \tag{13.20}$$

For any feasible point x of MOPSC and for each $i \in T^-$, $0 = -h_i(\bar{x}) = h_i(x)$, by quasiconvexity of h_i, we get

$$\langle \nabla h_i(\bar{x}), x - \bar{x} \rangle \geq 0,\ \forall i \in T^-. \tag{13.21}$$

Similarly, we have

$$\langle \nabla h_i(\bar{x}), x - \bar{x} \rangle \leq 0,\ \forall i \in T^+. \tag{13.22}$$

Also $G_i(x) \leq G_i(\bar{x}),\ \forall i \in \alpha^+ \cup \beta_H^+$, and $H_i(x) \leq H_i(\bar{x}),\ \forall i \in \gamma^+ \cup \beta_G^+$. Since all of these functions are quasiconvex, we get

$$\langle \nabla G_i(\bar{x}), x - \bar{x} \rangle \leq 0,\ \forall i \in \alpha^+ \cup \beta_H^+, \tag{13.23}$$

$$\langle \nabla H_i(\bar{x}), x - \bar{x} \rangle \leq 0,\ \forall i \in \gamma^+ \cup \beta_G^+. \tag{13.24}$$

From Eqs. (13.20)–(13.24), we have

$$\begin{aligned}
&\langle \nabla g_i(\bar{x}), x - \bar{x} \rangle \leq 0,\ \forall i \in I_g(\bar{x}),\\
&\langle \nabla h_i(\bar{x}), x - \bar{x} \rangle \leq 0,\ \forall i \in T^+,\\
&\langle \nabla h_i(\bar{x}), x - \bar{x} \rangle \geq 0,\ i \in T^-,\\
&\langle \nabla G_i(\bar{x}), x - \bar{x} \rangle \leq 0,\ \forall i \in \alpha^+ \cup \beta_H^+,\\
&\langle \nabla H_i(\bar{x}), x - \bar{x} \rangle \leq 0,\ \forall i \in \gamma^+ \cup \beta_G^+.
\end{aligned}$$

Since $\alpha^- \cup \gamma^- \cup \beta_G^- \cup \beta_H^- = \phi$, we get

$$\left\langle \sum_{\alpha \cup \beta} \lambda_i^G \nabla G_i(\bar{x}), x - \bar{x} \right\rangle \leq 0,\ \left\langle \sum_{\beta \cup \gamma} \lambda_i^H \nabla H_i(\bar{x}), x - \bar{x} \right\rangle \leq 0,$$

$$\left\langle \sum_{i \in I_g(\bar{x})} \lambda_i^g \nabla g_i(\bar{x}), x - \bar{x} \right\rangle \leq 0,\ \left\langle \sum_{i=1}^{q} \lambda_i^h \nabla h_i(\bar{x}), x - \bar{x} \right\rangle \leq 0.$$

So,

$$\left\langle \left(\sum_{i \in I_g(\bar{x})} \lambda_i^g \nabla g_i(\bar{x}) + \sum_{i=1}^{q} \lambda_i^h \nabla h_i(\bar{x}) + \sum_{\alpha \cup \beta} \lambda_i^G \nabla G_i(\bar{x}) + \sum_{\beta \cup \gamma} \lambda_i^H \nabla H_i(\bar{x}) \right), x - \bar{x} \right\rangle \leq 0,$$

which contradicts (13.19). Hence, $\bar{x}$ is a weakly efficient solution for MOPSC. This completes the proof. □

Theorem 13.5.2 *Let $\bar{x}$ be a feasible point of MOPSC and the M-stationarity conditions hold at $\bar{x}$. Suppose that each $f_i(i = 1, \ldots, m)$ is strictly pseudoconvex at $\bar{x}$, $g_i(i \in I_g(\bar{x}))$, $h_i(i \in T^+)$, $-h_i(i \in T^-)$, $G_i(i \in \alpha^+ \cup \beta_H^+ \cup \beta^+)$, $-G_i$ $(i \in \alpha^- \cup \beta_H^-)$, $H_i(i \in \gamma^+ \cup \beta_G^+ \cup \beta^+)$, $-H_i(i \in \gamma^- \cup \beta_G^-)$ are quasiconvex at $\bar{x}$. If $\alpha^- \cup \gamma^- \cup \beta_G^- \cup \beta_H^- = \phi$, then $\bar{x}$ is efficient solution for MOPSC.*

Proof The proof follows the lines of the proof of Theorem 13.5.1. □

13.6 Duality

In this section, we formulate and study a Wolfe-type dual problem for the MOPSC under the generalized convexity assumption. The Wolfe-type dual problem is formulated as follows:

$$WDMOPSC(\bar{x}) \quad \max_{u,\lambda} \; f(u) + \Big[\sum_{i \in I_g} \lambda_i^g g_i(u) + \sum_{i=1}^{q} \lambda_i^h h_i(u) + \sum_{i=1}^{l} [\lambda_i^G G_i(u) + \lambda_i^H H_i(u)]\Big] e$$

subject to:

$$0 \in \sum_{i=1}^{m} \rho_i \nabla f_i(u) + \sum_{i \in I_g} \lambda_i^g \nabla g_i(u) + \sum_{i=1}^{q} \lambda_i^h \nabla h_i(u) + \sum_{i=1}^{l} [\lambda_i^G \nabla G_i(u) + \lambda_i^H \nabla H_i(u)], \quad (13.25)$$

$$\forall i \in I^g(\bar{x}) : \lambda_i^g \geq 0,$$

$$\forall i \in \alpha(\bar{x}) : \lambda_i^H = 0,$$

$$\forall i \in \gamma(\bar{x}) : \lambda_i^G = 0,$$

$$\forall i \in \beta(\bar{x}) : \lambda_i^G \lambda_i^H = 0,$$

where, $e := (1, \ldots, 1) \in \mathbb{R}^m$, $\lambda = (\lambda^g, \lambda^h, \lambda^G, \lambda^H) \in \mathbb{R}^{k+p+2l}$, $\rho = (\rho_1, \ldots, \rho_m)$ ≥ 0 and $\sum_i^m \rho_i = 1$.

Theorem 13.6.3 (Weak Duality) *Let $\bar{x}$ be feasible for* MOPSC, *(u, ρ, λ) feasible for* WDMOPSC *$(\bar{x})$ and index sets $I_g, \alpha, \beta, \gamma$ defined accordingly. Suppose that each $f_i(i = 1, \ldots, m)$, $g_i(i \in I_g(\bar{x}))$, $h_i(i \in T^+)$, $-h_i(i \in T^-)$, $G_i(i \in \alpha^+ \cup \beta_H^+ \cup \beta^+)$, $-G_i$ $(i \in \alpha^- \cup \beta_H^-)$, $H_i(i \in \gamma^+ \cup \beta_G^+ \cup \beta^+)$ and $-H_i(i \in \gamma^- \cup \beta_G^-)$ are pseudoconvex at at u. If $\alpha^- \cup \gamma^- \cup \beta_G^- \cup \beta_H^- = \phi$, Then,*

$$f(x) \nleq f(u) + \Big[\sum_{i\in I_g} \lambda_i^g g_i(u) + \sum_{i=1}^{q} \lambda_i^h h_i(u) + \sum_{i=1}^{l} [\lambda_i^G G_i(u) + \lambda_i^H H_i(u)]\Big]e.$$

Proof Let

$$f(x) \leq f(u) + \Big[\sum_{i\in I_g} \lambda_i^g g_i(u) + \sum_{i=1}^{q} \lambda_i^h h_i(u) + \sum_{i=1}^{l} [\lambda_i^G G_i(u) + \lambda_i^H H_i(u)]\Big]e.$$

Then there exist n such that

$$f_n(x) < f_n(u) + \sum_{i\in I_g} \lambda_i^g g_i(u) + \sum_{i=1}^{q} \lambda_i^h h_i(u) + \sum_{i=1}^{l} [\lambda_i^G G_i(u) + \lambda_i^H H_i(u)]$$

and

$$f_i(x) \leqq f_i(u) + \sum_{i\in I_g} \lambda_i^g g_i(u) + \sum_{i=1}^{q} \lambda_i^h h_i(u) + \sum_{i=1}^{l} [\lambda_i^G G_i(u) + \lambda_i^H H_i(u)], \forall i \neq n.$$

From the Definition 13.9 and above inequality, we have

$$\left\langle\left(\sum_{i=1}^{m} \rho_i \nabla f_i(u) + \sum_{i\in I_g} \lambda_i^g \nabla g_i(u) + \sum_{i=1}^{q} \lambda_i^h \nabla h_i(u) + \sum_{i=1}^{l} [\lambda_i^G \nabla G_i(u) + \lambda_i^H \nabla H_i(u)]\right), x - u\right\rangle < 0.$$

Then,

$$\sum_{i=1}^{m} \rho_i \nabla f_i(u) + \sum_{i\in I_g} \lambda_i^g \nabla g_i(u) + \sum_{i=1}^{q} \lambda_i^h \nabla h_i(u) + \sum_{i=1}^{l} [\lambda_i^G \nabla G_i(u) + \lambda_i^H \nabla H_i(u)] < 0.$$

Which is a contradiction to the feasibility of the (u, ρ, λ) for the WDMOPSC, therefore

$$f(x) \nleq f(u) + \Big[\sum_{i\in I_g} \lambda_i^g g_i(u) + \sum_{i=1}^{q} \lambda_i^h h_i(u) + \sum_{i=1}^{l} [\lambda_i^G G_i(u) + \lambda_i^H H_i(u)]\Big]e.$$

This complete the proof. □

Theorem 13.6.4 (Strong Duality) *If $\bar{x}$ is a efficient solution of* MOPSC, *such that* NNAMCQ *is satisfied at $\bar{x}$ and index sets $I_g, \alpha, \beta, \gamma$ defined accordingly. Let f_i (i =*

$1,\ldots m)$, $g_i(i \in I_g)$, $h_i(i \in J^+)$, $-h_i(i \in J^-)$, $G_i(i \in \alpha^- \cup \beta_H^-)$, $-G_i(i \in \alpha^+ \cup \beta_H^+ \cup \beta^+)$, $H_i(i \in \gamma^- \cup \beta_G^-)$, $-H_i(i \in \gamma^+ \cup \beta_G^+ \cup \beta^+)$ *satisfy the assumption of the Theorem 13.6.3 and If* $\alpha^- \cup \gamma^- \cup \beta_G^- \cup \beta_H^- = \phi$. *Then, there exists* $(\bar{\rho}, \bar{\lambda})$, *such that* $(\bar{x}, \bar{\rho}, \bar{\lambda})$ *is an efficient solution of* WDMOPSC $(\bar{x})$ *and respective objective values are equal.*

Proof Since, $\bar{x}$ is an efficient solution of MOPSC and the NNAMCQ is satisfied at $\bar{x}$, hence, $\exists\,\bar{\lambda} = (\bar{\lambda}^g, \bar{\lambda}^h, \bar{\lambda}^G, \bar{\lambda}^H) \in \mathbb{R}^{p+q+2l}$, such that the M-stationarity conditions for MOPSC are satisfied, that is,

$$0 = \sum_{i=1}^{m} \bar{\rho}_i \nabla f_i(\bar{x}) + \sum_{i \in I_g} \bar{\lambda}_i^g \nabla g_i(\bar{x}) + \sum_{i=1}^{q} \bar{\lambda}_i^h \nabla h_i(\bar{x}) + \sum_{i=1}^{l} [\bar{\lambda}_i^G \nabla G_i(\bar{x}) + \bar{\lambda}_i^H \nabla H_i(\bar{x})].$$

$$\forall i \in I^g(\bar{x}) : \lambda_i^g \geq 0, \forall i \in \alpha(\bar{x}) : \lambda_i^H = 0, \forall i \in \gamma(\bar{x}) : \lambda_i^G = 0, \forall i \in \beta(\bar{x}) : \lambda_i^G \lambda_i^H = 0.$$

Therefore, $(\bar{x}, \bar{\rho}, \bar{\lambda})$ is feasible for WDMOPSC $(\bar{x})$. By Theorem 13.6.3, from the feasibility condition of MOPSC and WDMOPSC $(\bar{x})$, we have

$$f(\bar{x}) = f(\bar{x}) + \Big[\sum_{i \in I_g} \bar{\lambda}_i^g g_i(\bar{x}) + \sum_{i=1}^{q} \bar{\lambda}_i^h h_i(\bar{x}) + \sum_{i=1}^{l} [\bar{\lambda}_i^G G_i(\bar{x}) + \bar{\lambda}_i^H H_i(\bar{x})]\Big] e. \tag{13.26}$$

Using Theorem 13.6.3 and from Eq. (13.26), we have

$$f(\bar{x}) = f(\bar{x}) + \Big[\sum_{i \in I_g} \bar{\lambda}_i^g g_i(\bar{x}) + \sum_{i=1}^{q} \bar{\lambda}_i^h h_i(\bar{x}) + \sum_{i=1}^{l} [\bar{\lambda}_i^G G_i(\bar{x}) + \bar{\lambda}_i^H H_i(\bar{x})]\Big] e$$

$$\nless f(u) + \Big[\sum_{i \in I_g} \lambda_i^g g_i(u) + \sum_{i=1}^{q} \lambda_i^h h_i(u) + \sum_{i=1}^{l} [\lambda_i^G G_i(u) + \lambda_i^H H_i(u)]\Big] e.$$

Hence, $(\bar{x}, \bar{\rho}, \bar{\lambda})$ is an efficient solution for WDMOPSC $(\bar{x})$ and the respective objective values are equal. □

13.7 Future Research Work

In the future, the concept of weak stationarity, Mordukhovich stationarity, and strong stationarity, i.e., W-stationarity, M-stationarity, and S-stationarity may be extended for nonsmooth multiobjective optimization problems with switching constraint using Mordukhovich limiting subdifferential and Michel–Penot subdifferential (see, [33,

45, 47]). Bao et al. [6] established new weak and strong suboptimality conditions for the general MPEC problems in finite-dimensional and infinite-dimensional spaces that do not assume the existence of optimal solutions. Bao and Mordukhovich [7] established necessary optimality conditions to super efficiency using variational principles for multiobjective optimization problems with equilibrium constraints. It will be interesting to obtain super efficiency, strong suboptimality conditions, and established necessary conditions for nonsmooth multiobjective optimization problems with switching constraints in the future.

Duality is an important subject in the study of mathematical programming problems as the weak duality provides a lower bound to the objective function of the primal problem. Pandey and Mishra [54, 55] formulated a Mond–Weir-type dual problem and established weak duality theorems, strong duality theorems under generalized standard Abadie constraint qualification for nonsmooth optimization problems with equilibrium constraints and semi-infinite mathematical programming problems with equilibrium constraints, respectively. Further, Mishra et al. [43] obtained a several duality theorems for mathematical programs with vanishing constraints. Recently, Su and Dinh [62] introduced the Mangasarian–Fromovitz-type regularity condition and the two Wolfe and Mond–Weir dual models for interval-valued pseudoconvex optimization problem with equilibrium constraints, as well as provided weak and strong duality theorems for the same using the notion of contingent epiderivatives with pseudoconvex functions in real Banach spaces. It will be interesting to study duality results in real Banach spaces for nonsmooth multiobjective optimization problems with switching constraint.

Acknowledgements The authors are grateful to anonymous referees for careful reading of the manuscript, which improved the chapter in its present form. We are grateful to Prof. S. K. Mishra for his most valuable support to design this chapter. The second author is supported by the Science and Engineering Research Board, a statutory body of the Department of Science and Technology (DST), Government of India, through project reference no. EMR/2016/002756.

References

1. Abadie, J.M. (ed.): Nonlinear Programming. Wiley, New York (1967)
2. Achtziger, W., Kanzow, C.: Mathematical programs with vanishing constraints: optimality conditions and constraint qualifications. Math. Program. **114**, 69–99 (2008)
3. Ardakani, J.S., Farahmand Rad, S.H., Kanzi, N., Ardabili, P.R.: Necessary stationary conditions for multiobjective optimization problems with nondifferentiable convex vanishing constraints. Iran. J. Sci. Technol. Trans. A Sci. **43**, 2913–2919 (2019)
4. Ardali, A.A., Movahedian, N., Nobakhtian, S.: Optimality conditions for nonsmooth mathematical programs with equilibrium constraints, using convexificators. Optimization **65**, 67–85 (2016)
5. Bao, T.Q., Gupta, P., Mordukhovich, B.S.: Necessary conditions in multiobjective optimization with equilibrium constraints. J. Optim. Theory Appl. **135**, 179–203 (2007)
6. Bao, T.Q., Gupta, P., Mordukhovich, B.S.: Suboptimality conditions for mathematical programs with equilibrium constraints. Taiwan. J. Math. **12**(9), 2569–2592 (2008)

7. Bao, T.Q., Mordukhovich, B.S.: Necessary conditions for super minimizers in constrained multiobjective optimization. J. Global Optim. **43**, 533–552 (2009)
8. Bazaraa, M.S., Goode, J.J., Nashed, M.Z.: On the cones of tangents with applications to mathematical programming. J. Optim. Theory Appl. **13**, 389–426 (1974)
9. Bigi, G., Pappalardo, M.: Regularity conditions in vector optimization. J. Optim. Theory Appl. **102**(1), 83–96 (1999)
10. Chieu, N.H., Lee, G.M.: Constraint qualifications for mathematical programs with equilibrium constraints and their local preservation property. J. Optim. Theory Appl. **163**, 755–776 (2014)
11. Chieu, N.H., Lee, G.M.: A relaxed constant positive linear dependence constraint qualification for mathematical programs with equilibrium constraints. J. Optim. Theory Appl. **158**, 11–32 (2013)
12. Chinchuluun, A., Pardalos, P.M.: A survey of recent developments in multiobjective optimization. Ann. Oper. Res. **154**, 29–50 (2007)
13. Clason, C., Rund, A., Kunisch, K., Barnard, R.C.: A convex penalty for switching control of partial differential equations. Syst. Control Lett. **89**, 66–73 (2016)
14. Clason, C., Rund, A., Kunisch, K.: Nonconvex penalization of switching control of partial differential equations. Syst. Control Lett. **106**, 1–8 (2017)
15. Flegel, M.L., Kanzow, C.: A Fritz John approach to first order optimality conditions for mathematical programs with equilibrium constraints. Optimization **52**, 277–286 (2003)
16. Flegel, M.L., Kanzow, C.: Abadie-type constraint qualification for mathematical programs with equilibrium constraints. J. Optim. Theory Appl. **124**(3), 595–614 (2005)
17. Flegel, M.L., Kanzow, C.: On M-stationary points for mathematical programs with equilibrium constraints. J. Math. Anal. Appl. **310**(1), 286–302 (2005)
18. Flegel, M.L., Kanzow, C.: On the Guignard constraint qualification for mathematical programs with equilibrium constraints. Optimization **54**(6), 517–534 (2005)
19. Flegel, M.L., Kanzow, C.: A direct proof for M-stationarity under MPEC-GCQ for mathematical programs with equilibrium constraints. In: Dempe, S., Kalashnikov, V. (eds.) Optimization with Multivalued Mappings: Theory, Applications, and Algorithms, pp. 111–122. Springer, Boston (2006)
20. Flegel, M.L., Kanzow, C., Outrata, J.V.: Optimality conditions for disjunctive programs with application to mathematical programs with equilibrium constraints. Set Valued Anal. **15**(2), 139–162 (2007)
21. Gfrerer, H., Ye, J.: New constraint qualifications for mathematical programs with equilibrium constraints via variational analysis. SIAM J. Optim. **27**(2), 842–865 (2017)
22. Gould, F.J., Tolle, J.W.: A necessary and sufficient qualification for constrained optimization. SIAM J. Appl. Math. **20**, 164–172 (1971)
23. Gugat, M.: Optimal switching boundary control of a string to rest infinite time. ZAMMJ. Appl. Math. Mech. **88**(4), 283–305 (2008)
24. Guignard, M.: Generalized Kuhn-Tucker conditions for mathematical programming problems in a Banach space. SIAM J. Contr. **7**, 232–241 (1969)
25. Guo, L., Lin, G.H.: Notes on some constraint qualifications for mathematical programs with equilibrium constraints. J. Optim. Theory Appl. **156**(3), 600–616 (2013)
26. Guo, L., Lin, G.H., Ye, J.J.: Second-order optimality conditions for mathematical programs with equilibrium constraints. J. Optim. Theory Appl. **158**(1), 33–64 (2013)
27. Hante, F.M., Sager, S.: Relaxation methods for mixed-integer optimal control of partial differential equations. Comput. Optim. Appl. **55**(1), 197–225 (2013)
28. Hoheisel, T., Kanzow, C.: First and second order optimality conditions for mathematical programs with vanishing constraints. Appl. Math. **52**(6), 495–514 (2007)
29. Hoheisel, T., Kanzow, C.: Stationary conditions for mathematical programs with vanishing constraints using weak constraint qualifications. J. Math. Anal. Appl. **337**, 292–310 (2008)
30. Hoheisel, T., Kanzow, C.: On the Abadie and Guignard constraint qualifications for mathematical programmes with vanishing constraints. Optimization **58**(4), 431–448 (2009)
31. Hoheisel, T., Kanzow, C., Outrata, J.V.: Exact penalty results for mathematical programs with vanishing constraints. Nonlinear Anal. **72**, 2514–2526 (2010)

32. Izmailov, A.F., Solodov, M.V.: Mathematical programs with vanishing constraints: optimality conditions, sensitivity, and a relaxation method. J. Optim. Theory Appl. **142**, 501–532 (2009)
33. Jeyakumar, V., Luc, D.T.: Nonsmooth calculus, minimality, and monotonicity of convexificators. J. Optim. Theory Appl. **101**, 599–621 (1999)
34. Kanzow, C., Mehlitz, P., Steck, D.: Relaxation schemes for mathematical programs with switching constraints. J. Optim. Meth. Soft. https://doi.org/10.1080/10556788.2019.1663425
35. Liberzon, D.: Switching in Systems and Control. Birkhauser, Boston (2003)
36. Li, X.F.: Constraint qualifications in nonsmooth multiobjective optimization. J. Optim. Theory Appl. **106**(2), 373–398 (2000)
37. Liang, Z.-A., Huang, H.-X., Pardalos, P.M.: Efficiency conditions and duality for a class of multiobjective fractional programming problems. J. Global Optim. **27**, 447–471 (2003)
38. Luo, Z.-Q., Pang, J.-S., Ralph, D.: Mathematical Programs with Equilibrium Constraints. Cambridge University Press, Cambridge, UK (1996)
39. Maeda, T.: Constraint qualifications in multiobjective optimization problems: differentiable case. J. Optim. Theory Appl. **80**(3), 483–500 (1994)
40. Maeda, T.: Second order conditions for efficiency in nonsmooth multiobjective optimization problems. J. Optim. Theory Appl. **122**(3), 521–538 (2004)
41. Mangasarian, O.L.: Nonlinear Programming. Mcgraw Hill, New York (1969)
42. Mehlitz, P.: Stationarity conditions and constraint qualifications for mathematical programs with switching constraints. Math. Program. **181**, 149–186 (2020)
43. Mishra, S.K., Singh, V., Laha, V.: On duality for mathematical programs with vanishing constraints. Ann. Oper. Res. **243**(1–2), 249–272 (2016)
44. Mishra, S.K., Singh, V., Laha, V., Mohapatra, R.N.: On constraint qualifications for multiobjective optimization problems with vanishing constraints. In: Xu, H., Wang, S., Wu, S.Y. (eds.) Optimization Methods, Theory and Applications. Springer, Berlin, Heidelberg (2015)
45. Mordukhovich, B.S.: Variations Analysis and Generalized Differentiation, I: Basic Theory. Grundlehren Series (Fundamental Principles of Mathematical Sciences), vol. 330. Springer, Berlin (2006)
46. Mordukhovich, B.S.: Equilibrium problems with equilibrium constraints via multiobjective optimization. Optim. Methods Soft. **19**, 479–492 (2004)
47. Mordukhovich, B.S.: Variational Analysis and Generalized Differentiation, II: Applications. Grundlehren Series (Fundamental Principles of Mathematical Sciences), vol. 331. Springer, Berlin (2006)
48. Mordukhovich, B.S.: Multiobjective optimization problems with equilibrium constraints. Math. Program. Ser. B **117**, 331–354 (2009)
49. Movahedian, N., Nobakhtian, S.: Necessary and sufficient conditions for nonsmooth mathematical programs with equilibrium constraints. Nonlinear Anal. **72**, 2694–2705 (2010)
50. Outrata, J.V.: Optimality conditions for a class of mathematical programs with equilibrium constraints. Math. Oper. Res. **24**, 627–644 (1999)
51. Outrata, J., Kocvara, M., Zowe, J.: Nonsmooth Approach to Optimization Problems with Equilibrium Constraints. Kluwer Academic Publishers, Dordrecht (1998)
52. Pareto, V.: Course d'E' conomie Politique. Rouge, Lausanne (1896)
53. Pandey, Y., Mishra, S.K.: On strong KKT type sufficient optimality conditions for nonsmooth multiobjective semi-infinite mathematical programming problem with equilibrium constraints. Oper. Res. Lett. **44**, 148–151 (2016)
54. Pandey, Y., Mishra, S.K.: Duality for nonsmooth optimization problems with equilibrium constraints, using convexificators. J. Optim. Theory Appl. **17**, 694–707 (2016)
55. Pandey, Y., Mishra, S.K.: Optimality conditions and duality for semi-infinite mathematical programming problems with equilibrium constraints, using convexificators. Ann. Oper. Res. **269**, 549–564 (2018)
56. Peterson, D.W.: A review of constraint qualifications in finite-dimensional spaces. SIAM Rev. **15**, 639–654 (1973)
57. Preda, V., Chitescu, I.: On constraint qualifications in multiobjective optimization problems: semidifferentiable case. J. Optim. Theory Appl. **100**(2), 417–433 (1999)

58. Rockafellar, R.T.: Convex Analysis. Princeton University Press, Princeton, New Jersey (1970)
59. Sager, S.: Reformulations and algorithms for the optimization of switching decisions in nonlinear optimal control. J. Process Control **19**(8), 1238–1247 (2009)
60. Scheel, S., Scholtes, S.: Mathematical programs with complementarity constraints: stationarity, optimality, and sensitivity. Math. Oper. Res. **25**(1), 1–22 (2000)
61. Seidman, T.I.: Optimal control of a diffusion/reaction/switching system. Evolut. Equ. Control Theory **2**(4), 723–731 (2013)
62. Van Su, T., Dinh, D.H.: Duality results for interval-valued pseudoconvex optimization problem with equilibrium constraints with applications. Comp. Appl. Math. **39**, 127 (2020)
63. Wang, L., Yan, Q.: Time optimal controls of semilinear heat equation with switching control. J. Optim. Theory Appl. **165**(1), 263–278 (2015)
64. Ye, J.J.: Necessary and sufficient optimality conditions for mathematical programs with equilibrium constraints. J. Math. Anal. Appl. **307**(1), 350–369 (2005)
65. Ye, J.J., Zhu, D.L., Zhu, Q.J.: Exact penalization and necessary optimality conditions for generalized bilevel programming problems. SIAM J. Optim. **2**, 481–507 (1997)
66. Ye, J.J.: Constraint qualifications and necessary optimality conditions for optimization problems with variational inequality constraints. SIAM J. Optim. **10**, 943–962 (2000)
67. Zuazua, E.: Switching control. J. Eur. Math. Soc. **13**(1), 85–117 (2011)

Chapter 14
Optimization of Physico-Chemical Parameters for the Production of Endoxylanase Using Combined Response Surface Method and Genetic Algorithm

Vishal Kapoor and Devaki Nandan

Abstract Endoxylanase production by *Trichoderma reesei* Rut C-30 was optimized under solid-state fermentation using a mixture of waste paper and wheat bran. Most effective variables for the endoxylanase production in screening experiments were incubation day, substrate ratio, solid:liquid ratio, and pH of the medium. In this chapter, a quadratic model was developed through response surface method followed by genetic algorithm to optimize the operational conditions for maximum endoxylanase production. The predicted optimal parameter for hybrid RSM-GA was tested and the final endoxylanase activity obtained was assessed very close to the predicted value. Optimization leads to the enhancement of endoxylanase activity by ~2.5 fold.

Keywords Endoxylanase · Response surface method · Genetic algorithm · Optimization · *Trichoderma ressei* · Rut C-30

14.1 Introduction

Response Surface Methodology (RSM), a combination of mathematical and statistical techniques, is useful for analyzing the effects of several independent variables on the system response without the need for a predetermined relationship between the objective function and the variables [10, 11, 25]. Sharma and Kumar [29] applied response surface method on plasma arc cutting to minimize dross formation rate. A significant reduction was found in dross formation by the application of optimum solution obtained. Danmaliki et al. [8] focused on the optimization of the experi-

V. Kapoor (✉) · D. Nandan
Indian Institute of Technology Kanpur, Kanpur 208016, India
e-mail: vishal.262570@gmail.com

D. Nandan
e-mail: devakinandan1804@gmail.com

V. Laha et al. (eds.), *Optimization, Variational Analysis and Applications*,
Springer Proceedings in Mathematics & Statistics 355,
https://doi.org/10.1007/978-981-16-1819-2_14

mental factors affecting adsorptive desulfurization process in a continuous flow system using response surface methodology. A face-centered central composite design (CCD) was used to statistically visualize the complex interactions of concentration, column length, dosage, and flow rate on the adsorption of dibenzothiophene.

Yolmeh and Jafari [37] presented the state-of-the-art applications of RSM in the optimization of different food processes such as extraction, drying, blanching, enzymatic hydrolysis and clarification, production of microbial metabolites, and formulation. He concluded that the appropriate selection of RSM design, independent variables (screening), and levels of the factors significantly influences the successful application of RSM. Mourabet et al. [24] employed Response surface methodology for the removal of fluoride on Brushite and the process parameters were optimized. Four important process parameters including initial fluoride concentration, pH, temperature, and B dose were optimized to obtain the best response of fluoride removal using the statistical Box Behnken design. There had been some studies regarding enzyme production using *Trichoderma* strains, a complete optimization of operational conditions for the production of the enzyme has received little attention in the literature [13, 28]. The effects of moisture percentage, temperature, pH, inoculum, and nitrogen source on the production of endoxylanase from *T. longibrachiatum* were optimized through RSM [3].

The effect of incubation day, substrate ratio, solid:liquid ratio, and pH on the production of endoglucanase by *Trichoderma reesei* Rut C-30 using agro-residue were also optimized through RSM [19]. Optimization of such processes with RSM provided great benefits in the production of endoxylanase and endoglucanase.

The traditional methods of optimization and search do not perform well over a broad spectrum of problem domains. Traditional techniques are not efficient when practical search space is too large [16]. Genetic algorithms (GA) are computer search and optimization algorithms based on mechanics of natural genetic and natural selection and widely used in a wide range of problems due to their usability, ease of operation, minimum requirements, and global perspective [18, 30]. Specific work has shown that GA is a valuable technique for achieving optimal solutions to solve the problems [1, 5, 6].

There is one of the drawbacks of using a GA for optimization - since there is no guarantee of optimality, there is always the chance that there is a better chromosome lurking somewhere in the search space. Typically, the GA is coupled with a local search mechanism to find the optimal chromosome in a region. So, if a hybrid algorithm is used, the problem reduces which ensures that the GA could be run as many times as is needed to pick out all the good regions [21]. Genetic algorithms are good optimization methods and have the following advantages: (1) they do not need the objective function to be continuous, convex, or unimodal, and (2) they are very efficient due to their ability to perform parallel searches in the feasible space and the testing of small blocks of good solutions under multiple scenarios. These two advantages make them very suitable for optimization with RSM, particularly in cases of discontinuity or where spaces are very constrained or irregular [2]. In the present study, attempts have been made by employing RSM coupled GA approach to quantitatively evaluate the individual and combined interaction effect(s) of physico-chemical parameters on production of endoxylanase by *Trichoderma reesei* Rut C-30 under solid-state fermentation using a novel mixture of waste paper and wheat bran.

14.2 Materials and Methods

14.2.1 Microorganism and Material

Trichoderma reesei Rut C-30 was obtained from Regional Research Laboratory (RRL), Trivandrum. For stock culture maintenance, the strain was grown in 2% (w/v) malt extract-agar slants at around 28 °C and sub-cultured once in 2 weeks. Waste paper (WP) and wheat bran (WB), locally available from the market, were used as a substrate for enzyme production. All chemicals were procured from Merck, India, and Sigma Co., USA.

14.2.2 Inoculum Development for Endoxylanase Production

To obtain the spores, 6-day old culture slants of *T. reesei* Rut C-30 was used by adding 10 ml of sterile distilled water. The spores were scraped off with inoculating loop, aseptically. Spores disperse evenly by agitation in a vortex-cyclomixer for 5 min. Haemocytometer was used to determine the spore count. Inoculum containing $3.6x10^6$ spores/ml was used for subsequent fermentation.

14.2.3 Production Medium

To culture the organisms primary culture medium was used [9, 19]. The fermentation medium was sterilized at 121 °C or 15 psi for 20 min.

14.2.4 Substrate and Solid-State Fermentation

Solid-state fermentation for enzyme production was done using well mixed autoclaved substrate (WP and WB) (5 g) with 10 ml of modified Mandels medium [9] in the 250-ml Erlenmeyer flasks. 0.2 ml of spore suspension was used over sterilized substrate and mixed thoroughly. Various process parameters was studied in preliminary experiments to access their impact on enzyme production and most promising ones (incubation day (X_1) (4, 5, and 6), substrate ratio (X_2) (WP : WB) (1:4, 1:5, and 1:6), solid: liquid ratio (X_3) (1:0.5, 1:1, and 1:1.5) and medium pH (X_4) (4.5, 5.0, and 5.5)) were selected to asses further during statistical analysis (Table 14.1).

Table 14.1 Independent variables and their levels in the experimental design

Independent variables	Symbols	Code levels				
		−2	−1	0	1	2
Incubation day	(X_1)	3	4	5	6	7
Substrate ratio	(X_2)	3:1	4:1	5:1	6:1	7:1
Solid : Liquid ratio	(X_3)	1:0	1:0.5	1:1	1:1.5	1:2
pH	(X_4)	4.0	4.5	5.0	5.5	6.0

14.2.5 Enzyme Extraction and Assay

Ten milliliters of 5% glycerol–water solution were used for leaching out the extracellular enzyme produced by fermented biomass of *T. reesei* Rut C-30 in 2 h at room temperature. The biomass was filtered with cheese cloth under pressure and filtrate was centrifuged at 10,000 rpm at 4 °C for 10 min. The supernatant was used to assess the endoxylanase activity using 1% xylan. The dinitrosalicylic acid reagent [23] was used to determine the concentration of reducing sugar and the enzyme activity was expressed in International Units (IU). IU was calculated as number of μ moles of product (xylose) equivalents released per milliliter per minute.

14.2.6 Experimental Design

In order to ascribe the effect of factors on response surface in the region of investigation, a central composite design (CCD) with four factors at five levels was performed (Table 14.1). In order to obtain ratio for factor level of substrate, the value of one substrate (WP) and in solid: liquid ratio, solid substrate were made constant and other variable factors (WB and liquid medium) were entered in Design expert software. Enzyme activity (IU/g) of endoxylanase (Y) was taken as a response from the 30 sets analyzed (Table 14.2).

14.2.7 Statistical Analysis

Response surface methodology may be summarized as a collection of statistical tools and techniques for constructing and exploring an approximate functional relationship between a response variable and a set of design variables [35]. The response variable was fitted by a second-order model in order to correlate the response variable to the independent variables. The general form of the second-degree polynomial equation is:

$$Y_i = \beta_o + \sum \beta_i x_i \div \sum \beta_{ii} x_i^2 \div \sum \beta_{ij} X_i X_j \qquad (14.1)$$

Table 14.2 Experimental design and results of the central composite design

Run	Variables				Response (Y)	
	X_1	X_2	X_3	X_4	Endoxylanase activity (IU/g)	
	Incubation days	Substrate ratio	Solid:liquid ratio	pH	Actual value	Predicted value
1	5	1:5	1:1	5.0	161.16	169.36
2	5	1:5	1:1	5.0	162.10	169.36
3	4	1:6	1:1.5	5.5	92.65	91.36
4	4	1:6	1:0.5	5.5	61.51	59.74
5	4	1:6	1:1.5	4.5	229.16	184.15
6	4	1:6	1:0.5	4.5	170.51	152.22
7	6	1:6	1:0.5	5.5	71.10	97.51
8	6	1:6	1:0.5	4.5	94.31	91.60
9	6	1:4	1:1.5	5.5	146.89	179.90
10	6	1:4	1:0.5	5.5	121.26	114.14
11	6	1:4	1:1.5	4.5	163.58	143.22
12	6	1:6	1:1.5	4.5	80.35	95.59
13	6	1:6	1:1.5	5.5	131.28	101.20
14	4	1:4	1:1.5	4.5	215.69	208.0
15	5	1:5	1:1	5.0	163.32	169.36
16	6	1:4	1:0.5	4.5	91.13	107.15
17	5	1:5	1:1	5.0	161.23	169.36
18	4	1:4	1:1.5	5.5	161.72	142.29
19	4	1:4	1:0.5	4.5	132.05	140.0
20	4	1:4	1:0.5	5.5	79.11	78.59
21	5	1:5	1:2	5.0	38.46	74.56
22	5	1:3	1:1	5.0	209.63	191.99
23	7	1:5	1:1	5.0	90.76	56.85
24	5	1:5	1:0	5.0	35.56	6.87
25	5	1:5	1:1	4.0	152.20	177.92
26	5	1:7	1:1	5.0	100.46	125.51
27	5	1:5	1:1	5.0	163.32	148.03
28	5	1:5	1:1	5.0	162.39	148.03
29	3	1:5	1:1	5.0	38.54	79.86
30	5	1:5	1:1	6.0	140.42	122.11

Here, Y_i is the predicted response; $X_i X_j$ are input variables which influence the response variable Y; β_0 is the offset term; β_i is the ith linear coefficient; β_{ii} is the ith quadratic coefficient and β_{ij} is the ijth interaction coefficient.

Table 14.3 List of used GA parameters

S. No.	Parameter	Value
1.	Population size	100
2.	Length of chromosome	40
3.	Selection operator	Roulette method
4.	Crossover operator	Single point operator
5.	Crossover probability	0.9
6.	Mutation probability	0.01

The second-order polynomial coefficients were calculated using the statistical software Design-Expert 10.0 (Stat-Ease, Inc., Minneapolis, USA). The data obtained from RSM on endoxylanase production were subjected to the analysis of variance (ANOVA). Statistical significance of the model equation was determined by Fisher's test value, and the proportion of variance explained by the model was given by the multiple coefficients of determination, R squared (R^2) value. It also includes Student's t-value for the estimated coefficients and the associated probabilities p(t). For each variable, the quadratic models were represented as contour plots (2D).

14.2.8 Genetic Algorithm

The general optimization procedure using a genetic algorithm is shown in Fig. 14.1. For this reason, the software formulation was made using an objective function from RSM and the various functions of the GA toolbox on the MATLAB platform so that the GA can generate a population set that could reproduce and cross among itself in order to create the best possible solution for a given number of generations.

The program is executed after the program formulation has been completed in order to obtain optimized process parameters for the desired response endoxylanase production. The parameters used for GA are shown in Table 14.3. The fitness parameter is enzyme activity of endoxylanase obtained by RSM.

The practical constraints imposed during the Friction Stir Welding (FSW) operations are stated as follows:

Parameter bounds:

- Incubation day

$$IDL \leq ID \leq IDU \tag{14.2}$$

where IDL and IDU are the lower and upper bounds of incubation day, respectively.

- Substrate ratio

$$SRL \leq SR \leq SRU \tag{14.3}$$

where SRL and SRU are the lower and upper bounds of substrate ratio, respectively.

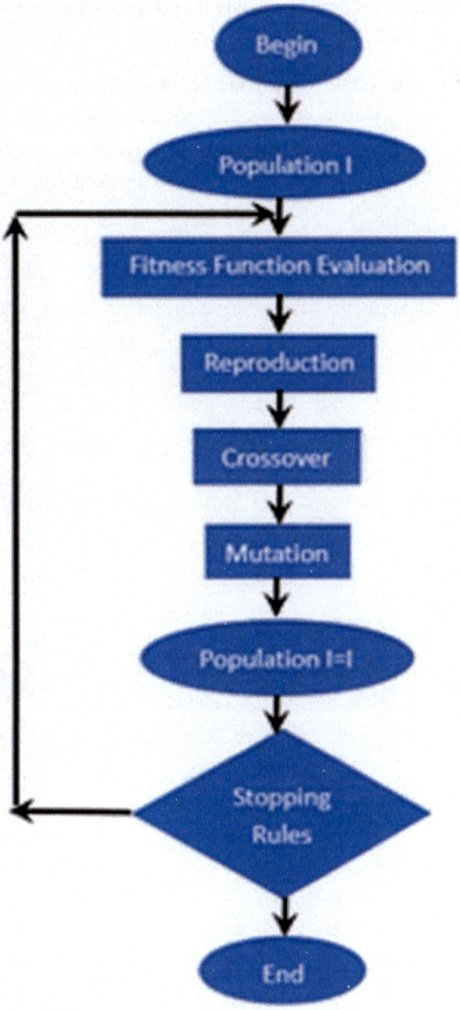

Fig. 14.1 General optimization procedures for genetic algorithm

- Solid liquid ratio

$$SLRL \leq SLR \leq SLRU \tag{14.4}$$

  where SLRL and SLRU are the lower and upper bounds of solid: liquid ratio, respectively.

- pH

$$pHL \leq pH \leq pHU \tag{14.5}$$

  where pHL and pHU are the lower and upper bounds of pH, respectively.

14.3 Results and Discussion

On the basis of initial results of endoxylanase production, the boundary limits of each variable were determined (Table 14.1). Data from the 30 sets were analyzed to yield regression equation and regression coefficient (R^2). The response Y was fitted with second-order polynomial equation (14.6).

$$Y(\text{endoxylanase activity}) = 158.70 - 5.75X_1 - 16.62X_2 + 16.92X_3 - 13.95X_4 - 19.92X_1^2 + 2.68X_2^2 + 26.83X_3^2 + 0.50X_4^2 - 6.92X_1X_2 - 6.98X_1X_3 + 24.60X_1X_4 - 8.02X_2X_3 - 7.77X_2X_4 - 0.076X_3X_4 \quad (14.6)$$

Perturbation plot (Fig. 14.2) shows the comparative effects of all the physico-chemical components on endoxylanase activity. In Fig. 14.2, a steep curvature in incubation days and solid: liquid ratio curve shows that the response of endoxylanase activity was very sensitive to these factors. The relatively flat lines of substrate ratio and pH shows insensitivity of the responses to change in these two components of the medium.

This regression equation is used as the fitness function for GA. The parameters used for GA are shown in Table 14.3. The result obtained by RSM is considered as an initial solution for performing GA. The fitness parameter is enzyme production

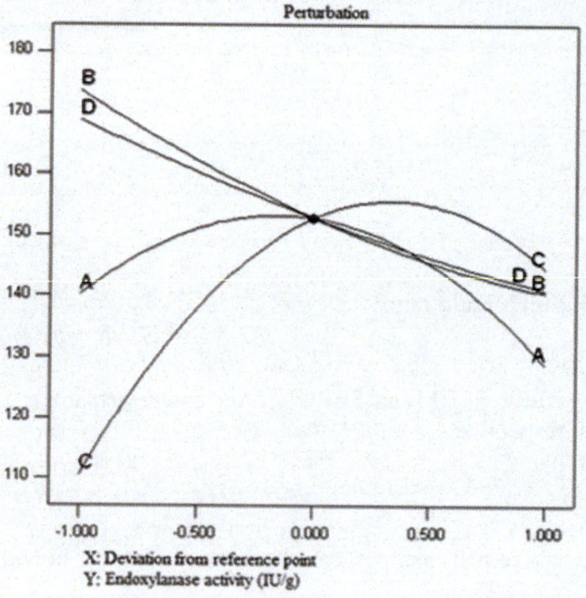

Fig. 14.2 Perturbation plots for endoxylanase activity by *Trichoderma reesei* Rut C-30; (A) Incubation day, (B) Substrate ratio, (C) Solid:Liquid ratio, and (D) pH

Table 14.4 Optimum results and confirmation test

Method applied	Optimum parameter				Response parameter	Actual Value	% Error
	ID	SR	SLR	pH			
RSM	4	6	1.5	4.5	184.15	202.37	9.003
RSM+ GA	4	6	1.5	4.5	207.42		2.49

Table 14.5 ANOVA analysis for responses (Y) endoxylanase activity (IU/g)

Source	Sum of squares	DF	Mean square	F-value	Prob >F
For Y					
Model	62089.05	14	4434.93	4.24	0.0054
Residual	14650.70	14	1046.48		
Lack of fit	14647.22	10	1464.7	1686.34	<0.0016
Pure error	3.47	4	0.87		
$R^2 = 0.8091$					
Adeq. precision = 8.344					

for both RSM and Hybrid RSM-GA methods. Results obtained by both methods are shown in Table 14.4.

The statistical significance of the model equation was evaluated by the F-test for analysis of variance (ANOVA). The ANOVA statistics for the response Y is shown in Table 14.5. The results of the quadratic model indicated that this could be used to navigate the design space. Table 14.5 evinces that the prob >F-values for the endoxylanase production is lower than 0.05 indicating that the quadratic model was significant. The coefficient of determination (R^2) that was found to be close to 1 (0.809 for Y) also advocated a high correlation between observed and predicted values. The "lack of fit test" compares the residual error to the "Pure Error" from replicated experimental design points. The p-value, lesser than 0.05, for the response indicates that lack of fit for the model was significant. Adequate precision measures the signal to noise ratio and a ratio greater than 4 is desirable. The adequate precision was 8.344. The high values of adequate precision demonstrated that the model is significant for the process (Table 14.5).

Usually, it is essential to ensure that the selected model provides an adequate approximation to the real system. By applying the diagnostic plots such as the predicted versus actual value plot, the model adequacy can be judged. The correlation coefficient between actual and predicted values for Y was 0.809. This R^2 value illustrates good agreement between the calculated and observed results within the range of experiment.

Table 14.6 Endoxylanase production using agro-residue from fungal strains

S. No.	Xylanase activity	Microorganism	Agro substrate	Optimum physical conditions	References
1.	92 IU/ml	*Trichoderma reesei* Rut C-30	Rice straw	pH 4.8; incubation temperature (29 °C); incubation period (5 days)	[7]
2.	4.62 U/ml	*Trichoderma viride*	Maize straw	Incubation period (14–17 days); pH 3.5–4.0; incubation temperature (25 °C); substrate concentration (5%)	[15]
3.	43.8 U/ml	*Penicillium* sp. SS1	Wheat bran, rice bran and sawdust	Incubation period (4 days); pH 9.0; incubation temperature (50 °C)	[4]
4.	1906.5 U/ml	*Aspergillus niger*	Palm leaf	Inoculum concentration (1 ml); incubation temperature (28 °C); moisture level (70%)	[26]
5.	14.44 U/ml	*Scytalidium thermophilum*	Soy flour	–	[17]
6.	42.5 U/g	*Aspergillus niger*	Barley bran	Incubation temperature (35 °C); pH 5.5; moisture level (75%)	[33]
7.	51.43 U/ml	*Penicillium glabrum*	Brewer's spent grain	pH 3.0; incubation temperature (60 °C)	[20]
8.	73.09 U/ml	*Trichoderma viride*	Pineapple peel	pH 7.5; incubation temperature (28 °C); substrate concentration (2%)	[12]
9.	73.0 U/ml	*Aspergillus fumigatus* RSP-8	Sorghum straw	pH 5.0–9.0; inoculum concentration (0.5–2.0%); incubation temperature (26–34 °C); substrate concentration (0.5–3%)	[27]
10.	202.37 U/ml	*Trichoderma reesei* Rut C-30	Wheat bran, waste paper	Incubation period (4 days); pH 4.5; incubation temperature (50 °C); inoculum concentration (0.2 ml)	Current study

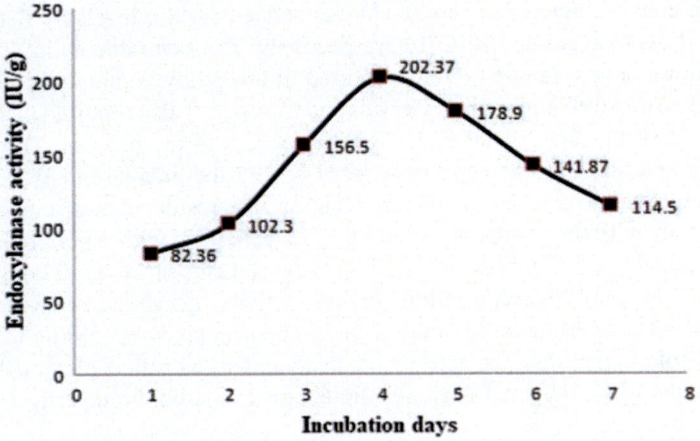

Fig. 14.3 Endoxylanase activity profile under the optimal conditions suggested by the model. endoxylanase activity: substrate ratio = 1:6, solid: liquid ratio = 1:1.5, and pH = 4.5

Fig. 14.4 Contour plot for the effect of solid: liquid ratio x incubation day on endoxylanase activity. Not plotted variables are fixed at zero level in the graph

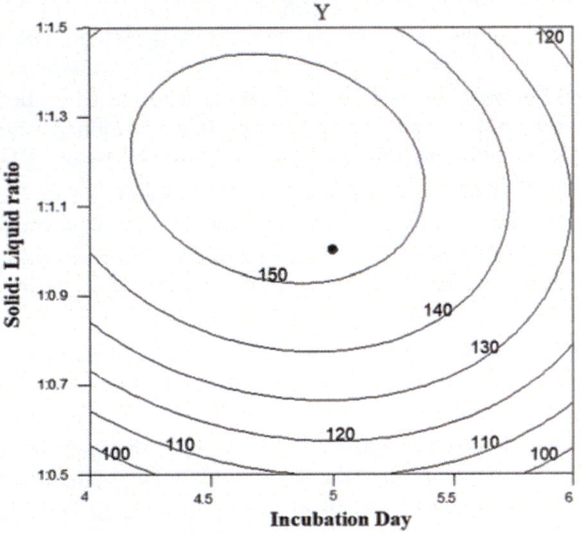

Experimental run 5 (Table 14.2) shows the highest endoxylanase activity (229.16 IU/g). The activity was obtained at incubation day (4), substrate ratio (1 WP: 6 WB), solid: liquid ratio (1:1.5), and pH (4.5). The time of the highest endoxylanase activity (4 days) corresponds to the mid-stationary growth phase which is in agreement with findings by Liu et al. [22], who reported optimum cultivation time for endoxylanase activity of *Trichoderma viride* in SSF between 4 and 5 days. The highest endoxylanase activity was found from the cultivation of mixed substrates of WP: WB in (1:6) ratio. This result was similar to the Thygesen et al. [34] and Singh et al. [31, 32] in which cultivation on agro-residual substrate was the favorable operating parameter for the enhancement of the endoxylanase and α-amylase levels by *T. reesei* Rut C-30 and *Streptomyces* sp. MSC702, respectively. The favorable solid: liquid ratio for maximum endoxylanase activity reported in this study is mainly at mid or low levels (1:1.5) of solid: liquid ratio. Gervais and Molin [14] also reported that moisture content in SSF plays a crucial factor for the success of the process. The optimum enzyme production was approximately at pH 4.5. Further increase in pH had no significant effect on the production of endoxylanase. The results obtained in low pH are in accordance with the results obtained by Colina et al. [7] who reported the highest xylanase activity by *T. reesei* Rut C-30 growing on rice straw at pH 4.8 and also by Xiong et al. [36] who reported the highest activity of xylanase-I at 4.0 pH. The highest activity experimentally obtained in the present study according to the CCD was ~2.5-fold higher than the activity obtained under conditions previously used in our laboratory (88.2 IU/g at 4 days, substrate ratio 1:5, solid: liquid ratio 1:1 and pH 4.8).

A validation of the model is given in Fig. 14.3, which shows the cultivation of *T. reesei* Rut C-30 for the endoxylanase production under optimal conditions of substrate ratio (1:6), solid: liquid ratio (1:1.5), and pH (4.5). The maximum endoxylanase activity obtained was 202.37 IU/g in 4 days. In this case, the model predicted endoxylanase activity of 184.15 IU/g in 4 days. The experimental value was found to be 9.8% higher than the predicted value, confirming the closeness of the model to the experimental result. To study the interaction between all the four components three dimensional curves were plotted. Combined effect of incubation day and solid: liquid ratio on endoxylanase production is shown in Fig. 14.4 as a contour plot. The endoxylanase activity tends to be the highest within the range of incubation days 4–5 and solid: liquid ratio 1.1–1.5. Incubation day played a critical role in fungal growth and it also showed a very strong interaction with solid: liquid ratio. Pairing the other factors produced a flat response surface showing that these factors had no significant effect on endoxylanase activity.

In the present study, we found that the high production of endoxylanase can be achieved by *T. reesei* Rut C-30, using a mixture of agricultural residues as substrate. Although many microorganisms have been reported to produce endoxylanase under solid-state fermentation, *T. reesei* Rut C-30 was found to have such activity appreciably in the present study. The optimized xylanase activity in the present study was comparatively higher from most of the earlier reported fungal species using low-value crude agriculture-based raw materials as a substrate and could be considered for vast biotechnological applications (Table 14.6). While differences observed in the production through various fungal strains are basically owing to the critical fermentation parameters and nature of substrates which determined xylanase catalytic activity.

14.4 Conclusion

In this work, evidences have been found to determine the optimal growth conditions for the production of endoxylanase by *T. reesei* Rut C-30 in SSF using response surface method and Hybrid RSM-GA method. The maximal activity of the enzyme produced was 202.37 IU/g for endoxylanase, when optimized conditions were employed. The enzyme activity predicted by the model at optimal conditions agreed fittingly with experimental data, thus confirming the model validity. The results obtained by hybrid RSM-GA are more accurate on the confirmation test. Hybrid RSM-GA method could be a powerful optimization for estimating optimal response of endoxylanase production by *T. reesei* Rut C-30. This study is expected to facilitate further work on the purification of the endoxylanase produced by *T. reesei* Rut C-30.

Acknowledgements We would like to thank the anonymous referees for their suggestions, which improved the original version of the chapter.

References

1. Ahmad, I., Jeenanunta, C., Chanvarasuth, P., Komolavanij, S.: Prediction of physical quality parameters of frozen shrimp (*Litopenaeus vannamei*): an artificial neural networks and genetic algorithm approach. Food Bioproc. Tech. **7**(5), 1433–1444 (2014)
2. Alvarez, M.J., Ilzarbe, L., Viles, E., Tanco, M.: The use of genetic algorithms in response surface methodology. Qual. Technol. Quant. Manag. **6**(3), 295–307 (2009)
3. Azin, M., Mravej, R., Zareh, D.: Production of xylanase by *Trichoderma longibrachiatum* on a mixture of wheat bran and wheat straw: optimization of culture condition by Taguchi method. Enzyme Microb. Technol. **40**, 801–805 (2007)
4. Bajaj, B.K., Sharma, M., Sharma, S.: Alkalistable endo-β-1,4-xylanase production from a newly isolated alkalitolerant *Penicillium* sp. SS1 using agro-residues. 3 Biotech. 1, 83–90 (2011)
5. Chatterjee, S., Bandopadhyay, S.: Reliability estimation using a genetic algorithm-based artificial neural network: an application to a load–haul–dump machine. Expert Syst. Appl. **39**(12), 10943–10951 (2012)
6. Chen, G.Y., Fu, K.Y., Liang, Z.W., Sema, T., Li, C., Tontiwachwuthikul, P., Idem, R.: The genetic algorithm based back propagation neural network for MMP prediction in CO_2-EOR process. Fuel **126**, 202–212 (2014)
7. Colina, A., Sulbarán-De-Ferrer, B., Aiello, C., Ferrer, A.: Xylanase production by *Trichoderma reesei* Rut C-30 on rice straw. Appl. Biochem. Biotechnol. **108**, 715–724 (2003)
8. Danmaliki, G.I., Saleh, T.A., Shamsuddeen, A.A.: Response surface methodology optimization of adsorptive desulfurization on nickel/activated carbon. Chem. Eng. J. **313**, 993–1003 (2017)
9. Das, M., Banerjee, R., Bal, S.: Multivariable parameter optimization for endoglucanase production by *Trichoderma reesei* Rut C-30 from *Ocimum gratissimum* seed. Braz. Arch. Biol. Technol. **51**, 35–41 (2008)
10. Draper, N.R., John, J.A.: Response-surface design for quantitative and qualitative variables. Technometrics **30**(4), 423–8 (1988)
11. Draper, N.R., Lin, D.K.J.: Small response-surface designs. Technometrics **32**(2), 187–194 (1990)
12. Fortkamp, D., Knob, A.: High xylanase production by *Trichoderma viride* using pineapple peel as substrate and its application in pulp biobleaching. Afr. J. Biotech. **13**(22), 2248–2259 (2014)

13. Gerber, P.J., Heitmann, J.A., Joyce, T.W.: Purification and characterization of xylanases form *Trichoderma*. Bioresour. Technol. **61**, 127–40 (1997)
14. Gervais, P., Molin, P.: The role of water in solid-state fermentation. J. Biochem. Eng. **13**, 85–101 (2003)
15. Goyal, M., Kalra, K.L., Sarren, V.K., Soni, G.: Xylanase production with xylan rich lignocellulosic wastes by a local soil isolate of *Trichoderma viride*. Braz. J. Microbiol. **39**(3), 535–541 (2008)
16. He, J., Sato, M. (eds.): Advances in Computing Science-ASIAN 2000: 6th Asian Computing Science Conference Penang, Malaysia, November 25–27, 2000 Proceedings (No. 1961). Springer Science & Business Media (2000)
17. Joshi, C., Khare, S.K.: Induction of xylanase in thermophilic fungi *Scytalidium thermophilum* and *Sporotrichum thermophile*. Braz. Arch. Biol. Biotechnol. **55**(1), 21–27 (2012)
18. Kalyanmoy, D.: Optimizations for Engineering Design- Algorithm and Examples, pp. 290–333. Prentice Hall of India, New Delhi (1996)
19. Kapoor, V., Singh, R., Banerjee, R., Kumar, V.: Statistical optimization of production parameters for endoglucanase by *Trichoderma reesei* Rut C-30 employing agro-residue. Dyn. Biochem. Process Biotech. Mol. Biol. **5**, 35–40 (2011)
20. Knob, A., Beitel, S.M., Fortkamp, D., Terrasan, C.R., de Almeida, A.F.: Production, purification, and characterization of a major *Penicillium glabrum* xylanase using Brewer's spent grain as substrate. Biomed Res. Int. **1–8** (2013)
21. Lakshmanan, V.: Using a genetic algorithm to tune a bounded weak echo region detection algorithm. J. Appl. Meteorol. **39**, 222–230 (1999)
22. Liu, J., Youn, X., Zeng, G., Shi, J., Chen, S.: Effect of biosurfactant on cellulase and xylanase production by *Trichoderma viride* in solid substrate fermentation. Process Biochem. **41**, 2347–2351 (2006)
23. Miller, G.L.: Use of dinitrosalicylic acid reagent for determining reducing sugars. Anal. Chem. **31**, 426–428 (1959)
24. Mourabet, M., El Rhilassi, A., El Boujaady, H., Bennani-Ziatni, M., Taitai, A.: Use of response surface methodology for optimization of fluoride adsorption in an aqueous solution by Brushite. Arab. J. Chem. **10**, S3292–S3302 (2017)
25. Myers, R.H., Montgomery, D.C.: Response Surface Methodology: Process and Product Optimization Using Designed Experiments, p. 43. Wiley, New York (2002)
26. Norazlina, I., Pushpahvalli, B., Ku Halim, K..H.., Norakma, M..N.: Comparable study of xylanase production from *Aspergillus niger* via solid state culture. J. Chem. Chemical Eng. **6**(12), 1106–1113 (2012)
27. Ravichandra, K., Yaswanth, V.V.N., Nikhila, B., Ahmad, J., Srinivasa Rao, P., Uma, A., Ravindrababu, V., Prakasham, R.S.: Xylanase production by isolated fungal strain, *Aspergillus fumigatus* RSP-8 (MTCC 12039): Impact of agro-industrial material as substrate. Sugar Tech. **18**(1), 29–38 (2016)
28. Reczey, K., Szengyel, Zs., Eklund, R., Zacchi, G.: Cellulase production by *Trichoderma reesei*. Bioresour. Technol. **57**, 25–30 (1996)
29. Sharma, D.N., Kumar, J.R.: Optimization of dross formation rate in plasma arc cutting process by response surface method. Materials Today: Proceedings (2020)
30. Sharma, D.N., Tewari, M.: Optimization of Friction Stir Welding parameters using combined Taguchi L9 and Genetic Algorithm. International Conference an artificial intelligence and application (IEEE-COER-ICAIA-2019) (2019)
31. Singh, R., Kapoor, V., Kumar, V.: Production of thermostable, Ca^{+2}-independent, maltose producing α-amylase by *Streptomyces* sp. MSC702 (MTCC 10772) in submerged fermentation using agro-residues as sole carbon source. Ann. Microbiol. **62**, 1003–1012 (2012)
32. Singh, R., Kapoor, V., Kumar, V.: Influence of carbon and nitrogen sources on the α-amylase production by a newly isolated thermotolerant *Streptomyces* sp. MSC702 (MTCC 10772). Asian J. Biotechnol. **3**(6), 540–553 (2011)
33. Soliman, H.M., Sherief, A.A., Tanash, A.B.E.: Production of Xylanase by *Aspergillus niger* and *Trichoderma viride* using some agriculture residues. Int. J. Agric. Res. **7**, 46–57 (2012)

34. Thygesen, A., Thomsen, A.B., Schmidt, A.S., Jorgensen, H., Ahring, B.K., Olsson, L.: Production of cellulose and hemicelluloses degrading enzymes by filamentous fungi cultivated on wet oxidized wheat straw. Enzyme Microb. Technol. **32**, 606–615 (2003)
35. Venter, G.: Non-dimensional response surfaces for structural optimization with uncertainty. Ph.D. thesis, University of Florida, USA (1998)
36. Xiong, H., Weymarn, N.V., Leisola, M., Turunen, O.: Influence of pH on the production of xylanases by *Trichoderma reesei* Rut C-30. Process Biochem. **39**, 731–736 (2004)
37. Yolmeh, M., Jafari, S.M.: Applications of response surface methodology in the food industry processes. Food Bioproc. Tech. **10**(3), 413–433 (2017)

Chapter 15
Optimal Duration of Integrated Segment Specific and Mass Promotion Activities for Durable Technology Products: A Differential Evolution Approach

A. Kaul, Anshu Gupta, S. Aggarwal, P. C. Jha, and R. Ramanathan

Abstract Promotion is carried out by firms for effective communication with potential customers so that response is achieved at different levels namely, awareness, interest, evaluation, trial, adoption, and market growth for the products. Firms have limited financial resources and time to market any of their products. Promotion activities on the other hand show diminishing returns. It is imperative for firms to use their resources judiciously and use scientific methods for related decisions. In this chapter, we propose an optimization model to determine the optimal duration of a promotion campaign for durable technology products marketed in a segmented market with an integrated segment-specific and mass promotion strategy. The proposed model at the same time incorporates the growth in the market potential due to promotional activities. There is limited scholarly research available in this domain and aspects of promotion and marketing environment considered in this study are not considered

A. Kaul (✉)
ASMSOC, NMIMS University, Mumbai, 7th Floor, V.L. Mehta Road, Vile-Parle (West), Mumbai 400056, India
e-mail: kaularshia25@gmail.com

A. Gupta
School of Business, Public Policy and Social Entrepreneurship, Dr. B.R. Ambedkar University Delhi, Delhi 110006, India
e-mail: guptaanshu.or@gmail.com

S. Aggarwal
LBSIM, 11/07, Dwarka Sector 11, Near Dwarka Sector 11, Metro Station, New Delhi, Delhi 110075, India
e-mail: sugandha_or@yahoo.com

P. C. Jha
Department of Operational Research, University of Delhi, Delhi 110007, India
e-mail: pcjhadu@gmail.com

R. Ramanathan
Business and Management Research Institute (BMRI), University of Bedfordshire, University Square, Luton, Bedfordshire LU1 3JU, UK
e-mail: ram.ramanathan@beds.ac.uk

V. Laha et al. (eds.), *Optimization, Variational Analysis and Applications*,
Springer Proceedings in Mathematics & Statistics 355,
https://doi.org/10.1007/978-981-16-1819-2_15

in any other research. Solution methodology based on nature-inspired optimization algorithm differential evolution is proposed, given the NP-hard nature of the proposed model and the suitability of the method to solve problems with real-valued decision variable with convergence to a global solution. A real-life case study is presented to illustrate model application, and test and compare performance with a similar recent study developed on the assumption of static market size. The proposed model shows fair results over the comparative study.

15.1 Introduction

Promotion is an important element of the marketing mix used by firms to communicate with the target markets to create awareness, build interest, and disseminate information about the features and the value delivered by the product(s). Promotion also serves to stimulate buying decisions of potential customers, product reinforcement, and expand the size of the potential customer base [1]. While planning for promotion, a firm is required to focus on several aspects such as budgeting for the promotion, identification of forms and types of promotion channels, resource allocation, and strategies to be adopted in the different channels and time and duration of the promotion. In the literature, several scholars have discussed and conducted studies related to these aspects see [2–6]. In relation to time and duration of the promotion, mainly three types of promotion strategies are used, namely, continuous, flighting, and pulsing [7]. The correct strategy to be followed depends on several factors such as the type of product, availability of promotion resources, and the competitor's strategy. Products which fall in the category of durable technology have a typical Product Life Cycle (PLC) characterized by a bell-shaped curve, wherein the actual shape is dependent upon product type and other marketing variables [8, 9]. Irrespective of the type of promotion strategy used by a firm all products in this category follow a set pattern where they reach the maturity stage subsequent to the introduction and growth, after which inevitably sales decline continuously, leading to the end of PLC. The marginal return from promotion is a decreasing function of time which is likely to become negligible towards the end of the lifecycle [3, 10–12]. This makes it essential to study the PLC and determine the optimal time for which promotion activities must be continued. This study aims to analyze the optimal duration of promotion started at the launch phase of the PLC for a durable technology product with respect to the continuous promotion strategy. Increasing heterogeneity in the customer preferences owing to the availability of several product choices, differences in the disposable income levels, access to information, and emergence of new retail formats calls for segmentation of the potential market into smaller homogeneous segments [13, 14] and conducting promotion targeted to those segments [14]. Availability of several choices for a product also increases the competition among the firms offering these products. The firms competing in this kind of market try to create product differentiation in the perception of their potential customers through promotion and try to gain competitive advantage. For every segment, the promotion plan is tailored to cater to the preferences of the segment, termed as Below the Line (BTL) promotion. While BTL promotion customized for segments, serves to target the segment potential, firms also conduct Above the Line (ATL) promotion comprising the promotion activities that are largely non-targeted and conducted with an objective of wider reach and focused on building the brand. Promotion channels such as national television, print media, and online advertising which have a wider reach are among the preferred channels for ATL promotion activities, while local media including local television, print, sponsorships, brand activation, and in-store promotions are some of the preferred channels for BTL promotion activities [15–17]. A promotion strategy that uses a combination of ATL and BTL promotion activities is termed as Through the Line (TTL) promotion [2, 15]. Many of the firms in present times adopt a TTL promotion strategy with the allocation of promotion resources among the ATL and BTL activities varying across industries and firms. ATL promotions are expected to influence the brand value of a firm along with a wider reach to the audience. Largely as a result of the influence of ATL promotions and to a lesser extent due to BTL promotions, the firms also try to reach and influence a larger potential market. Promotion is carried out not only to target the potential market but also influence the size of the target market [1, 15, 18, 19]. It is expected that as the products move through their PLC adoption, the size of the potential market will grow [16].

There is a fair amount of literature on issues relating to timing and duration of promotion [20–23]. In this study, we have developed a profit optimization model to determine the duration of a continuous promotion carried out with TTL strategy simultaneously considering the growing market potential and tested through a real-life case study of a durable technology product. To the best of our knowledge there is no research in the literature that addresses these aspects [21, 22, 24]. The proposed model is developed on the innovation diffusion model given by [16] which analyzes the adoption growth of the product under consideration using the TTL promotion strategy. There are other innovation diffusion models [23] that consider the diffusion of innovation under the TTL promotion strategy; however these models assume that the size of the potential market is static and remains constant over the PLC. The study of [16] has proposed two models considering growing market size (dynamic), one assuming linear and the other assuming exponential market growth, respectively. The authors have called the growing market size a dynamic nature of the market.

In this study, the solution methodology based on nature-inspired optimization algorithm Differential Evolution (DE) is used to solve the optimization models considering the NP-hard nature of the proposed model. DE algorithms are a class of nature-inspired optimization algorithms that can be used to solve models involving real-valued decision variables with faster convergence and certainty of global solution [25–27]. The model is tested with a real-life application and the results are also compared to a previous study by [28].

The chapter is organized as follows: in Sect. 15.2, the detailed background of the study through review of literature is discussed and the research gap is highlighted along with the contribution of our study. In Sect. 15.3, conceptual framework and model development is discussed. In Sect. 15.4, the solution methodology is described.

In Sect. 15.5, a real-life case study is presented to validate the application of the theoretical model. The results and discussions are presented in Sect. 15.6. The managerial and theoretical implications of the model are highlighted in Sect. 15.7. Section 15.8 concludes the chapter and provides the future scope of research in this field.

15.2 Literature Review

Over the years, researchers have attempted to study the various aspects related to the development of promotion strategies including the time dimension of promotion planning. One of the initial studies in this direction is by [29], where the author proposed a dynamic optimization model to determine the optimal long-run equilibrium level of advertising using the sales advertising response model by [30]. The findings suggested that the level of advertising is a non-monotone function of the rate of decay parameter. Reference [31] followed a two-stage approach to make an assessment of the promotion policy in medium and long terms for mature product categories. In this research, authors studied the long-term effects of different types of advertising and promotion orientation (price and non-price) on consumers' price sensitivity in the packaged food industry. The results showed that the size of consumer population sensitive to price and promotion increases over time. Increased expenditure on promotion may not achieve long-term trends in terms of market share. Mahajan and Muller [10] conducted a comparative analysis between the pulsing and continuous advertising strategies. The research determined the optimal timing and number of pulses for a particular data setting. The findings of the study suggested that a shaped advertising response function is preferable in case of the pulsing strategy. In line with the results of the above studies, several researchers discussed the diminishing effect of returns of advertisement expenditure as an important phenomenon in determining the level and timing of advertisements [3, 11, 12]. Smith [32] discussed that product differentiation as well as market segmentation are important for a successful marketing strategy. The study also highlighted consideration of cost as an important factor in determining the level of product differentiation and segmentation. In marketing literature, several studies discuss the fact that market segmentation facilitates management of the consumer heterogeneity and that promotion strategies in segments are to be customized to the needs of the segment [33–36]. The researchers have also highlighted the implementation issues in segmentation strategy for promotion such as those related to cost, financial and market benefits, practical constraints, etc., and suggested the use of a structured procedure and planning. Jobber [37] studied the effectiveness of BTL promotions. The study discussed that when the sale of products is conducted in large supermarkets which stock several product choices for buyers it becomes important for marketing firms to invest in BTL promotions for local sales along with ATL promotions. Schultz [2] discussed the growth of BTL promotion activities by marketers in the United States. The study highlighted that traditionally firms used ATL promotion to manage brands through the traditional media, however with the growth of sales promotion the use of BTL promotion activities has increased.

Over the years, marketers have used different combinations of ATL and BTL promotions known as TTL promotion strategies customized to their products and markets [38, 39]. Burrato et al. [40] conducted a study that analyzed the advertising of new products in segmented markets and developed a profit optimization model based on advertising intensity in segments and goodwill created. The study assumed two types of advertising—mass advertising that influences the entire market and creates an effectiveness spectrum and target advertising for segments. Jha et al. [23] proposed an innovation diffusion model to describe the adoption level of durable technology products over the PLC incorporating the effect of a TTL promotion strategy. The study assumed that the market share of such products remains constant over the PLC and also ignored the repeat purchase behavior like similar studies in this area such as [41]. The marketing literature discusses that along with creating awareness and motivating a target market for purchase decision effective promotion brings growth in the size of the target market and develops customer loyalty [42]. Aggarwal et al. [16] incorporated these aspects in their study and proposed diffusion models considering dynamic market size and repeat purchasing behavior. Given the different promotion strategies available; it is noteworthy that firms do not have infinite resources. The resources both in terms of time and finances are limited in nature. Firstly, considering the diminishing returns of promotion over the product life cycle firms are required to determine the time up to which the product is to be marketed [3, 11, 12]. Secondly, according to the literature on technology substitution in marketing in the case of durable technology products firms constantly innovate and introduce new products before their earlier products become obsolete to remain competitive. It is important for firms to introduce new products at a strategic time such that the time to market the new product matches or exceeds the industry target and they are able to make desirable returns from the previous products [43]. Once the new products are introduced the earlier products diffuse in the market mainly through word of mouth effects and price promotions. Thus it is essential to determine the duration of direct promotion and the optimal resources to be utilized in promotion [44–46]. Various authors have proposed different strategies for the determination of the duration of different forms of promotion. Aggarwal and Vaidyanathan [47] in their research conducted two studies. In the first study for fast-moving consumer goods, authors determined that short duration (time-limited promotion) has a greater impact on the purchase of products as compared to long-term promotions (time-independent promotion). The second study in relation to durable technology products discussed that limiting the validity of the promotion has an impact on purchase behavior. The discussion by Esteban-Bravo et al. [21] focused on the frequency and interval of promotions. The researchers highlighted the importance of planning for non-price promotions from the point of view of the duration. The objectives were to maximize the profit taking into account the decay in economic returns with time. The authors proposed a dynamic optimization model to determine the optimal duration of a promotion campaign assuming that customer decision follows a state-dependent Markov process at the aggregated level. Cetin [20] proposed a mathematical model to establish the optimal time duration of an advertising campaign for a technology innovation such that the profit is maximized. The revenue is taken as a function of adoption level, measured using a pure external influence adoption model, fixed cash flow of advertising cost, and fringe cost independent of time. Beltov et al. [22] aimed to study the price promotions, time, and duration of promotion for sales in retail stores. A dynamic optimization model is proposed assuming two competitive brands under retail sales and price promotions are implemented on one brand at a time. Devlin et al. [48] discussed the effect of time-limited price promotions on consumers. An experimental comparative assessment is made to assess the effect of time-limited price and non-time-limited price promotions on consumers, which helped in future decision making. Results show that no direct relationship exists between time-limited promotions and purchase behavior. Lin and Lin [24] discussed a model in which they establish the duration of the promotion campaign before reaching a steady-state assuming two competitive products in the market. A Markovian profit maximization model is proposed integrating entropy and diffusion theory. The study contributes by implementing the theoretical theories to model the problem for a case of the canned coffee market in Taiwan. Some of the models cannot be generalized. Duran et al. [49] developed a model to determine the optimal stopping time of seasonal ticket sales for a sports and entertainment event and start time of single event ticket sales. It was assumed that demand follows a Poisson process. The objective was to maximize the profit resulting from the sales of the tickets so that sufficient demand for the seasonal tickets is achieved giving adequate time for the sale of single tickets. Similar to the study of Cetin [20] that uses a pure external influence model for determining optimal duration for a technological innovation, Aggarwal et al. [44] developed an optimization model based on a mixed influence model. The use of mixed influence model is supported as social influence affects product adoption significantly along with the promotion activities as many adopters wait to see the response from the social influences before adopting the product. The proposed solution methodology uses DE algorithm to solve the model. The study by Lo et al. [50] looked at the problem of optimal duration from the point of view of determining the time frame of promotion (time-lapsed and time to go) for group buying deals for the tourism and hospitality industry similar to Duran et al. [49] using analysis of covariance. The combined effect of time lapsed and time to go helps to achieve the optimum sales of group-buying deals given a 5-day promotional period. Through this study restaurateurs and group-buying websites can determine the ideal time for promotion duration to generate sales. Danaher et al. [51] in their research focus on the effect of mobile-coupons (m-coupons) on the response of the customers. The response is considered in terms of time, place at which the coupons are delivered, and the duration for which the coupons are valid. It is observed that the length of expiration must be shortened to persuade urgent purchase. Although m-coupons exist, the traditional coupons still dominate over the m-coupons. Kaul et al. [28] studied the optimal duration of promotion campaign for durable technology products promoted in a segmented market under the TTL promotion strategy. The objective of the proposed model is to maximize the profits based on adoption level using a mixed influence diffusion model under the TTL strategy proposed by Jha et al. [23]. The authors assumed that market size remains constant (static) throughout the PLC and ignore the growth of market due to promotion. The proposed study caters to this research gap by developing an optimization model for determining the optimal duration of promotion of durable technology products under the TTL promotion strategy considering a growing potential market. The proposed optimization model is an NP-hard problem due to the complex non-linear mathematical form of the diffusion model used to describe the product adoption level. For a given problem having functions with mathematical forms for which it is difficult to establish mathematical nature (convexity), and involve multiple parameters and variables, it is important to decide a suitable method for handling such a problem. Nature-inspired optimization algorithms, also called soft computing algorithms, find applications when the global optimization methods are not applicable. In the literature, several soft computing algorithms and their versions are available. This research uses DE algorithm because the algorithm can handle real-valued decision variables, shows faster convergence, is easy to implement and computations are simple, fast, and reliable. The algorithm can be implemented without knowing the mathematical nature of the functions involved in the model [25, 26, 52, 53]. From the basic DE method by Storn and Price [25] to the various new hybrid versions as discussed by Das and Suganthan [27], there has been a significant development in research in relation to the DE algorithm. Storn and Price [25] discussed the differential evolution approach for non-linear and non-differentiable continuous space functions. The authors established that the method converges faster with more certainty compared to several well-known and established global optimization methods. Liu and Lampinen [54] discussed the new version of the differential evolution algorithm for fuzzy logic controllers for adapting to the search parameters for mutation and crossover operations. The experimental results have been shown for standard test functions and showed the superiority of the new algorithm. Das et al. [55] discussed the case of the performance of the DE algorithm for the case when the fitness functions are noisy and continuously changing. The authors have proposed two improved DE algorithms for achieving the global optima for noisy functions. Das and Suganthan [27] discussed an extensive literature for the period 1995–2010 on the different variations in DE and the various applications which have been studied. Tasgetiren et al. [56], presented a Discrete DE (DDE) algorithm for solving the no-wait flowshop scheduling problems. Tasgetiren et al. [57] used DDE algorithm to obtain the solution of the single machine total weighted tardiness problem with sequence-dependent setup times. The DE algorithm is useful to find a global optimal solution for highly non-linear and non-convex problems (Tsafarakis) [58].

15.2.1 Literature Gap and Research Motivation

Initial studies in the area of promotion duration are discussed from the point of view of pulsing, see Mahajan and Muller [10]; Balakrishnan and Hall [11]; Sethi [29] or price promotions [22]. Some others have considered the specific cases of duration for the validity of different types of promotion coupons and their effects on the purchase intentions of the potential adopters [51]. In other cases, the time limits on the duration of the promotions and the effects on purchase are considered

[47, 50]. There has been an emphasis on the fact that, as time progresses, there is a diminishing effect on the marginal rate of return due to promotion. Therefore, it becomes imperative for marketers to analyze the duration for which the promotion must be continued so that the promotion remains lucrative for them. Several authors have underscored the importance of finding out the optimal time and duration for a promotion with respect to different products or markets [20, 21, 24, 44]. Though there has been research to determine the optimal duration of promotion in the extant literature, there is limited research that has considered the segmented nature of the market. As discussed earlier, segmentation of the potential market is increasingly carried out by marketers as an important element of promotion mix so as to customize the positioning of their products in the various segments. It allows the development of promotional mix with respect to segments and targets the potential segment with segment-specific promotion strategy along with promotion carried out by means of mass media catering to the mass market, known as TTL promotion strategy [15, 17, 44]. Distinguishing features of this research are as follows:

- Profit optimization models are proposed to determine the optimal length of the promotion period for durable technology products over the PLC.
- The proposed model is developed considering the market segmentation strategy followed by marketers currently to increase the effectiveness of its promotion activities. Under the segmented promotion strategy, given the individualistic preferences and increasing heterogeneity of customers in target markets, the promotion activities are customized to cater to smaller segments along with mass promotions for brand establishment and accelerate market growth. Though the marketing literature discuses such integrated TTL promotion activities, the analytical studies are very limited in this area.
- The model incorporates the coefficient of growth in segment potentials due to the effect of promotion. The proposed study seems to be considering this aspect in promotion duration studies for the first time. The earlier studies in this area are based on static market size assumption.
- The proposed model finds practical application to study the effectiveness of both ATL and BTL promotion strategies, evaluation and allocation of resources among ATL and BTL promotions, facilitates decisions related to duration of promotion activities, and decisions related to new product introductions and technology substitutions.
- The proposed models are tested on real-life data using nature-inspired optimization algorithm DE to solve the models, compared with previous research, and sensitivity with respect to resource availability is conducted on the decision variable.

15.3 Conceptual Framework and Model Development

The conceptual framework of the proposed study is shown in Fig. 15.1.

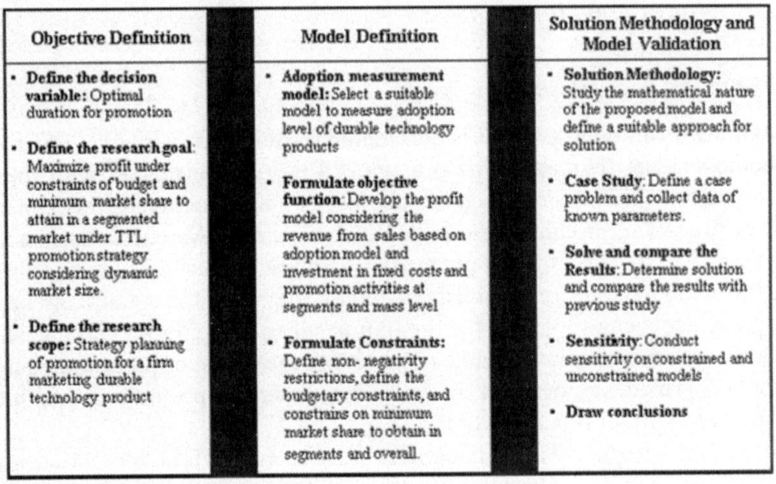

Fig. 15.1 Research framework

15.3.1 System of Notations

S	Number of market segments
i	Index for segments; $i = 1, 2, \ldots, S$
$\overline{N}_i(.)$	Expected number of potential adopters in i^{th} segment by time t
$\overline{N}_i$	Expected initial market size in i^{th} segment by time t
$\overline{N}(.)$	Expected total potential adopters by time t
b_i, β_i	Diffusion model parameters for external and internal influence in i^{th} segment
$x_i(t)$	Instantaneous rate of promotion in the ith segment
$X_i(t)$	Cumulative segment specific promotion efforts in i^{th} segment by time t
$X(t)$	Cumulative mass promotion efforts by time t
$N_i(.)$	Expected number of adopters in i^{th} segment by time t
$R_i(t)$	Revenue earned in i^{th} segment by time t
$C_i(t)$	Variable cost incurred in i^{th} segment by time t
FC_i	Fixed cost in i^{th} segment
a_i	Fixed cash flow on promotion per time unit in i^{th} segment
A	Fixed cash flow on mass promotion per time unit
$\omega_i\prime, \omega_i\prime\prime$	Unit sale, cost price of the product in the ith segment
ω_i	Profit per unit in the ith segment; $(\omega_i = \omega_i\prime - \omega_i\prime\prime)$
α_i	Coefficient of mass promotion in i^{th} segment ; $\alpha_i \in [0, 1)$
g_i	Coefficient of dynamic market size in i^{th} segment; $g_i \in [0, 1)$
N_i^*	Minimum market share to be achieved in i^{th} segment
N^*	Minimum market share to be achieved in total market
r	Present value factor
Z	Promotion budget
$\phi(t)$	Profit function

15.3.2 Model Development

Thus, the model is developed as follows:

1. A durable technology product is considered, promoted in a market divided into S homogeneous segments. The product diffuses in the market with time and follows a typical PLC described by four stages—introduction, growth, maturity, and decline. The product is promoted to spread the awareness, accelerate the adoption, and increase market potential. The marginal revenue due to promotion decreases on account of diminishing returns that call for a trade-off between adoption level achievable and the promotion expenditure. Profit maximization in such a case can be achieved while determining the extent of promotion duration.
2. Adoption growth is governed by the external (due to promotion) and internal influences (word of mouth) over time.
3. The product is marketed in each segment using a targeted BTL promotion strategy tailored for segments according to the segment characteristics and promotion media preferences. The segment-driven promotion strategies are supported with ATL promotion strategy for mass promotion to cater to the wider potential market.
4. With the spread of product awareness due to promotion, internal influences, and other factors such as population growth and economic changes the market potential grows in each segment.
5. Firms have finite promotion resources to spend.
6. The total profit realized from all the segments by any time (say t) is calculated as the difference between revenue obtained from adoption by time t and the expenditure on fixed costs and mass and segmented promotions.

The profit function based on above considerations is formulated as follows:

$$\max \phi(t) = \sum_{i=1}^{S} e^{-rt}(R_i(t) - C_i(t) - a_i(t)) - FC_i - AX(t)e^{-rt}$$

$$= \sum_{i=1}^{S} e^{-rt}(\omega_i\prime N_i(X_i(t)) - \omega_i\prime\prime N_i(X_i(t), X(t)) - a_i X_i(t)) - FC_i - AX(t)e^{-rt}$$

$$= \sum_{i=1}^{S} e^{-rt}(\omega_i N_i(X_i(t), X(t)) - a_i X_i(t)) - FC_i - AX(t)e^{-rt}$$

In equation (1), $N_i(X_i(t), X(t))$ represents the adoption level of the product by time t satisfying assumptions (1–4). The adoption measurement model proposed in Aggarwal et al. [16] satisfies diffusion environment defined in assumptions (1–4), in which is developed on the assumption that the *rate of adoption with respect to promotional effort intensity is proportional to the remaining number of non-adopters for durable technology products marketed in the segmented market under internal and external influences. The external influence is the result of joint effect of mass and segment-specific promotion activities.* The authors proposed different mathematical functions to describe the promotional intensity functions and to describe the growth of market potential. The adoption models in Aggarwal et al. [16] are briefly described below.

Diffusion Model 1 (M1): Assumes market size grows exponentially in response to the external and internal influences, i.e.

$$N_i(X_i(t), X(t)) = \overline{N_i} e^{(g_i X_i(t) + \alpha_i X_i(t))} \tag{15.1}$$

The adoption model is given as

$$N_i(X_i(t), X(t)) = \frac{N_i b_i}{b_i + g_i} \left[\frac{e^{g_i (X_i(t) + \alpha_i X(t))} - e^{b_i (X_i(t) + \alpha_i X(t))}}{1 + \beta_i e^{-b_i (X_i(t) + \alpha_i X(t))}} \right] \quad \forall (i = 1, 2, \ldots, S) \tag{15.2}$$

Diffusion Model 2 (M2): Assumes the market size grows linearly in response to the external and internal influence, i.e.

$$N_i(X_i(t), X(t)) = \overline{N_i} g_i(X_i(t), \alpha_i(X(t))) \tag{15.3}$$

The adoption model is given as

$$N_i(X_i(t), X(t)) = \frac{\overline{N_i}}{1 + \beta_i e^{-b_i (X_i(t) + \alpha_i X(t))}} \left[g_i(X_i(t) + \alpha_i X(t)) + (1 - e^{(-b_i (X_i(t) + \alpha_i X(t))))}) \left(1 - \frac{g_i}{b_i} \right) \right], \tag{15.4}$$

$\forall\, i = 1, 2, \ldots, S$. In this study, we assume that the product adoption level in segments can be described according to diffusion models M1 and M2. The study by Aggarwal et al. [16] assumes that given the dynamic nature of the potential market, the nature of adoption in each segment follows a similar mathematical form with different parameter values. For determining the promotion duration, we can continue with the assumption and use either model M1 or M2 for all the segments in the profit maximization model. On the other hand, the best-fit dynamic market size adoption model could be established for each segment, which is then used in the profit optimization model. The results are presented for all cases in the case study section.

In the literature, mathematical functions are proposed to describe the promotion effort expenditure as a function of time assuming *instantaneous rate of promotion expenditure is proportional to the available balance of promotion resources.* Under this assumption exponential ($\mu_i(1 - e^{-\gamma_i t})$), Rayleigh ($\mu_i(1 - exp(-\gamma_i \frac{t^2}{2}))$), Weibull ($\mu_i(1 - exp(-\gamma_i t^{m_i}))$), and logistic ($\frac{\mu_i}{1+\gamma_i (e^{(-\gamma_i t)})}$) forms of promotion effort functions were proposed with usual meaning of notations [16]. We assume the segment specific as well as mass promotion effort functions are described by one of the above best-fit forms estimated from the data collected. The parameters of adoption growth models and Promotion Effort Functions (PEFs) can be estimated from the observed adoption and promotion expenditure data for some initial periods or past data of similar products. The values of cost and sales price, parameters of fixed cash flow on differentiated and mass promotions, and fixed cost in each segment are provided by the firm. Substituting the values of the parameters, the profit function can be optimized to obtain the optimal value of promotion duration. The optimal value of promotion duration as determined from profit function (1), which imposes no restriction on the promotion effort expenditure. Further, it is observed that the firm has limited financial resources available. In such a case, a constrained profit optimization model is defined as budgetary restrictions as given in constraint (15.5) below:

$$\sum_{i=1}^{S} a_i X_i(t) + AX(t) \leq Z \tag{15.5}$$

The potential customers in every segment possess unique characteristics, each segment responds uniquely to the external and internal influences in terms of product adoption. Product diffusion may be fast in some segments and comparatively slower in other segments. Profit maximization under budgetary restrictions could result in a solution such that a high market potential is captured in segments with higher adoption rates and relatively low in segments with a slow response to the diffusion process. The firm may want to ensure a certain minimum market share that it might want to achieve in the respective segments and the market as a whole. Under such a situation additional constraints on the minimum level of adoption to be attained in each segment could be imposed along with the constraint on the minimum level of combined adoption attainable by the optimal time. Equations (15.6) and (15.7) below defines these constraints, respectively.

$$N_i(X_i(t), X(t)) \geq N_i^* \ \forall \ i = 1, 2, , S \tag{15.6}$$

$$\sum_{i=1}^{S} N_i(X_i(t), X(t)) \geq N^* \tag{15.7}$$

Additional constraints can be imposed on the proposed models depending on a specific application such as budgetary restrictions on a particular segment and/or strategy.

15.4 Solution Methodology

The adoption growth models as well as PEFs in the objective function are described by non-linear functions. The constraints in the proposed model are in the form of non-linear inequalities. Due to the non-linear nature of component functions, the overall objective function is a complex non-linear function. The convexity of the proposed optimization model is difficult to establish. Nature-inspired optimization algorithms [62] are widely studied, successfully used, and accepted methods in the recent years for solving optimization models. Researchers have developed methods in this class of solution methodology that converge to a nearly global solution. We propose to use DE algorithm as methodology to solve the models formulated in this study. The proposition is supported from the literature wherein the authors Price and Storn [26] established that the method is suitable to solve non-linear optimization problems with real-valued decision variables. The algorithm converges faster with more certainty compared to several well-known and established global optimization methods. The step-by-step approach to solve the proposed model using DE algorithm is given in Appendix A. Solving an optimization model with DE requires the value of several parameters of the methodology apart from the model parameters. The literature of DE has discussed the values of the model parameters and based on the studies we have chosen the values (for details on DE algorithm and determining values of its parameters reader can refer to [52, 59, 60, 62].

In the literature (refer Storn and Price [26]; Ali and Törn [61], Vesterstrom and Thomsen [62] it is suggested that F which is the scaling factor lies in the range $F \in [0.5, 1.2]$; Cr defined as crossover probability lies in the range $Cr \in [0.8, 1]$; $NP = 10 * D$ (size of population) and D is defined as the dimension of the model, equal to the number of decision variables, here $D = 1$ (corresponding to $t =$ duration of promotion). The chosen values of the parameters in the current study are taken as $F = 0.7$, $Cr = 0.9$ and the roulette wheel process is used for selection of the base vector. The size of the population in the case is taken to be $NP = 15$, $15 \gneqq 10 * D$. The DE algorithm is coded on the DEV C++ on an Intel(R) Core(TM) i3 CPU @2.13 GHz, 2 GB RAM, Windows 7 operating system for solving the case study.

15.5 Case Study

15.5.1 Data Description

In this section, we validate and test the performance of the proposed model through numerical illustrations. The results of case illustrations are also compared to a similar study Kaul et al. [28] for verification of results and establishing the performance of the proposed model. The models proposed in Kaul et al. [28] are developed assuming a static market size under an integrated segment-specific and mass promotion strategy. The adoption data of a new durable technology product marketed in four segments with respect to mass and segment-specific promotion (TTL) over a period of an initial 24 months is used. The data and parameters of the model are adopted from the studies [16, 28]. Estimates of the parameters of the adoption models (M1 and M2; equation 3 and 5, respectively) are taken from Aggarwal et al. [16] and the remaining parameters of the objective function (equation 1) are taken from (Kaul et al.) [28]. The data is shown in Table 15.1.

The results are computed based on the proposed model for three different cases and compared with [28]. The cases are described in Table 15.2 as illustrations 1–

4. In illustrations 1 and 2 diffusion models (M1 and M2), respectively, describes the adoption level in all segments. In illustration 3, the best-fit dynamic market size models are taken for measuring the adoption level in all segments. The best-fit adoption model is determined based on the Mean Square Errors (MSE) of estimates reported in Aggarwal et al. [16] and calculated using the non-linear regression module in SPSS software [63]. It may be noted here that the form of the promotion effort function is also based on the best-fit model from the models described in Sect. 15.2. Illustration 4 corresponds to the optimization model for promotion duration in Kaul et al. [28] used to draw comparison.

15.6 Results and Discussions

The unconstrained profit maximization model is solved for all the illustrations as given in Table 15.1 using DE algorithm with parameter values discussed in Sect. 15.4. The results are as shown in Table 15.3.

It can be seen from Table 15.3 that optimal duration of promotion campaign for illustrations 1–4 (refer Table 15.2) is 46.97, 35.68, 46.74, and 37.76 months ,respectively. Comparing the results of the proposed model with the study Kaul et al. [28], it can be seen that in case of illustrations 1 and 3 corresponding to the adoption measurement models with exponential market growth in each segment and best-fit models, respectively, higher expected profits are achievable. That is the study Kaul et al. [28] underestimates the profit. However, the expected profit achievable in illustration 2 underestimates the results. The underestimation in this case is accounted for by the assumption of linear growth model for market size, that doesn't fit well on the data under consideration. Further continuing the promotion for optimal times as shown in Table 15.3, raises the requirement of promotion budget, which is at a much higher level as compared to the proposed budget of INR160 million. To control the promotion spending budgetary restriction of INR160 million as given in constraint (6) is imposed on the model and solution is obtained as given in Table 15.4.

The optimal promotion duration (t*) determined under a budgetary constraint of INR160 million is 24.19 months for all the illustrations. This could be attributed to the similar mathematical nature of the models. However, the results in Table 15.4 shows significant variation in the expected adoption level achievable and expected profit between the models, with the highest profit of INR 21,081,102,172 corresponding to illustration 1 (exponential market growth model in adoption measurement model). The profit achievable according to Kaul et al. [28] is INR 13,411,166,872 at the level of INR 160 million expenditure in promotion which clearly shows underestimation of results as expected with the optimization model Kaul et al. [28] developed on the assumption of static market size. The results of unconstrained as well as constrained models with budgetary restriction show that the results of the Kaul et al. [28] model underestimates the results for the data under consideration. This suggests that the proposed model performs better than the Kaul et al. [28] model for the case under consideration. Further, the application of the proposed model is explored imposing

Table 15.1 Parameters of the model

Segment	S1	S2	S3	S4	Mass promotion
Sales price (in INR)	457000	468000	446000	453700	4 million
Variable cost (in INR)	388000	403000	378200	393500	4 million
Unit promotion cost (in INR)	1300000	1020000	1820000	1720000	4 million
Fixed cost	40 million	40 million	40 million	40 million	
Proposed promotion budget	160 million				
Present value factor	0.02				
Parameters of diffusion model (M1)					
$\overline{N}_i$	41330	66633	57232	162318	
b_i	0.123077	0.428514	0.476789	0.336248	
β_i	31.7	176.76	220.09	396.09	
α_i	0.339	0.2	0.213	0.24	
g_i	0.0453	0.0662	0.0509	0.015	
MSE	1128.17	5690.43	8144.91	143545.00	
Parameters of diffusion model (M2)					
$\overline{N}_i$	98435	94427	50631	154853	
b_i	0.116946	0.445413	0.481654	0.296856	
β_i	58.91	259.51	208.54	399.71	
α_i	0.25	0.2	0.2	0.33	
g_i	0.0597	0.05	0.0964	0.0202	
MSE	59571.97	7242.82	5787.91	113393.58	
Best-fit PEF	Weibull	Exponential	Exponential	Exponential	Exponential
PEF parameters					
μ_i	41.56	14.66	12.27	33.41	77.89
γ_i	0.0034	0.0265	0.0274	0.0155	0.01485
m_i	1.72	1	1	1	1

Source Aggarwal et al. [16]; Kaul et al. [28]

Table 15.2 Description of numerical illustrations

Diffusion model used to describe the adoption level in segments

Segments	Illustration 1	Illustration 2	Illustration 3	Illustration 4
S1	Proposed model with model M1 to describe adoption level (equation 3)	Proposed model with model M2 to describe adoption level (equation 5)	Proposed model with model M1 to describe adoption level (equation 3)	Results of Kaul et al. [28] for comparison of results
S2	Proposed model with model M1 to describe adoption level (equation 3)	Proposed model with model M2 to describe adoption level (equation 5)	Proposed model with model M1 to describe adoption level (equation 3)	Results of Kaul et al. [28] for comparison of results
S3	Proposed model with model M1 to describe adoption level (equation 3)	Proposed model with model M2 to describe adoption level (equation 5)	Proposed model with model M2 to describe adoption level (equation 5)	Results of Kaul et al. [28] for comparison of results
S4	Proposed model with model M1 to describe adoption level (equation 3)	Proposed model with model M2 to describe adoption level (equation 5)	Proposed model with model M2 to describe adoption level (equation 5)	Results of Kaul et al. [28] for comparison of results

additional restrictions. Restriction of budget to INR 160 million cuts down the percentage adoption level achievable by $t^* = 24.19$ months. For example in illustration 1, the expected percentage adoption level by $t^* = 24.19$ months in segments S1–S4 is 28.85%, 55.79%, 65.59% and 81.57% respectively and total adoption level is 51.97% compared to 73.03, 86.6, 90.34, and 95.72% in segments and 83.64% in total according to the unconstrained model. Similar results are obtained for the other cases as shown in Tables 15.3 and 15.4. Referring to earlier discussion in Sect. 15.3, firms may also set the minimum adoption level to be achieved in each segment with constraint (15.6) as well as restriction on the total market share (constraint (15.7)). The budget restriction and minimum achievable market size restrictions are contradictory in nature and may lead to infeasibility of the optimization model. DE provides a compromised solution in case of infeasibility. Here we also present the sensitivity analysis on the proposed model (for illustration 1–3) with different values of budget constraint (15.5). For sensitivity analysis of Kaul et al. [28] model (illustration 4) reader may refer to original study. The budget is increased by 5% of the previous value in the iterations of the sensitivity analysis. The results of sensitivity are illustrated graphically here only for illustration 1 in Fig. 15.2. Fig. 15.2 shows the expected adoption levels (in %) achievable in each segment and total market for different values of budget for illustration 1. Further Fig. 15.3 shows the comparative analysis of the profits achievable for illustrations 1–3 for different levels of budgetary restrictions. Figure 15.4 shows the corresponding percentage change in profit for every 5% increase in the budget. From Fig. 15.2 it can be inferred that the expected adoption level and profit increase on increasing the promotion resources for all the illustrations (1–3). However, Fig. 15.4 depicts that the increase in profit is

Table 15.3 Results of unconstrained model

Optimal promotion duration (t^*) (in months)	Segment	Expected adoption level by t^*	Expected % adoption level by t^*	Expected level profit (in INR)	Expected promotion expenditure (in INR)
Illustration 1					
46.97	S1	312,586.10	73.08		
46.97	S2	193,319.29	86.6		
46.97	S3	124,237.80	90.34		
46.97	S4	231,767.01	95.72		
	Total	861,910.20	83.64	21,822,746,005	262,826,610
Illustration 2					
35.68	S1	195,326.94	57.45		
35.68	S2	122,547.35	73.38		
35.68	S3	88,169.16	73.90		
35.68	S4	176,384.89	75.93		
	Total	582,428.36	67.83	18,365,804,761	218,675,849
Illustration 3					
46.74	S1	310,914.14	73.08		
46.74	S2	192,631.71	86.61		
46.74	S3	114,052.53	86.42		
46.74	S4	226,380.30	90.94		
	Total	843,978.69	82.04	21,465,844,625	262,018,777
Illustration 4					
37.76	S1	200,350.11	69.57		
37.76	S2	155,113.23	99.05		
37.76	S3	102,743.12	96.04		
37.76	S4	179,509.19	80.39		
	Total	637,715.64	82.30	19,317,800,000	227,722,672

at a decreasing rate due to the diminishing rate of return of promotion activities. The sensitivity analysis provides insight to the decision-makers in deciding an optimal level of spending on promotion and the time duration of promotion activities. The figures have been presented for some of the illustrations; results for other illustrations can be observed similarly and not shown here because of similarity in results.

Further results are obtained by imposing the restriction in constraints (15.6) and (15.7), setting $N_i^* = 50\%$ and $N^* = 60\%$ to obtain a trade-off between the budget and the market size aspiration restrictions for the proposed models. Detailed results for this case are shown in Table 15.5. The results for all models are obtained compromising budget and the market share aspiration constraints are satisfied. Here, again the proposed model with exponential market growth in adoption measurement model

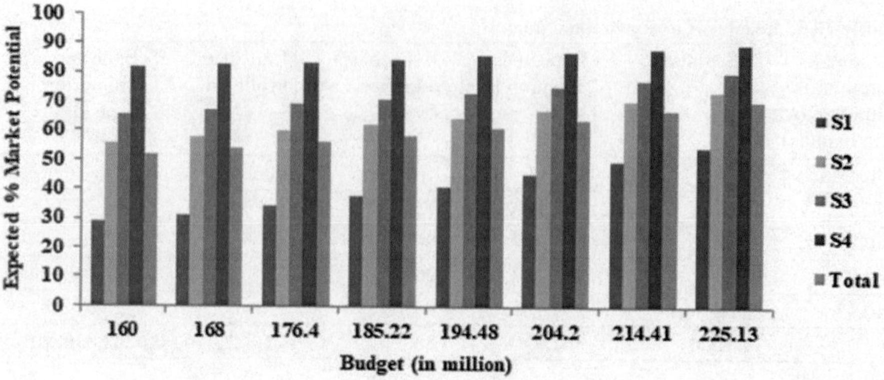

Fig. 15.2 Sensitivity analysis for profit model in illustration 1

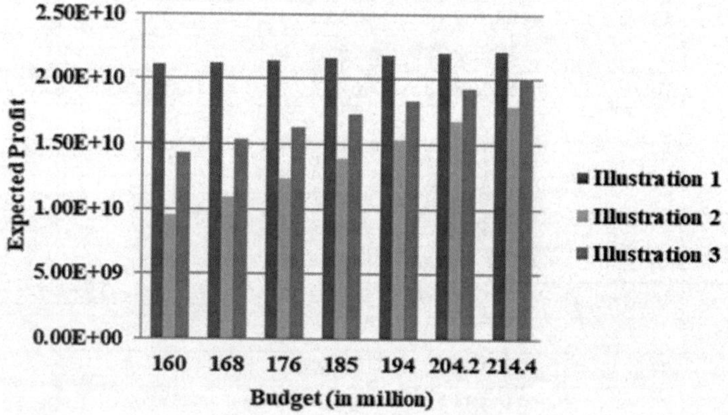

Fig. 15.3 Comparative analysis for profit achievable for illustration 1–3

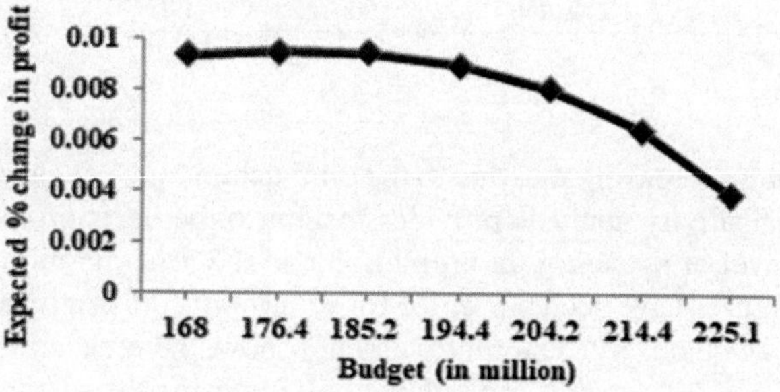

Fig. 15.4 Analysis for marginal change in profit for illustration 1

Table 15.4 Results with budget constraint (Z = INR 160 Million)

Optimal promotion duration (t^*) (in months)	Segment	Expected adoption level by t^*	Expected % adoption level by t^*	Expected profit (in INR)
I Illustration 1				
24.19	S1	123,407.11	28.85	
24.19	S2	124,531.78	55.79	
24.19	S3	90,203.028	65.59	
24.19	S4	197,524.19	81.57	
	Total	535,666.09	51.97	21,081,102,172
I Illustration 2				
24.19	S1	73,077.17	21.49	
24.19	S2	56,336.93	33.73	
24.19	S3	41,323.40	34.63	
24.19	S4	71,581.24	30.81	
	Total	242,318.75	28.22	9,490,825,709
I Illustration 3				
24.19	S1	123,407.11	29.00	
24.19	S2	124,531.79	55.99	
24.19	S3	41,323.40	31.31	
24.19	S4	71,581.24	28.76	
	Total	360,843.53	35.07	14,364,236,054
I Illustration 4				
24.19	S1	70,060.29	24.33	
24.19	S2	131,832.95	84.18	
24.19	S3	70,634.27	66.02	
24.19	S4	66,150.18	29.62	
	Total	338,677.70	43.70	13,411,166,872

gives the best performance at the same level of promotion expenditure and duration of promotion.

15.7 Managerial and Theoretical Implications

- Profit maximization for the firm by obtaining optimal duration of the promotion campaign: As firms do not have unlimited financial resources which they can invest on the promotion of products, they need to decide the optimal duration of the promotion campaign such that the profit is also maximized. Apart from the constraint on financial resources, there is also a diminishing rate of return of

Table 15.5 Results of unconstrained model

Optimal promotion duration (t^*) (in months)	Segment	Expected adoption level by t^*	Expected % adoption level by t^*	Expected profit (in INR)	Expected promotion expenditure (in INR)
Illustration 1					
34.74	S1	212,810.92	50.00		
34.74	S2	156,759.12	70.23		
34.74	S3	106,638.00	77.55		
34.74	S4	214,235.80	88.48		
	Total	690,443.85	66.99	22,195,628,124	214,415,302
Illustration 2					
34.74	S1	186,296.84	54.79		
34.74	S2	118,214.72	70.78		
34.74	S3	85,082.75	71.31		
34.74	S4	169,631.34	73.02		
	Total	559,225.66	65.13	17,961,953,101	214,415,302
Illustration 3					
34.74	S1	212,810.93	50.02		
34.74	S2	156,759.12	70.48		
34.74	S3	85,082.75	64.47		
34.74	S4	169,631.35	68.14		
	Total	624,284.15	60.68	20,125,768,054	214,415,302

promotion. The promotion efforts cease to have an effect after a certain amount of time and it becomes futile to continue the promotion campaign. Through the proposed model the expected time and financial resources the firm needs to invest for the promotion of its product in case of durable technology products can be determined.

- Trade-off between budget and market aspiration level: The proposed model and solution methodology allows decision-maker to conduct sensitivity analysis at different levels of budget and market share aspirations and take an appropriate decision rather than getting a fixed solution.
- Flexibility to modify the model with respect to a particular nature of market growth and promotional effort function. The proposed model is not limited to the use of market growth and PEFs discussed in the study. A generalized optimization is developed that can easily be modified according to more mathematical forms of these functions with respect to a specific situation.
- Comparative analysis between models of literature: A comparison is made of the proposed model with a recent model in the literature to test the performance of the proposed model for real-life situations. The comparative model Kaul et al. [28] was developed on the adoption measurement model that assumes static market

size. It was expected that with this assumption results would be underestimated. The results show that the proposed model developed on the adoption model with exponential market growth performs better than the Kaul et al. [28] model for the chosen case study establishing the performance of the proposed model.
- Application of nature-inspired optimization algorithm Differential Evolution: The proposed research comes under the category of an application research of DE algorithm in the class of nature-inspired optimization. The methodology presented in the paper finds implication for academic researchers and practitioners to apply the method in practice and future research.

15.8 Conclusion and Future Scope

The study proposes a profit optimization model to determine the optimal duration of promotion for durable technology products marketed in segmented market under an integrated BTL and ATL promotion strategy. In recent time, this integrated strategy is increasingly followed by the marketers. Through BTL promotion activities firms target segment potentials and with ATL promotion strategy the objective is to target mass market along with influencing the market size. The model is developed on a recent adoption measurement model that describes the adoption level of PLC under an integrated ATL and BTL strategy along with incorporating the market growth due to promotions. There are limited studies incorporating the effect of an integrated promotion strategy in the literature. The proposed model is tested on a real-life case study and results are compared with a similar study that is developed on an adoption measurement model to describe the adoption level with an assumption of static market potential. The assumption of static market size is not realistic as the firms carry out promotions not only to spread product awareness and motivate purchase decision but also to establish the brand and stimulate market growth. The proposed model thus finds more practical application compared to similar studies available in the literature. The results of the case study also prove the performance of the proposed model. Further, the model development is carried out adding constraints with an incremental approach. The decision-makers can easily add and remove constraints with respect to a particular situation. Nature-inspired optimization methodology, differential evolution algorithm adds further flexibility in the model by allowing sensitivity of results for different constraints and levels of restrictions. The immediate scope of further research identified in the study is to test the model performance with other soft computing algorithms and conduct sensitivity of DE parameters. Future research can explore the development of adoption measure model based on other forms of market growth and promotional effort functions and show applicability of these models for determining the optimal duration.

Acknowledgements The authors would like to express their gratitude to the referees for the valuable comments.

Steps for Differential Evolution
The following are the steps for the Differential Evolution Algorithm[Eooktistov, 2006[53]]:
Start of Steps
Step 1: Random individuals are chosen for generating the initial population.
Step 2: New population can be created by repeating the following steps till the stopping criterion is achieved
(i) **[Selection]** Random individuals are selected for reproduction
(ii)**[Reproduction]** New individuals are created by mutation and crossover
(iii)**[Evolution]** The fitness function values of the individuals are computed
(iv)**[Advanced Population]** The new population is selected from the target individual and trial individual.
End of Steps
Details of Steps
Initialization
In case one wants to optimize the function of D number of real parameters, one must select a population with size NP. The parameter vectors of NP have the form

$$X_{i,G} = (x_{1,i,G}, x_{2,i,G}, \ldots, x_{D,i,G})$$

D is the defined as dimension, i is the index of individual and G represents the number of generations.
In the first step, all solution vectors are randomly initialized. The lower and upper bound for generating the initial solution vector are $l = \{l_1, l_2, \ldots, l_D\}$ and $u = \{u_1, u_2, \ldots, u_D\}$. The equation given below is used to generate the initial population.

$$x_{j,i,0} = l_j + \text{rand}_{i,j}[0,1] \times (u_j - l_j)$$

In the above equation i the index for individual, j represents the component index and $\text{rand}_{i,j}[0,1]$ is a random number lying between 0 and 1 and is uniformly distributed. The population vectors $X_{i,0} = (x_{1,i,0}, x_{2,i,0}, \ldots, x_{D,i,0})$ are defined as the target vectors.
Mutation
Steps of mutation, recombination and selection is undergone by each of the NP parameter vector. Given a parameter vector $X_{i,G}$, first three vectors are randomly selected $X_{r1,G}, X_{r2,G}, X_{r3,G}$ where the indices are distinct. The particular individual which is perturbed is given by $V_{i,G}$ and is given based on the three randomly selected individuals.

$$V_{i,G} = X_{r1,G} + F \times (X_{r2,G} - X_{r3,G})$$

in the above equation $r1, r2, r3 \in \{1, 2, \ldots, NP\}$ and are selected randomly where $r1 \neq r2 \neq r3 \neq i$, $F \in [0, 1.2]$ and $V_{i,G}$ is defined as the mutation vector.
Crossover
The individual which has been perturbed $V_{i,G} = (v_{1,i,G}, v_{2,i,G}, \ldots, v_{D,i,G})$ and the current population $X_{i,G} = (x_{1,i,G}, x_{2,i,G}, \ldots, x_{D,i,G})$, then undergo crossover operation, generating the population of candidates or the trial vectors $U_{i,G} = (u_{1,i,G}, u_{2,i,G}, \ldots, u_{D,i,G})$. $u_{j,i,G}$ is defined as given below

$$u_{1,i,G} = \begin{cases} v_{1,i,G} \text{ if } \text{rand}_{i,j}[0,1] \leq C_r \; \forall j = j_{rand} \\ x_{1,i,G} \text{ otherwise} \end{cases}$$

In the above expression $C_r \in [0.1]$ and is defined as the crossover probability j_{rand} is the random parameter chosen once for each i.
Selection
From the individuals in the current population, the population for the next generation is selected and the corresponding trial vector is determined according to the rule below.

$$X_{i,G+1} = \begin{cases} U_{i,G} \text{ if } f(U_{i,G}) \geq f(X_{i,G}) \\ X_{i,G} \text{ otherwise} \end{cases}$$

In the above expression $f(\cdot)$ is the value of the objective function. Individuals of the temporary population are compared with their counterparts of the current population. The steps of mutation, recombination and selection are repeatedly carried out till the stopping criterion for the algorithm is reached.
Handling of Constraints in DE
The method used for handling constraints is Pareto ranking. The value of constraints is calculated at both the target and trial vectors. The Pareto ranking methods follows the following three rules:
(i) The preferred feasible vector between two feasible vectors (target and trial) is the one which gives the best objective function value.
(ii) Given two vectors (target and trial), the vector which is feasible is preferred and not the infeasible one.
(iii) Given two feasible vectors, the one with the lowest sum of constraint violation is preferred.
Criterion for stopping computation by algorithm
DE algorithm can be stopped under the following conditions:
(i) A maximum number of generations have been reached
(ii) The accuracy level of results has been obtained that is $|f_{max} - f_{min}| \leq \epsilon$

Fig. 15.5 Steps of Differential Evolution

Appendix (Steps of Differential Evolution)

See Fig. 15.5.

References

1. Kotler, P., Keller, K.L., Koshy, A., Jha, M.: Marketing Management-a South Asian Perspective. Pearson India, New Delhi (2013)
2. Schultz, D.E.: Above or below the line? Growth of sales promotion in the United States. Int. J. Advert. **6**(1), 17–27 (1987)
3. Hahn, M., Hyun, J.S.: Advertising cost interactions and the optimality of pulsing. Manag. Sci. **37**(2), 157–169 (1991)
4. Mihiotis, A., Tsakiris, I.: A mathematical programming study of advertising allocation problem. Appl. Math. Comput. **148**(2), 373–379 (2004)
5. Kwak, N.K., Lee, C.W., Kim, J.H.: An MCDM model for media selection in the dual consumer/industrial market. Eur. J. Oper. Res. **166**(1), 255–265 (2005)
6. Coulter, K., Sarkis, J.: Development of a media selection model using the analytic network process. Int. J. Advert. **24**(2), 193–215 (2005)
7. Egan, J.: Marketing Communications. Sage Publications Ltd., London (2015)
8. Rogers, E.: Diffusion of Innovations. Free Press, New York (1962)
9. Exploit the product lifecycle (1965). https://hbr.org/1965/11/exploit-the-product-life-cycle. Accessed 13 Jan 2020
10. Mahajan, V., Muller, E.: Advertising pulsing policies for generating awareness for new products. Mark. Sci. **5**(2), 89–106 (1986)
11. Balakrishnan, S., Hall, N.G.: Maximin procedure for the optimal insertion timing of ad executions. Eur. J. Oper. Res. **85**(2), 368–382 (1995)
12. Freimer, M., Horsky, D.: Periodic advertising pulsing in a competitive market. Mark. Sci. **31**(4), 637–648 (2012)
13. Lin, C.F.: Segmenting customer brand preference: demographic or psychographic. J. Prod. Brand Manag. **11**(4), 249–268 (2002)
14. McDonald, M., Dunbar, I.: Market Segmentation: How to Do It, How to Profit from It. Elsevier Butterworth-Heinemann, Oxford (2004)
15. The double jeopardy of sales promotion (1990). https://hbr.org/1990/09/the-double-jeopardy-of-sales-promotions. Accessed 13 Jan 2020
16. Aggarwal, S., Gupta, A., Govindan, K., Jha, P.C., Meidutė, I.: Effect of repeat purchase and dynamic market size on diffusion of an innovative technological consumer product in a segmented market. Technol. Econ. Dev. Econ. **20**(1), 97–115 (2014)
17. Sridhar, S., Germann, F., Kang, C., Grewal, R.: Relating online, regional, and national advertising to firm value. J. Mark. **80**(4), 39–55 (2016)
18. Above the line (ATL), below the line (BTL) & through the line (TTL) marketing (2018). https://www.feedough.com/atl-btl-ttl-marketing/. Accessed 13 Jan 2020
19. Arora, N.: ATL, BTL and TTL marketing in education industry. Int. J. Res. Innov. Soc. Sci. **2**(1), 13–15 (2018)
20. Cetin, E.: Determining the optimal duration of an advertising campaign using diffusion of information. Appl. Math. Comput. **173**(1), 430–442 (2006)
21. Esteban-Bravo, M., Múgica, J.M., Vidal-Sanz, J.M.: Optimal duration of magazine promotions. Mark. Lett. **16**(2), 99–114 (2005)
22. Beltov, T., Jorgensen, S., Zaccour, G.: Optimal retail price promotions. Anales de estudiosecconómicos y empresariales, Servicio de Publicaciones **16**, 9–36 (2006)
23. Jha, P.C., Aggarwal, S., Gupta, A., Kumar, U.D., Govindan, K.: Innovation diffusion model for a product incorporating segment-specific strategy and the spectrum effect of promotion. J. Stat. Manag. Syst. **17**(2), 165–182 (2014)
24. Lin, C., Lin, Y.T.: Robust analysis on promotion duration for two competitive brands. J. Oper. Res. Soc. **59**(4), 548–555 (2008)
25. Price, K., Storn, R.: Differential evolution-a simple and efficient adaptive scheme for global optimization over continuous spaces, Technical report, International Computer Science Institute, Berkley (1995)

26. Storn, R., Price, K.: Differential evolution - a simple and efficient heuristic for global optimization over continuous spaces. J. Global Optim. **11**, 341–359 (1997)
27. Das, S., Suganthan, P.N.: Differential evolution: a survey of the state-of-the-art. IEEE Trans. Evol. Comput. **15**(1), 4–31 (2011)
28. Kaul, A., Gupta, A., Aggarwal, S., Jha, P.C.: Differential evolution approach to determine the promotion duration for durable technology product under the effect of mass and segment-driven strategies. In: Pant, M., Deep, K., Bansal, J., Nagar, A., Das, K. (eds.) Advances in Intelligent Systems and Computing, vol. 437, pp. 947–960. Springer, Singapore
29. Sethi, S.P.: Note-optimal long-run equilibrium advertising level for the Blattberg-Jeuland model. Manag. Sci. **29**(12), 1436–1443 (1983)
30. Blattberg, R.C., Jeuland, A.P.: A micromodeling approach to investigate the advertising-sales relationship. Manag. Sci. **27**(9), 988–1005 (1981)
31. Mela, C.F., Gupta, S., Lehmann, D.R.: The long-term impact of promotion and advertising on consumer brand choice. J. Mark. Res. **34**(2), 248–261 (1997)
32. Smith, W.R.: Product differentiation and market segmentation as alternative marketing strategies. J. Mark. **21**(1), 3–8 (1956)
33. Mahajan, V., Jain, A.K.: An approach to normative segmentation. J. Mark. Res. **15**, 338–345 (1978)
34. McBurnie, T., Clutterbuck, D.: ive Your Company the Marketing Edge. Penguin Books, London (1988)
35. Meadows, M., Dibb, S.: Assessing the implementation of market segmentation in retail financial services. Int. J. Serv. Ind. Manag. **9**(3), 266–285 (1998)
36. Dibb, S.: Market segmentation: strategies for success. Mark. Intell. Plan. **16**(7), 394–406 (1998)
37. Jobber, D.: Evaluating the effectiveness of below-the-line promotion: a critique. Eur. J. Mark. **7**(1), 64–69 (1973)
38. Gautam, A.: The impact of above the line promotion tools used in the telecom sector - a case study of reliance communications in Western Uttar Pradesh circle, India. Eur. J. Bus. Soc. Sci. **3**(4), 29–38 (2014)
39. Lancaster, G., Reynolds, P.: Above and below-the-line promotion. In: Lancaster, G., Reynolds, P. (eds.) Marketing, Macmillan Business Masters, pp. 235–265. Palgrave, London (1998)
40. Buratto, A., Grosset, L., Viscolani, B.: Advertising a new product in a segmented market. Eur. J. Oper. Res. **175**(2), 1262–1267 (2006)
41. Bass, F.M.: A new product growth model for consumer durables. Manag. Sci. **15**, 215–227 (1969)
42. Simkin, L., Dibb, S.: Prioritising target markets. Mark. Intell. Plan. **16**(7), 407–417 (1998)
43. Mahajan, V., Muller, E.: Timing, diffusion, and substitution of successive generations of technological innovations: the IBM mainframe case. Technol. Forecast. Soc. Chang. **51**(2), 109–132 (1996)
44. Aggarwal, S., Gupta, A., Singh, Y., Jha, P.C.: Optimal duration and control of promotional campaign for durable technology product. In: Proceedings of 2012 the IEEE IEEM, December 2012, Hong Kong, China (2012)
45. Manik, P., Gupta, A., Jha, P.C.: Multi stage promotional resource allocation for segment specific and spectrum effect of promotion for a product incorporating repeat purchase behavior. Int. Game Theory Rev. **17**(02), 1540021(1-21) (2015)
46. Chanda, U., Bardhan, A.K.: Modelling innovation and imitation sales of products with multiple technological generations. J. High Technol. Manag. Res. **18**(2), 173–190 (2008)
47. Aggarwal, P., Vaidyanathan, R.: Use it or lose it: purchase acceleration effects of time-limited promotions. J. Consum. Behav. **2**(4), 393–403 (2003)
48. Devlin, J., Ennew, C., McKechnie, S., Smith, A.: A study of time limited price promotions. J. Prod. Brand Manag. **16**(4), 280–285 (2007)
49. Duran, S., Swann, J.L., Yakici, E.: Dynamic switching times from season to single tickets in sports and entertainment. Optim. Lett. **6**(6), 1185–1206 (2012)
50. Lo, A., Wu, J., Law, R., Au, N.: Which promotion time frame works best for restaurant group-buying deals? Tour. Recreat. Res. **39**(2), 203–219 (2014)

51. Danaher, P.J., Smith, M.S., Ranasinghe, K., Danaher, T.S.: Where, when, and how long: factors that influence the redemption of mobile phone coupons. J. Mark. Res. **52**(5), 710–725 (2015)
52. Price, K.V., Storn, R.M., Lampinen, J.A.: Differential Evolution: A Practical Approach to Global Optimization. Springer, Berlin (2005)
53. Feoktistov, V.: Differential Evolution: in Search of Solutions. Springer, Berlin (2006)
54. Liu, J., Lampinen, J.: A fuzzy adaptive differential evolution algorithm. Soft Comput. Fusion Found. Methodol. Appl. **9**(6), 448–462 (2005)
55. Das, S., Konar, A., Chakraborty, U.: Improved differential evolution algorithms for handling noisy optimization problems. Proc. IEEE Congr. Evol. Comput. **2**, 1691–1698 (2005)
56. Tasgetiren, M.F., Pan, Q.K., Suganthan, P.N., Liang, Y.C.: A discrete differential evolution algorithm for the no-wait flowshop scheduling problem with total flow time criterion. In: IEEE Symposium on Computational Intelligence in Scheduling, Hawaii, pp. 251–258 (2007)
57. Tasgetiren, M.F., Pan, Q., Liang, Y.C.: Discrete differential evolution algorithm for the single machine total weighted tardiness problem with sequence dependent setup times. Comput. Oper. Res. **36**(6), 1900–1915 (2009)
58. Tsafarakis, S., Saridakis, C., Matsatsinis, N., Baltas, G.: Private labels and retail assortment planning: a differential evolution approach. Ann. Oper. Res. **247**(2), 677–692 (2016)
59. Das, S., Mullick, S.S., Suganthan, P.N.: Recent advances in differential evolution-an updated survey. Swarm Evol. Comput. **27**, 1–30 (2016)
60. Piotrowski, A.P.: Review of differential evolution population size. Swarm Evol. Comput. **32**, 1–24 (2017)
61. Ali, M., Törn, A.: Population set-based global optimization algorithms: some modifications and numerical studies. Comput. Oper. Res. **31**, 1703–1725 (2004)
62. Vesterstrom, J., Thomsen, R.: A comparative study of differential evolution, particle swarm optimization, and evolutionary algorithms on numerical benchmark problems. In: IEEE Congress on Evolutionary Computation CEC2004, Portland OR, USA (2004)
63. Meyers, L.S., Gamst, G.C., Guarino, A.J.: Performing Data Analysis Using IBM SPSS. Wiley, New York (2013)

Chapter 16
A Secure RGB Image Encryption Algorithm in Optimized Virtual Planet Domain

Manish Kumar

Abstract The primary aim of this chapter is to provide an optimized, secure RGB image encryption algorithm using 4D hyper-chaotic system in Virtual Planet Domain (VPD). We have constructed a new keyspace, and it shows that the keyspace can resist brute force attacks infeasible. We have tested the proposed algorithm on standard test images. We have successfully verified the robustness of the proposed algorithm by using commonly known attacks such as the differential, cropped, noise, and entropy attacks. Finally, we compared the proposed technique with existing algorithms, and the data (shown in tables) confirms that the proposed algorithm is competitive and can resist exhaustive attacks efficiently.

16.1 Introduction

With advances in the digital era, communication between devices has become more prominent. In today's world of cut-throat competition, there is always a better technique available round the clock with new encryption schemes enabling secure transmission of image data through a network. Every day we transfer much important information over the Internet; data mostly comprise images. Images have become one of the most useful information carriers, which is often used for military, medical science, biometric authentication, and online personal photographs [38]. The most widely known chaos theory was first introduced by Edward Lorenz in 1963 and states chaos when the present determines the future. However, the approximate present does not approximately determine the future. It means that a small change in initial conditions would lead to complete desperate outcomes. From the past decade, many chaos-based image encryption techniques have been proposed [2, 3, 9, 11, 13, 14, 16–19, 21, 24, 26–29, 31, 33, 39, 40, 43–45]. Chaos-based cryptography has received significant attention because of noise-like signals, ergodicity, mixing and

M. Kumar (✉)
Department of Mathematics, Birla Institute of Technology and Science-Pilani, Hyderabad Campus, Hyderabad 500078, India
e-mail: manish.math.bhu@gmail.com

V. Laha et al. (eds.), *Optimization, Variational Analysis and Applications*, Springer Proceedings in Mathematics & Statistics 355,
https://doi.org/10.1007/978-981-16-1819-2_16

sensitivity to initial conditions, which are often connected with those of good ciphers, such as confusion and diffusion. One can find a very close relationship between the chaos system and in terms of diffusion and sensitivity to initial conditions, as well as randomness. To generate the random number sequence, author in [31] used the 1D chaotic map. The technique in [9] describes an algorithm for the rapid numerical application of the class of linear operation to arbitrary operation by using the orthogonal wavelet transform, and due to restriction of the only 1D case provides small keyspace. The security limitation of the 1D chaotic cryptography leads to the nonlinear chaotic map, which uses tangent function and algorithm and iterated many times to provide a high level of security. However, the keystream becomes less sensitive to see [19]. In [39], the authors have suggested that the 1D chaotic map renders small keyspace and weak security. The shuffling process of the image in [33] offers the real-time secure 3D chaotic map, with XOR and modulo operator, and can resist brute force to some extent as compared to fast encryption technique [21] with limited keyspace [29].

The challenges faced in chaos-based encryption have gained much attention to researcher's, to invent secure image encryption techniques [2, 3, 27, 28]. In [26], an encryption algorithm is based on the Lorenz system. Aiming at the protection of information many image encryption techniques have been evolved [11, 14, 40, 44].

The two basic concepts of the cryptosystem are confusion and diffusion [45], the chaotic map generates the pseudo-randomized sequence in [43]. In the first step, confusion was achieved by scrambling the image pixel, and likewise, diffusion was achieved by modifying scrambled pixel values forming the cipher image. Commonly used chaotic maps (i.e., Logistic, Tent, Sine, and so on) suffer from nonlinear distribution. In [18], the authors claim that the discrete chaotic map provides difficulty in the permutation stage of an image. Further, in [13, 17], the algorithm proposed works only on square images. In order to overcome the drawback of the conventional permutation-only-type image cipher, the authors have introduced a new significant diffusion effect in permutation procedure through a two-stage bit-level shuffling algorithm [16]. A bunch of encryption schemes [7, 10, 25, 36, 37, 41] have ended up being extremely weak and show deep security flaws that make them sensitive to various attacks. The keyspace generated from low-dimensional chaotic map gives a shorter periodicity.

In order to overcome the abovementioned flaws, we have proposed a *new, secure* encryption scheme using 4D hyper-chaotic system in the VPD domain. A new formula for a keyspace is derived in such a way that it resists commonly known attacks. The proposed algorithm provides a secure platform for lossless data transmission.

The rest of the work is organized as follows: in Sect. 16.2, we discussed the key generation process by using the TLBO algorithm on the 4D hyper-chaotic system. A virtual planet encoding and decoding scheme is explained. We have elaborated on the proposed optimized encryption algorithm in Sect. 16.3. In Sect. 16.4, the proposed algorithm has been tested with standard test images, and the security analysis has been performed successfully. In Sect. 16.5, a comparison between the proposed algorithm with the methods given in [1, 4–6, 8, 12, 15, 19, 20, 22, 29, 30, 35, 42] has been done. In Sect. 16.6, conclusion drawn from the present work is mentioned.

16.2 Optimized 4D Hyper-chaotic System and Virtual Planet Domain

16.2.1 4D Hyper-chaotic System

A 4D hyper-chaotic system provides chaotic behavior with a minimum of two positive Lyapunov exponents. The main highlighting feature of the 4D hyper-chaotic system is the sensitive dependence on the initial conditions and the parameters. In the proposed algorithm, one can use any 4D hyper-chaotic system to generate keyspace. For simulation purposes, we have used a 4D Rössler hyper-chaotic system to generate keyspace. The Rössler 4D hyper-chaotic system has been given in Eq. (16.1). The chaotic behavior of this system exists for the initial conditions $x = -10$, $y = -6$, $z = 0$, $w = 10$ and the control parameters $a = 0.25$, $b = 3$, $c = 0$, $d = 0.05$, and $e = 0.0025$ as shown in Fig. 16.1. The sensitivity plots for this system are given in Fig. 16.2.

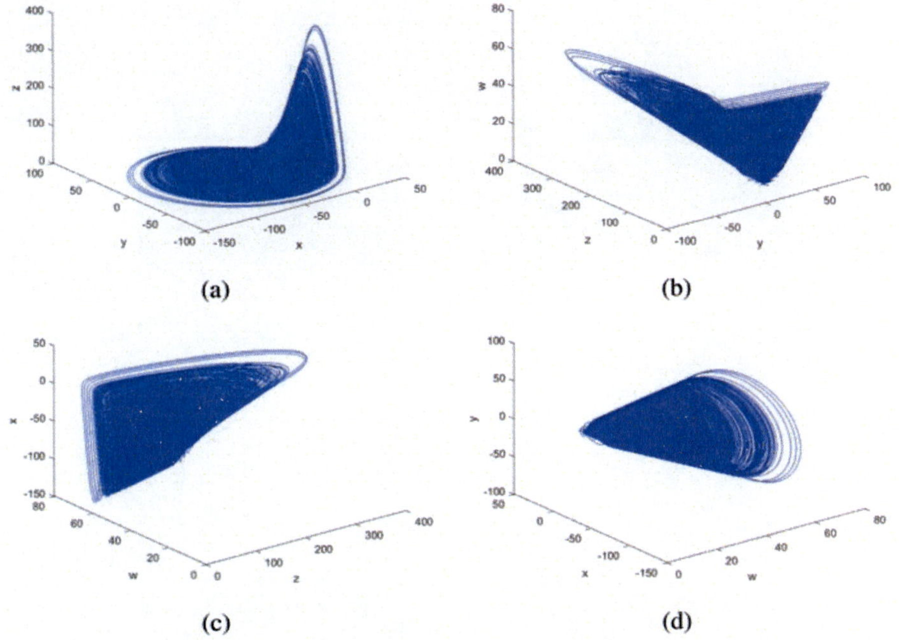

Fig. 16.1 3D phase portraits of 4D hyper-chaotic system (16.1) in **a** xyz-plane, **b** yzw-plane, **c** zwx-plane, **d** wxy-plane

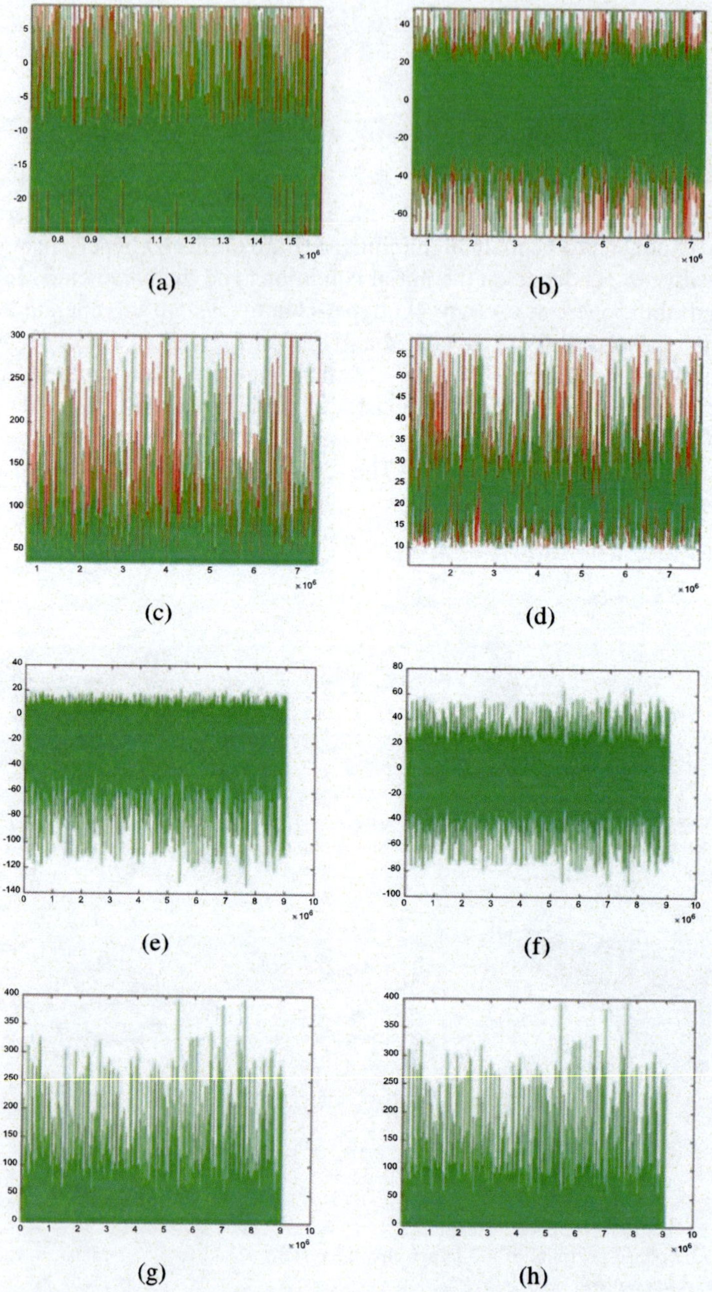

Fig. 16.2 Key sensitivity plots for the small change in x, y, z, and w sequences: **a** $x = 10$ and $x^\star = 10.000000000000001$, **b** $y = -6$ and $y^\star = -6.000000000000001$, **c** $z = 0$ and $z^\star = 0.000000000000001$, and **d** $w = 10$ and $w^\star = 10.000000000000001$, **e** $x = 10$ and $x^\star = 10.0000000000000001$, **f** $y = -6$ and $y^\star = -6.0000000000000001$, **g** $z = 0$ and $z^\star = 0.0000000000000001$, and **h** $w = 10$ and $w^\star = 10.0000000000000001$

$$\left.\begin{aligned} \dot{x} &= x + e(-y - z) \\ \dot{y} &= y + e(x + ay + w) \\ \dot{z} &= z + e(b + xz) \\ \dot{w} &= w + e(-cz + dw) \end{aligned}\right\} \tag{16.1}$$

16.2.2 Teaching–learning-Based Optimization (TLBO) Algorithm

The teaching–learning-based optimization algorithm was first proposed by Rao et al. in [34], which simulates the process of teaching–learning in the classroom. The main concept of TLBO is to reach toward the best solution and leave from the worst solution in every iteration, as proposed in sequence, which only needs the common controlling parameters and do not need any specific parameters. The TLBO algorithm is widely accepted by researchers working in the optimization field. We are not exploring the TLBO algorithm in more detail, but we may refer to see [48]. In this method, the population consists of learners in a class, and design variables are courses offered. The process of working of TLBO is divided into two parts: *Teacher Phase* and *Learner Phase*. The *Teacher Phase* means learning from the teacher, and the *Learner Phase* means learning through the interaction between learners. The above optimization method yields parameters of the 4D hyper-chaotic system that lead to the lowest correlation among adjacent pixels or the highest entropy in the encrypted image.

16.2.3 Virtual Planet Domain Encoding Process

Motivated from [23], we have used a virtual planet encoding scheme to provide a high level of security to the proposed algorithm in terms of commonly known attacks (such as Brute force, cropping, known plain image, cipher image, differential attacks).

16.2.4 Virtual Planet Domain Encoding Process

Motivated from [23], we have used a virtual planet encoding scheme to provide a high level of security to the proposed algorithm in terms of commonly known attacks (such as Brute force, cropping, known plain image, cipher image, differential attacks).

16.2.4.1 Planet Encoding

The RGB image consists of three-channel matrices, namely, red channel, green channel, and the blue channel. The RGB image is encrypted using the planet domain. The first pixel of every channel matrix is taken out as location (1,1,1) for red channel pixel, (1,1,2) corresponds to the second channel pixel, and lastly (1,1,3) corresponds to the third channel pixel. Each planet of the coding domain is uniquely represented by a set of a 3-bit binary number, as explained in the block diagram of encoding (as shown in Fig. 16.3). Now we take every channel pixel value and convert it to an 8-bit binary number, which constitutes and defines the 3-bit grouped binary number, which gives the planet order used by rule-20075 as discussed in Table 16.1.

16.2.4.2 Virtual Planet Diffusion Scheme

To diffuse the entire RGB image in the virtual planet domain, we use the sequence U which was obtained through the generated keyspace. Now, we sort the key sequence U to get the XOR table, which depicts the XOR operation performed between every

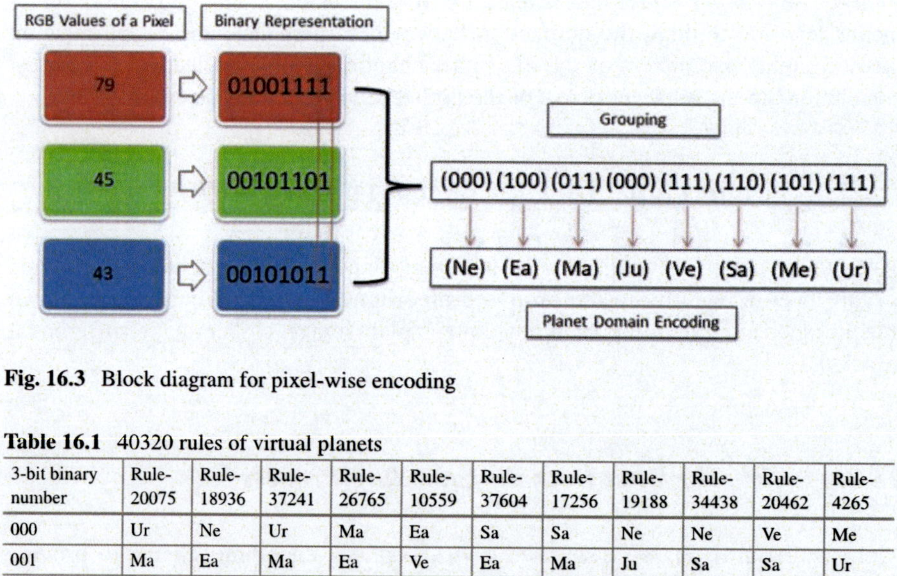

Fig. 16.3 Block diagram for pixel-wise encoding

Table 16.1 40320 rules of virtual planets

3-bit binary number	Rule-20075	Rule-18936	Rule-37241	Rule-26765	Rule-10559	Rule-37604	Rule-17256	Rule-19188	Rule-34438	Rule-20462	Rule-4265
000	Ur	Ne	Ur	Ma	Ea	Sa	Sa	Ne	Ne	Ve	Me
001	Ma	Ea	Ma	Ea	Ve	Ea	Ma	Ju	Sa	Sa	Ur
010	Ju	Ur	Ve	Ve	Me	Ma	Ur	Ur	Ur	Ne	Ne
011	Me	Me	Ne	Ju	Ma	Ve	Me	Me	Ju	Ea	Ma
100	Ne	Sa	Ju	Ur	Ne	Ne	Ne	Ea	Ea	Me	Ju
101	Sa	Ma	Ea	Me	Ju	Ur	Ve	Sa	Ve	Ma	Sa
110	Ea	Ve	Sa	Ne	Sa	Ju	Ju	Ve	Me	Ju	Ve
111	Ve	Ju	Me	Sa	Ur	Me	Ea	Ma	Ma	Ur	Ea

Table 16.2 Virtual planet encoding schemes

Virtual planet	Ur (000)	Ma (001)	Ju (010)	Me (011)	Ne (100)	Sa (101)	Ea (110)	Ve (111)
Scheme 1	178	187	166	174	191	180	199	179
Scheme 2	170	200	168	199	192	181	187	122

Table 16.3 Planet diffusion

row/col	Ur	Ma	Ju	Me	Ne	Sa	Ea	Ve
Ur=000	178	187	166	174	191	180	199	179
Ma=001	187	178	174	166	180	191	179	199
Ju=010	166	174	178	187	199	179	191	180
Me=011	174	166	187	178	179	199	180	191
Ne=100	191	180	199	179	178	187	166	174
Sa=101	180	191	179	199	187	178	174	166
Ea=110	199	179	191	180	166	174	178	187
Ve=111	179	199	180	191	174	166	187	178

encoded planet by using the Rule-20075, as shown in Table 16.3. For instance, any planet (represented by a 3-bit number) can be encoded to a new decimal number between 0 and 255 by either Scheme 1 or Scheme 2, and the process is shown in Table 16.2 as follows:

Ur corresponds to 000 is encoded as 178, Ma corresponds to 001 is encoded as 187, Ju corresponds to 010 is encoded as 166, Me corresponds to 011 is encoded as 174, Ne corresponds to 100 is encoded as 191, Sa corresponds to 101 is encoded as 180, Ea corresponds to 110 is encoded as 199, and Ve corresponds to 111 is encoded as 179. We sort the key sequence V to get the XOR table, which depicts the XOR operation performed between every encoded planet by utilizing Rule-20075 and 178 encoded as 170, 187 encoded as 200, 166 encoded as 168, 174 encoded as 199, 191 encoded as 192, 180 encoded as 181, 199 encoded as 187, and 179 encoded as 122 (described in Table 16.2).

16.2.4.3 Virtual Planet Transform

On extracting numbers from the sequence, we generate an array that helps to form a key, used to transform the planet using this key, and the final output is now transformed and has substantially increased entropy.

16.2.4.4 Virtual Planet Decoding

We initialize a matrix for a dummy image of the size compatible with the size of the plane RGB image. The new 2D matrices obtained have columns that are divided by 8 to get the exact number of columns like that of the plane image, and "3" is on account of three channels. The elements of the decoded image are extracted by comparing them with the encoded domain, and its equivalent binary representation is fed into the transpose of the matrix. Finally, we convert binary to decimal, and then all the planes are extracted and reshaped to get the RGB planet cipher image.

16.3 Proposed Encryption and Decryption Algorithm

The proposed encryption and decryption process is explained in a detailed manner in the following subsections. Further, encryption and decryption results are shown in Fig. 16.5.

16.3.1 Encryption Procedure

The flowchart of the proposed encryption process is shown in Fig. 16.4 and encryption results are provided in Fig. 16.5. The process of each step is explained as follows:

Step 1: The selection of best initial parameters for the 4D hyper-chaotic system is made by using the TLBO algorithm keeping in the view that it provides the lowest correlation among adjacent pixels or the highest entropy.

Step 2: Iterate Eq. (16.1) by using initial parameters obtained from Step 1 for (x, y, z, w), with n_0 times consistently to stay away from the unsafe result of the transitional procedure.

Step 3: To bring all values of each sequence in the range [0, 255], we apply modulo 256 operations on each sequence and round it off to the nearest integer.

Step 4: Perform bit-wise XOR between the first two hyper-chaotic sequences (x, y) and the last two hyper-chaotic sequences (z, w) to produce two sequences U and V, respectively.

Step 5: Read the three channels of the RGB image, mark the index, and accordingly shuffle all the three channels.

Step 6: Define the virtual planet domain using rule-20075 in terms of a 3-bit binary number.

Step 7: Take the pixel value of the shuffled image of size $(m \times n \times 3)$ obtained from Step 2, which is a 3-bit binary number group to an 8-bit binary number so that the whole image is converted to the image of size $(m \times n \times 8)$.

Step 8: Sort the key sequence U using VPD scheme 1 as given in Table 16.2 and diffuse the image with respect to the planet rule used above in Step 5.

Step 9: Sort the key sequence V using VPD scheme 2 as given in Table 16.2 and diffuse the image with respect to the planet rule used above in Step 5.

Step 10: Sort the key sequence V and displace the image pixels with respect to the VPD used.

Step 11: Now get the RGB encrypted image by dividing the column by 8.

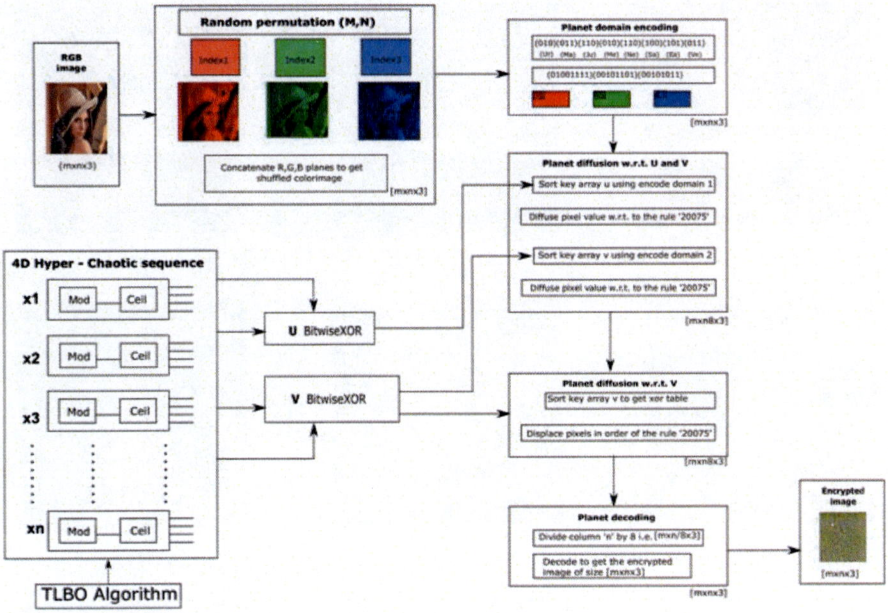

Fig. 16.4 Block diagram for encryption

16.3.2 Decryption Procedure

The decryption process is carried out in a reverse manner.

16.4 Statistical Analysis

Compressing a picture is entirely different from compressing raw binary information. Of course, the general-purpose compression can be used to compress images. Moreover, the results would be optimal. This is due to the applied statistical analysis of an image that may be exploited by the encryption algorithm significantly designed for it. Also, a number of the finer details within the image may be sacrificed for saving a lot of information measure. The images have to be reproduced once decompressed. Two of the error metrics properties used to compare the compression techniques are the Mean Square Error (MSE) and, therefore, the Peak Signal-to-Noise Ratio (PSNR). The statistical properties of pictures are set by the priority of adapting secondary conducts like filtering, restoring, cryptography, and form recognition to the image signal. The essential techniques are enforced to suppress noise or increase a weak signal.

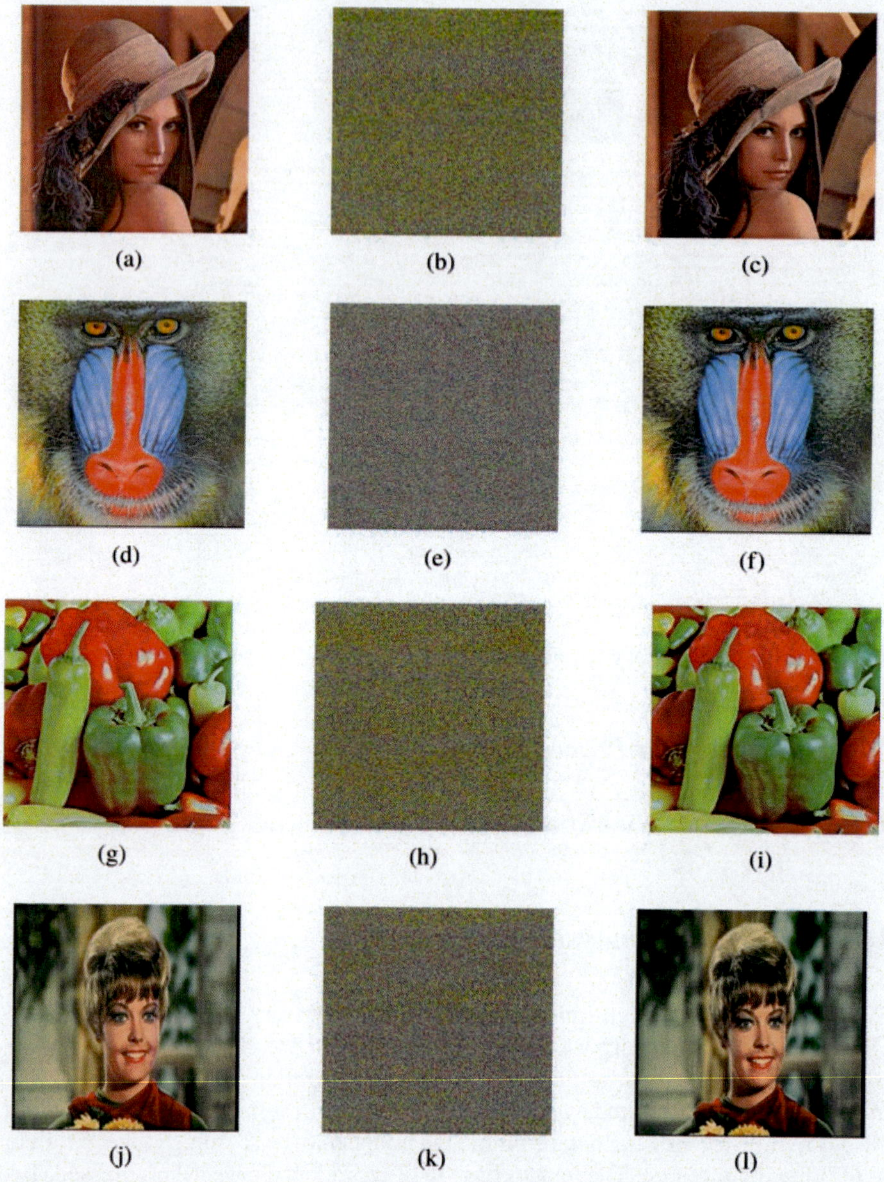

Fig. 16.5 Encryption and decryption results by using the Rule-20075: **a** plane Lena image, **b** cipher Lena image, **c** perfectly decrypted Lena image, **d** plane Baboon image, **e** cipher Baboon image, **f** perfectly decrypted Baboon image, **g** plane Peppers image, **h** cipher Peppers image, **i** perfectly decrypted Peppers image, **j** plane Zelda image, **k** cipher Zelda image, and **l** perfectly decrypted Zelda image

16.4.1 Mean Square Error

The MSE of each RGB component is a measure of similarity between the plane and the decrypted image. A low value of MSE testifies the efficiency of the algorithm to regenerate the encoded image by showing how far residuals are from the regression line data points. The formula for the same is given by

$$MSE = \frac{1}{M \times N} \sum_{m} \sum_{n} \left[|f(m\Delta x, n\Delta y) - f_0(m\Delta x, n\Delta y)|^2 \right],$$

Where Δx and Δy are the pixel sizes, f and f_0 are the intensities of the decrypted and plane image separately that demonstrate the MSE of the plane image and the decrypted image parts.

16.4.2 Peak Signal-to-Noise Ratio

It is used to measure the quality of reconstruction of lossy and lossless compressions like the ones we are using, which involve wavelet packet transform and planet encoding. Any kind of error introduced during the transform becomes the noise for the input signal data. The formula for PSNR is given by

$$PSNR = 10 \times log10 \frac{256 \times 256}{MSE}.$$

It is generally related to the MSE values and a higher value would normally correspond to the higher quality of image decryption, Table 16.4 represents the MSE and PSNR between the plane image and decrypted image.

Table 16.4 MSE and PSNR between the plain image and decrypted image of Lena, Baboon, pepper, and Zelda

RGB elements of image	MSE	PSNR
Red	0	∞
Green	0	∞
Blue	0	∞

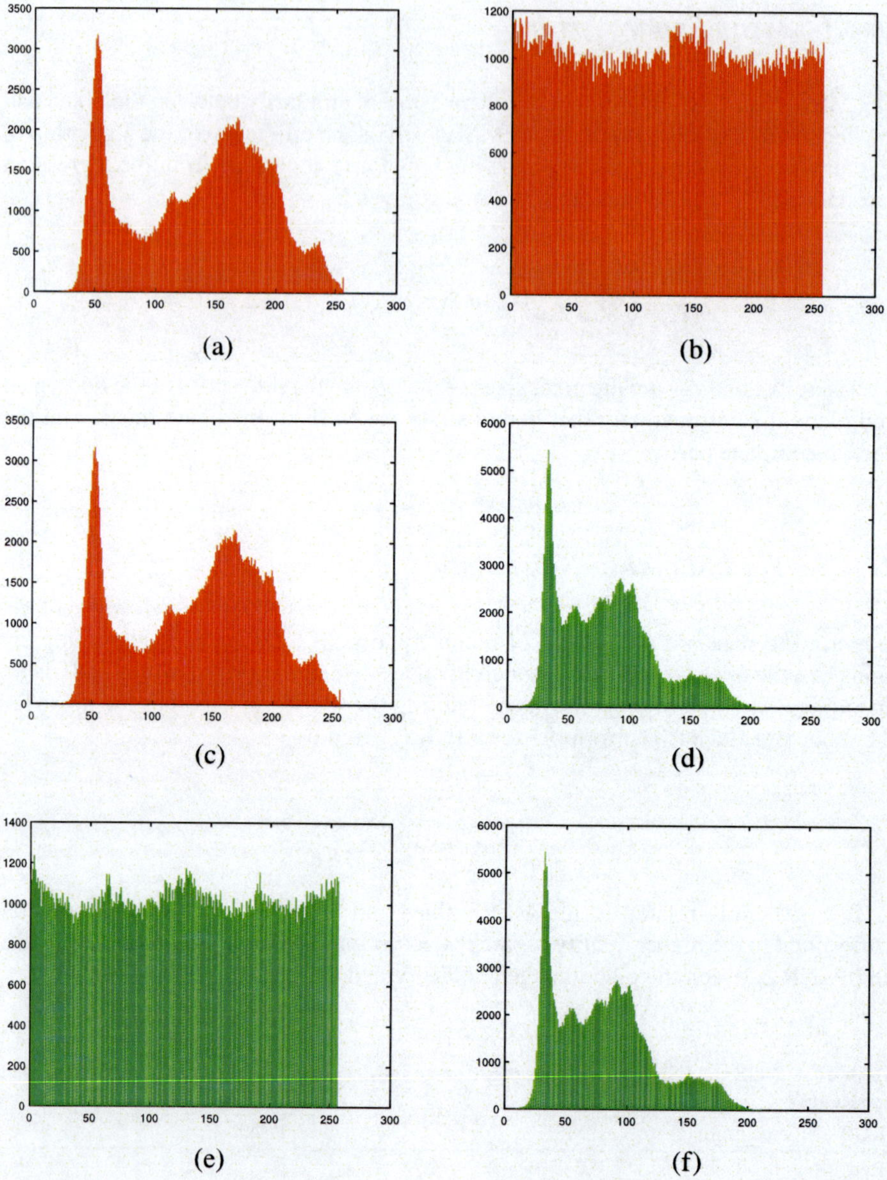

Fig. 16.6 Histogram plots: **a** plane Lena red component, **b** cipher Lena red component, **c** decrypted Lena red component, **d** plane Lena green component, **e** cipher Lena green component, **f** decrypted Lena green component, **g** plane Lena blue component, **h** cipher Lena blue component, **i** decrypted Lena blue component

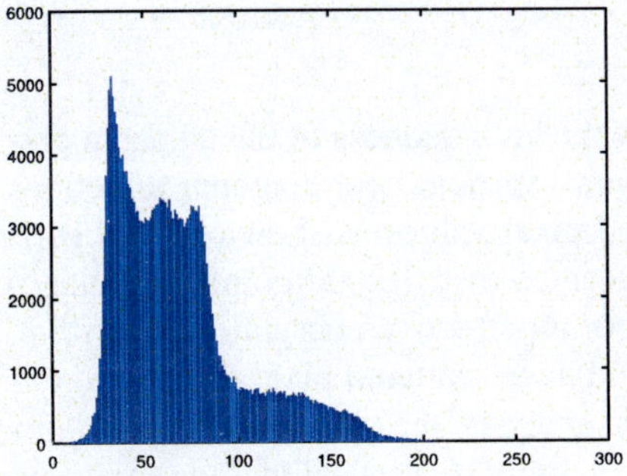

(g)

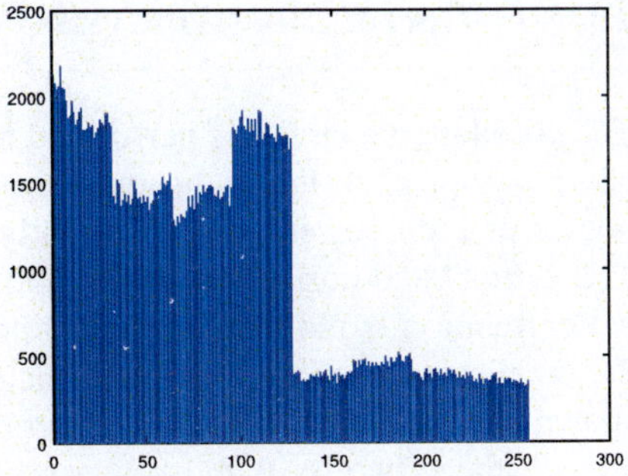

(h)

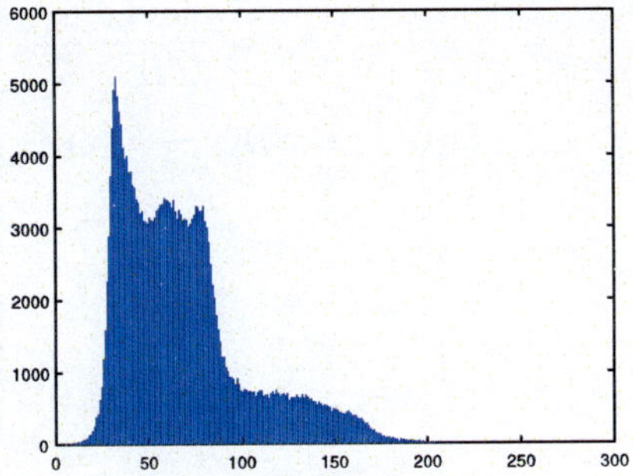

(i)

Fig. 16.6 (continued)

16.4.3 Histogram Analysis

An image is a very distinct structure that makes perfect sense for a human being, and it does not mean that a computer can understand and process the image in the same way. In this case, we can transform the image in different ways, which are more reliable for the machine and can extract different features, one of these representations is a histogram. The histogram of an image shows the significant characteristic to analyze the statistical feature, which gives the pixel distribution for an image, Fig. 16.6, showing the histogram for various RGB images and the cipher images. To prevent a histogram analysis attack, it is necessary to confirm that the distribution of the cipher image must hide the redundancy of the plain image and should not leak any data pertaining to the plain image or the connection between plain image and the cipher image.

16.4.4 Correlation Analysis

The correlation analysis is performed to know the boundedness of the adjacent pixels. Every pixel in the plain image in horizontal, vertical, and diagonal directions possesses high correlation with tightly packed pixel values as shown in Fig. 16.7. The vertical and horizontal correlation is performed over 10,000 points and diagonal correlation is plotted over 1,000 points. For an effective encryption algorithm, the pixel correlation of the cipher image should have scattered pixel distribution as shown in Fig. 16.7. To express such correlation analysis, the correlation coefficients are calculated by (16.2) and values presented in Table 16.5 reveal that the proposed algorithm is good in terms for the correlation coefficient.

$$\left.\begin{aligned} r_{xy} &= \frac{cov(x, y)}{\sqrt{D_x}\sqrt{D_y}}, \\ cov(x, y) &= E\left[(x - E(x))(y - E(y))\right], \\ E(x) &= \frac{1}{L}\sum_{i=1}^{L} x_i, \\ D_x &= \frac{1}{L}\sum_{i=1}^{L}(x_i - E(x))^2. \end{aligned}\right\} \tag{16.2}$$

16.4.5 Cropping Analysis

The cropping analysis is performed on the cipher image to ensure the robustness of the algorithm while transferring the encrypted image to other ends on the communication channel over the network. Figure 16.8a, c, e, g shows the encrypted image is cropped to 50% and the respective decrypted images give some amount of perceptual data (i.e., partial image can be recovered).

16.4.6 Pixel Sensitivity and Key Sensitivity

Any marginal change in a single value of the pixel of an encrypted image results in an entirely distorted decrypted image that can be easily witnessed in Fig. 16.9. It is in relation to the complexity of the encryption and decryption algorithm and the particular importance of every pixel value, even in a large image.

Similarly, any change in the key also results in an entirely distorted image, as shown in the figure, which can be easily witnessed in the following plot. The robustness of any cryptography systems mainly depends on the keyspace. Regardless of

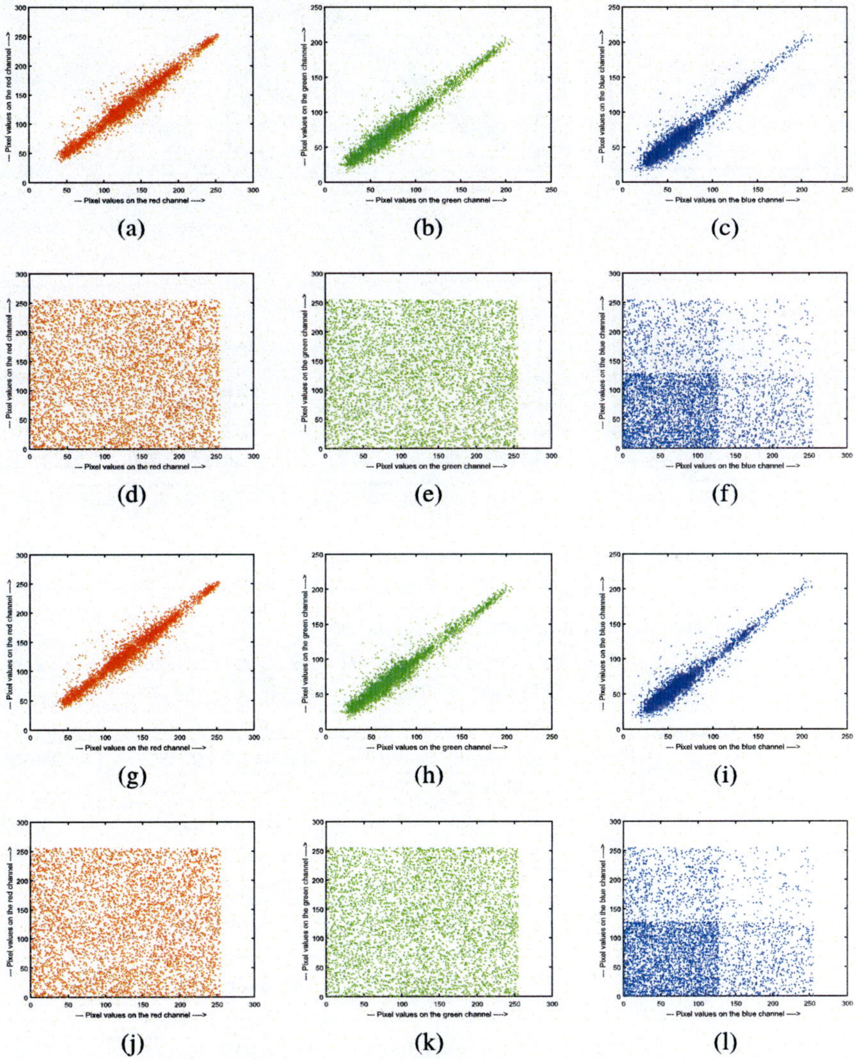

Fig. 16.7 Correlation plots: **Diagonal correlation a** plane red plane, **b** plane green plane, **c** plane blue plane, **d** cipher red plane, **e** cipher green plane, **f** cipher blue plane, **Horizontal correlation g** plane red plane, **h** plane green plane, **i** plane blue plane, **j** cipher red plane, **k** cipher green plane, **l** cipher blue plane, **Vertical correlation m** plane red plane, **n** plane green plane, **o** plane blue plane, **p** cipher red plane, **q** cipher green plane, **r** cipher blue plane

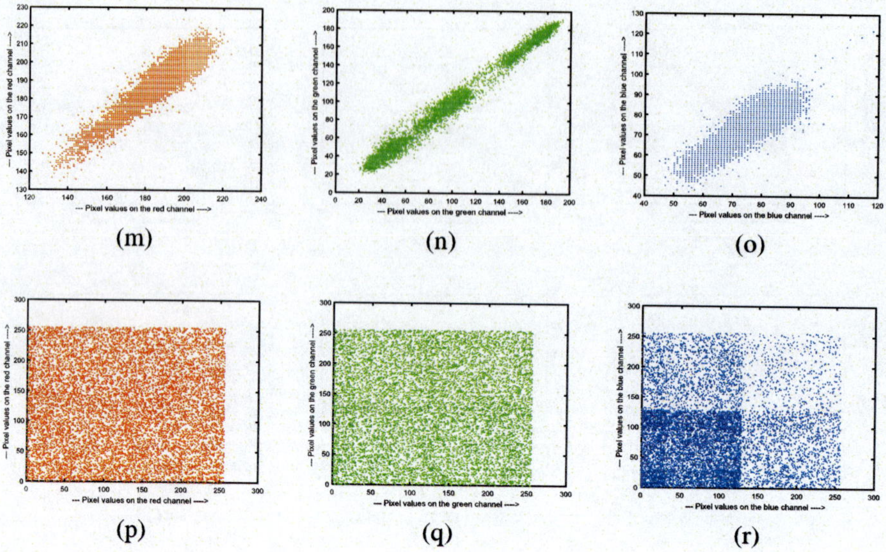

Fig. 16.7 (continued)

Table 16.5 Correlation coefficient for different test images

Image	Channel	Horizontal correlation		Vertical correlation		Diagonal correlation	
		Plane image	Cipher image	Plane image	Cipher image	Plane image	Cipher image
Zelda	Red	0.9948	4.019e-06	0.9672	2.094e-06	0.9924	1.408e-06
	Green	0.9718	2.881e-06	0.8949	5.934e-06	0.9860	1.181e-06
	Blue	0.9965	7.934e-06	0.9562	2.521e-06	0.9810	7.934e-06
Lena	Red	0.9678	6.123e-06	0.9712	1.153e-06	0.9783	2.807e-06
	Green	0.9591	3.729e-06	0.9906	2.005e-06	0.9651	2.452e-06
	Blue	0.9996	6.103e-06	0.9293	4.080e-06	0.9183	6.135e-06
Baboon	Red	0.8930	1.194e-06	0.9301	1.045e-06	0.7040	7.158e-06
	Green	0.8854	3.006e-06	0.8499	1.822e-06	0.5708	8.115e-06
	Blue	0.8766	0.175e-06	0.8807	1.779e-06	0.8024	1.724e-06
Pepper	Red	0.9663	2.552e-06	0.9612	1.237e-06	0.9375	1.041e-06
	Green	0.9720	2.071e-06	0.9722	7.026e-06	0.9662	2.406e-06
	Blue	0.9309	4.278e-06	0.9542	3.241e-06	0.9305	1.100e-06
Flower	Red	0.9682	1.008e-06	0.9851	2.523e-06	0.9692	1.094e-06
	Green	0.9666	9.784e-06	0.9660	1.106e-06	0.9389	8.127e-06
	Blue	0.9791	5.571e-06	0.9713	3.482e-06	0.9392	9.466e-06

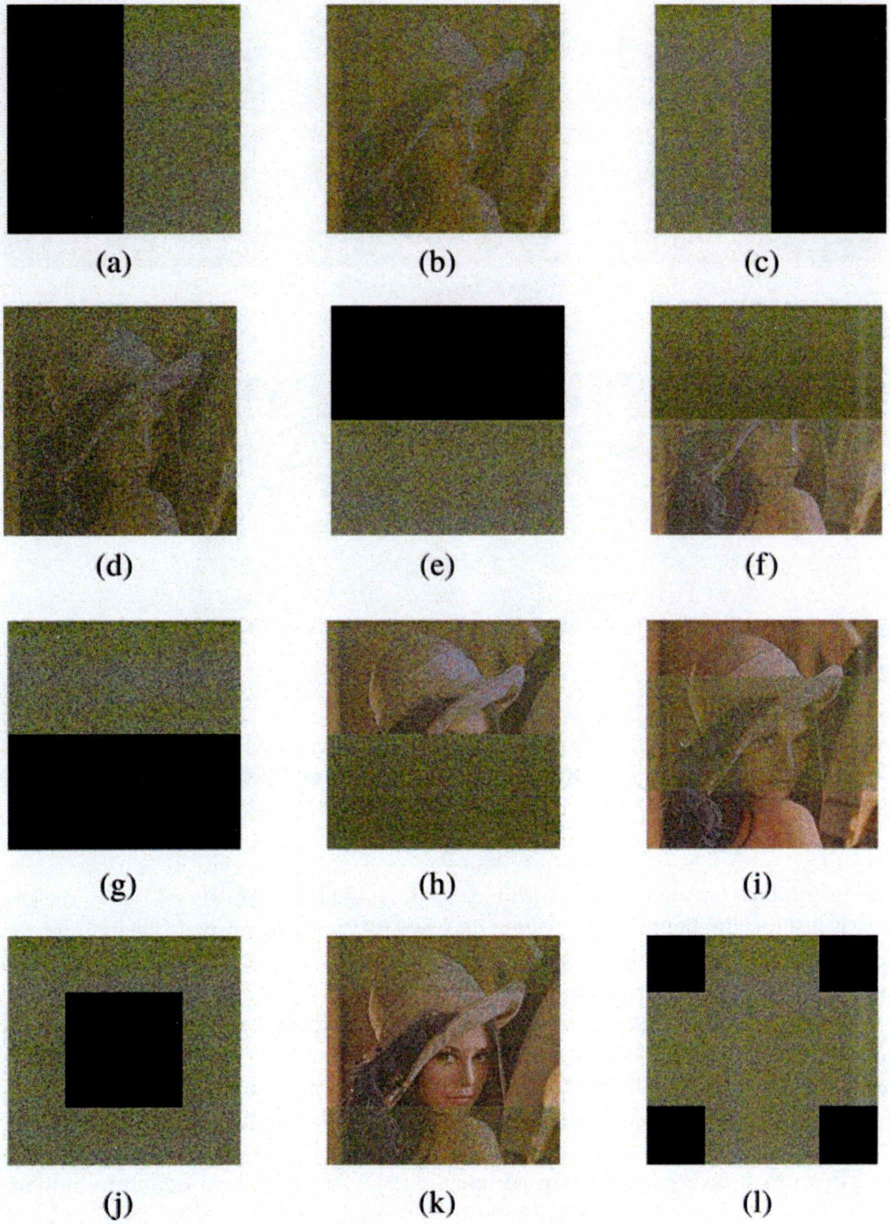

Fig. 16.8 Cropping analysis plots: **a** Lena red plane, **b** encrypted Lena red plane, **c** decrypted Lena red plane, **d** Lena green plane, **e** encrypted Lena green plane, **f** decrypted Lena green plane, **g** Lena blue plane, **h** encrypted Lena blue plane, **i** decrypted Lena blue plane

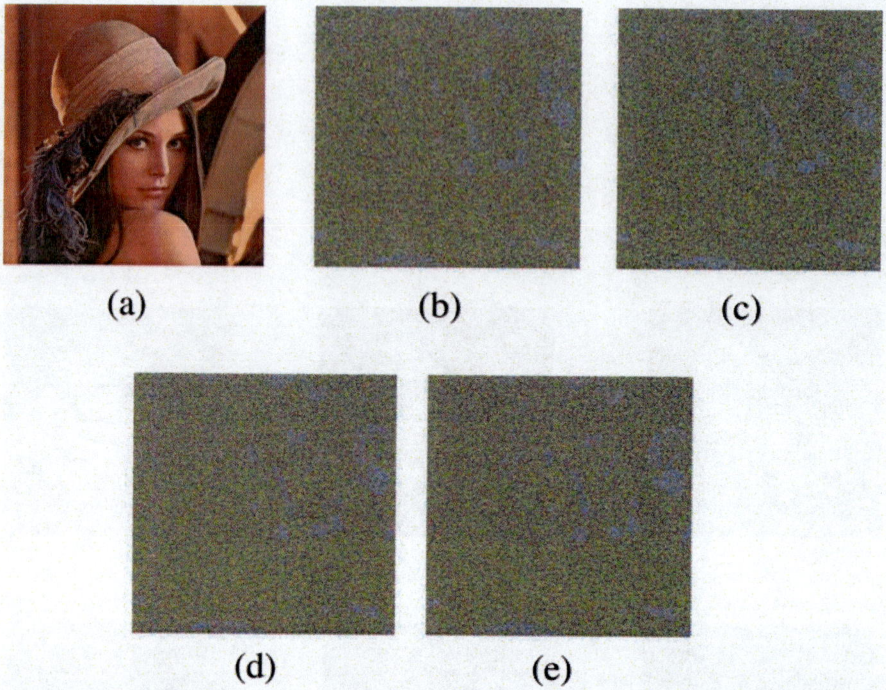

Fig. 16.9 Key sensitivity plots: **a** plane image, **b** cipher image ($x_0 = -10.000000000000001$, $y_0 = -6$, $z_0 = 0$, and $w_0 = 10$), **c** cipher image ($x_0 = -10$, $y_0 = -6.000000000000001$, $z_0 = 0$, and $w_0 = 10$), **d** cipher image ($x_0 = -10$, $y_0 = -6$, $z_0 = 0.000000000000001$, and $w_0 = 10$), **e** cipher image ($x_0 = -10$, $y_0 = -6$, $z_0 = 0$, and $w_0 = 10.000000000000001$)

however robust and elegant, the formula can be. If the key is poorly chosen or the key size is just too small, the cryptosystem may be broken. In the proposed technique, the key consists of n key sequences, i.e., $x_1, x_2, x_3, x_4, x_5, x_6, x_7, x_8, \ldots, x_n$ as shown in Fig. 16.4. Every x_n yields to four key parameters x, y, z, w of 4D hyper-chaotic system. To assess the key sensitivity, we tend to perform the encryption with keys ($x^{\star} = -10.000000000000001$, $y^{\star} = -6.000000000000001$, $z^{\star} = 0.000000000000001$, and $w^{\star} = 10.000000000000001$) to form x_1 by keeping the other respective parameter as constant. The cipher images for initial parameters with added 10^{15} values are shown in Fig. 16.9.

The encryption is performed by adding the $10^{15 \times 4}$ to all the key sequences individually. The overall keyspace includes the 40302 possibilities of planet rules and hyper-chaotic key sensitivity for x_n is $10^{60} \times 10^{60} \times \cdots \times 10^{60}$ (n times)$= 10^{60n}$. To choose U and V key streams from the x_n sequences we have $(4n - 1)!$ distinct options. So the overall keyspace is of variable size given by $40302 \times (4n - 1)! \times 10^{60n}$, which is large enough to resist the possible brute force attacks.

For simulation purpose, we take $n = 1$ for which the key size is $(3)! \times 10^{60} \times 40302 = 2.41812e + 65 \approx 10^{65}$.

16.4.7 Robustness Against Known Plain-Image and Chosen Cipher-Image Attacks

The cipher image is also known as encrypted information as it is comprised of plain-image information in encoded form (i.e., unreadable form without knowing the exact cipher to decrypt image). The reverse of encryption is turning out the cipher information to plain image (Table 16.6).

16.4.7.1 Known Plain-Image Attack

In addition to a brute force attack, another standard attack is known as a known plain-image attack. In this attack, the attacker has a copy of a plain image and a cipher image. It allows the attacker to analyze the relationship between the plain image and the cipher image. It would be unusual for an attacker to have information, and there is some compromise that has already occurred to fetch the original plain image. An

Table 16.6 Keyspace compared with existing techniques

Technique	Keyspace
[19]	10^{45}
[29]	2^{128}
[1]	2^{96}
[4]	2^{203}
[5]	10^{112}
[6]	10^{30}
[8]	2^{260}
[12]	2^{36}
[15]	2^{230}
[20]	10^{70}
[22]	10^{124}
[30]	2^{157}
[35]	2^{80}
[42]	10^{114}
Proposed algorithm	$40302 \times (4n - 1)! \times 10^{60n}$

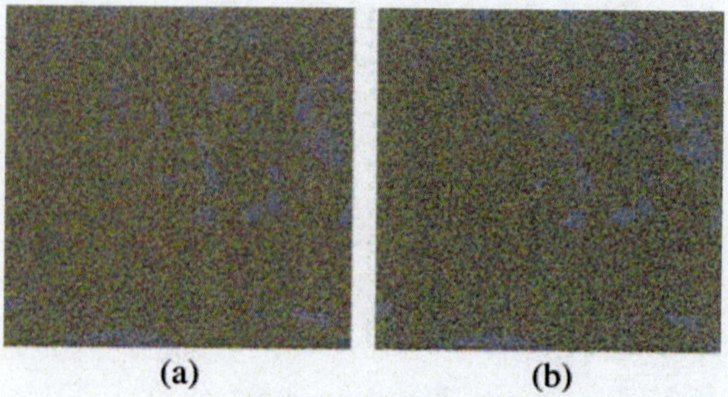

Fig. 16.10 Robustness against attacks: **a** Plain-image attack, **b** cipher-image attack

adversary can figure out the frequency distribution for pixel patterns in the cipher image. The more sophisticated algorithm makes a constant level in the frequency distribution of cipher images so that patterns are not revealed to get any information for the plain image. However, the known plain-image attacks can work on simpler algorithms. Figure 16.10a shows that the proposed algorithm is free from this attack.

16.4.7.2 Cipher-Image Attack

Cipher-image attack is attacking model for crypt-analysis which gathers the information by obtaining the decryption by chosen cipher image from these pieces of information with a goal to acquire the secret key. By the formal definition of Elgamal cryptosystem [32] is semantically secure under the chosen cipher-image attack. Figure 16.10b demonstrates that the proposed algorithm is free from this attack.

16.4.8 Differential Analysis

The differential analysis is very prominent in evaluating and comparing the similarities between any two images.

$$NPCR = \frac{\sum_{i=1}^{N}\sum_{j=1}^{M} D(i,j)}{N \times M} \times 100. \tag{16.3}$$

If $C^1(i,j) = C^2(i,j)$ then $D(i,j) = 0$. Otherwise $D(i,j) = 1$.

$$UACI = \frac{\sum_{i=1}^{N}\sum_{j=1}^{M} \frac{|C^1(i,j)-C^2(i,j)|}{255}}{N \times M} \times 100, \tag{16.4}$$

Table 16.7 NPCR and UACI for different images

Zelda		Lena		Baboon		Pepper		Barbara	
NPCR	UACI	NPCR	UACI	NPCR	UACI	NPCR	UACI	NPCR	UACI
99.6138	33.4710	99.61331	33.5300	99.6151	33.4804	99.6036	33.4890	99.6088	33.4653

where C^1 and C^2 are cipher images before and after one pixel change in original image. NPCR and UACI values for different test images are shown in Table 16.7.

16.4.8.1 Statistical Test for NPCR

From [46], we can demonstrate the proposed algorithm is good by using statistical test for NPCR. Suppose we have two cipher images C^1 and C^2 of size $512 \times 512 \times 3$ each, then hypotheses (H_0 and H_1) with significance level α for $N(C^1, C^2)$ are

$$H_0 : N(C^1, C^2) = \mu_N, \tag{16.5}$$

$$H_1 : N(C^1, C^2) < \mu_N. \tag{16.6}$$

Reject H_0, if $N(C^1, C^2) < N^*_\alpha$, otherwise accept H_0, where

$$N^*_\alpha = \mu_N - \phi^{-1}(\alpha)\sigma_N = \frac{\left(F - \phi^{-1}(\alpha)\sqrt{\frac{F}{MN}}\right)}{F+1}, \tag{16.7}$$

$$\mu_N = \frac{F}{F+1}, \tag{16.8}$$

$$\sigma_N^2 = \frac{F}{(F+1)^2 MN}, \tag{16.9}$$

where F is the largest pixel value in the original image.

Observe from Table 16.8, $N(C^1, C^2)$ values for Baboon, Lena, and Peppers exceed N^*_α values for $\alpha = 0.05, 0.01$, and 0.001. So we can accept the null hypothesis (H_0). Hence, the NPCR values confirm that the proposed algorithm is good.

16.4.8.2 Statistical Test for UACI

Likewise, again from [46], we can demonstrate the proposed algorithm is good by using statistical test for UACI. Assuming that, we have two cipher images C^1 and C^2 of size $512 \times 512 \times 3$ each, then hypotheses (H_0 and H_1) with significance level α for $U(C^1, C^2)$ are

Table 16.8 Statistical test for NPCR

Testing cipher image	F = 255				
	μ_N	σ_N	$N^*_{0.05}$	$N^*_{0.01}$	$N^*_{0.001}$
Numerical values	99.6094	0.0122	99.5893	99.5810	99.5717
Zelda (99.6138)			PASS	PASS	PASS
Lena (99.61331)			PASS	PASS	PASS
Baboon (99.6151)			PASS	PASS	PASS
Peppers (99.6036)			PASS	PASS	PASS
Barbara (99.6088)			PASS	PASS	PASS

$$H_0 : U(C^1, C^2) = \mu_U, \tag{16.10}$$

$$H_1 : N(C^1, C^2) < \mu_U. \tag{16.11}$$

Reject H_0, if $U(C^1, C^2)(U^{*+}_\alpha, U^{*-}_\alpha)$, otherwise accept H_0, where

$$U^{*+}_\alpha = \mu_U + \phi^{-1}(\alpha/2)\sigma_U, \tag{16.12}$$

$$U^{*-}_\alpha = \mu_U - \phi^{-1}(\alpha/2)\sigma_U, \tag{16.13}$$

$$\mu_U = \frac{F+2}{3F+3}, \tag{16.14}$$

$$\sigma_U^2 = \frac{(F+2)(F^2+2F+3)}{18(F+1)^2 MNF}, \tag{16.15}$$

Table 16.9 Statistical test for UACI

Testing cipher image	F = 255				
	μ_U	σ_U	$U^{*+}_{0.05}/U^{*-}_{0.05}$	$U^{*+}_{0.01}/U^{*-}_{0.01}$	$U^{*+}_{0.001}/U^{*-}_{0.001}$
Numerical values	33.4635	0.0462	33.3730/33.5541	33.3445/33.5826	33.3115/33.6156
Zelda (33.4710)			PASS	PASS	PASS
Lena (33.5300)			PASS	PASS	PASS
Baboon (33.4804)			PASS	PASS	PASS
Peppers (33.4890)			PASS	PASS	PASS
Barbara (33.4653)			PASS	PASS	PASS

Table 16.10 Information entropy for different standard images

Information entropy						
Image	Zelda	Lena	Baboon	Pepper	Barbara	Tulip
Plain image	7.6587	7.4958	7.7621	7.70473	7.71537	7.662
Cipher image	7.9997	7.9998	7.9997	7.9996	7.9997	7.9998

Notice from Table 16.9, $U(C^1, C^2)$ values for Baboon, Lena, and Peppers belong to the interval $(U_{\alpha}^{*+}, U_{\alpha}^{*-})$ for $\alpha = 0.05, 0.01$, and 0.001. So we can accept the null hypothesis (H_0). Hence, the UACI values confirm that the proposed algorithm is good.

16.4.9 Information Entropy

The information entropy for a gray image or RGB images is outlined as follows:

$$H(s) = \sum_{i=0}^{2^N-1} p(s_i) \log_2 \left(\frac{1}{p(s_i)} \right), \tag{16.16}$$

where N is that the range of bits to represent the pixel value, s_i represents the probability of image pixels, log represents the base log of 8-bit image, and the theoretical value tends to be eight. Table 16.10 shows the information entropy values of different images. For an information source with 2^N states, $H(s) = N$ bits.

16.4.10 Efficiency of Proposed Encoding Technique

The shuffling process in DNA encoded domain discussed in [47] fails when every pixel is same (i.e., either 0, or 85, or 170 or 255) which reveals the secret keys. For instance, if we take $R = R_{ij}$, $G = G_{ij}$, and $B = B_{ij}$ layers and let the binary representation of one pixel (say R_{11}, B_{11}, and G_{11}) from each layer be $r_1r_2r_3r_4r_5r_6r_7r_8$, $g_1g_2g_3g_4g_5g_6g_7g_8$, and $b_1b_2b_3b_4b_5b_6b_7b_8$, respectively. If we design our proposed algorithm based on [47], then attack is possible if all planets in encoded image are equal, i.e., $r_1g_1b_1 = r_2g_2b_2 = r_3g_3b_3 = \cdots = r_8g_8b_8$ for all pixels. So there are eight possible pictures that are $\forall i, j, [R_{ij}, G_{ij}, B_{ij}] =$ [0 0 0], [0 0 255], [0 255 0], [0 255 255], [255 0 0], [255 0 255], [255 255 0], [255 255 255]. However, in our proposed encoding scheme, we have introduced a series of complex diffusion processes that

Table 16.11 Run time for the algorithm in seconds for image of size 512 × 512 × 3

Image	Key generation	Encryption	Decryption
Lena	3.100102	10.703014	12.955467

resist against vulnerabilities of chosen plain/cipher-image attacks. The encrypted images for abovementioned eight pictures are demonstrated in Fig. 16.11.

16.4.11 Run Time of the Algorithm

This section provides the run time of every phase (i.e., key generation, the run time for the encryption phase, and the run time for the decryption phase). It is elaborated in Table 16.11 which is performed on Windows 10 Pro and on MATLAB 2018a.

16.5 Comparison

There exist several hyper-chaotic-based image encryption techniques in the literature. We draw the comparison between some of the existing algorithms and proposed algorithm, as shown in Table 16.12 on the basis of some performance analysis parameters. For a secure cryptosystem, the huge keyspace is required, which is highlighted in Table 16.6. For the encryption algorithm to be robust, we proposed a large keyspace of variable size based on the possible selection of key streams from the generated n number of hyper-chaotic sequences. The correlation coefficients of neighboring pixels of plain image and cipher image are compared with other techniques, as listed in Table 16.12. The information entropy of cipher is very close to 8, which clearly states the randomness of cipher and its unpredictability is guaranteed. NPCR is about 99%, and the UACI is about 33%, which predicts the encryption algorithm is sensitive to any change in the plain image.

16.6 Conclusion

A new, robust, and optimized secure encryption algorithm has been proposed by using a 4D hyper-chaotic system in VPD. A new formula for keyspace has been designed to protect algorithms from commonly known attacks, as can be seen in Sect. 16.4. The proposed algorithm has been tested and verified successfully on standard RGB images (Zelda, Lena, Baboon, Pepper, and Barbara). The value presented in Table 16.6 reveals that the proposed algorithm has better keyspace and also user-

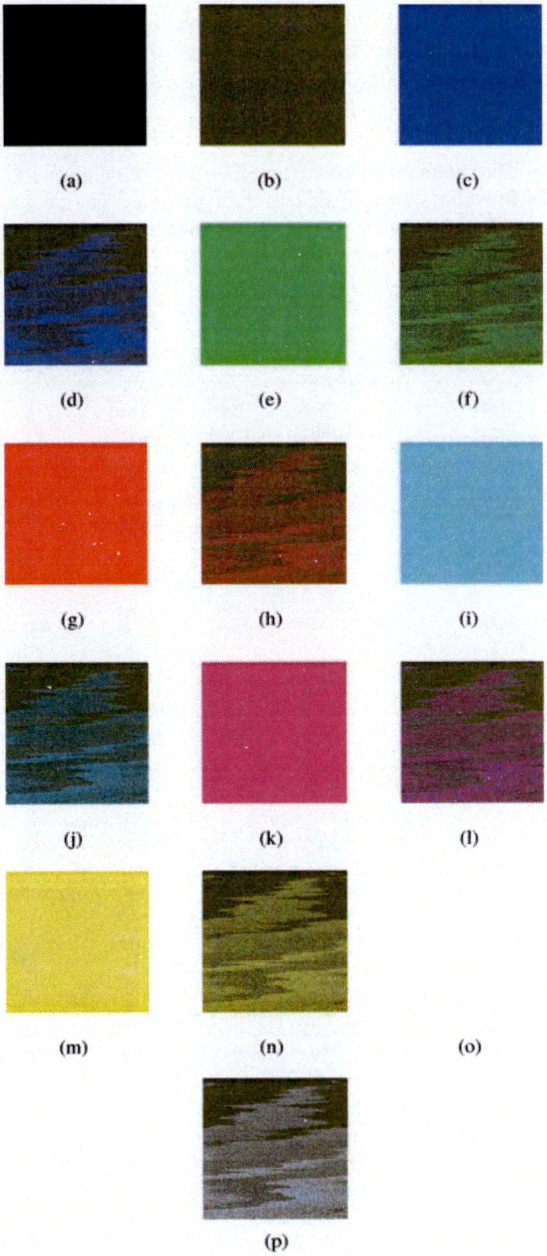

Fig. 16.11 $\forall i, j\ [R_{ij}, G_{ij}, B_{ij}]$: **a** [0,0,0], **b** encrypted [0,0,0], **c** [0,0,255], **d** encrypted [0,0,255], **e** [0,255,0], **f** encrypted [0,255,0], **g** [255,0,0], **h** encrypted [255,0,0], **i** [0,255,255], **j** encrypted [0,255,255], **k** [255,0,255], **l** encrypted [255,0,255], **m** [255,255,0], **n** encrypted [255,255,0], **o** [255,255,255], **p** encrypted [255,255,255]

friendly. The data presented in Table 16.12 suggests that the proposed algorithm is superior to other existing algorithms. Further, the efficiency of the proposed algorithm can be viewed from Fig. 16.11. Hence, the proposed algorithm can opt for the secure transmission of RGB images effectively.

Acknowledgements The author is thankful to the Science and Engineering Research Board, Government of India, for providing financial support through the project file no: YSS/2015/000930. The author is also grateful to the anonymous referees for their constructive comments and valuable suggestions, which have helped to improve the chapter.

Table 16.12 Comparison table

Techniques	Horizontal correlation		Vertical correlation		Diagonal correlation		NPCR	UACI	Entropy
	Plain image	Cipher image	Plain image	Cipher image	Plain image	Cipher image			
[19]	0.9240	–0.158	0.9561	–0.0653	0.9265	0.03231	–	–	–
[29]	0.9765	0.0445	0.9796	0.0284	0.9502	0.0206	99.600	33.40	–
[1]	0.9471	–0.0159	0.9665	–0.0195	0.8985	0.0135	99.629	28.50	7.9975
[4]	0.9341	0.0041	0.9634	0.0036	0.9402	0.0027	25.000	19.00	–
[5]	0.9535	0.0095	0.9616	0.0106	0.9503	0.0048	–	–	7.9891
[6]	0.9603	–0.0030	0.9257	0.0085	0.9055	0.0003	99.5956	33.60	7.9912
[8]	0.9574	0.0038	0.9399	0.0023	0.9183	0.0004	41.9620	33.25	7.9968
[12]	0.9176	0.01183	0.9541	0.00016	0.9020	0.0148	50.300	25.20	–
[15]	0.9537	0.0047	0.9792	0.0030	0.9245	0.0047	99.5100	33.45	7.9997
[20]	0.9241	–0.0142	0.9524	–0.0074	0.9017	–0.0183	–	–	–
[22]	0.9700	–0.0043	0.9409	0.00141	–	–	99.6048	33.50	7.9890
[30]	0.9411	–0.0003	0.9702	0.0014	0.9153	0.0001	99.600	33.54	–
[35]	0.9187	0.005230	0.9557	0.00612	0.8877	–0.007312	99.61	33.48	7.9981
[42]	0.9721	–0.0029	0.9739	–0.0017	0.9705	0.0004	99.59	33.45	7.9971
Proposed algorithm	0.9877	5.3183e-06	0.9394	2.4126e-06	0.9864	3.7980e-06	99.613	33.53	7.9998

References

1. Abdlrudha, H.H., Nasir, Q.: Low complexity high security image encryption based on nested PWLCM chaotic map. In: International Conference for Internet Technology and Secured Transactions (ICITST), pp. 220–225. IEEE (2011)
2. Aguilar-Bustos, A., Cruz-Hernández, C.: Synchronization of discrete-time hyperchaotic systems: an application in communications. Chaos Solitons Fractals **41**(3), 1301–1310 (2009)
3. Aguilar-Bustos, A., Cruz-Hernández, C., López-Gutiérrez, R., Posadas-Castillo, C.: Synchronization of different hyperchaotic maps for encryption. Nonlinear Dyn. Syst. Theory **8**(3), 221–236 (2008)
4. Ahadpour, S., Sadra, Y.: A chaos-based image encryption scheme using chaotic coupled map lattices (2012). arXiv:1211.0090
5. Ahmad, M., Alam, M.S.: A new algorithm of encryption and decryption of images using chaotic mapping. Int. J. Comput. Sci. Eng. **2**(1), 46–50 (2009)
6. Al-Najjar, H., et al.: Digital image encryption algorithm based on a linear independence scheme and the logistic map. In: Proceedings of ACIT-2011 (2011)
7. Arroyo, D., Alvarez, G., Amigó, J.M., Li, S.: Cryptanalysis of a family of self-synchronizing chaotic stream ciphers. Commun. Nonlinear Sci. Numer. Simul. **16**(2), 805–813 (2011)
8. Behnia, S., Akhshani, A., Mahmodi, H., Akhavan, A.: A novel algorithm for image encryption based on mixture of chaotic maps. Chaos Solitons Fractals **35**(2), 408–419 (2008)
9. Beylkin, G., Coifman, R., Rokhlin, V.: Fast wavelet transforms and numerical algorithms i. Commun Pure Appl. Math. **44**(2), 141–183 (1991)
10. Caragata, D., Tutanescu, I.: On the security of a new image encryption scheme based on a chaotic function. Signal, Image Video Process. **8**(4), 641–646 (2014)
11. Chang, C.C., Hwang, M.S., Chen, T.S.: A new encryption algorithm for image cryptosystems. J. Syst. Softw. **58**(2), 83–91 (2001)
12. Chen, G., Mao, Y., Chui, C.K.: A symmetric image encryption scheme based on 3d chaotic cat maps. Chaos Solitons Fractals **21**(3), 749–761 (2004)
13. Chen, J.X., Zhu, Z.L., Fu, C., Yu, H., Zhang, Y.: Reusing the permutation matrix dynamically for efficient image cryptographic algorithm. Signal Process. **111**, 294–307 (2015)
14. Coppersmith, D.: The data encryption standard (DES) and its strength against attacks. IBM J. Res. Dev. **38**(3), 243–250 (1994)
15. Faraoun, K.: Chaos-based key stream generator based on multiple maps combinations and its application to images encryption. Int. Arab J. Inf. Technol. **7**(3), 231–240 (2010)
16. Fu, C., Lin, B.B., Miao, Y.S., Liu, X., Chen, J.J.: A novel chaos-based bit-level permutation scheme for digital image encryption. Opt. Commun. **284**(23), 5415–5423 (2011)
17. Fu, C., Meng, W.H., Zhan, Y.F., Zhu, Z.L., Lau, F.C., Chi, K.T., Ma, H.F.: An efficient and secure medical image protection scheme based on chaotic maps. Comput. Biol. Med. **43**(8), 1000–1010 (2013)
18. Fu, C., Zhang, G.Y., Zhu, M., Chen, Z., Lei, W.M.: A new chaos-based color image encryption scheme with an efficient substitution keystream generation strategy. Secur. Commun. Netw. (2018)
19. Gao, H., Zhang, Y., Liang, S., Li, D.: A new chaotic algorithm for image encryption. Chaos Solitons Fractals **29**(2), 393–399 (2006)
20. Gao, T., Chen, Z.: A new image encryption algorithm based on hyper-chaos. Phys. Lett. A **372**(4), 394–400 (2008)
21. Han, F., Hu, J., Yu, X., Wang, Y.: Fingerprint images encryption via multi-scroll chaotic attractors. Appl. Math. Comput. **185**(2), 931–939 (2007)
22. Hossain, M.B., Rahman, M.T., Rahman, A.S., Islam, S.: A new approach of image encryption using 3d chaotic map to enhance security of multimedia component. In: 2014 International Conference on Informatics, Electronics and Vision (ICIEV), pp. 1–6. IEEE (2014)
23. Kumar, M., Mohapatra, R., Agarwal, S., Sathish, G., Raw, S.: A new RGB image encryption using generalized Vigenére-type table over symmetric group associated with virtual planet domain. Multimed. Tools Appl. **78**(8), 10227–10263 (2019)

24. Kwok, H., Tang, W.K.: A fast image encryption system based on chaotic maps with finite precision representation. Chaos Solitons Fractals **32**(4), 1518–1529 (2007)
25. Li, C., Arroyo, D., Lo, K.T.: Breaking a chaotic cryptographic scheme based on composition maps. Int. J. Bifurc. Chaos **20**(08), 2561–2568 (2010)
26. Liu, H., Wang, X.: Triple-image encryption scheme based on one-time key stream generated by chaos and plain images. J. Syst. Softw. **86**(3), 826–834 (2013)
27. López-Mancilla, D., Cruz-Hernández, C.: Output synchronization of chaotic systems: model-matching approach with application to secure communication. Nonlinear Dyn. Syst. Theory **5**(2), 141–156 (2005)
28. López-Mancilla, D., Cruz-Hernández, C.: Output synchronization of chaotic systems under nonvanishing perturbations. Chaos Solitons Fractals **37**(4), 1172–1186 (2008)
29. Mao, Y., Chen, G., Lian, S.: A novel fast image encryption scheme based on 3d chaotic baker maps. Int. J. Bifurc. Chaos **14**(10), 3613–3624 (2004)
30. Masmoudi, A., Bouhlel, M.S., Puech, W.: A new image cryptosystem based on chaotic map and continued fractions. In: 18th European Signal Processing Conference, pp. 1504–1508. IEEE (2010)
31. Matthews, R.: On the derivation of a chaotic encryption algorithm. Cryptologia **13**(1), 29–42 (1989)
32. Meier, A.V.: The ElGamal cryptosystem (2005)
33. Pareek, N.K., Patidar, V., Sud, K.K.: Image encryption using chaotic logistic map. Image Vis. Comput. **24**(9), 926–934 (2006)
34. Rao, R.V., Savsani, V.J., Vakharia, D.: Teaching-learning-based optimization: a novel method for constrained mechanical design optimization problems. Comput.-Aided Des. **43**(3), 303–315 (2011)
35. Slimane, N.B., Bouallegue, K., Machhout, M.: A novel image encryption scheme using chaos, hyper-chaos systems and the secure Hash algorithm SHA-1. In: International Conference on Control, Automation and Diagnosis (ICCAD), pp. 141–145. IEEE (2017)
36. Solak, E., Çokal, C.: Algebraic break of image ciphers based on discretized chaotic map lattices. Inf. Sci. **181**(1), 227–233 (2011)
37. Solak, E., Çokal, C., Yildiz, O.T., Biyikoğlu, T.: Cryptanalysis of Fridrich's chaotic image encryption. Int. J. Bifurc. Chaos **20**(05), 1405–1413 (2010)
38. Srivastava, A.: A survey report on different techniques of image encryption. Int. J. Emerg. Technol. Adv. Eng. **2**(6), 163–167 (2012)
39. Sun, F., Liu, S., Li, Z., Lü, Z.: A novel image encryption scheme based on spatial chaos map. Chaos Solitons Fractals **38**(3), 631–640 (2008)
40. Tang, Y., Wang, Z., Fang, J.A.: Image encryption using chaotic coupled map lattices with time-varying delays. Commun. Nonlinear Sci. Numer. Simul. **15**(9), 2456–2468 (2010)
41. Wang, X., Liu, L.: Cryptanalysis of a parallel sub-image encryption method with high-dimensional chaos. Nonlinear Dyn. **73**(1–2), 795–800 (2013)
42. Wang, X., Zhu, X., Zhang, Y.: An image encryption algorithm based on Josephus traversing and mixed chaotic map. IEEE Access **6**, 23733–23746 (2018)
43. Wang, X.Y., Gu, S.X.: New chaotic encryption algorithm based on chaotic sequence and plain text. IET Inf. Secur. **8**(3), 213–216 (2014)
44. Wei, X., Guo, L., Zhang, Q., Zhang, J., Lian, S.: A novel color image encryption algorithm based on DNA sequence operation and hyper-chaotic system. J. Syst. Softw. **85**(2), 290–299 (2012)
45. Wong, K.W., Kwok, B.S.H., Law, W.S.: A fast image encryption scheme based on chaotic standard map. Phys. Lett. A **372**(15), 2645–2652 (2008)
46. Wu, Y., Noonan, J.P., Agaian, S.: NPCR and UACI randomness tests for image encryption. Cyber J.: Multidiscip. J. Sci. Technol. J. Sel. Areas Telecommun. (JSAT) **1**(2), 31–38 (2011)
47. Xie, T., Liu, Y., Tang, J.: Breaking a novel image fusion encryption algorithm based on DNA sequence operation and hyper-chaotic system. Opt.-Int. J. Light Electron Opt. **125**(24), 7166–7169 (2014)
48. Zou, F., Chen, D., Xu, Q.: A survey of teaching-learning-based optimization. Neurocomputing **335**, 366–383 (2019)

Chapter 17
Identification and Analysis of Key Sustainable Criteria for Third Party Reverse Logistics Provider Selection Using the Best Worst Method

Jyoti Dhingra Darbari, Shiwani Sharma, and Mark Christhian Barrueta Pinto

Abstract Growing environmental issues, social concerns, enforced regulations and intense competition have motivated electronic companies to inculcate Reverse Logistics (RL) practices in action for sustainable Reverse Supply Chain (RSC). Due to lack of expertise and the heavy costs associated with the setting up of reverse logistics system, RL practices are widely embraced by most companies through Third Party Reverse Logistics Providers (3PRLPs). Due to the dependency of companies on 3PRLPs, the evaluation and selection of 3PRLP is a matter of strategic concern and requires critical decision-making. The main challenge in this regard that the companies face is to identify the appropriate criteria for assessing the performance of 3PRLP under a sustainable environment. In this sense, the main intent of the current study is to provide a systematic framework for an electronics company to (i) identify the most relevant 3PRLP performance evaluation criteria under three sustainability dimensions namely, economic, environmental and social, (ii) extract the most influential list of sustainable criteria and (iii) determine the weights of importance of the influential criteria. In order to attain this objective, a decision-making model is proposed in which firstly, the economic, environmental and social criteria are derived from an extensive literature survey. Secondly, Delphi technique is used to shortlist the most influential criteria. Thirdly, the Best Worst Method (BWM) is used to determine the importance of the shortlisted criteria. The result analysis shows that environmental sustainability is the primary focus of the companies for the implementation of RL, contrary to the assumption that economic performance is always

J. D. Darbari (✉)
Department of Mathematics, Lady Shri Ram College for Women, University of Delhi, Delhi 110024, India
e-mail: jydbr@hotmail.com

S. Sharma
Department of Operational Research, Faculty of Mathematical Sciences, University of Delhi, Delhi 110007, India
e-mail: shiwani.sharma.8668@gmail.com

M. C. B. Pinto
School of Business, Universidad Peruana de Ciencias Aplicadas (UPC), Lima 15023, Peru
e-mail: mbarruetapinto@gmail.com

V. Laha et al. (eds.), *Optimization, Variational Analysis and Applications*,
Springer Proceedings in Mathematics & Statistics 355,
https://doi.org/10.1007/978-981-16-1819-2_17

the major motivation. 'Quality', 'RL Practices' and 'Health and Safety' are accorded the highest ranking under economic, environmental and social dimensions, respectively. The proposed model can assist electronic companies in determining the most important criteria for sustainable 3PRLP selection for outsourcing RL activities.

17.1 Introduction

The process of Reverse Logistics (RL) involves activities aimed at the appropriate backward flow of products which are considered as reached their end-of use/end-of-life stage by the consumers [40]. Rogers and Tibben-Lembke [83] defines RL as, 'the trend of design, schedule, planning, controlling and warehousing and also information for returned products in reverse flow of classical supply chain in order to recover value and get the competitive advantage'. Figure 17.1 provides a schematic view of the forward and reverse flow of goods and the activities involved in a generic Supply Chain (SC). RL has gained immense attention in the past two decades as a result of the environmental sensitization of consumers and governments. For businesses, RL proves to be a key strategy in managing a sustainable SC [32]. Companies are inclined towards RL nowadays due to decrease in availability of raw materials and consequently rise in their prices [46]. The specific activities of RL such as repair, remanufacture, refurbish help gain monetary benefits in terms of reselling of refurbished products while recycling, disassemble and proper disposal help in reducing the ill effects of the dumping of unused products [103]. Moreover, the backward channel provides opportunities of jobs to various marginalized workers, specifically in developing nations such as India, Bangladesh and Taiwan. Hence, all the three dimensions of sustainability are covered naturally under the umbrella of RL activities [48, 101].

Most logistics systems fail to manage the concurrent flows as they have different necessities and are managed under different constraints [27]. The forward flow is customer demand driven, while the reverse flow is driven by the quantity of products returned. Each RL process requires a different considered focus, hence companies need to plan and design RL network which is an uphill task [34]. Additionally in RL the amount of returned products is uncertain, the backward flow is untimely and the condition of the products is unknown, which adds further complexity in scheduling and planning the RL activities [89]. Organizations, particularly, in India, although are legally bound to implement RL but do not have a suitable structure in place. There are many hindering factors such as lack of knowledge, lack of government support, lack of awareness amongst consumers and other financial and organizational constraints [75]. Consequently, most organizations prefer outsourcing the complex task of managing RL activities to reduce the cost of implementation, for streamlining the recovery and redistribution process and for focusing on their core competencies [2, 28].

Outsourcing of RL activities has its own challenges and choosing a reliable and sustainable Third Party Reverse Logistics Provider (3PRLP) is a daunting task. The associated financial as well as operational impact along with the long-term effect of the partnership needs to be recognized [102]. Accordingly, the organization may choose to outsource all or some of its RL activities. Consequently, 3PRLP selection process sometimes involves choosing a single 3PRLP and other times choosing multiple 3PRLPs. Moreover, the outsourcing may be done under partial or full collaboration with 3PRLPs [24]. The strategic decision of choosing the 3PRLP, the activities to be outsourced and the nature of the partnership must be based on a critical analysis of the operational, financial, sustainable capabilities of the reverse logistics provider by the Reverse Supply Chain (RSC) managers.

Within the context of 3PRLP evaluation and selection process, identification of appropriate criteria of evaluation is of prime importance as they echo the organization's requirements and expectations from the partnership with the 3PRLP. In addition to the traditional criteria such as cost, quality, flexibility and responsiveness, assessing the capabilities of 3PRLPs with regard to environmental and social concerns has become imperative for organizations focusing on managing sustainable SC practices [11, 15, 31]. Clearly, unlike the evaluation criteria for forward logistics provider, which is more economically driven, the evaluation criteria of 3PRLP must include environmental and social performance indicators. Hence, 3PRLP evaluation process requires a more detailed list of attributes and criteria, most of which may be difficult to quantify and involves a more difficult process of data collection [39]. Moreover, the filtration of the criteria to extract the most significant ones and the sorting of the criteria in the order of their weights of importance are essential parts of the 3PRLP selection process. Selection of criteria ideally should be company specific, case specific and industry specific as they impact the decision selection of 3PRLPs. Moreover, the process of selection of performance criteria for 3PRLP evaluation with sustainable perspective is dominated by the presence of conflicting opinions of different stakeholders of the SC of the organization, which adds to the complexity in the decision-making environment. Multiple-criteria decision-making (MCDM) techniques promise to be very effective in this regard for simultaneously evaluating various criteria based on sustainability dimensions in group decision-making environment [63].

Although there has been ample research on the need for outsourcing to 3PRLPs for achieving a sustainable RSC and the type of criteria to be considered for evaluation of 3PRLPs, however, most of these studies are theoretical in nature. Very few studies have developed mathematical models for the identification and selection of criteria in a systemized manner. This study focuses on identification and selection of key performance criteria for the evaluation of 3PRLPs based on all three dimensions of sustainability, by developing a decision-making model for an electronic company based in India. The company XYZ is looking for a partnership venture with a suitable 3PRLP with the aim of achieving a sustainable RSC. In the first stage, an exhaustive list of criteria based on economic, environmental and social dimensions is prepared through an extensive literature survey. The criteria are identified specifically for the evaluation of 3PRLPs who are providing services in the electronics industry. In order to extract the most relevant criteria as per the company's requirements, Delphi technique is employed to gather opinions of the Decision Makers (DMs)

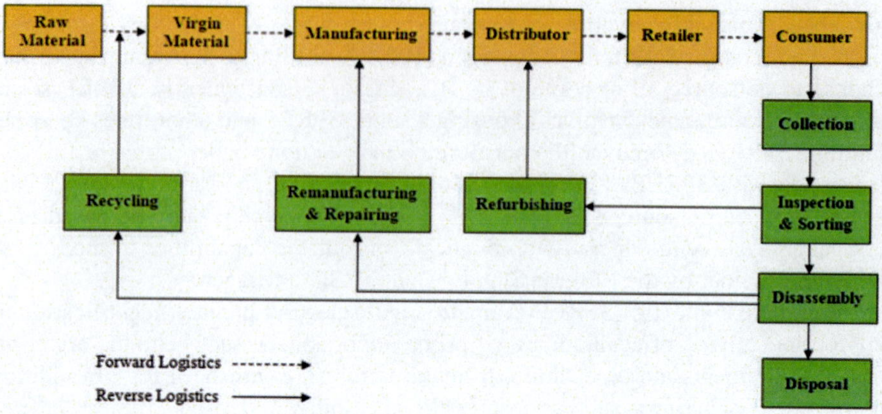

Fig. 17.1 Flow of physical goods for forward and reverse logistics [1, 95, 104]

through a structured questionnaire and semi-structured interviews. The data analysis of the information gathered through the Delphi technique helps in the first level of filtration of the criteria. In the second stage, Best Worst Method (BWM) technique is employed to rank the importance of economic, environmental and social dimensions and also to rank the criteria under the three dimensions of sustainability as per the decision-making team.

The remainder of this chapter is organized as follows: Sect. 17.2 provides literature review on the need for outsourcing in RL and the importance of sustainability related factors for evaluation and selection of 3PRLP. Section 17.3 explains the proposed methodology developed for the identification, evaluation and selection of criteria with regard to all the three dimensions of sustainability. The application of the proposed methodology is presented in Sect. 17.4. Section 17.5 provides the result discussion of the study. Section 17.6 concludes the paper and includes suggestions for future research scope.

17.2 Literature Review

The focus of the study is on the analysis of key sustainable criteria for the evaluation of third-party logistics provider in RL. The literature review presented in this section discusses the work done by researchers over the years in that direction. The literature review section is divided into three sections: Sect. 17.2.1 discusses the need for outsourcing in RL in SC; Sect. 17.2.2 demonstrates the plethora of work with regard to identification, evaluation and selection of sustainable performance criteria for provider selection in RL; Sect. 17.2.3 highlights the research gap and provides the significant contribution of the present study.

17.2.1 Outsourcing in Reverse Logistics

Forward logistics in SC refers to all activities with regard to the flow of product and information from the suppliers to the customers for satisfying customer's needs and meeting their expectations [16]. Contrary to this, RL refers to all activities of SC aimed at managing the reverse flow of returned product from the consumption point to the origin point for the purpose of capturing value and proper disposal [83]. However, it does not imply that RL is just reversing the forward logistics [31, 64]. RL faces many complexities and its effective implementation requires suitable RL network configuration to carry the broad range of activities such as collection, sorting, inspection, disassembly, remanufacture, recycling and disposal [35]. Due to the lack of knowledge and infrastructure, most firms prefer to outsource RL activities to specialized 3PRLP for advantages such as reduced costs, advanced technology and better performance [4]. However, the problem of third party provider selection faces greater complexity for outsourcing activities related to RL in comparison to traditional forward logistics because of the major difference in their scope of work and expertise [41]. Even the most successful third-party logistics providers are not able to manage the reverse flow of products efficiently and effectively [27]. 3PRLPs must be specialized in handling the value-added activities for the reverse flow of returned products [2]. They must be well equipped to carry these activities following proper environmental guidelines [8]. Therefore, dependency of the firms on 3PRLPs is huge in terms of achievement of sustainable business practices [19]. Due to these differentiators with regard to the objective of outsourcing, 3PRLPs play a strategic role in aiding firms to attain sustainable competitive advantage, government support and customer satisfaction [40]. Hence, suitable 3PRLP selection for outsourcing in RL is a crucial decision for RSC managers and has emerged as an important research area [102].

17.2.2 Sustainable Performance Criteria for Provider Selection in Reverse Logistics

The decision of provider selection, while considering complete or partial outsourcing of the RL activities, needs the development of a comprehensive conceptual framework based on various performance metrics [1, 26]. The framework is broadly influenced by the set of criteria and the evaluation approach [105]. Identification of an appropriate set of performance criteria is a critical stage of the decision-making process, as it significantly impacts the evaluation rankings of the alternatives [13]. Hence, 3PRLP selection problem must be characterized by exhaustive research on the selection of performance criteria of evaluation of 3PRLPs. In the literature, traditional economic criteria such as cost of services, financial position, asset ownership are considered essential criteria by most authors [6, 77, 87]. Further, process-based criteria such as resource capacity, network capacity, skilled manpower, service capability, flexibility and quality of service have always been considered important for evaluation of 3PRLPs [50, 95]. Moreover, 3PRLPs offering complete RL services must be equipped with advanced equipment, specialized infrastructure and secure IT and tracking system [2]. Most organizations seek to implement RL for pursuing sustainability goals as RL activities majorly cover all sustainability dimensions [59, 74, 96, 100]. Hence, sustainability performance metrics of 3PRLPs are extremely important for a effective RSC [3, 19]. A review of research on 3PRLP evaluation and selection demonstrates that evaluation criteria based on all three dimensions of sustainability—economic, environmental and social—are dominant in the recent literature [12, 15, 31]. However, there is a lack of studies focusing on the critical analysis of 3PRLP evaluation criteria and development of mathematical models for selection of criteria with regard to industry specific requirements.

17.2.3 Research Contribution

The literature analysis presented above demonstrates that sustainability related factors are essential for the evaluation of 3PRLPs. However, none of the studies have discussed all the performance evaluation criteria in a systematic way. It is evident from the above discussion that most of the studies with regard to developing 3PRLP evaluation criteria are based on the triple bottom approach. The motivation of researchers is more on developing models for the evaluation of 3PRLPs while less emphasis is laid on the systematic identification and selection of the key criteria. Moreover, most of the studies focusing on the need for developing the criteria for evaluation of 3PRLP are based on theoretical findings and lack development of analytic models. This gap is considered in the study. Most importantly, organizations need to consider the criteria which match their requirements [73]. In this direction, this chapter aims to develop a 3PRLP selection model for an Indian electronics company for the selection of key evaluation criteria identified from the plethora of criteria in the literature and practice. The novelty of the study is to provide a systematic framework for an electronics company to achieve the following objectives:

1. To identify a broad set of 3PRLP sustainable performance evaluation criteria through an extensive literature survey.
2. To prioritize the key sustainable criteria based on deliberations amongst the team of experts from an electronics company using Delphi technique which is very effective in managerial decision-making.
3. To determine the rank of importance of the key criteria under each sustainability dimension using BWM, an efficient MCDM technique.

17.3 Methodology used for Selection and Evaluation of Criteria

The selection process of 3PRLP ideally must involve a thorough evaluation of the performance of 3PRLPs based on key criteria based on all three sustainability dimensions. Hence, the focus of the study is to develop a systematic model which can provide guidance to the case company in (i) identifying the most relevant 3PRLP performance evaluation criteria under three sustainability dimensions namely, economic, environmental and social, (ii) extracting the most influential list of sustainable criteria and (iii) determining the weights of importance of the influential criteria. In order to attain this objective, a decision-making model is proposed, in which, firstly, the economic, environmental and social criteria are derived from an extensive literature survey. Secondly, Delphi technique is used to shortlist the most influential criteria. Thirdly, in accordance with the above evaluated results, BWM is used to determine the importance of the shortlisted criteria. The steps of the proposed methodology are described in the following sections:

17.3.1 Identification of Criteria

For the purpose of evaluation of 3PRLP, identification of relevant criteria is carried out with the aid of an extensive research analysis of studies on 3PRLP evaluation. On the basis of the broad literature review, a total of thirteen economic criteria and eleven environmental and eleven social criteria are identified. The relevant criteria have been briefly described in Tables 17.1, 17.2 and 17.3.

17.3.2 Delphi Technique for Identification of Key Sustainable Criteria for 3PRLP Evaluation

The Delphi technique is used with consideration to varying outlooks of DMs in evaluating the importance of each criterion under the three dimensions considered in this study namely, economic, environmental and social. The decision-making team included 7 members of the company each with a minimum experience of six years. They were designated as Manager Supply Chain Operations, Manager Business Operations, Manager Human Resources, Senior Manager Information and Security, General Manager CSR and Sustainability, Chief Financial Officer.

The Delphi technique can be elaborated in the following steps [60]:

Step 1: The principal step is to identify the possible criteria for each of the three dimensions through a broad literature review. For our evaluation, we have thirteen criteria for the economic dimension, and eleven criteria each for environmental and social dimensions, respectively.

Step 2: Post the identification of the criteria, the DMs scrutinize each and every criterion based on the sustainability impact they put on outsourcing the logistics. The dependency amongst the identified criteria is also checked.

Table 17.1 Economic criteria for evaluation of 3PRLP

Notation	Criteria	Description	References
1	Cost	Per unit cost of RL processes-collection, inspection, storage, disassembly, remanufacturing, disposal, service and other associated logistics costs	[37, 47, 57, 85, 95]
2	Reputation and market share	It refers to the opinion of the customers about how well the logistics organization is satisfying their needs	[4, 95, 99]
3	Delivery and services	Reliability of quality assurance in carrying out recovery process, documentation and transportation	[4, 56, 99]
4	Technological expertise	Investment in strong technical development ability to implement RL activities, level of advanced equipment	[4, 41, 76]
5	Geographical reach	Geographical location, distribution coverage, market coverage	[5, 71, 91]
6	RL capacity	Financial capacity to invest in all RL operations, network capacity, transport capacity, specialized infrastructure	[4, 44, 47, 66, 92]
7	Financial stability/position	RSC performance, mutual commitment towards business needs, market share, liquidity, profitability	[4, 7, 14, 37, 38, 57]
8	Management capability	Warehouse management, transportation management, manpower, capacity of facilities	[30, 38, 62, 90]
9	Technique level	Range of services, inventory management, manpower planning, space utilization, resource allocation, demand forecasting, equipment handling	[30, 38, 62]
10	Service capability	Quality service, configuration flexibility, adaptation to change in market	[4, 17, 23, 45, 51, 58, 76, 90]
11	Communication and IT system	Investment in logistics information system, IT and information security system	[4, 7, 49, 62]
12	Relationship	Mutual commitment, trust and fairness, channel relationship	[44, 94]
13	Strategic fit	Attitude, ability to match its resources and capabilities with opportunities in the external environment	[22]

Table 17.2 Environmental criteria for evaluation of 3PRLP

Notation	Criteria	Description	References
1	Reverse logistics	Developing efficient logistics system for carrying all RL practices such as collection, sorting, recycle, remanufacture and redistribution with emphasis on maximizing value creation	[2, 18, 20, 80, 88]
2	Green design	Use of environmentally-efficient logistics system, green design of facilities to factor in short-term as well as long-term impact on the environment	[47, 77]
3	Environmental management practices	Monitoring of environmental level of RL activities, environmental credentials earned, employee training	[11, 25, 33, 54, 95]
4	Pollution prevention	Measures adopted and efforts made for reduction, elimination, or prevention of pollutant emissions	[10, 22, 52]
5	Resource consumption	Reduction in the consumption of resources-energy, raw material and water	[10, 22, 25, 52, 54, 78]
6	Degree of closure/safe recycling	Impact of recycling on the outside environment	[21, 70]
7	Pollution control	Waste minimisation and reduction of carbon footprint in every stage of the SC	[10, 22, 25, 54]
8	Green practices	Green technology, green packaging using bio-degradable materials, employees training	[98]
9	Customer satisfaction	Matching degree of customer expectation regarding environment safety	[16]
10	Environmental protection compliance and commitment	ISO compliance, respect for environmental protection laws and environmental policies, commitment and alignment towards environmental objectives	[9, 38, 61]
11	Disposal capability	Capability of disposal of wastes in order to protect environment	[53, 55, 78, 93]

Step 3: Post analysis, the criteria are ranked on basis of their importance which is assessed through a developed questionnaire with the panel of experts. The DMs rank the criteria on the following scale: 'very poor'-1, 'poor'-2, 'medium'-3, 'good'-4 and 'very good'-5.

Step 4: The specified ranks are then collected and the mean of the ranks for each criteria is calculated. Further, normalization is done to obtain the final ranking.

Step 5: The top six out of thirteen economic criteria, five out of eleven environmental criteria and four out of eleven social criteria are selected as per the DMs opinion.

Table 17.3 Social criteria for evaluation of 3PRLP

Notation	Criteria	Description	References
1	Cooperation with government agencies	Compliance with various ILO laws relating to employee welfare and compliance with government employment law	[62]
2	Stakeholder satisfaction	Health, education, housing, security, grants and donations, supporting community projects and economic welfare and growth	[22]
3	Employment practices	Building relationship with the staff, employment compensation, and flexible working arrangements	[10, 22, 25, 54, 95]
4	Health and safety	Respect for policies with regard to employee health and safety, workplace safety, security and safety procedural complains	[10, 36, 63]
5	Employment stability	Career development, employee contracts	[20, 29, 34, 63]
6	Local community influence/publicity	Promotions for betterment of society	[10, 38]
7	Supporting education	Educating people about importance of reuse, recycle, remanufacture	[2]
8	Equity labour sources	Policies towards labour equity	[22, 38]
9	Corporate image	Market reputation, image among public	[67, 72, 79]
10	Job opportunities	Opportunities for employment by the organization	[22, 43, 68, 69, 86]
11	Value to customer	Consumer education, customer satisfaction and responsiveness	[63]

17.3.3 Best Worst Method for Ranking of Key Sustainable Criteria for 3PRLP Evaluation

The BWM technique was developed by Rezaei, 2015 and has since been applied to numerous multi-criteria-based modelling problems [42, 84, 97]. The major advantages of using BWM over other multi-criteria-based evaluation techniques are: (i) the number of pairwise comparisons is less resulting in less time, cost and effort; (ii) it results in better consistency of the judgement matrix.

Consider the set of 'k' criteria $\{C_1, C_2, \ldots, C_k\}$ and the set of 'm' DMs $\{DM_1, DM_2, \ldots, DM_m\}$. The BWM technique to find the weights of importance of the 'k' criteria is briefly described below [81]:

Step 1: Each DM is asked to select his/her best (most desirable) and the worst (least desirable) criteria.

Let C_B^i be the best criteria and C_W^i be the worst criteria of the ith DM ($i = 1, 2, \ldots, m$).

Step 2: For each DM, the preference of the best criteria over the other criteria is calculated.
A numerical scale of 1–9 is used in this study, where a value of '1' represents equal preference and a value of '9' represents the extreme preference of the best criteria over the other criteria. This results in the Best-to-Others (BO) vector given by

$$\{a_{B1}^i, a_{B2}^i, \ldots, a_{Bk}^i\}$$

Where, a_{Bj}^i indicates the preference of the best criteria over jth criteria.
Also, $a_{Bj}^i \geqslant 1 \quad \forall j = 1, 2, \ldots, k \quad$ and $\quad a_{BB}^i = 1$.

Step 3: For each DM, the preference of each criterion with the worst criteria is calculated. This results in the Others-to-Worst (OW) vector given by

$$\{a_{1W}^i, a_{2W}^i, \ldots, a_{kW}^i\}$$

Where, a_{jW}^i indicates the preference of the jth criteria over the worst criteria.
Also, $a_{jW}^i \geqslant 1 \quad \forall j = 1, 2, \ldots, k \quad$ and $\quad a_{WW}^i = 1$.

Step 4: Calculate the optimal weights $(v_1^i, v_2^i, \ldots, v_k^i)$ of the criteria as per the judgement of ith DM. The objective is to ascertain the optimal weights of the criteria in order to minimize the maximum of the absolute differences $|v_B^i - a_{Bj}^i v_j^i|$ and $|v_j^i - a_{jW}^i v_W^i|$ for $j = 1, 2, \ldots, k$.

Step 5: Formulate the min-max model as follows [82]:

$$\min \max_j \{|v_B^i - a_{Bj}^i v_j^i|, |v_j^i - a_{jW}^i v_W^i|\}$$

Subject to

$$\sum_{j=1}^{k} v_j^i = 1$$

$$v_j^i \geqslant 0 \quad \forall j = 1, 2, \ldots, k$$

Step 6: Using α^i to denote the maximum absolute difference, formulate the following equivalent linear model for calculating weights of criteria as per the ith DM [82]:

$$\min \alpha^i$$

Subject to

$$|v_B^i - a_{Bj}^i v_j^i| \leqslant \alpha^i \quad \forall j = 1, 2, \ldots, k$$

$$|v_j^i - a_{jW}^i v_W^i| \leqslant \alpha^i \quad \forall j = 1, 2, \ldots, k$$

$$\sum_{j=1}^{k} v_j^i = 1$$

$$v_j^i \geqslant 0 \quad \forall j = 1, 2, \ldots, k$$

α^i can be considered as an indicator of the consistency of the comparisons. Its value close to zero shows a high level of consistency. The reliability of the model also relies on the value of α^i. The greater the value, the less reliable the comparisons are [65].

Step 7: Solve the linear model of BWM to get the optimal weights.
Let the optimal solution of model formulated in Step 6 be given by $(v_1^{i*}, v_2^{i*}, \ldots, v_k^{i*})$ and the optimal objective value be α^{i*}.

Step 8: Calculate the final weights $w_1, w_2, \ldots, w_k$ of criteria by taking average of the optimal weights obtained for each DM as follows:

$$w_j = \frac{\sum_{i=1}^{m} v_j^{i*}}{m} \qquad \forall j = 1, 2, \ldots, k$$

17.4 Application of the Proposed Methodology

17.4.1 Identification of Key Criteria Using Delphi Technique

The objective of using the Delphi technique is to select the most important criteria according to the DMs from a list of thirteen criteria in economic dimension and eleven in environmental and social dimensions respectively. The criteria must be shortlisted on the basis of their importance in evaluating the capabilities of 3PRLPs in sustainably managing the RL operations. The Delphi technique aids in identifying the critical criteria, the inter-dependency amongst the criteria and the irrelevant criteria as per the DMs opinions and end goals. Henceforth, the key sustainable criteria are extracted as shown in Fig. 17.2.

This has resulted in finalization of six key economic criteria: (1) *Financial Performance (FNP)* refers to the financial capability of the 3PRLP in providing the RL services at minimum cost and its mutual commitment towards achieving liquidity and profitability for organization; (2) *Resource Capacity (RCP)* which refers to the capacity of the 3PRLP to invest in RL operations, facility development and other infrastructure development; (3) *Quality (QL)* corresponds to the quality of the service provided by the 3PRLP and the quality of the final remanufactured product, recovered parts and material; (4) *Assets Management (ASSM)* refers to management of the facilities and vehicles, transportation activities, manpower engaged by the 3PRLP; (5) *Technology Innovation (TI)* incorporates the ability of the 3PRLP to invest in technical development in order to fulfil the RL service level, provide information security system for a better communication between the facilities and advanced components and equipment for better working conditions; (6) *Optimization Capabilities (OPC)* refers to the technique level and the range of services provided by the 3PRLP. It also includes the inventory management, space utilization, demand forecasting and equipment handling skills of 3PRLP.

The evaluation of eleven environmental criteria using Delphi technique resulted in clustering the criteria and identifying five key criteria with the aim towards selecting 3PRLP who will be able to carry RL activities with reduced environmental degradation. The five combined environmental criteria are (1) *RL Practices (RLP)* which

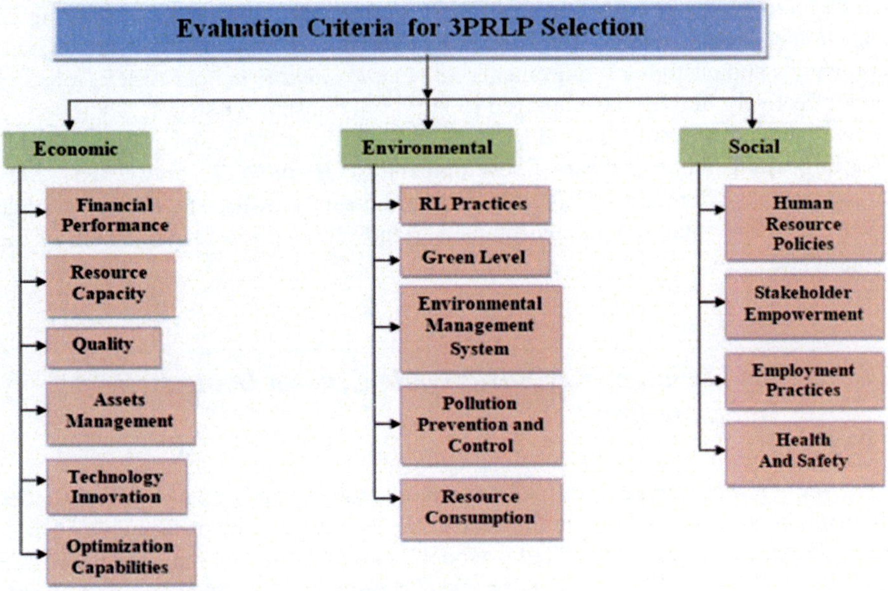

Fig. 17.2 Key sustainable criteria for 3PRLP evaluation

includes developing efficient logistics system by 3PRLP for carrying all RL activities such as collection, sorting, recycling, remanufacture and redistribution with emphasis on maximizing value creation and minimizing the deterioration of the environment. *RLP* also includes the capability of disposal of wastes in order to protect the environment; (2) *Green Level (GRL)* of 3PRLP is measured in terms of the green practices adopted by the 3PRLP such as green packaging using biodegradable materials and training of employees is an unavoidable practice for the safety of the environment. It also involves green design of 3PRLP's facilities to factor in short-term as well as long-term impact on the environment; (3) *Environmental Management System (EMS)* refers to the commitment and alignment of 3PRLP towards the environmental objectives of the organization. Its compliance towards the environmental protection laws and environmental policies. Its efforts towards reduction of carbon footprint in every stage of the RSC; (4) *Pollution Prevention and Control (PP&C)* relates to measures adopted and efforts made by 3PRLPs for reduction, elimination, or prevention of pollutant emissions; (5) *Resource Consumption (RCN)* refers to the ability of 3PRLP to reduce the consumption of resources such as energy, raw material and water.

Eleven social criteria are evaluated and combined in the following four key criteria: (1) *Human Resource Policies (HRP)* which is to check compliance of 3PRLP with various ILO laws related to employee welfare and transparency towards labour equity. Compliance and transparency with regards to employment laws is very important as most RL activities in India are still conducted in an unorganized manner involving women and children to work in hazardous conditions; (2) *Stakeholder Empowerment*

(STE) refers to the contribution of 3PRLP towards educating and empowering its stakeholders. It also refers to the ability of the 3PRLP to respond effectively towards company's and customer requirements; (3) *Employment Practices (EMP)* refers to how effectively 3PRLP has managed to build relationship with staff. Additionally, it also includes the attitude of 3PRLP towards employment compensation, flexible working arrangements and career development; (4) *Health and Safety (H&S)* refers to the policies adopted by 3PRLP to ensure the safety of the employees, provide security, and maintaining an environment friendly workplace for the health of the employees.

17.4.2 Evaluation of Key Sustainable Criteria Using Best Worst Method

Next, the BWM technique is utilized to prioritize the key performance criteria under the triple bottom line approach and reduce the existence of inconsistency of DMs. Four BWM models are formulated—Model 1 is for finding the rank of importance of the three sustainability dimensions viz. economic, environmental and social. Table 17.4 below provides the weights of each dimension obtained from solving model 1 on the basis of preferences given by DM 1. It can be seen from Table 17.4 the value of α^1 for model 1 is 0.045, which is closer to zero. Hence, the evaluation of DM1 is consistent.

Similarly, evaluation of weights of the other six DMs are determined and the final average weights of the three sustainability dimensions are calculated. The result is shown graphically in Fig. 17.3. It can be seen that the environmental dimension gains the highest average weight with the economic dimension following at second number and the social dimension achieves the third rank.

BWM technique is also used for evaluating the criteria under the three sustainability dimensions as illustrated in Fig. 17.2. The results of the three BWM models are presented in Tables 17.5, 17.6 and 17.7. Table 17.5 represents the weights of the top six shortlisted economic criteria obtained from solving model 2 on the basis of preferences given by DM 1. The value of α^1 for model 2, in this case, is 0.094, which means the comparison of criteria for DM1 is consistent.

Table 17.4 BO and OW vectors and weights of sustainability dimensions derived from model 1 (DM 1)

Criteria	Best/BO	Worst/OW	Weight
Economic	2	4	0.332
Environmental	1	6	0.571
Social	7	1	0.097
α^1			0.045

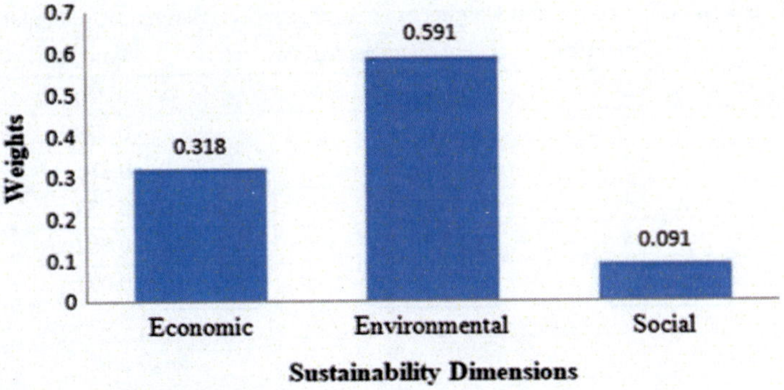

Fig. 17.3 Graphical representation of weights of sustainability dimensions

Similarly, evaluation of weights of the other six DMs are determined and the final average weights of the criteria under economic dimension are calculated. The result is shown graphically in Fig. 17.4. The top six amongst the thirteen criteria in the descending order of their average weights are; *'Quality' (QL)* (0.375), *'Financial Performance' (FNP)* (0.234), *'Resource Capacity' (RCP)* (0.156), *'Technology Innovation' (TI)* (0.094), *'Optimization Capabilities' (OPC)* (0.094) and *'Assets Management' (ASSM)* (0.047).

Table 17.6 represents the weights of the top five shortlisted environmental criteria obtained from model 3 on the basis of preferences given by DM 1. The value of α^1 for model 3 is obtained as 0.095, which shows the comparison is consistent for DM1.

Similarly, evaluation of weights of the other six DMs are determined and the final average weights of the criteria under environmental dimension are calculated. The result is shown graphically in Fig. 17.5. The top five criteria in descending order of their average weights are: *'RL Practices' (RLP)* (0.437), *'Environmental Management System' (EMS)* (0.266), *'Green Level' (GRL)* (0.133), *'Pollution Prevention and Control' (PP&C)* (0.106), *'Resource Consumption' (RCN)* (0.057).

Table 17.7 represents the weights of the top four shortlisted social criteria obtained from model 4 on the basis of preferences given by DM 1. The value of α^1 for model 4 is 0.044, the consistency ratio is very close to zero, hence the result is reliable.

Similarly, evaluation of weights of the other six DMs are determined and the final average weights of the criteria under social dimension are calculated. The result is shown graphically in Fig. 17.6. In today's era, an organization needs to have respect for policies with regard to employee health and safety, workplace safety, security and safety procedural compliance. Hence, it must also give emphasis on the same aspects while evaluation of 3PRLP. The criteria in the descending order of their average weights are; *'Health and Safety' (H&S)* (0.485), *'Employment Practices' (EMP)* (0.265), *'Human Resource Policies' (HRP)* (0.176), *'Stakeholder Empowerment' (STE)* (0.074).

Table 17.5 BO and OW vectors and weights of economic criteria derived from model 2 (DM 1)

Criteria	Best/BO	Worst/OW	Weight
FNP	2	5	0.252
RCP	3	3	0.178
QL	1	6	0.381
ASSM	7	1	0.037
TI	5	3	0.081
OPC	5	4	0.071
α^1			0.094

Table 17.6 BO and OW vectors and weights of environmental criteria derived from model 3 (DM 1)

Criteria	Best/BO	Worst/OW	Weight
RLP	1	6	0.384
GRL	4	4	0.213
EMS	2	5	0.284
PP&C	5	3	0.094
RCN	7	1	0.025
α^1			0.095

Table 17.7 BO vector, OW vector and weights of social criteria derived from model 4 (DM 1)

Criteria	Best/BO	Worst/OW	Weight
HRP	3	3	0.211
STE	6	1	0.062
EMP	2	4	0.186
H&S	1	6	0.541
α^1			0.044

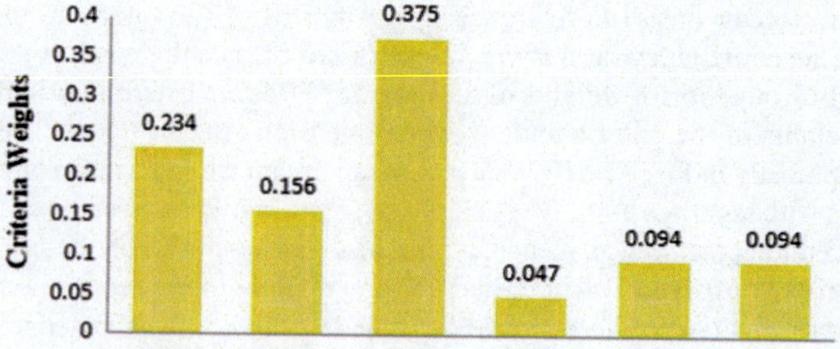

Fig. 17.4 Graphical representation of weights of economic criteria

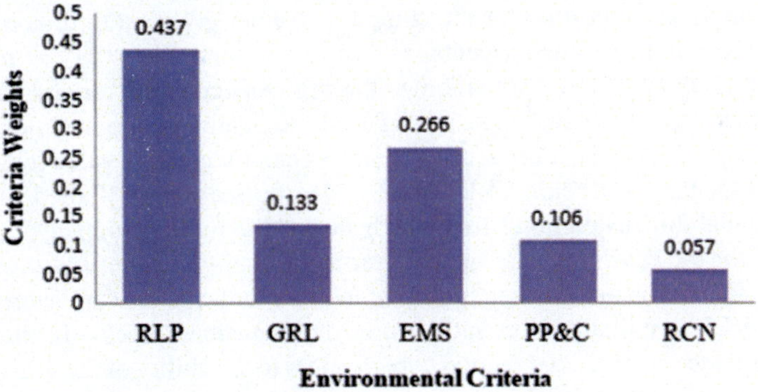

Fig. 17.5 Graphical representation of weights of environmental criteria

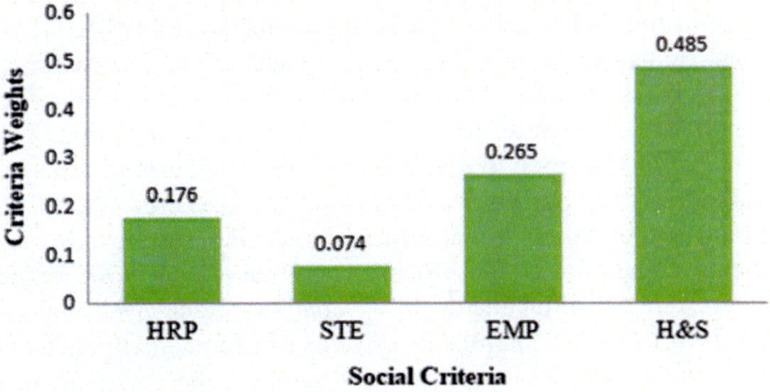

Fig. 17.6 Graphical representation of weights of social criteria

17.5 Result Discussion

The sustainability criteria shortlisted using Delphi technique are evaluated under each dimension using BWM. Next, the weights of importance of the three dimensions of sustainability are derived using BWM. The results of the four BWM models have been presented in Tables 17.4, 17.5, 17.6 and 17.7. Table 17.4 provides the weights of each dimension obtained from solving model 1. It can be seen that the environmental dimension gains the highest weight with the economic dimension following at second number and the social dimension achieves the third rank. The environmental dimension being ranked first is reflective of the DMs opinion that the primary objective of the organization in choosing to outsource to 3PRLP is to manage the returned flow of products and associated activities in an environmentally safe manner. The second rank of economic dimension shows that financial performance and RL associated costs hold importance for bringing profit to the organization. The social dimension is ranked third, which implies that workplace safety and employment practices although important for the company, are not given more importance than environmental and economic aspects.

Table 17.5 shows the top six amongst the thirteen economic criteria. *'QL'* plays an important role in outsourcing to 3PRLP as the quality of the recycled material, refurbished product and quality service are important for creating value for customers and which is the idea behind RL. Also, it can be seen from Table 17.4 that the environmental dimension ranks first which shows that for the company, the focus is on *'RLP'* for the environmental gains in terms of quality recovery of products and materials. Further, in outsourcing logistics, an important concern for the organization is that 3PRLP is mutually committed towards its business needs. In this context, *'FNP'* has hence been ranked second, which refers to the ability of the 3PRLP to gain economic benefits from RSC performance for the organization. Next, *'RCP'* which ranks third measures the capability of 3PRLP to invest in RL network operations and specialized infrastructure. The criteria *'TI'* and *'OPC'* hold the same level of importance. Both criteria have relevance in measuring the ability of 3PRLP to invest in strong technical development and efficiently manage RSC processes.

Table 17.6 represents the weights of the top five shortlisted environmental criteria obtained from model 3. Due to increase in environmental pollution, stakeholders demand for reduction of carbon footprint in every stage of the RSC. This justifies the obtained rankings of the criteria based on judgments of the DMs. *'RLP'* is the highest ranked criteria under the environmental dimension. In RL, sustainability is of utmost importance and for that 3PRLP must focus on execution of all *'RLP'* efficiently and enhancement of safe recycling and disposal capability. Ranked second is *'EMS'*, as the company is strict about compliance towards environmental policies. Hence, it wants to associate with 3PRLP who actively monitors the environmental level of their *'RLP'* and adheres to all the environmental protection laws and environmental policies. *'GRL'* is ranked third, which measures the capability of 3PRLP to focus on the green design of facilities to factor in short-term as well as long-term impact on the environment, in order to enhance the environmental performance of the RSC network.

Table 17.7 represents the weights of the top four shortlisted social criteria obtained from model 4. The criteria *'H&S'* has received the first rank, which shows that the company is concerned towards not only maintaining safety standards for their organization, but also expect the same from the 3PRLP. Ranked second is *'EMP'*, which means the 3PRLP must have the ability to contribute towards career development of their employees while also providing opportunities to the local people for the development of regional sustainability. Ranked third is *'HRP'* as compliance with various ILO laws relating to employee welfare is needed in RL. It is essential as many unorganized sectors use unscientific methods to recycle and recover full value from the returned products.

A comparative ranking of weights of the three dimensions of sustainability for 3PRLP evaluation and within each dimension the importance of criteria as per the DMs has been shown graphically in Fig. 17.7. It gives a clear picture to the RSC managers regarding how much emphasis must be laid on the criteria for the evaluation of 3PRLP for achieving sustainability.

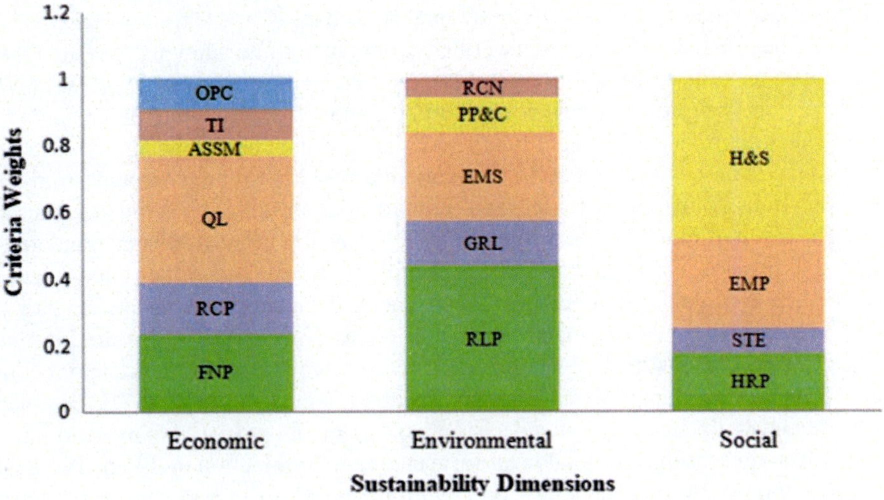

Fig. 17.7 Graphical representation of the importance of criteria under each sustainability dimension

17.6 Conclusion

Concerning the result analysis, the conclusion of the study is presented. The identification of key performance criteria for 3PRLPs is complex. This research has attempted an integrated MCDM model, which combines Delphi and BWM techniques to evaluate and select the appropriate key performance criteria for the selection of sustainable 3PRLPs. The proposed model is applied in the electronic industry to check the applicability and validity of the model. From the list of sustainable criteria derived from a broad literature review, few are shortlisted and weighted in order to acquire the main aspects for the assessment of sustainable performance of 3PRLPs. Delphi technique has been applied for the first level of screening of the criteria which is based on the results of various levels of questionnaires sent to a panel of experts and choosing the most prominent criteria of economic, environmental and social dimensions according to the DMs. The criteria have been shortlisted on the basis of their importance in evaluating the sustainable performance of 3PRLPs. Post the Delphi technique, the inter-dependent key criteria have been clubbed together to acquire a smaller number of criteria to ease the complexity of the decision-making. Next, the methodology involves the BWM technique to prioritize the key performance criteria under the triple bottom line approach. Four BWM models have been formulated. First model is for finding the rank of importance of the three sustainability dimensions viz. economic, environmental and social. Next, three BWM models are utilized to find the rank of importance of all criteria under each sustainability dimension. The result of model 1 shows that the environmental dimension has achieved the highest preference since the motivation behind RL is to achieve reduce the negative impact of the SC activities, and hence environmental sustainability is the foremost responsibility of the 3PRLPs. The importance of environmental dimension justifies the DMs outlook on the criteria selection. As per the result derived from model 2, *QL* and *FNP* have been the topmost key performance criteria under the economic dimension. *QL* is of great importance as the quality of service and product is of high significance for a customer, whereas *FNP* refers to the economic benefits for the company from the RL operations. Value for customers and economic benefits have been the idea behind RL. The result of model 3 yields *RLP* and *EMS* as the top two criteria under environmental dimension. This ascertains that reduction of carbon footprint and compliance towards environmental policies in RSC are the major goals towards sustainability. Lastly, model 4 solved for social dimension yields *H&S* and *EMP* as the top two social criteria. It is justifiable as safety and opportunity for employees is a major consideration for the organization and hence expect the same from 3PRLP. The obtained results validate that the integrated decision-making model proposed in the study successfully addresses the sustainable performance criteria selection problem. The final list of criteria derived in the study along with their rank of importance, sustainable 3PRLP selection problem can prove to be very useful in sustainable 3PRLP selection problem. There are a few limitations of the study. The criteria identified in the study broadly covers all the sustainable aspects of the evaluation of 3PRLPs. However, the study is limited to electronic industries. Further, the criteria selection is based on the opinions of the DMs of a specific industry which can vary when applied to other case studies. However, it has a lot of scope for modification by researchers and practitioners with regard to the change in the decision-making environment. This study can also be expanded by incorporating the risk dimension, as risk is an important factor to be taken under control while performing the RL. Risk factors like financial risk, operational risk and organizational risk can be considered while selecting the 3PRLPs.

Acknowledgements We would like to show our gratitude to the anonymous reviewers for their valuable comments and suggestions to improve the chapter.

References

1. Agrawal, S., Singh, R.K., Murtaza, Q.: A literature review and perspectives in reverse logistics. Resour. Conserv. Recycl. **97**, 76–92 (2015)
2. Agrawal, S., Singh, R.K., Murtaza, Q.: Outsourcing decisions in reverse logistics: sustainable balanced scorecard and graph theoretic approach. Resour. Conserv. Recycl. **108**, 41–53 (2016)
3. Agrawal, S., Singh, R.K., Murtaza, Q.: Reverse supply chain issues in Indian electronics industry: a case study. J. Remanuf. **8**(3), 115–129 (2018)
4. Aguezzoul, A.: Third-party logistics selection problem: a literature review on criteria and methods. Omega **49**, 69–78 (2014)
5. Aktas, E., Agaran, B., Ulengin, F., Onsel, S.: The use of outsourcing logistics activities: the case of turkey. Transp. Res. Part C: Emerg. Technol. **19**(5), 833–852 (2011)

6. Ali, S.M., Arafin, A., Moktadir, M.A., Rahman, T., Zahan, N.: Barriers to reverse logistics in the computer supply chain using interpretive structural model. Global J. Flex. Syst. Manag. **19**(1), 53–68 (2018)
7. Andersson, D., Norrman, A.: Procurement of logistics services - a minutes work or a multi-year project? Eur. J. Purch. Supply Manag. **8**(1), 3–14 (2002)
8. Anttonen, M., Halme, M., Houtbeckers, E., Nurkka, J.: The other side of sustainable innovation: is there a demand for innovative services? J. Clean. Prod. **45**, 89–103 (2013)
9. Awasthi, A., Chauhan, S.S., Goyal, S.K.: A fuzzy multicriteria approach for evaluating environmental performance of suppliers. Int. J. Prod. Econ. **126**(2), 370–378 (2010)
10. Bai, C., Sarkis, J.: Integrating sustainability into supplier selection with grey system and rough set methodologies. Int. J. Prod. Econ. **124**(1), 252–264 (2010)
11. Bai, C., Sarkis, J.: Flexibility in reverse logistics: a framework and evaluation approach. J. Clean. Prod. **47**, 306–318 (2013)
12. Bai, C., Sarkis, J.: Integrating and extending data and decision tools for sustainable third-party reverse logistics provider selection. Comput. Oper. Res. **110**, 188–207 (2019)
13. Bouzon, M., Govindan, K., Rodriguez, C.M.T., Campos, L.M.: Identification and analysis of reverse logistics barriers using fuzzy Delphi method and AHP. Resour. Conserv. Recycl. **108**, 182–197 (2016)
14. Boyson, S., Corsi, T., Dresner, M., Rabinovich, E.: Managing effective third party logistics relationships: what does it take? J. Bus. Logist. **20**(1), 73 (1999)
15. Centobelli, P., Cerchione, R., Esposito, E.: Environmental sustainability in the service industry of transportation and logistics service providers: systematic literature review and research directions. Transp. Res. Part D: Transport Environ. **53**, 454–470 (2017)
16. Chopra, S., Meindl, P., Kalra, D.V.: Supply Chain Management: Strategy, Planning, and Operation, vol. 232. Pearson, Boston (2013)
17. Choy, K.L., Chow, H.K., Tan, K.H., Chan, C.K., Mok, E.C., Wang, Q.: Leveraging the supply chain flexibility of third party logistics - hybrid knowledge-based system approach. Expert Syst. Appl. **35**(4), 1998–2016 (2008)
18. Cochran, J.K., Ramanujam, B.: Carrier-mode logistics optimization of inbound supply chains for electronics manufacturing. Int. J. Prod. Econ. **103**(2), 826–840 (2006)
19. Colicchia, C., Marchet, G., Melacini, M., Perotti, S.: Building environmental sustainability: empirical evidence from logistics service providers. J. Clean. Prod. **59**, 197–209 (2013)
20. da Silveira Guimarães, J.L., Salomon, V.A.P.: ANP applied to the evaluation of performance indicators of reverse logistics in footwear industry. Procedia Comput. Sci. **55**, 139–148 (2015)
21. Deng, W.J., Giesy, J.P., So, C.S., Zheng, H.L.: End-of-life (EoL) mobile phone management in Hong Kong households. J. Environ. Manag. **200**, 22–28 (2017)
22. Dou, Y., Sarkis, J.: A joint location and outsourcing sustainability analysis for a strategic offshoring decision. Int. J. Prod. Res. **48**(2), 567–592 (2010)
23. Efendigil, T., Önüt, S., Kongar, E.: A holistic approach for selecting a third-party reverse logistics provider in the presence of vagueness. Comput. Ind. Eng. **54**(2), 269–287 (2008)
24. Fawcett, S.E., Fawcett, A.M., Watson, B.J., Magnan, G.M.: Peeking inside the black box: toward an understanding of supply chain collaboration dynamics. J. Supply Chain Manag. **48**(1), 44–72 (2012)
25. Gauthier, C.: Measuring corporate social and environmental performance: the extended life-cycle assessment. J. Bus. Ethics **59**(1–2), 199–206 (2005)
26. Geethan, K.A.V., Jose, S., Chandar, C.S.: Methodology for performance evaluation of reverse supply chain. Int. J. Eng. Technol. **3**(3), 213–224 (2011)
27. Genchev, S.E., Richey, R.G., Gabler, C.B.: Evaluating reverse logistics programs: a suggested process formalization. Int. J. Logist. Manag. (2011)
28. Giri, B.C., Sarker, B.R.: Improving performance by coordinating a supply chain with third party logistics outsourcing under production disruption. Comput. Ind. Eng. **103**, 168–177 (2017)
29. Goebel, P., Reuter, C., Pibernik, R., Sichtmann, C.: The influence of ethical culture on supplier selection in the context of sustainable sourcing. Int. J. Prod. Econ. **140**(1), 7–17 (2012)

30. Göl, H., Çatay, B.: Third-party logistics provider selection: insights from a Turkish automotive company. Supply Chain Manag.: Int. J. (2007)
31. Govindan, K., Cheng, T.C.E.: Sustainable supply chain management. Comput. Oper. Res. **54**(C), 177–179 (2015)
32. Govindan, K., Soleimani, H.: A review of reverse logistics and closed-loop supply chains: a journal of cleaner production focus. J. Clean. Prod. **142**, 371–384 (2017)
33. Govindan, K., Pokharel, S., Sasikumar, P.: A hybrid approach using ISM and fuzzy TOPSIS for the selection of reverse logistics provider. Resour. Conserv. Recycl. **54**, 28–36 (2009)
34. Govindan, K., Palaniappan, M., Zhu, Q., Kannan, D.: Analysis of third party reverse logistics provider using interpretive structural modeling. Int. J. Prod. Econ. **140**(1), 204–211 (2012)
35. Govindan, K., Sarkis, J., Palaniappan, M.: An analytic network process-based multicriteria decision making model for a reverse supply chain. Int. J. Adv. Manuf. Technol. **68**(1–4), 863–880 (2013)
36. Govindan, K., Khodaverdi, R., Jafarian, A.: A fuzzy multi criteria approach for measuring sustainability performance of a supplier based on triple bottom line approach. J. Clean. Prod. **47**, 345–354 (2013)
37. Govindan, K., Khodaverdi, R., Vafadarnikjoo, A.: A grey DEMATEL approach to develop third-party logistics provider selection criteria. Ind. Manag. Data Syst. (2016)
38. Govindan, K., Kadziński, M., Sivakumar, R.: Application of a novel PROMETHEE-based method for construction of a group compromise ranking to prioritization of green suppliers in food supply chain. Omega **71**, 129–145 (2017)
39. Govindan, K., Agarwal, V., Darbari, J.D., Jha, P.C.: An integrated decision making model for the selection of sustainable forward and reverse logistic providers. Ann. Oper. Res. **273**(1–2), 607–650 (2019)
40. Govindan, K., Kadziński, M., Ehling, R., Miebs, G.: Selection of a sustainable third-party reverse logistics provider based on the robustness analysis of an outranking graph kernel conducted with ELECTRE I and SMAA. Omega **85**, 1–15 (2019)
41. Guarnieri, P., Sobreiro, V.A., Nagano, M.S., Serrano, A.L.M.: The challenge of selecting and evaluating third-party reverse logistics providers in a multicriteria perspective: a Brazilian case. J. Clean. Prod. **96**, 209–219 (2015)
42. Gupta, H., Barua, M.K.: Supplier selection among SMEs on the basis of their green innovation ability using BWM and fuzzy TOPSIS. J. Clean. Prod. **152**, 242–258 (2017)
43. Hasan, M.: Sustainable supply chain management practices and operational performance (2013)
44. Hong, J., Chin, A.T., Liu, B.: Logistics outsourcing by manufacturers in China: a survey of the industry. Transp. J. 17–25 (2004)
45. Hwang, B.N., Shen, Y.C.: Decision making for third party logistics supplier selection in semiconductor manufacturing industry: a nonadditive fuzzy integral approach. Math. Probl. Eng. (2015)
46. John, S.T., Sridharan, R., Kumar, P.R.: Reverse logistics network design: a case of mobile phones and digital cameras. Int. J. Adv. Manuf. Technol. **94**(1–4), 615–631 (2018)
47. Kafa, N., Hani, Y., El Mhamedi, A.: A fuzzy multi criteria approach for evaluating sustainability performance of third-party reverse logistics providers. In: IFIP International Conference on Advances in Production Management Systems, pp. 270–277. Springer, Berlin (2014)
48. Kannan, D.: Role of multiple stakeholders and the critical success factor theory for the sustainable supplier selection process. Int. J. Prod. Econ. **195**, 391–418 (2018)
49. Kannan, G., Haq, A.N.: Analysis of interactions of criteria and sub-criteria for the selection of supplier in the built-in-order supply chain environment. Int. J. Prod. Res. **45**(17), 3831–3852 (2007)
50. Kannan, G., Murugesan, P., Senthil, P., Noorul Haq, A.: Multicriteria group decision making for the third party reverse logistics service provider in the supply chain model using fuzzy TOPSIS for transportation services. Int. J. Serv. Technol. Manag. **11**(2), 162–181 (2009)
51. Keshavarz Ghorabaee, M., Amiri, M., Kazimieras Zavadskas, E., Antuchevičienė, J.: Assessment of third-party logistics providers using a CRITIC–WASPAS approach with interval type-2 fuzzy sets. Transport **32**(1), 66–78 (2017)

52. Klassen, R.D., Whybark, D.C.: The impact of environmental technologies on manufacturing performance. Acad. Manag. J. **42**(6), 599–615 (1999)
53. Knemeyer, A.M., Ponzurick, T.G., Logar, C.M.: A qualitative examination of factors affecting reverse logistics systems for end-of-life computers. Int. J. Phys. Distrib. Logist. Manag. (2002)
54. Labuschagne, C., Brent, A.C., Van Erck, R.P.: Assessing the sustainability performances of industries. J. Clean. Prod. **13**(4), 373–385 (2005)
55. Lai, K.H., Wu, S.J., Wong, C.W.: Did reverse logistics practices hit the triple bottom line of Chinese manufacturers? Int. J. Prod. Econ. **146**(1), 106–117 (2013)
56. Lambert, S., Riopel, D., Abdul-Kader, W.: A reverse logistics decisions conceptual framework. Comput. Ind. Eng. **61**(3), 561–581 (2011)
57. Lao, S.I., Choy, K.L., Ho, G.T.S., Tsim, Y.C., Chung, N.S.H.: Determination of the success factors in supply chain networks: a Hong Kong-based manufacturer's perspective. Meas. Bus. Excell. (2011)
58. Li, F., Li, L., Jin, C., Wang, R., Wang, H., Yang, L.: A 3PL supplier selection model based on fuzzy sets. Comput. Oper. Res. **39**(8), 1879–1884 (2012)
59. Li, J., Wang, Z., Jiang, B.: Managing economic and social profit of cooperative models in three-echelon reverse supply chain for waste electrical and electronic equipment. Front. Environ. Sci. Eng. **11**(5), 12 (2017)
60. Linstone, H.A., Turoff, M. (eds.): The Delphi Method, pp. 3–12. Addison-Wesley, Reading (1975)
61. Liou, J.J., Tamošaitienė, J., Zavadskas, E.K., Tzeng, G.H.: New hybrid COPRAS-G MADM model for improving and selecting suppliers in green supply chain management. Int. J. Prod. Res. **54**(1), 114–134 (2016)
62. Liu, H.T., Wang, W.K.: An integrated fuzzy approach for provider evaluation and selection in third-party logistics. Expert Syst. Appl. **36**(3), 4387–4398 (2009)
63. Mavi, R.K., Goh, M., Zarbakhshnia, N.: Sustainable third-party reverse logistic provider selection with fuzzy SWARA and fuzzy MOORA in plastic industry. Int. J. Adv. Manuf. Technol. **91**(5–8), 2401–2418 (2017)
64. Meade, L., Sarkis, J., Presley, A.: The theory and practice of reverse logistics. Int. J. Logist. Syst. Manag. **3**(1), 56–84 (2007)
65. Mi, X., Tang, M., Liao, H., Shen, W., Lev, B.: The state-of-the-art survey on integrations and applications of the best worst method in decision making: why, what, what for and what's next? Omega **87**, 205–225 (2019)
66. Mothilal, S., Gunasekaran, A., Nachiappan, S.P., Jayaram, J.: Key success factors and their performance implications in the Indian third-party logistics (3PL) industry. Int. J. Prod. Res. **50**(9), 2407–2422 (2012)
67. Muller, A., Kolk, A.: CSR performance in emerging markets evidence from Mexico. J. Bus. Ethics **85**(2), 325–337 (2009)
68. Nikolaou, I.E., Evangelinos, K.I.: A framework for evaluating the social responsibility quality of reverse logistics. Quality Management in Reverse Logistics, pp. 53–72. Springer, London (2013)
69. Nikolaou, I.E., Evangelinos, K.I., Allan, S.: A reverse logistics social responsibility evaluation framework based on the triple bottom line approach. J. Clean. Prod. **56**, 173–184 (2013)
70. Nnorom, I.C., Osibanjo, O.: Toxicity characterization of waste mobile phone plastics. J. Hazard. Mater. **161**(1), 183–188 (2009)
71. Pamučar, D., Stević, Ž., Zavadskas, E.K.: Integration of interval rough AHP and interval rough MABAC methods for evaluating university web pages. Appl. Soft Comput. **67**, 141–163 (2018)
72. Parast, M.M., Adams, S.G.: Corporate social responsibility, benchmarking, and organizational performance in the petroleum industry: a quality management perspective. Int. J. Prod. Econ. **139**(2), 447–458 (2012)
73. Perçin, S.: An integrated fuzzy SWARA and fuzzy AD approach for outsourcing provider selection. J. Manuf. Technol. Manag. (2019)

74. Prajapati, H., Kant, R., Shankar, R.: Bequeath life to death: state-of-art review on reverse logistics. J. Clean. Prod. **211**, 503–520 (2019)
75. Prakash, C., Barua, M.K.: Integration of AHP-TOPSIS method for prioritizing the solutions of reverse logistics adoption to overcome its barriers under fuzzy environment. J. Manuf. Syst. **37**, 599–615 (2015)
76. Prakash, C., Barua, M.K.: A combined MCDM approach for evaluation and selection of third-party reverse logistics partner for Indian electronics industry. Sustain. Prod. Consum. **7**, 66–78 (2016)
77. Prakash, C., Barua, M.K.: An analysis of integrated robust hybrid model for third-party reverse logistics partner selection under fuzzy environment. Resour. Conserv. Recycl. **108**, 63–81 (2016)
78. Presley, A., Meade, L., Sarkis, J.: A strategic sustainability justification methodology for organizational decisions: a reverse logistics illustration. Int. J. Prod. Res. **45**(18–19), 4595–4620 (2007)
79. Ravi, V., Shankar, R., Tiwari, M.K.: Productivity improvement of a computer hardware supply chain. Int. J. Product. Perform. Manag. (2005)
80. Razzaque, M.A., Sheng, C.C.: Outsourcing of logistics functions: a literature survey. Int. J. Phys. Distrib. Logist. Manag. (1998)
81. Rezaei, J.: Best-worst multi-criteria decision-making method. Omega **53**, 49–57 (2015)
82. Rezaei, J.: Best-worst multi-criteria decision-making method: some properties and a linear model. Omega **64**, 126–130 (2016)
83. Rogers, D.S., Tibben-Lembke, R.S.: Going backwards: reverse logistics trends and practices. The University of Nevada, Reno. Center for Logistics Management, Reverse Logistics Council (1998)
84. Salimi, N., Rezaei, J.: Evaluating firms' R&D performance using best worst method. Eval. Program Plan. **66**, 147–155 (2018)
85. Sarkis, J., Talluri, S.: A model for strategic supplier selection. J. Supply Chain Manag. **38**(4), 18–28 (2002)
86. Sarkis, J., Helms, M.M., Hervani, A.A.: Reverse logistics and social sustainability. Corp. Soc. Responsib. Environ. Manag. **17**(6), 337–354 (2010)
87. Sasikumar, P., Haq, A.N.: Analysing interactions among battery recycling barriers in the reverse supply chain. Enterprise Networks and Logistics for Agile Manufacturing, pp. 249–269. Springer, London (2010)
88. Sasikumar, P., Haq, A.N.: Integration of closed loop distribution supply chain network and 3PRLP selection for the case of battery recycling. Int. J. Prod. Res. **49**(11), 3363–3385 (2011)
89. Serrato, M.A., Ryan, S.M., Gaytán, J.: A Markov decision model to evaluate outsourcing in reverse logistics. Int. J. Prod. Res. **45**(18–19), 4289–4315 (2007)
90. Sharma, S.K., Kumar, V.: Optimal selection of third-party logistics service providers using quality function deployment and Taguchi loss function. Benchmarking: Int. J. (2015)
91. Singh, R., Shankar, R., Kumar, P., Singh, R.K.: A fuzzy AHP and TOPSIS methodology to evaluate 3PL in a supply chain. J. Model. Manag. (2012)
92. Stock, G.N., Greis, N.P., Kasarda, J.D.: Logistics, strategy and structure. Int. J. Oper. Prod. Manag. (1998)
93. Tan, A.W.K., Yu, W.S., Arun, K.: Improving the performance of a computer company in supporting its reverse logistics operations in the Asia-Pacific region. Int. J. Phys. Distrib. Logist. Manag. (2003)
94. Tate, K.: The elements of a successful logistics partnership. Int. J. Phys. Distrib. Logist. Manag. (1996)
95. Tavana, M., Zareinejad, M., Santos-Arteaga, F.J., Kaviani, M.A.: A conceptual analytic network model for evaluating and selecting third-party reverse logistics providers. Int. J. Adv. Manuf. Technol. **86**(5–8), 1705–1721 (2016)
96. Turki, S., Sauvey, C., Rezg, N.: Modelling and optimization of a manufacturing/remanufacturing system with storage facility under carbon cap and trade policy. J. Clean. Prod. **193**, 441–458 (2018)

97. van de Kaa, G., Kamp, L., Rezaei, J.: Selection of biomass thermochemical conversion technology in the Netherlands: a best worst method approach. J. Clean. Prod. **166**, 32–39 (2017)
98. Wu, C., Barnes, D.: An integrated model for green partner selection and supply chain construction. J. Clean. Prod. **112**, 2114–2132 (2016)
99. Yayla, A.Y., Oztekin, A., Gumus, A.T., Gunasekaran, A.: A hybrid data analytic methodology for 3PL transportation provider evaluation using fuzzy multi-criteria decision making. Int. J. Prod. Res. **53**(20), 6097–6113 (2015)
100. Yu, H., Solvang, W.D.: Incorporating flexible capacity in the planning of a multi-product multi-echelon sustainable reverse logistics network under uncertainty. J. Clean. Prod. **198**, 285–303 (2018)
101. Zarbakhshnia, N., Jaghdani, T.J.: Sustainable supplier evaluation and selection with a novel two-stage DEA model in the presence of uncontrollable inputs and undesirable outputs: a plastic case study. Int. J. Adv. Manuf. Technol. **97**(5–8), 2933–2945 (2018)
102. Zarbakhshnia, N., Soleimani, H., Ghaderi, H.: Sustainable third-party reverse logistics provider evaluation and selection using fuzzy SWARA and developed fuzzy COPRAS in the presence of risk criteria. Appl. Soft Comput. **65**, 307–319 (2018)
103. Zarbakhshnia, N., Soleimani, H., Goh, M., Razavi, S.S.: A novel multi-objective model for green forward and reverse logistics network design. J. Clean. Prod. **208**, 1304–1316 (2019)
104. Zarbakhshnia, N., Wu, Y., Govindan, K., Soleimani, H.: A novel hybrid multiple attribute decision-making approach for outsourcing sustainable reverse logistics. J. Clean. Prod. **242**, 118461 (2020)
105. Zhang, R., Zhang, H., Liu, B.: Selection of reverse-logistics servicer for electronic products with fuzzy comprehensive evaluation method. Grey Syst.: Theory Appl. (2012)

Chapter 18
Efficiency Assessment Through Peer Evaluation and Benchmarking: A Case Study of a Retail Chain Using DEA

Anshu Gupta, Nomita Pachar, and Mark Christhian Barrueta Pinto

Abstract Retail industry in developing countries like India has observed immense growth in the past two decades and has marked a significant position in the global retail market due to technological advancements, globalization, rise in customer expenditure, emergence of multiple retail formats and increasing interest of investors in this sector. The growth in the retail sector is coupled with intense competition, shrinking revenues and rising expenditure on promotional activities, drawing attention of the decision-makers towards efficient operations. It is imperative to develop a robust approach for efficiency measurement for retail stores to support planning and implementation of efficient operations and expand the supply chain capabilities. The existing literature for retail stores efficiency assessment has mostly considered the self-appraisal approach, limiting its practical application due to inherent issues of total weight flexibility and pseudo-efficiency. In this study, we have presented an approach for efficiency assessment of retail stores through peer evaluation using the cross-efficiency models of Data Envelopment Analysis (DEA) to address these issues. The study also identifies pseudo-efficient stores using the concept of maverick index and defines benchmarks for all inefficient stores including maverick stores for developing improvement strategies. A case study of Indian electronic retail chain is presented to demonstrate the application.

Keywords Retail performance assessment · Cross efficiency · DEA · Benchmarking · Maverick stores · Pseudo-efficiency

---

A. Gupta (✉)
School of Business, Public Policy and Social Entrepreneurship, Dr. B. R. Ambedkar University Delhi, Delhi 110006, India
e-mail: guptaanshu.or@gmail.com

N. Pachar
Department of Operational Research, University of Delhi, Delhi 110007, India
e-mail: nomita.or.du@gmail.com

M. C. B. Pinto
School of Business, Universidad Peruana de Ciencias Aplicadas (UPC), Lima 15023, Peru
e-mail: mbarruetapinto@gmail.com

V. Laha et al. (eds.), *Optimization, Variational Analysis and Applications*, Springer Proceedings in Mathematics & Statistics 355,
https://doi.org/10.1007/978-981-16-1819-2_18

18.1 Introduction

Indian retail industry has inhabited a phenomenal position in global retail ranking with its emergence as a dynamic industry, accounting for more than 10% of GDP and around 8% employment in the country [12]. Fostered by high market potential and low economic risk, the retail sector has witnessed enhanced profitability in the highly competitive and ever-changing marketplace. Competitiveness and complexity are continuously soaring in this industry due to overabundance of consumer choice, fast changing technology and blooming of multi-format retailing [41]. To survive in the competitive marketplace and meet the challenges of today's business environment, retailers are evolving continuously with improved operational efficiency and supply chain capabilities [18, 20]. Retail chains can manage the flow of goods in an efficient and effective way by ensuring availability of the right product in the right place at the right time and satisfying constantly changing market demand [40]. Sustained performance and continuous improvement are key for long-term sustainability of any business including the retail trade. Along with devising strategies in this direction it is imperative for the retail firms to develop an approach for efficiency measurement scientifically. An efficiency measurement approach is useful for businesses, for monitoring and evaluating the performance of its several business units and its stakeholders accounting the input resource utilization to yield well-defined outputs [30]. When a business involves multiple comparable units such as stores in a retail chain in such a case apart from measurement of efficiency of the individual units (commonly known as Decision-Making Units (DMUs)), firms also need to identify the best practices group [46] for benchmarking the inefficient units. In this direction, our study presents a data envelopment analysis (DEA)-based efficiency measurement approach through peer assessment of multiple stores of an electronic retail (ER) chain. Further benchmarking reference sets are derived using multiple correlation clustering (MCC) to help decision-makers deal with the inefficiencies of the inefficient and pseudo-efficient DMUs in comparison to the best performers.

DEA and some of its extensions including cross-efficiency DEA model are well-accepted approaches for relative efficiency measurement of comparable DMUs [15]. Several characteristics of DEA and cross-efficiency DEA model encourage the use of this methodology for efficiency measurement of a group of retail stores operated by a centralized management. These include—(1) measurement of efficiency based on multiple dimensions of performance, (2) objective assessment of efficiency as no subjective scoring is required form the decision-maker or based on qualitative criteria, (3) dimensions of performance measured on non-homogeneous scales can be included and (4) input and output (I/O) dimensions could be differentiated which is beneficial for further use of the results for devising improvement strategies [5, 16]. Introduced from the seminal work of Charnes, Cooper and Rhodes [6] (commonly called as CCR model), the traditional CCR model provides the relative efficiencies of DMUs in comparison to others based on self-appraisal assuming constant return to scale. The CCR [6] model is then extended by Banker, Charnes and Cooper [3] (commonly called as BCC model) under variable return to scale assumption. The conventional approach has two major issues—(1) it does not provide a ranking for the best performers in the set of DMUs under consideration [14] and (2) the problem of pseudo-efficient DMUs [39]. Given the characteristic of total weight flexibility in the conventional model several DMUs may be identified as efficient gaining the highest level of efficiency (equal to 1) leading to the issue of pseudo-efficient DMUs and also the requirement of ranking the efficient units [5]. The self-evaluation model could not eliminate unrealistic weights without collecting the weight restriction from decision-makers [27]. The cross-efficiency model helps to overcome these issues [14, 27, 28].

The idea of self/peer-evaluation is often related with the performance assessment of an individual as in personnel management. However, the application of cross-efficiency through peer evaluation is not limited to people and has been used in the literature in different contexts such as for measuring the efficiency of nursing homes, coastal cities, public procurement and portfolio selection [16, 27, 28, 36]. Studies in the literature have discussed the efficiency evaluation of retail stores applying DEA models. Most of these studies are based on the conventional DEA models (CCR or BCC), there is no notable article in the literature in the context of retail stores, in general, and an Indian electronic retail chain, in particular. Our study presents an application of the cross-efficiency DEA models for objective efficiency evaluation of stores of an Indian electronic retail chain. Further our study also identifies the pseudo-efficient stores and benchmarks for improvement of inefficient as well as pseudo-efficient stores.

Structure of the remaining chapter is as follows: Sect. 18.2 elaborates the relevant review of literature; Sect. 18.3 defines the problem of the study; Sect. 18.4 explains the methodology; the results and findings are discussed in Sect. 18.5 and Sect. 18.6 demonstrates the conclusion.

18.2 Literature Review

The focus of the study is to analyse the peer efficiency of retail stores of an Indian ER chain based on DEA through peer evaluation. The existing studies related to efficiency evaluation in retail have generally considered conventional DEA (self-appraisal) approaches ignoring the peer evaluation. Our manuscript considers this issue and presents a DEA-based methodology for evaluating peer efficiency of multiple retail stores. DEA is a well-known technique and the conventional model of self-appraisal has two formulations: CCR [6] and BCC [3]; various theoretical extensions have been discussed in the literature for different contexts including cross-efficiency [36], super efficiency [2], attractiveness [37], variable benchmark model [11] and slack-based model [7]. These models have been widely employed in different fields including transport [4], banking [8], retail [24] and health [23] in the literature. The efficiency evaluation in the retail sector based on basic DEA has been explored by some researchers [17, 22, 24, 25, 44]. The following paragraph provides a glimpse of the last 10 years of research related to performance evaluation in retail based on basic DEA.

Yu and Ramanathan [44] evaluated Chinese retail organization's economic efficiency based on two inputs (carpet area and staff) and two outputs (profit and sales) employing CCR DEA model. Authors used Malmquist Productivity Index (MPI) to examine the changes in efficiency with respect to different years for the period 2000–2003 and the influence of environmental variables applying bootstrapped Tobit Regression (TR). Gupta and Mittal [21] measured the productivity of grocery retail firms located in National Capital Region (NCR), India through CCR model of DEA, based on six inputs (store area, check points, SKUs, number of employees, employees cost and working hours) and two outputs (sales and customers conversion ratio). Lau [26] investigated the retail distribution network's efficiency measurement approach as an alternative to conventional optimization approach using transportation cost as I/O defined in terms of sales data in the basic DEA model. Pande and Patel [32] examined cost efficiency of retail stores of a pharmacy company in NCR, India and derived the effect of footfalls, sales and operating expenses on efficiency using the TR model. Gandhi and Shankar [17] followed the approach of [44] and measured the economic efficiency of Indian retail firms of the period 2008–2010. In a similar study, Xavier et al. [43] presented efficiency analysis for retail stores of a clothing retail firm of Portugal. Ko et al. [25] demonstrated the measurement of efficiency for a household retail chain in Korea and examined the effect of competitive environment and assortment on efficiency values using the TR model. Gupta et al. [22] demonstrated DEA-based methodology for analytical selection of performance dimensions and efficiency measurement of multiple retail stores with a case study of an Indian ER chain again based on the CCR model. It is evident that the major focus of the researchers for efficiency measurement related to retail sectors remained on the conventional models employing self-appraisal models.

As discussed in the introduction section, the DEA self-appraisal model has some practical issues limiting the application of the models including—no ranking of efficient DMUs and pseudo-efficiency. To deal with these issues, Sexton et al. [36] proposed an extension of the basic DEA model which is known as a cross-efficiency approach for measuring peer efficiency of DMUs based on multiple I/O. Cross-efficiency models provide a solution to the issue of unrealistic weights without collecting prior information on weight constraints and also provide unique ranking of all DMUs [14, 27, 28]. Doyle and Green [14] proposed extension of the concept of cross-efficiency and developed aggressive and benevolent formulations considering secondary objectives for resolving ambiguity, and also discussed the concept of maverick index to deal with the issue of pseudo-efficiency. Higher values of this index indicate overestimation of efficiency of the concerned DMU through self-appraisal. The concept of cross-efficiency in DEA has gained a lot of attention by researchers and practitioners [1, 16, 34, 36]. This section reviews and identifies gaps in literature and highlights the contributions of our study.

Talluri and Sarkis [39] illustrated the use of cross-efficiency approach in DEA for evaluating layout of cellular manufacturing systems with two inputs (number of workers and number of machines) and three outputs considered as average (flow time, work in process levels and labour utilization). Sarkis [34] presented an analysis of different DEA ranking techniques (basic DEA model, cross-efficiency model, super efficiency model, ranked efficiency, radii of classification rankings) and Multiple Criteria Decision-Making (MCDM) methods (PROMETHE, ELECTRE and SMART). A case study of solid waste management of Finland is used with five inputs (cost, health effects, global effects, surface water releases and acidificative releases) and three outputs (employees, technical feasibility and resource recovery). The results demonstrated that judgement of DMUs in the DEA technique provided the better results than given by the MCDM techniques. Adler et al. [1] reviewed the ranking approaches in the DEA which are cross-efficiency, maverick index, super efficiency, benchmarking, multivariate statistical techniques, ranked inefficient units through proportional inefficiency. The results of analysis are demonstrated through a numerical illustration of a nursing home as given in [36]. Braglia et al. [5] presented an approach of the efficiency evaluation based on cross-efficiency of Iranian steel plants with 5 inputs and 12 outputs. Further, authors computed the maverick index for determining pseudo-efficiency plants and also employed cluster analysis for benchmarking. Talluri and Narasimhan [38] proposed a framework to identify suppliers for strategic sourcing and calculated efficiency scores of suppliers by using the DEA model. Authors also conducted the peer evaluation of suppliers to overcome the weight flexibility issue of CCR model. Liang et al. [27] extended the model of cross-efficiency that was given by [14] and introduced an alternative secondary goal in cross-efficiency evaluation. The model was illustrated through a numerical example. It selected 13 open coastal cities and 5 special economic zones in 1989 of China based on two inputs and three outputs. Yu et al. [45] measured the SC performance based on different information sharing scenarios through cross-efficiency DEA approach. The result of the study demonstrated that sharing demand information is the most efficient scenario for efficient supply chains. Falagario et al. [16] presented a decision-making tool for selecting the best supplier using aggressive and benevolent formulations of cross-efficiency in DEA. The validity of the approach is supported through a case study of an Italian public procurement agency with two inputs and two outputs which are execution time and price, and enhancement plants and free maintenance after post-delivery, respectively. Lim et al. [28] proposed a strategy for selecting the portfolio by using DEA cross-efficiency technique. Further this study addressed the variation in cross-efficiencies through mean variance framework. The applicability of the approach is demonstrated through Korean stock market with nine inputs and seven outputs. Wu et al. [42] proposed the idea of satisfaction degree in cross-efficiency technique through a max min mode and gave two algorithms to solve the models. Liu et al. [29] evaluated the eco-efficiency of 23 coal-fired power plants of China using a cross-efficiency approach and considered the idea of undesirable output in the model. Omrani et al. [31] evaluated the energy efficiency of 20 zones of Iranian transportation sector with five inputs and four outputs based on cross-efficiency and cooperative game approach. Chen et al. [9] assessed environmental efficiency with undesirable outputs of China during 2006–2015 using DEA cross-efficiency approach. Further, authors proposed the three strategies which are environmental protection, economic development and win-win strategies based on the objective of decision-makers. Goswami and Ghadge [19] developed a DEA-based model considering undesirable and desirable outputs for evaluating supplier efficiency and also measured the cross-efficiency to accomplish peer evaluation. Authors demonstrated the validity of the approach through application of Hyundai Steel Company.

It is noticeable from the above Review of Literature (ROL) that there are ample number of studies that discussed issues related to peer efficiency measurement with different applications. However, the application in the context of retail is limiting.

18.2.1 Contribution of the Study

From the ROL, it is evident that the concept of peer evaluation is explored by a lot of researchers in the literature with diverse applications; however, application of the efficiency measurement approach through peer evaluation is yet to be explored in the context of the retail sector as demonstrated in Table 18.1. With respect to the case study in consideration the decision-makers were interested in exploring the relative efficiency assessment though peer evaluation and comparison with the CCR efficiency.

1. In the literature, peer evaluation of efficiency measurement is discussed with respect to various fields and different countries [9, 16, 28, 29, 31] while no

Table 18.1 Existing gap in the literature

Studies	Methodology			Application of retail sector	Indian case study
	Cross-efficiency	Maverick index	Benchmarking		
Talluri and Sarkis [39]	✓	✓	✓	×	×
Sarkis [34]	✓	×	×	×	×
Adler et al. [1]	✓	✓	✓	×	×
Braglia et al. [5]	✓	✓	✓	×	×
Talluri and Narasimhan [38]	✓	×	×	×	×
Liang et al. [27]	✓	×	×	×	×
Yu et al. [45]	✓	×	×	×	×
Falagario et al. [16]	✓	×	×	×	×
Lim et al. [28]	✓	×	×	×	×
Liu et al. [29]	✓	×	×	×	×
Omrani et al. [31]	✓	×	×	×	×
Chen et al. [9]	✓	×	×	×	×
Goswami and Ghadge [19]	✓	×	×	×	×
Our study	✓	✓	✓	✓	✓

significant study exists for peer evaluation for efficiency assessment in the retail sector and in particular related to Indian retail context.

2. Another limitation of the existing research is that the discussion of pseudo-efficient DMUs and determination of benchmarks for further improvement of inefficient units are explored limitedly [5, 14, 39]. In this manuscript, through comparison of CCR and aggressive (and benevolent) efficiency, we have also identified the pseudo-efficient units and have computed benchmarks for further improvement of inefficient units as well as for pseudo-efficient units.

Considering these arguments specific contributions of the our study are as follows:

1. The study assesses the efficiency of retail stores of an Indian electronic retail chain based on peer appraisal through different models of cross-efficiency and compares the results.
2. Cross-efficiency assessments are compared with the CCR efficiency using Maverick index to identify pseudo-efficiency such that improvement strategies may be devised for inefficient as well as pseudo-efficient stores.
3. Benchmarking reference sets are derived using MCC to determine the closest benchmarks for all inefficient stores.

18.3 Problem Definition

The case company in focus is an Indian ER chain that offers a large assortment of consumer electronic goods [22]. The retail chain manages several stores spread over Delhi NCR area. For efficacious management of the retail chain, the decision-maker is looking for an effective analytical approach for continuous monitoring of the performance of the stores and devises strategies to enhance their performance. The performance of the stores depends on several factors such as size of the store, location and number of personnel and product assortment. The study presents a cross-efficiency-based DEA approach for efficiency assessment through peer evaluation. Subsequently, pseudo-efficient units are identified and benchmark sets are determined for all inefficient and pseudo-efficient stores. The case study discussed here presents the results based on analysis of data related to 24 stores selected for demonstration of results by the decision-maker.

18.4 Methodology

The assessment of efficiency using DEA is conducted identifying the performance measures as I/O, and efficiency is defined by the ratio weighted sum of output and input. In the basic CCR model of DEA [6], each DMU is enabled to propose its own weights in order to maximize its outputs with respect to certain constraints on the inputs of all the DMUs [10]. In this scenario, a unit under evaluation may achieve the status of an efficient unit through a set of I/O weights wherein some I/O achieves nearly zero value and few or just single inputs and outputs get significant non-negative value. A retail store consumes several inputs such as monetary expenses on day-to-day operations, inventory cost, promotional expenses, store area and number of staff to generate outputs in the form of sales and customer satisfaction. Computation of efficiency with positive weights only for some I/O limits the practical applications of the classical DEA model leading to the problem of pseudo-efficiency. Since the cross-efficiency model enables peer evaluation of efficiency and overcomes this problem, this study uses the cross-efficiency model of DEA [36] along with the aggressive and benevolent formulations [14] for assessment efficiency of a group of retail stores. Following the efficiency assessment pseudo-efficient DMUs are identified using Maverick index and benchmarking sets are determined based on MCC [14].

18.4.1 Cross-Efficiency Assessment

The classical model of cross-efficiency assessment enables computation of efficiency through peer appraisal by a two-stage process. In the first stage (known as the 'self-appraisal' stage), for each DMU its CCR efficiency score [6] is computed. In the second stage, efficiency scores are calculated for each DMU using the weights of the other DMUs, resulting in the computation of a Cross-Efficiency Matrix (CEM) (as shown in Fig. 18.1). The efficiency of a DMU is then calculated by aggregation of efficiency scores computed in the second stage. In the CEM, the element at ith row and jth column is the efficiency of DMU j with the optimal weights of DMU i. The diagonal of the CEM represent the CCR efficiency for each DMU. The mathematical formulation of the model is as follows. Assuming there are n DMUs consuming m inputs to generate s outputs, the ith input of jth DMU ($j = 1, 2, \ldots,$ k, $\ldots,$ n) represented by x_{ij} and rth output of jth DMU denoted by y_{rj}. The basic formulation of cross-efficiency evaluation is the input-oriented CCR DEA model [6] for computing optimal weights of I/O. In the first stage, the weights of kth DMU are computed using the following CCR DEA model:

$$\begin{aligned} \theta_k = \max \ & \textstyle\sum_{r=1}^{s} u_{rk} y_{rk} \\ \text{s.t.} \quad & \textstyle\sum_{i=1}^{m} v_{ik} x_{ik} = 1 \\ \textstyle\sum_{r=1}^{s} u_{rj} y_{rj} - & \textstyle\sum_{i=1}^{m} v_{ij} x_{ij} \le 0 \ \forall j \\ & u_{rj}, v_{ij} \ge 0 \end{aligned} \tag{M1}$$

where u_{rj} is the weight associated with rth output of jth DMU and v_{ij} is the weight associated with ith input of jth DMU. The solution of the model (1) provides optimal values of the I/O weights and efficiency value of DMU k. Using the optimal solution of stage 1 the cross-efficiencies of DMU l for all $l = 1, 2, \ldots,$ n can be calculated from the following equation in stage 2:

$$\theta_{kl} = \frac{\sum_{r=1}^{s} u_{rk}^{*} y_{rl}}{\sum_{i=1}^{m} v_{ik}^{*} x_{il}} \tag{M2}$$

where u_{rk}^{*} and v_{rk}^{*} denote optimal weights of DMU k according to model (M1). Cross-efficiency of lth DMU is calculated by averaging the efficiency over row/columns as depicted in Eq. (1).

$$\theta_l = \frac{1}{n-1} \sum_{k=1, k\neq j}^{n} \theta_{kl}, \forall l \tag{1}$$

Along the rows each θ_{kl} is interpreted as efficiency that DMU k accords to DMU $l \, \forall \, j$, averaging over rows is DMU k's average appraisal of peers against which it would like to compare itself and along the columns θ_{kl} represent the peer appraisal of DMU l and averaging down column l is the average peer appraisal of DMU l.

Cross-efficiency scores obtained from the basic DEA (as in stage 2) are often not unique as it depends on which of the optimal solutions of the linear programming model of CCR DEA model is employed, again limiting the usefulness of the approach [27]. To deal with this issue, Doyle and Green [14] proposed aggressive and benevolent formulation of cross-efficiency. In the aggressive approach, DMU k determines weights that minimize efficiency of peers and the benevolent approach maximizes the efficiency of DMU k and also the efficiency of peers. These formulations are as follows:

$$\min \sum_{r=1}^{s} u_r \sum_{j=1, j\neq k}^{n} y_{rj} \qquad (M3)$$

or

$$\max \sum_{r=1}^{s} u_r \sum_{j=1, j\neq k}^{n} y_{rj} \qquad (M4)$$

$$\text{s.t.} \sum_{i=1}^{m} v_i \sum_{j=1, j\neq k}^{n} x_{ij} = 1 \qquad (2)$$

$$\sum_{r=1}^{s} u_r y_{rk} - \theta_k \sum_{i=1}^{m} v_i x_{ik} = 0 \forall k \qquad (3)$$

$$\sum_{r=1}^{s} u_r y_{rj} - \sum_{i=1}^{m} v_i x_{ij} = 0 \forall j;\; j \neq k \qquad (4)$$

$$u_r, v_i \geq 0 \qquad (5)$$

The objective functions (M3) and (M4) are the secondary objective functions of the cross-efficiency model and represent the aggressive and benevolent formulations, respectively, along with the constraints (2)–(5).

18.4.2 Maverick Index

Maverick index represents the deviation between CCR and cross-efficiency [36] and is a measure of the deviation when moving from self-appraisal to the corresponding peer appraisal, i.e. the Maverick index for DMUk is defined as

$$M_k = \frac{\theta_{kk} - \theta_k}{\theta_k} \tag{18.1}$$

Rating DMU	Rated DMU						Averaged appraisal of peers
	1	2	.	.	.	n	
1	θ_{11}	θ_{12}	.	.	.	θ_{1n}	A_1
2	θ_{21}	θ_{22}	.	.	.	θ_{2n}	A_2
.	.	.				.	
.	.	.				.	
.	.	.				.	
n	θ_{n1}	θ_{n2}	.	.	.	θ_{nn}	A_n
Averaged appraisal by peers	θ_1	θ_2	.	.	.	θ_n	

Fig. 18.1 Cross-efficiency matrix

where θ_{kk} and θ_k are the CCR and cross-efficiency, respectively, of DMU k. A DMU may become efficient during self-evaluation while obtaining low efficiency in peer evaluation. Higher value of maverick index indicates that an efficient DMU is over-estimated because of poor discrimination.

18.4.3 Identification of Benchmarks

Benchmarking is a technique employed by organizations for improvement of the low performers in comparison to the best performers. Benchmarking measures the performance of a low-performing unit against the best-performing unit [35]. The benchmarking approach based on classical DEA has some limitations. Firstly, the inefficient DMUs and their corresponding reference sets may not be similarly inherent, and may represent an unachievable target for inefficient DMUs [39]. Secondly, benchmarking set for pseudo-efficient could not be determined. In the literature, researchers have discussed benchmarks for the inefficient/pseudo-DMUs through the cluster analysis [5, 35, 39].

Here we have used the Multiple Correlation Clustering (MCC) method developed by [13] to form benchmarking clusters. It is a technique based on an iterative procedure that partitions the set of DMUs in two subsets; these subsets are further segregated until homogeneity is obtained between DMUs. The steps for MCC are as follows:

1. The correlation matrix is computed based on the cross-efficiency matrix of DMUs.
2. If the values of the correlations in step 1 are close to +1 or −1 (approximation taken up to a suitable precision) go to step 3 otherwise compute higher order correlation matrices until a correlation matrix with all values +1/−1 is obtained.
3. Dataset is partitioned based on negative and positive correlations to form two clusters of DMUs.

4. Steps 1–3 are repeated until one of the condition is satisfied.
   a. The resultant correlation matrix has all values close to +1 implying the units in consideration are alike.
   b. The maximum number of possible iterations is reached.
   c. The number of units in the partitioned matrix is too small to further divide into separate clusters.

This MCC approach has several characteristics that favours its application for clustering including—the method can form clusters of highly intercorrelated units, can detect even small noise signals from large noise, missing data could be inferred from higher order correlations and is impervious to multicollinearity [35]. Within each cluster, the DMU with the highest cross-efficiency score can be considered as a benchmark for the other members of the same cluster. The methodology is demonstrated in the following section with a case study.

18.5 Case Study

This section demonstrates application of the cross-efficiency model for estimating the efficiency of 24 stores of an ER chain. Through ROL and discussion with the decision-maker key I/O performance measures to be used for efficiency computation are identified. The study considers the five inputs: operating expenses [43], average inventory cost [32], number of employees [25], promotional expenses [22] and store size [32]; and two outputs: profits [17] and customer satisfaction [22] to compute the peer efficiency of an ER chain's 24 stores. Rescaled data is obtained from the case organization. The results presented here are computed using the 'MultiplierDEA' package [33] in R software on the PC with 4 GB RAM and intel Core i3-5020U CPU @ 2.20 GHZ.

18.5.1 Analysis and Results

The values of the CCR, cross (CE), aggressive (AE) and benevolent (BE) efficiency are computed using the data obtained from the organization for I/O mentioned above, using models (equations) (M1), (M2), (M3) and (M4), respectively. The efficiency values, maverick indices (MI) and ranking of the stores obtained using these models are listed in Table 18.2.

According to the CCR model, retail stores R3, R4, R15, R17 and R20 have the perfect efficiency score of 1, while if we compare the results of efficiency values of the four different models, store R4 attains the highest efficiency according to all the models and could be considered as leader in terms of its performance. The low value of the maverick indices for this store also supports its leader status. The results are also consistent for the stores R3 and R20 that attain the II and III highest values of efficiency, respectively, in all models along with low values of maverick index. Low value of the maverick index reasserts the performance of these stores as high value of this index is indicative of the pseudo-efficient status of a productive unit. Store R21 though did not attain perfect efficiency in CCR model is ranked above the store R15 due to the higher value of maverick index associated with R15. The retail store R17 is also efficient according to the self-appraisal model while it has attained 7th rank through peer appraisal, again this could be attributed to the pseudo-efficiency issue. The store has been overestimated due to poor discrimination associated with the CCR model as represented by the maverick index for the store. The cross-efficiency models surface the pseudo-efficient units and hence could provide better results for benchmarking in terms of providing benchmarks for the pseudo-efficient as well as inefficient stores.

The results of cross-efficiency are used as input for the MCC to form the clusters of alike stores wherein a store having highest efficiency in a cluster acts as benchmark for other stores in that group. Comparing the I/O measures of a low-performing unit with the best performer, decision-makers can devise strategies for improvement of the inefficient units. The benchmarks are defined through the MCC clustering approach as described in the methodology section. First-order and higher order correlations are computed using the CEM to obtain correlation values close to ± 1 and I level clusters are obtained as shown in Fig. 18.2. Further clusters are derived following the stopping criteria of MCC and a total of six clusters of stores are obtained. Within a cluster the store with highest efficiency acts as the closest benchmark [39] for the other stores. Retail store R15 is benchmark for R1, R12 and R17 in the I cluster (C1); store R10 is benchmark for R11 in II cluster (C2); store R4 is benchmark for stores R7, R14, R18 and R24 in III cluster (C3); R3 is benchmark for R2, R13, R16 and R22 in C4; R6 is benchmark for R5, R8, R19 and R23 in C5 and store R20 is benchmark for R21 in C6. This analysis gives realistic insights to management for forming improvement strategies for inefficient/maverick stores. For the maverick retail store R17, the closest benchmark is the store R15. Similarly we can define benchmarks for all pseudo-efficient stores following this method. It may be noted here that if benchmarking reference sets are defined based on the CCR efficiency [22] then benchmarks for maverick stores could not be defined and hence DEA efficiency measures could not be used to define the improvement strategies for these stores. Further the results of MCC can also be used for stepwise benchmarking [5]. For example, R11 can benchmark R10 in step one following incremental improvement strategies and later it can follow a higher order benchmark.

Following implications can be drawn based on the analysis and results presented above:

1. Rapid growth of the retail industry in developing economies like India has fueled tough competition in the market with several players striving to capture a notable size of market share and attain a competitive positioning. Firms in the market are required to devise strategies for gaining competitive advantage and nourish their businesses for long-term sustainability and remain profitable. In order to achieve these goals, decision-makers are required to focus on continuous

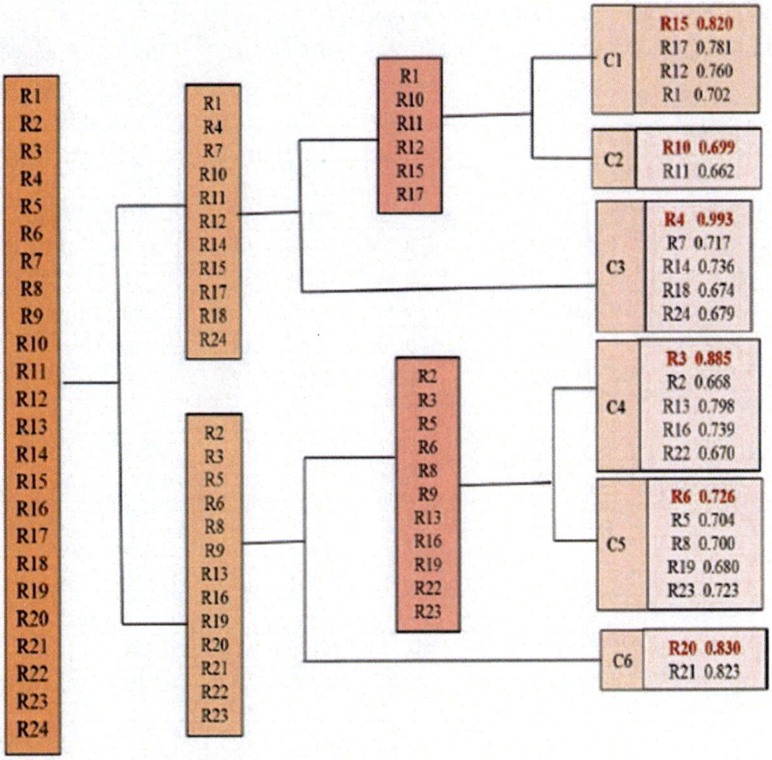

Fig. 18.2 Clusters for benchmarks

improvement of its products and processes, improving supply chain capabilities and adding more value for its potential customers [18]. It is imperative for the firm to devise improvement strategies and monitor their performance. The formulation and implementation of effective improvement strategies require firms to follow robust methods of performance measurement and identification of weaknesses, strengths and opportunities. DEA is a widely accepted and practically implemented methodology in this context and has been developed for different contexts. The existing literature related to efficiency measurement related to retail trade mostly considers the assessment of efficiency through self-appraisal (CCR model of [6]). The self-appraisal approach has some of the inherent issues such as overestimation of efficiency leaving to pseudo-efficient DMUs, poor discrimination and non-uniqueness of I/O weights. The peer assessment of efficiency based on cross-efficiency DEA models helps to resolve these issues. Our study discusses the application of the cross-efficiency DEA models in relation to retail and the case study provides a guideline for application for the practitioners.

2. In the direction of devising strategies for improvement and identifying opportunities for improvement, benchmarking is an effective approach. High-performing

Table 18.2 Efficiency scores of the retail stores

Retail stores	CCR	Conventional CE model			Aggressive model			Benevolent model		
		CE	MI	Rank	AE	MI	Rank	BE	MI	Rank
R1	0.794	0.702	0.131	16	0.683	0.162	16	0.706	0.124	18
R2	0.842	0.668	0.260	23	0.649	0.297	23	0.696	0.210	20
R3	1.000	0.885	0.130	2	0.872	0.147	2	0.905	0.104	2
R4	1.000	0.993	0.007	1	0.974	0.027	1	1.000	0.000	1
R5	0.798	0.704	0.134	15	0.689	0.158	15	0.711	0.122	16
R6	0.817	0.726	0.125	12	0.709	0.152	12	0.739	0.105	13
R7	0.851	0.717	0.187	14	0.704	0.209	13	0.725	0.174	14
R8	0.757	0.700	0.081	17	0.682	0.109	17	0.710	0.066	17
R9	0.868	0.737	0.178	10	0.720	0.205	9	0.757	0.147	10
R10	0.832	0.699	0.190	18	0.673	0.236	18	0.719	0.156	15
R11	0.741	0.662	0.119	24	0.645	0.149	24	0.677	0.095	24
R12	0.973	0.760	0.280	8	0.742	0.311	8	0.757	0.286	9
R13	0.878	0.798	0.100	6	0.775	0.133	6	0.805	0.091	6
R14	0.818	0.736	0.111	11	0.717	0.141	10	0.742	0.103	11
R15	1.000	0.820	0.220	5	0.802	0.247	5	0.819	0.221	5
R16	0.853	0.739	0.154	9	0.713	0.196	11	0.758	0.125	8
R17	1.000	0.781	0.280	7	0.762	0.313	7	0.778	0.285	7
R18	0.762	0.674	0.131	21	0.656	0.161	21	0.684	0.115	23
R19	0.739	0.680	0.087	19	0.660	0.119	20	0.696	0.062	21
R20	1.000	0.830	0.205	3	0.810	0.235	3	0.833	0.201	3
R21	0.978	0.823	0.188	4	0.806	0.214	4	0.827	0.183	4
R22	0.837	0.670	0.249	22	0.650	0.288	22	0.696	0.203	19
R23	0.832	0.723	0.151	13	0.702	0.186	14	0.742	0.122	12
R24	0.741	0.679	0.091	20	0.663	0.118	19	0.691	0.072	22

productive units of an organization can act as benchmarks for low performers, wherein an inefficient/maverick DMU can follow an efficient unit to gain competitiveness [39]. Our study presents the MCC method to form clusters of homogeneous units and determines the closest benchmark targets for inefficient units. The clustering method for deriving the benchmarking set also finds implications for incremental benchmarking [5].

3. The results of the study can also be used for devising strategies for optimal reallocation of the centralized resources of an organization according to efficiency targets. Organizations can also use the ranking of DMUs obtained using the cross-efficiency assessment for further decision-making.

18.6 Conclusion

The Indian retail industry has boomed enormously in the past two decades, contributing to the growth of the nation's economy and creating several employment opportunities. While several factors such as globalization, technology advancements and emergence of multiple retail formats have contributed to the growth of the industry, it has also attracted new entrants in the market intensifying competition. Thus, retail firms are required to focus more than ever on their competitive positioning, efficient operations and improvement. A robust method for measuring and monitoring performance is an important prerequisite in this direction. This chapter presents an application of the cross-efficiency DEA models for peer assessment of efficiency of stores of a retail chain with an application of an electronic retail chain. Efficiency measures are computed using the conventional CCR, cross-efficiency, aggressive and benevolent models, and comparison is drawn between different measures. The concept of maverick index is used to identify the maverick stores and benchmarks are obtained for all inefficient/maverick stores based on multiple correlation clustering. Application of the peer assessment methodology proposed in the study is validated with a single case study of an electronic retail chain. To overcome this limitation, future research should extend the applications for other retail firms and industry sectors. Another limitation of the study is that benchmark clusters are determined using the MCC method to establish the consistency of results one need to triangulate the results with other methods. The future work in this area can focus on identifying the dimensions for improvement for inefficiency stores and develop optimization models for centralized distribution of resources.

Acknowledgements The authors are indebted to the anonymous referees whose comments helped a lot to improve the quality of the chapter.

References

1. Adler, N., Friedman, L., Sinuany-Stern, Z.: Review of ranking methods in the data envelopment analysis context. Eur. J. Oper. Res. **140**(2), 249–265 (2002)
2. Andersen, P., Petersen, N.C.: A procedure for ranking efficient units in data envelopment analysis. Manag. Sci. **39**(10), 1261–1264 (1993)
3. Banker, R.D., Charnes, A., Cooper, W.W.: Some models for estimating technical and scale inefficiencies in data envelopment analysis. Manag. Sci **30**(9), 1078–1092 (1984)
4. Bazargan, M., Vasigh, B.: Size versus efficiency: a case study of US commercial airports. J. Air Transp. Manag. **9**(3), 187–193 (2003)
5. Braglia, M., Zanoni, S., Zavanella, L.: Measuring and benchmarking productive systems performances using DEA: an industrial cases. Prod. Plan. Control **14**(6), 542–554 (2003)
6. Charnes, A., Cooper, W.W., Rhodes, E.: Measuring the efficiency of decision making units. Eur. J. Oper. Res. **2**(6), 429–444 (1978)
7. Charnes, A., Cooper, W.W., Golany, B., Seiford, L., Stutz, J.: Foundations of data envelopment analysis for Pareto-Koopmans efficient empirical production functions. J. Econ. **30**(1–2), 91–107 (1985)

8. Chen, Y., Cook, W.D., Li, N., Zhu, J., Rhodes, E.: Additive efficiency decomposition in two-stage DEA. Eur. J. Oper. Res. **196**(3), 1170–1176 (2009)
9. Chen, L., Wu, F.M., Wang, Y.M., Li, M.J.: Analysis of the environmental efficiency in China based on the DEA cross-efficiency approach under different policy objectives. Expert Syst. (e12461) (2019)
10. Cook, W.D., Kress, M.: A data envelopment model for aggregating preference rankings. Manag. Sci. **36**(11), 1302–1310 (1990)
11. Cook, W.D., Seiford, L.M., Zhu, J.: Models for performance benchmarking: measuring the effect of e-business activities on banking performance. Omega **32**(4), 313–322 (2004)
12. Dhanabhakyam, M.: Indian retail industry - its growth, challenges and opportunities (online). https://www.fibre2fashion.com/industry-article/2203/indian-retail-industry-its-growth-challenges-and-opportunities (2007). Accessed 15 July 2020
13. Doyle, J.R.: MCC-multiple correlation clustering. Int. J. Man-Mach. Stud. **37**(6), 751–765 (1992)
14. Doyle, J., Green, R.: Efficiency and cross-efficiency in DEA: derivations, meanings and uses. J. Oper. Res. Soc. **45**(5), 567–578 (1994)
15. Emrouznejad, A., Yang, G.L.: A survey and analysis of the first 40 years of scholarly literature in DEA: 1978–2016. Socio-Econ. Plan. Sci. **61**, 4–8 (2018)
16. Falagario, M., Sciancalepore, F., Costantino, N., Pietroforte, R.: Using a DEA-cross efficiency approach in public procurement tenders. Eur. J. Oper. Res. **218**(2), 523–52 (2012)
17. Gandhi, A., Shankar, R.: Efficiency measurement of Indian retailers using data envelopment analysis. Int. J. Retail. Distrib. Manag. **42**(6), 500–520 (2014)
18. Ganesan, S., George, M., Palmatier, R.W., Weitz, B.: Supply chain management and retailer performance: emerging trends, issues, and implications for research and practice. J. Retail. **85**(1), 84–94 (2009)
19. Goswami, M., Ghadge, A.: A supplier performance evaluation framework using single and bi-objective DEA efficiency modelling approach: individual and cross-efficiency perspective. Int. J. Prod. Res. **58**(10), 3066–3089 (2020)
20. Grewal, D., Levy, M., Kumar, V.: Customer experience management in retailing: an organizing framework. J. Retail. **85**(1), 1–14 (2009)
21. Gupta, A., Mittal, S.: Measuring retail productivity of food and grocery retail outlets using the DEA technique. J. Strat. Mark. **18**(4), 277–289 (2010)
22. Gupta, A., Pachar, N., Darbari, J.D., Jha, P.C.: Efficiency assessment of Indian electronics retail stores using DEA. Int. J. Bus. Perform. Supply Chain Model. **10**(4), 386–414 (2019)
23. Kabasakal, A., Kutlar, A., Sarikaya, M.: Efficiency determinations of the worldwide railway companies via DEA and contributions of the outputs to the efficiency and TFP by panel regression. Cent. Eur. J. Oper. Res. **23**(1), 69–88 (2015)
24. Keh, H.T., Chu, S.: Retail productivity and scale economies at the firm level: a DEA approach. Omega **31**(2), 75–82 (2003)
25. Ko, K., Chang, M., Bae, E.S., Kim, D.: Efficiency analysis of retail chain stores in Korea. Sustainability **9**(9), 1629 (2017)
26. Lau, K.H.: Measuring distribution efficiency of a retail network through data envelopment analyses. Int. J. Prod. Econ. **146**(2), 598–611 (2013)
27. Liang, L., Wu, J., Cook, W.D., Zhu, J.: Alternative secondary goals in DEA cross-efficiency evaluation. Int. J. Prod. Econ. **113**(2), 1025–1030 (2008)
28. Lim, S., Oh, K.W., Zhu, J.: Use of DEA cross-efficiency evaluation in portfolio selection: an application to Korean stock market. Eur. J. Oper. Res. **236**(1), 361–368 (2014)
29. Liu, X., Chu, J., Yin, P., Sun, J.: DEA cross-efficiency evaluation considering undesirable output and ranking priority: a case study of eco-efficiency analysis of coal-fired power plants. J. Clean. Prod. **142**, 877–885 (2017)
30. Magnussen, J.: Efficiency measurement and the operationalization of hospital production. Health Serv. Res. **31**(1), 21 (1996)
31. Omrani, H., Shafaat, K., Alizadeh, A.: Integrated data envelopment analysis and cooperative game for evaluating energy efficiency of transportation sector: a case of Iran. Ann. Oper. Res. **274**(1–2), 471–499 (2019)

32. Pande, S., Patel, G.N.: Assessing cost efficiency of pharmacy retail stores and identification of efficiency drivers. Int. J. Bus. Perform. Manag. **14**(4), 368–385 (2013)
33. Puthanpura, A.K.: Multiplier data envelopment analysis and cross efficiency (online). https://cran.r-project.org/web/packages/MultiplierDEA/MultiplierDEA.pdf (2018). Accessed 16 August 2020
34. Sarkis, J.: A comparative analysis of DEA as a discrete alternative multiple criteria decision tool. Eur. J. Oper. Res. **123**(3), 543–557 (2000)
35. Sarkis, J., Talluri, S.: Performance based clustering for benchmarking of US airports. Transp. Res. Part A: Policy Pract. **38**(5), 329–346 (2004)
36. Sexton, T.R., Silkman, R.H., Hogan, A.J.: Data envelopment analysis: critique and extensions. New Dir. Program Eval. **1986**(32), 73–105 (1986)
37. Simonson, I., Tversky, A.: Choice in context: tradeoff contrast and extremeness aversion. JMR J. Mark. Res. **29**(3), 281 (1992)
38. Talluri, S., Narasimhan, R.: A methodology for strategic sourcing. Eur. J. Oper. Res. **154**(1), 236–250 (2004)
39. Talluri, S., Sarkis, J.: Extensions in efficiency measurement of alternate machine component grouping solutions via data envelopment analysis. IEEE Trans. Eng. Manag. **44**(3), 299–304 (1997)
40. Thomas, R.W., Esper, T.L., Stank, T.P.: Testing the negative effects of time pressure in retail supply chain relationships. J. Retail. **86**(4), 386–400 (2010)
41. Verhoef, P.C., Kannan, P.K., Inman, J.J.: From multi-channel retailing to omni-channel retailing: introduction to the special issue on multi-channel retailing. J. Retail. **91**(2), 174–181 (2015)
42. Wu, J., Chu, J., Zhu, Q., Yin, P., Liang, L.: DEA cross-efficiency evaluation based on satisfaction degree: an application to technology selection. Int. J. Prod. Res. **54**(20), 5990–6007 (2016)
43. Xavier, J.M., Moutinho, V.F., Moreira, A.C.: An empirical examination of performance in the clothing retailing industry: a case study. J. Retail. Consum. Serv. **25**, 96–105 (2015)
44. Yu, W., Ramanathan, R.: An assessment of operational efficiency of retail firms in China. J. Retail. Consum. Serv. **16**(2), 109–122 (2009)
45. Yu, M.M., Ting, S.C., Chen, M.C.: Evaluating the cross-efficiency of information sharing in supply chains. Expert Syst. Appl. **37**(4), 2891–2897 (2010)
46. Zhu, J.: Quantitative Models for Performance and Benchmarking: Data Envelopment Analysis with Spreadsheets, vol. 213. Springer, New York (2009)

Chapter 19
Spherical Search Algorithm: A Metaheuristic for Bound-Constrained Optimization

Rakesh Kumar Misra, Devender Singh, and Abhishek Kumar

Abstract This chapter is based on a recently published paper [9] of authors of this chapter in which a method for solving bound-constrained non-linear global optimization problems has been proposed. The algorithm obtains a sphere and then generates new trial solutions on its surface. Hence, this algorithm has been named as Spherical Search (SS) algorithm. This chapter starts with an introduction to the SS algorithm and then discusses different components and steps of the algorithm, viz., initialization of population, the concept of a spherical surface, the procedure of generation of trial solutions, selection of new population using greedy selection, stopping criteria, steps of the algorithm, and space and time complexity of the algorithm. Then, the algorithm has been applied to solve 30 bound-constrained global optimization benchmark problems of IEEE CEC 2014 suite and the results of the spherical search algorithm on these benchmark problems have been compared with the results of variants of well-known algorithms such as particle swarm optimization, genetic algorithm, covariance matrix adapted evolution strategy, and Differential Evolution on these problems to demonstrate its performance. Further, the SS algorithm has been applied to solve a model order reduction problem, an example of a real-life complex optimization problem.

Keywords Spherical search algorithm · Real-life optimization problems · Bound constrained optimization problem · Optimization algorithm · Global optimization

Nomenclature

$N \in \mathbb{N}$	Number of solutions in a population, Pop.

R. K. Misra (✉) · D. Singh · A. Kumar
Department of Electrical Engineering, Indian Institute of Technology (BHU), Varanasi 221005, India
e-mail: rkmisra.eee@iitbhu.ac.in

D. Singh
e-mail: dsingh.eee@iitbhu.ac.in

A. Kumar
e-mail: abhishek.kumar.eee13@iitbhu.ac.in

V. Laha et al. (eds.), *Optimization, Variational Analysis and Applications*, Springer Proceedings in Mathematics & Statistics 355,
https://doi.org/10.1007/978-981-16-1819-2_19

$k \in \mathbb{N}$	Iteration index.
$D \in \mathbb{N}$	Dimension of search space.
$\bar{x_i}^{(k)} \in \mathbb{R}^{(D\times 1)}$	ith solution from Pop at kth iteration.
$\bar{y_i}^{(k)} \in \mathbb{R}^{(D\times 1)}$	Trial solution corresponding to $\bar{x_i}^{(k)}$.
$\bar{z_i}^{(k)} \in \mathbb{R}^{(D\times 1)}$	Search direction corresponding to $\bar{x_i}^{(k)}$.
$A^{(k)} \in \mathbb{R}^{(D\times D)}$	An orthogonal matrix at kth iteration.
$\bar{c}^{(k)} \in \mathbb{R}^{(N\times 1)}_{>0}$	A step-size control vector at kth iteration.
$c_i^{(k)} \in \mathbb{R}_{>0}$	ith element of step-size control vector $\bar{c}^{(k)}$ corresponding to $\bar{x_i}^{(k)}$.
$B^{(k)}$	A binary diagonal matrix consisting of elements of binary vector $\bar{b}^{(k)}$ at kth iteration.
$\bar{b}^{(k)}$	A binary vector of dimension D.
$b_i^{(k)}$	Element of $\bar{b}^{(k)}$ corresponding to $\bar{x_i}^{(k)}$.

19.1 Introduction

In the literature, different optimization algorithms have been proposed to solve complex real-life global optimization problems. It has been experienced that meta-heuristics are efficient and effective as compared to the deterministic algorithm for solving complex real-life global optimization problems. The advantages of meta-heuristics are as follows. Meta-heuristics are algorithms are simple; they do not need derivatives of the objective function, can be easily applied to a variety of problems with minor changes in its structure, and have in-built ability to avoid local minima. However, as per No-Free-Lunch theorem [17] there cannot exist a universal method capable of solving all the problems efficiently.

There are two classes of meta-heuristics, viz., single-agent based and population based. Population-based meta-heuristics start with multiple initial solutions which improve over iterations. These algorithms explore large search space in each iteration and during exploration they keep sharing information among the individual solutions of the population and hence can avoid local minima; however, they need more function evaluations per iteration as compared to the single-agent class of meta-heuristics. As per the source of inspiration, meta-heuristics can be classified into two categories: (i) Evolutionary Algorithms (EAs) [4, 8, 13, 15, 16, 19] and (ii) Swarm-Based Algorithms (SAs) [3, 6, 7, 11, 18]. For a meta-heuristic to have improved performance, it must strike balance between two major characteristics, exploration and exploitation. To achieve the required balance, various parameters are introduced in a meta-heuristics. These parameters need to be tuned employing some rule-of-thumb operations which may make a particular meta-heuristic to perform better for a particular class of problems. The spherical search algorithm is having single parameter which is self-adaptive (i.e., it does not need tuning) due to which it gives improved performance for most of the problems. Further, the spherical search algorithm has very good balance between exploration and exploitation, is rotation-invariant, is able to map the contour of search space, and maintains high diversity during its run.

19.2 Spherical Search Optimization Algorithm [9]

This section develops the mathematical framework of the SS algorithm. In the SS algorithm, the search space is a vector space in which each candidate solution is represented as a vector.

Search space is assumed to be of D-dimension. In this search space, a target direction is decided using individual location and a target location. In every iteration, SS algorithm creates a $(D-1)$-spherical boundary keeping the target direction as its main axis. Trial solutions corresponding to each individual are generated on the surface of the spherical boundary. Figure 19.1 demonstrates the idea of spherical boundary for a two-dimensional search space. The figure illustrates the different components of SS algorithm related to spherical boundary as listed below:

- –: 1-spherical boundary for each of individual.
- ☆: target location.
- ★: individual location.
- O: trial location.

In Fig. 19.1, for each individual vector, trial solutions are generated on the 1-spherical boundary. Fitness of the trial solutions are evaluated on the basis of the objective function and better locations pass on into the next iteration.

When target location of an individual is far-off, resulting $(D-1)$-spherical boundary is large and the trail solutions naturally explore the search space. Whereas in case of nearby target, the resulting $(D-1)$-spherical boundary is small, which results in the exploitation of the search space. This phenomenon creates a balance of exploration and exploitation in the spherical search algorithm.

For every iteration, best solution is the location with the best fitness value. The iterations continue until either function evaluations become equal to the maximum number of function evaluations or the solution does not update for a specified number of iterations.

Flowchart of the SS algorithm is given in Fig. 19.2. Different steps of the flowchart are explained in the following subsections.

19.2.1 Initialization of Population

Let us assume that $Pop^{(k)}$ is population at kth iteration.

$$Pop^{(k)} = [\bar{x_1}^{(k)}, \bar{x_2}^{(k)}, \ldots \bar{x_N}^{(k)}]. \tag{19.1}$$

In this population, each of the element, $\bar{x_i}^{(k)}$, is a D-dimensional vector representing a $D-$dimensional point (solution) in search space as follows:

$$\bar{x_i}^{(k)} = [x_{i1}^{(k)}, x_{i2}^{(k)}, \ldots x_{iD}^{(k)}]^T. \tag{19.2}$$

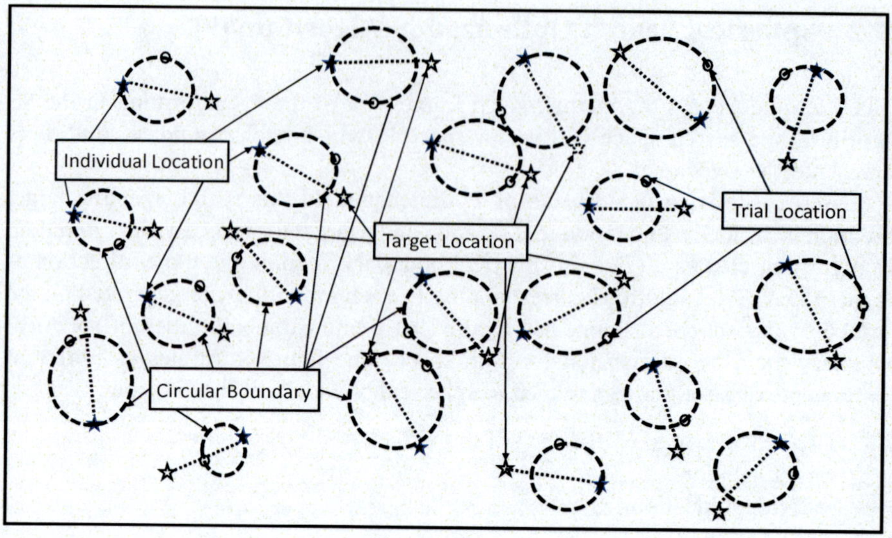

Fig. 19.1 Demonstrating the 1-spherical (circular) boundary of individuals of SS algorithm in 2-D search space [9]

Fig. 19.2 Simple framework of SS algorithm [9]

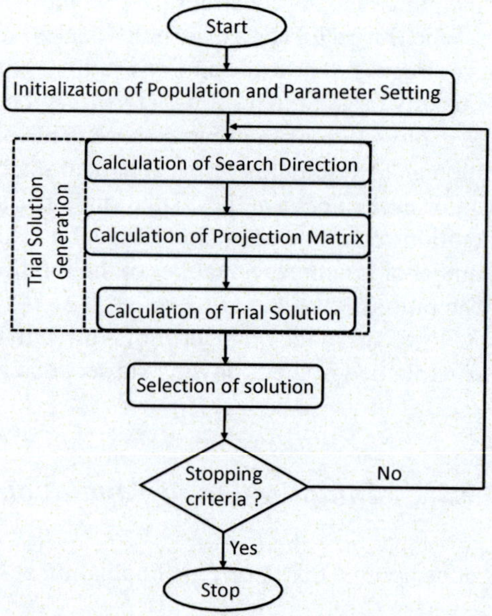

For $k = 0$, x_{ij}^{k} is initialized randomly within the limit (0, 1] using $rand(0, 1]$ which generates random number from uniform distribution keeping the initialization values between x_{hj} (upper bound) and x_{lj} (lower bound) of jth element as follows:

$$x_{ij}^{0} = (x_{hj} - x_{lj}) * rand(0, 1] + x_{lj} \tag{19.3}$$

19.2.2 Spherical Surface and Trial Solutions

As spherical search algorithm is a population-based optimization algorithm, in every iteration, new potential solutions are computed. Some of these potential solutions may become part of the population in the next iterations.

For each solution, a $(D - 1)$-spherical boundary is generated as an intermediate step toward evaluation of potential new solution. Method of generating these potential new solutions called as trial solutions, $\bar{y}_i$, has been demonstrated on a 2-D search space in Fig. 19.3 where locus of $\bar{y}_i$ is shown as $(D - 1)$-spherical boundary which becomes a circle (1-sphere) having diameter of $c_i\bar{z}_i$ in case of 2-D search space. In this case, search direction, $\bar{z}_i$, passes through the center of $(D - 1)$-spherical boundary. Method to evaluate $\bar{z}_i$ has been discussed in subsequent sections.

Trial solutions, $\bar{y}_i$, for kth iteration are obtained using following expression:

$$\bar{y_i}^{(k)} = \bar{x_i}^{(k)} + c_i^{(k)} P_i^{(k)} \bar{z_i}^{(k)}. \tag{19.4}$$

Value of $\bar{y_i}^{k}$ is decided with the help of a projection matrix P_i.

Search direction, $\bar{z_i}^{(k)}$, should be such that $i-$th solution moves toward the better solutions. Calculation of search direction has been graphically demonstrated in

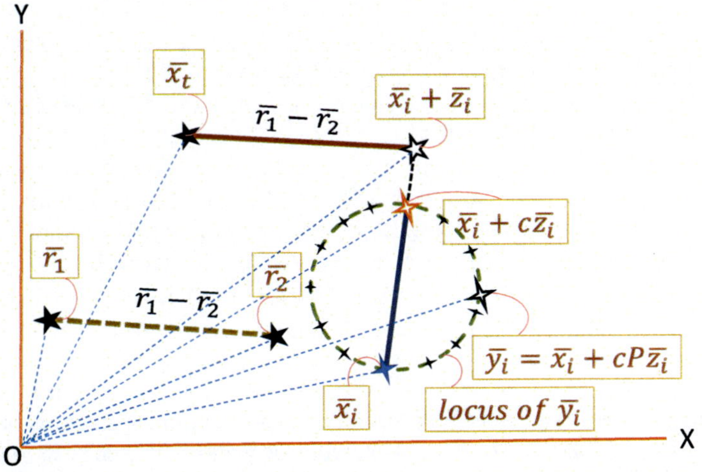

Fig. 19.3 Demonstrating the solution update scheme of SS algorithm in 2-D search space [9]

Fig. 19.3 which shows that two random vectors, r_1 and r_2, along with a target point (a vector), $\bar{x_t}$, need to be conceived and used in the following manner to obtain search direction, $\bar{z_i}^{(k)}$.

$$\bar{z_i}^{(k)} = (\bar{x_t}^{(k)} + \bar{r}_2^{(k)} - \bar{r}_1^{(k)}) - \bar{x_i}^{(k)}. \quad (19.5)$$

Vectors r_1 and r_2 are two random solutions from the population.

In SS algorithm, to calculate the search direction, $\bar{z_i}^{(k)}$, two methods, namely, *towards-rand* and *towards-best* have been used. The former implements exploration component and the latter implements the exploitation component in the algorithm. To strike a balance between the exploration and exploitation, population is sorted in the decreasing order of their fitness values and then the population is divided into two parts. First part of the population containing better solutions is used to evaluate the search direction, $\bar{z_i}^{(k)}$ for ith solution at kth iteration implementing *towards-rand* in the following manner:

$$\bar{z}_i^{(k)} = \bar{x}_{p_i}^{(k)} + \bar{x}_{q_i}^{(k)} - \bar{x}_{r_i}^{(k)} - \bar{x}_i^{(k)}, \quad (19.6)$$

where p_i, q_i, and r_i are randomly selected integer values between 1 and N such that $p_i \neq q_i \neq r_i \neq i$. When this equation is compared with Eq. (19.5), then $\bar{x}_{p_i}$ corresponds to the target points $\bar{x_t}$ and difference term $(\bar{x_q} - \bar{x_r})$ corresponds to $\bar{r}_2 - \bar{r}_1$ which is an approximation of distribution of difference of solutions in a population and helps algorithm maintain solution diversity iteration-by-iteration avoiding convergence to local minima.

The second part of population is used to evaluate the search direction, $\bar{z_i}^{(k)}$ for ith solution at kth iteration implementing *towards-best* in the following manner:

$$\bar{z}_i^{(k)} = \bar{x}_{pbest_i}^{(k)} + \bar{x}_{q_i}^{(k)} - \bar{x}_{r_i}^{(k)} - \bar{x}_i^{(k)}, \quad (19.7)$$

where $x_{pbest_i}^{(k)}$ is a solution selected randomly from top p best solutions out of population at iteration k and corresponds to the target points $\bar{x_t}$ when compared with (Eq. 19.5). Here also, difference term $(\bar{x_q} - \bar{x_r})$ corresponds to $\bar{r}_2 - \bar{r}_1$.

So far evaluation of $\bar{z_i}$ has been discussed; however, to evaluate $\bar{y_i}$ as per Eq. (19.4), besides $\bar{z_i}$, c_i, and projection matrix, P_i also need to be evaluated.

In Eq. (19.4), projection matrix, $P = A'B_iA$, which is a symmetrical matrix returns projections of $c_i\bar{z_i} + \bar{x_i}$ by linearly transforming it and thereby creating a $(D-1)$-spherical boundary. A being an orthogonal matrix produces infinite number of combinations whereas $\bar{b_i}$ being a binary vector can produce only finite number of combinations. Figure 19.4 plots locus of 10,000 randomly generated samples out of all possible projections of point (1, 1) given by projection matrix, P, on a 2-D search space which is a circular ring of diameter $\sqrt{2}$ having center at (0.5, 0.5).

In the beginning of every iteration, following two terms are initialized randomly (i) an orthogonal matrix, A, such that $AA' = I$ and (ii) binary diagonal matrix, B_i such that $0 < \text{rank}(B_i) < D$. A step-size control vector, $\bar{c}^{(k)}$, consists of N step-

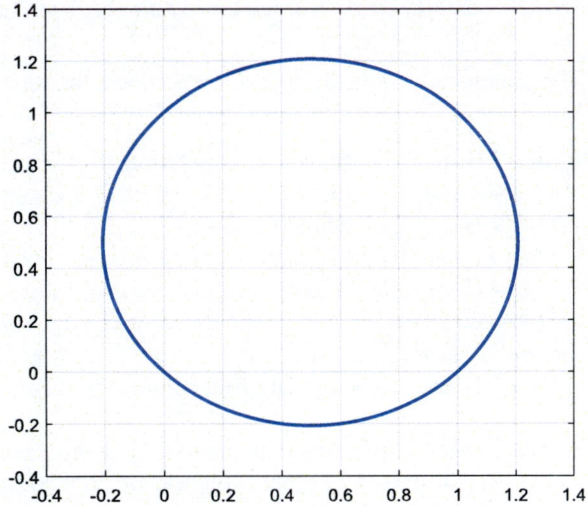

Fig. 19.4 Illustrating the locus of projection of point (1,1)[9]

size control parameter at kth iteration. Element $c_i^{(k)}$ is a step-size control parameter corresponding to ith trial solution and is calculated randomly in range of [0.5 0.7] at the start of kth iteration. This range has been decided by experimentation.

19.2.3 Selection of New Population for Next Iteration

Each of the solutions, $\bar{x_i}$, of the population, $Pop^{(k)}$, is updated using greedy selection procedure in the following manner:

$$\bar{x_i}^{(k+1)} = \begin{cases} \bar{y_i}^{(k)}, & \text{if} f(\bar{y_i}^{(k)}) \leq f(\bar{x_i}^{(k)}) \\ \bar{x_i}^{(k)}, & \text{otherwise} \end{cases} \tag{19.8}$$

Here, $f(\bar{y_i}^{(k)})$ is the objective function value of trial solution $\bar{y_i}^{(k)}$, and $f(\bar{x_i}^{(k)})$ is objective function value of solution $\bar{x_i}^{(k)}$ at kth iteration.

19.2.4 Stopping Criteria

The algorithm terminates when either the function evaluations, FEs, become equal to the specified maximum number of function evaluations, FE_{max}, or the solution does not update for a specified number of iterations.

19.2.5 Steps of Spherical Search Algorithm

The pseudocode of the SS algorithm is shown in Algorithm 4 and the steps are given below:

- **Step 1 (Line 2):** Initialize the population Pop and calculate objective function for each solution of Pop.
- **Step 2 (Line 3):** Calculate parameters: c_i.
- **Step 3 (Line 5):** Calculate the orthogonal matrix, A.
- **Step 4 (Line 7):** Calculate matrix B for each solution vector of population, Pop.
- **Step 5 (Lines 8 to 12):** Calculate the search direction for each solution vector of population, Pop.
- **Step 6 (Line 13):** Calculate trial solution for each solution vector of population Pop.
- **Step 7 (Line 16):** Update the population using greedy selection procedure.
- **Step 8 (Line 18):** The best solution of population is selected on the basis of minimum objective function value as *best solution*.
- **Step 9 (Line 19):** If the stopping criterion is met then go to **Step 10** else go to **Step 3**.
- **Step 10:** Return the best solutions.

```
 1: procedure SPHERICAL SEARCH ALGORITHM
 2: Initialize the Population, Pop
 3: c_i ← rand[0.5, 0.7]
 4: while FEs < FE_max do
 5:   A ← ComputeOrthogonalMatrix()
 6:   for i = 1 to N do
 7:     B_i ← ComputeBinaryVector()
 8:     if i < 0.5 * N then
 9:       z̄_i ← TowardsRand(i)
10:     else
11:       z̄_i ← TowardsBest(i)
12:     end if
13:     ȳ_i ← x̄_i + c_i A′ B_i A z̄_i
14:     f(ȳ_i) ← ObjectiveFunction(ȳ_i)
15:     FEs ← FEs + 1
16:     x̄_i ← GreedySelection(x̄_i, ȳ_i)
17:   end for
18:   Pop ← Sort(Pop)
19:   best solution ← Pop(i)
20: end while
21: end procedure
```

Algorithm 4: SS Algorithm

19.2.6 Space and Time Complexity of SS Algorithm

The space complexity of SS algorithm, i.e., maximum amount of space required during the optimization process, is $O(N \times D)$, where N is the size of population Pop and D is the dimension of solution space. Various steps of SS algorithm and their time complexities are as follows:

1. Initialization of population: $O(N \times D)$.
2. Calculation of objective function of each solution: $O(FE_{max} \times D)$=$O(Max_{iter} \times N \times D)$.
3. Calculation of orthogonal matrix: $O(Max_{iter} \times D \times log(D))$.
4. Calculation of trial solutions: $O(Max_{iter} \times N)$.

Hence, the overall time complexity of SS algorithm is
$O(Max_{iter} \times N \times D \times log(D)) = O(FE_{max} \times D \times log(D))$.

19.2.7 Validation of Performance of SS on Benchmark Problems

In general practice before applying to real-world problems, new algorithms are tested on the artificial problems, called benchmark problem, to analyze the effectiveness and efficiency with comparison to existing popular algorithms. The main reason behind that is the cost of implementation of the artificial problems is lower than real-life problems. In this chapter, problems of IEEE CEC 2014 benchmark suite [10] have been selected to test the performance of SS algorithm. Problems of this benchmark suite are hard to solve and characteristics of these problems closely resemble with real-life problems [10]. Four popular algorithms, viz., Particle Swarm Optimization (PSO) [7], Genetic Algorithm (GA)[16], Covariance Matrix Adaptation Evolution Strategy (CMAES) [5], and Differential Evolution (DE) [15] are selected for comparative analysis of performance of SS algorithm on the problems of IEEE CEC benchmark suite. To demonstrate the statistical differences between the performance of SS algorithm and other algorithms, Wilcoxon's signed rank test at 0.05 significance level has also been implemented and reported.

Table 19.1 shows the results of all algorithms on 30 problems of IEEE CEC 2014 benchmark suite. For each of the benchmark problems, the outcomes of Wilcoxon's signed rank test for SS algorithm as compared with each of the other four algorithms are presented in columns indicated as "W." The symbols "+," "=," and "−" used in Table 19.1 mean that performance of SS algorithm is better, comparable, or worse, respectively, as compared to the chosen algorithm. The summary of Wilcoxon's signed rank test is also reported at the bottom of this table for each of the four algorithms chosen for comparison. It is seen from Table 19.1 that the performance of SS algorithm is significantly better than PSO, GA, CMA-ES, and DE on 25, 19, 22, and 14 problems out of 30 problems, respectively. However, the other algorithms perform better than SS algorithm on 4, 12, 8, and 13 problems, respectively.

It can be concluded from above analysis that the performance of SS algorithms is better than the other existing popular algorithms on benchmark problems of IEEE CEC 2014 suite.

19.3 Application to Real-Life Complex Optimization Problems: Model Order Reduction

To demonstrate the effectiveness of the SS algorithm on real-life complex optimization problems, problem of Model Order Reduction (MOR) has been chosen as a representative problem.

Order reduction of a high-order linear time-invariant dynamic Single-Input and Single-Output system (SISO) applying SS algorithm is considered in this chapter. Following n-order SISO system having transfer function, $G(s)$, is considered for order reduction.

$$G(s) = \frac{N(s)}{D(s)} = \frac{\sum_{i=0}^{i=n-1} a_i s^i}{\sum_{i=0}^{n} b_i s^i}, \tag{19.9}$$

where a_i and b_i are the known parameters of the system. The objective in a MOR problem is to obtain a lower order (reduced order) model of a given SISO system preserving its important characteristics. It is expected that the step response of the reduced order model should match the step response of the original SISO system as closely as possible while minimizing Integral Square Error (ISE) and Impulse Response Energy (IRE). Reduced order model of the given SISO system (19.9) may be expressed as follows:

$$R(s) = \frac{N_r(s)}{D_r(s)} = \frac{\sum_{i=0}^{r-1} a_i' s^i}{\sum_{i=0}^{r} b_i' s^i}, \tag{19.10}$$

where r is the order of reduced model and $r \leq n$; a_i' and b_i' are the parameters of the reduced order model which shall be determined using SS optimization algorithm.

ISE is an error index expressed in terms of the time-domain unit step response of the original system, $y(t)$, and time-domain unit step response of the reduced order system,$y_r(t)$, which can be expressed mathematically as follows:

$$ISE = \int_{-\infty}^{\infty} \{y(t) - y_r(t)\}^2 dt. \tag{19.11}$$

The IRE of any system $G(s)$ is expressed in terms of its corresponding time-domain unit impulse response, $g(t)$, which can be mathematically represented as follows:

$$IRE = \int_{-\infty}^{\infty} g(t)^2 dt. \tag{19.12}$$

Table 19.1 Comparison of SS with other popular state-of-the-art algorithms on 30 problems of IEEE CEC 2014 benchmark suite with 30 dimension [9]

Prob	SS		PSO			GA			CMA-ES			DE		
	Mean	SD	Mean	SD	W	Mean	SD	W	Mean	SD	W	Mean	SD	W
1	3.31E+05	2.53E+05	5.14E+07	2.24E+07	+	1.09E+06	5.87E+05	−	8.16E+04	2.66E+04	−	7.06E+04	7.58E+04	−
2	0.00E+00	0.00E+00	2.42E+08	4.23E+08	+	7.57E+06	2.51E+04	+	4.59E+10	9.72E+09	+	0.00E+00	0.00E+00	=
3	3.68E−04	1.09E−03	5.97E+04	1.07E+04	+	2.14E+04	8.27E+03	+	2.96E+03	1.29E+03	+	0.00E+00	0.00E+00	−
4	7.01E+00	1.99E+01	1.66E+02	8.29E+01	+	3.62E+00	9.54E−01	−	4.71E+03	1.33E+03	+	1.83E−01	1.10E−01	−
5	2.09E+01	1.43E−01	2.08E+01	9.01E+01	+	2.09E+01	6.84E−02	+	2.00E+01	3.70E−04	+	2.09E+01	5.08E−02	+
6	1.13E+01	8.05E+00	2.58E+01	6.40E+01	+	2.15E+01	1.06E+00	+	6.50E+01	1.57E+00	+	7.08E−02	2.92E−01	−
7	3.72E−03	8.04E−03	9.28E+01	1.23E+01	+	1.28E+00	1.24E−02	+	1.12E+02	1.16E+01	+	0.00E+00	0.00E+00	−
8	2.52E+01	6.52E+00	5.09E−02	1.11E+00	−	1.01E+00	9.40E+00	−	8.98E+02	1.83E+02	+	1.22E+02	2.30E+01	+
9	1.22E+02	7.33E+01	3.18E+01	1.28E+02	−	2.83E+01	6.37E+00	−	9.14E+02	4.24E+00	+	1.77E+02	1.04E+01	+
10	7.11E+02	3.30E+02	7.61E+02	7.61E+02	+	8.85E+02	4.27E+02	+	1.09E+03	5.78E+02	+	5.43E+03	6.37E+02	+
11	4.81E+03	1.59E+03	6.83E+03	1.93E+03	+	7.53E+03	4.25E+02	+	2.16E+02	1.11E+02	−	6.68E+03	3.03E+02	+
12	8.79E−01	1.13E+00	2.86E+00	2.77E+01	+	5.86E−01	6.81E−02	−	1.20E−01	1.16E+00	−	2.39E+00	2.77E−01	+
13	5.28E−01	8.94E−02	6.40E−01	4.44E−01	+	2.79E−01	6.24E−03	−	1.31E+00	1.70E−01	+	3.43E−01	4.68E−02	=
14	5.56E−01	2.25E−01	4.21E−01	1.57E−01	+	2.54E−01	4.14E−02	−	1.57E+01	4.61E+00	+	2.71E−01	3.26E−02	−
15	1.64E+01	3.87E+00	4.47E+00	1.92E+00	−	1.06E+00	2.59E−02	−	1.51E+03	1.48E+00	+	1.55E+01	1.10E+00	−
16	1.28E+01	3.39E−01	1.29E+01	4.82E−01	=	1.30E+01	6.27E−01	+	1.62E+01	6.85E+00	+	1.23E+01	2.43E−01	=
17	9.94E+03	9.53E+03	3.26E+05	2.57E+05	−	3.04E+05	3.72E+04	−	4.42E+03	9.19E+02	−	1.30E+03	2.12E+02	−
18	1.55E+02	2.67E+02	1.02E+06	2.95E+06	+	5.85E+04	6.43E+04	+	2.50E+03	2.66E+02	−	5.33E+01	6.58E+00	−
19	5.68E+00	1.34E+00	2.82E+02	1.98E+01	+	1.62E+01	3.52E+01	+	2.04E+02	2.19E+01	+	4.45E+00	3.01E−01	−
20	9.30E+01	8.10E+01	1.08E+04	2.18E+03	+	2.36E+03	3.14E+03	−	2.76E+03	2.45E+02	−	3.39E+01	5.98E+00	−

(continued)

Table 19.1 (continued)

Prob	SS		PSO			GA			CMA-ES			DE		
	Mean	SD	Mean	SD	W	Mean	SD	W	Mean	SD	W	Mean	SD	W
21	2.43E+03	4.91E+03	7.54E+05	4.21E+05	+	1.34E+05	5.59E+04	–	3.53E+03	1.22E+03	–	6.48E+02	1.50E+02	–
22	1.34E+02	1.27E+02	8.24E+02	5.24E+02	+	1.38E+03	1.28E+02	+	5.17E+03	1.94E+02	+	4.92E+01	5.27E+01	–
23	2.00E+02	0.00E+00	3.42E+02	6.24E+00	+	3.01E+02	3.87E−02	+	2.81E+02	1.43E+01	+	3.15E+02	2.30E−13	+
24	2.00E+02	0.00E+00	2.05E+02	2.41E−01	+	2.51E+02	1.40E+01	+	2.76E+02	1.22E−01	+	2.14E+02	1.10E+01	+
25	2.00E+02	0.00E+00	2.20E+02	4.51E+00	+	2.18E+02	8.47E+00	+	2.09E+02	9.14E+00	+	2.03E+02	7.90E−02	+
26	1.01E+02	8.83E−02	1.00E+02	2.42E−01	+	1.56E+02	4.44E+01	+	2.75E+02	4.28E+01	+	1.00E+02	3.97E−02	+
27	2.11E+02	6.26E+01	2.51E+03	4.82E+02	+	7.89E+02	2.06E+02	+	4.50E+03	8.76E+02	+	3.60E+02	4.93E+01	+
28	2.00E+02	0.00E+00	1.81E+03	4.91E+02	+	3.46E+03	8.02E+03	+	5.98E+03	1.36E+02	+	8.03E+02	2.72E+01	+
29	8.80E+02	5.72E+02	8.77E+07	3.24E+07	+	1.39E+04	1.97E+05	+	1.27E+07	6.28E+06	+	5.52E+02	2.62E+02	+
30	2.59E+02	1.75E+02	4.11E+05	1.87E+04	+	3.51E+03	2.08E+03	+	7.18E+05	2.96E+05	+	4.94E+02	1.08E+02	+
+/=/–			25	1	4	19	0	12	22	0	8	14	3	13

Table 19.2 Comparison of performance of SS with other existing methods for test systems $G_1(s)$, $G_2(s)$, $G_3(s)$, and $G_4(s)$ (a0, a1, a2, and a3 are unknown parameters of reduced system, ISE: integral square error, IRE: impulse response energy, OFV: objective function values, NA: not available)

	Original	SS	MBDE	DE	FBDE	LICLDE	ABC
a0	G1(s)	8.0319E+01	7.2524E+01	2.2082E+02	8.5335E+01	1.0132E+02	4.8500E+02
a1		3.0175E+02	2.5251E+02	3.5012E+04	4.6230E+02	8.6789E+02	5.0000E+04
a2		9.9107E+01	8.9583E+01	1.2295E+03	1.1366E+02	1.6941E+02	4.1870E+03
a3		3.0175E+02	2.5251E+02	3.5012E+04	4.6230E+02	8.6789E+02	5.0000E+04
ISE		1.5834E−03	1.5932E−03	4.4376E−03	1.7827E−03	3.9344E−03	1.1626E−02
IRE	3.4068E+01	3.4068E+01	3.0766E+01	3.4069E+01	3.4069E+01	3.2862E+01	3.4061E+01
OFV		1.5845E−03	5.2531E−02	4.4495E−03	1.7891E−03	2.1960E−02	1.1738E−02
a0	G2(s)	−5.2651E−03	−4.5958E−03	2.9600E−02		−1.9500E−02	3.1800E−02
a1		1.1902E−01	8.2605E−02	2.1750E−01		2.8840E−01	4.0074E+00
a2		5.9424E+00	4.0210E+00	1.2395E+01	NA	1.4981E+01	1.3741E+01
a3		4.4631E+00	3.0989E+00	8.1560E+00		1.0820E+01	1.5028E+02
ISE		1.2141E−06	1.5581E−07	1.4514E−05	NA	4.3242E−06	5.3450E−04
IRE	2.6938E−04	2.6938E−04	2.7643E−04	2.6931E−04	NA	2.6925E−04	3.9253E−03
OFV		3.2702E−06	1.2927E−02	1.3525E−04	NA	2.4327E−04	8.7210E−01
a0	G3(s)	8.1133E−01	7.5984E−01	1.0759E+00		7.8530E−01	2.0340E−01
a1		2.6718E+00	2.9433E+00	9.5670E+00		2.9490E+00	8.9940E+00
a2		2.9469E+00	3.1529E+00	9.4527E+00	NA	3.1515E+00	7.9249E+00
a3		2.7927E+00	3.0799E+00	1.0000E+01		3.0823E+00	9.4008E+00
ISE		3.3766E−02	3.3701E−02	3.6423E−02	NA	3.3797E−02	3.5014E−02
IRE	5.4537E−01	5.4537E−01	5.3763E−01	5.4536E−01	NA	5.4548E−01	5.4551E−01
OFV		3.3770E−02	4.0849E−02	3.6426E−02	NA	3.3901E−02	3.5143E−02
a0	G4(s)	1.7324E+01	1.4887E+01	2.0000E+01	1.7322E+01	1.7203E+01	1.7387E+01
a1		5.3681E+00	4.7852E+00	5.6158E+00	5.3660E+00	5.3633E+00	5.3743E+00
a2		7.0261E+00	5.9883E+00	9.2566E+00	7.0240E+00	6.9298E+00	7.0910E+00
a3		5.3681E+00	4.7852E+00	5.6158E+00	5.3660E+00	5.3633E+00	5.3743E+00
ISE		8.0670E−04	4.0564E−03	3.7296E−02	8.0759E−04	9.0631E−04	8.5412E−04
IRE	2.1739E+01	2.1739E+01	1.8903E+01	2.1910E+01	2.1740E+01	2.1740E+01	2.1695E+01
OFV		8.0670E−04	7.3829E−02	4.1203E−02	8.4123E−04	9.2726E−04	1.8614E−03

In this chapter, the objective function, which is to be minimized, has been defined as follows in terms of both, ISE and IRE.

$$f = ISE + \frac{|IRE - IRE_r|}{IRE + IRE_r}, \tag{19.13}$$

where ISE is the error index calculated by Eq. (19.11), IRE and IRE_r are calculated by Eq. (19.12) for the original system and reduced order system, respectively.

19.3.1 Reduced Second-Order Model

High-order system, defined in Eq. (19.9), can also be represented as follows in terms of its eigenvalues (poles) λ_1, λ_2,..., λ_n.

$$G(s) = \frac{b_{n-1}s^{n-1} + b_{n-2}s^{n-2} + \cdots + b_0}{(s-\lambda_1)(s-\lambda_2)\ldots(s-\lambda_n)}. \tag{19.14}$$

The unit step response of the above system can be determined after a partial fraction of $\frac{G(s)}{s}$ in the following manner:

$$Y(s) = \frac{G(s)}{s} = \frac{k_0}{s} + \frac{k_1}{s-\lambda_1} + \frac{k_2}{s-\lambda_2} + \cdots + \frac{k_n}{s-\lambda_n}, \tag{19.15}$$

where k_i are real constants (residues).
Time-domain unit step response of $Y(s)$ can be obtained in the following manner by taking inverse Laplace transformation of Eq. (19.15).

$$y(t) = \mathscr{L}^{-1}\{Y(s)\} = k_0 + k_1e^{\lambda_1 t} + k_2e^{\lambda_2 t}\ldots k_ne^{\lambda_n t} = \sum_{i=0}^{n} k_ie^{\lambda_i t}. \tag{19.16}$$

Here, first term, k_0, is the steady-state response of the system. Let us consider that reduced order system is desired to be a system of second order as follows:

$$R(s) = \frac{a_0s + a_1}{a_2s^2 + a_3s + a_4}. \tag{19.17}$$

Equation (19.17) may also be written in terms of eigenvalues as follows:

$$R(s) = \frac{a_0s + a_1}{(s-\mu_1)(s-\mu_2)}, \tag{19.18}$$

where μ_1 and μ_2 are eigenvalues of the desired reduced order system. The step response of reduced system in Laplace transform domain shall be given as follows:

$$Y_r(s) = \frac{R(s)}{s} = \frac{k_0'}{s} + \frac{k_1'}{s-\mu_1} + \frac{k_2'}{s-\mu_2}, \tag{19.19}$$

where k_0', k_1', and k_2' are real constants. Step response in time domain shall be as follows:

$$y_r(t) = k_0' + k_1'e^{\mu_1 t} + k_2'e^{\mu_2 t}, \tag{19.20}$$

where k_0' is the steady-state response of reduced order system. Condition for perfectly matched steady-state responses of original system and reduced model is given by

$$k_0 = k'_1$$

$$a_4 = \frac{a'_0 a_1}{b'_0}. \tag{19.21}$$

Hence, the second-order MOR optimization problem has three unknown variables, $[a_0, a_1, a_2]$. These values of these variables must be obtained in such a manner that the objective function given in (19.13) is minimized.

19.3.2 Test Systems, Results, and Discussions

SS algorithm has been applied on following four test systems to study the effectiveness of performance:

$$G_1(s) = \frac{8169.13s^3 + 50664.97s^2 + 9984.32s + 500}{100s^4 + 10520s^3 + 52101s^2 + 10105s + 500}, \tag{19.22}$$

$$G_2(s) = \frac{s + 4}{s^4 + 19s^3 + 113s^2 + 245s + 150}, \tag{19.23}$$

$$G_3(s) = \frac{4.269s^3 + 5.10s^2 + 3.9672s + 0.9567}{4.3992s^4 + 9.0635s^3 + 8.021s^2 + 5.362s + 1}, \tag{19.24}$$

$$G_4(s) = \frac{18s^7 + 514s^6 + 5982s^5 + 36380s^4 + 122664s^3 + 222088s^2 + 185760s + 40320}{s^8 + 36s^7 + 546s^6 + 4536s^5 + 22449s^4 + 67284s^3 + 118124s^2 + 109584s + 40320}. \tag{19.25}$$

All these systems have different real poles. SS algorithm has been employed to obtain corresponding reduced second-order systems by minimizing the objective function defined in Eq. (19.13) in terms of ISE and IRE of original and reduced order systems. The results obtained by applying the SS algorithm for all four systems have been compared with four popular differential evolution-based algorithms reported in the literature, viz., Memory-Based Differential Evolution algorithm (MBDE) [12], Classical Differential Evolution (DE) [15], Fitness-Based Differential Evolution (FBDE) [14], Cognitive Learning in Differential Evolution (LICLDE) [1], and Artificial Bee Colony (ABC) [2]. Some of the parameters of SS algorithms like population size, independent runs, and stopping criterion are set according to LICLDE algorithm.

The best-known parameters of the reduced second-order system obtained by all algorithms are reported in Table 19.2. The best-reported parameters as per the objective function given in Eq. (19.13) are given in boldface in Table 19.2. The unit step responses and impulse responses of the original system and corresponding reduced order systems obtained using SS algorithms and five popular differential evolution-based algorithms considered in this chapter are shown in Figs. 19.5, 19.6, 19.7, and 19.8, respectively, for four test systems given in Eqs. (19.22–19.25).

From Table 19.2, it can be observed that for test systems $G_1 - G_4$, objective function values achieved by SS algorithm are significantly less than those obtained

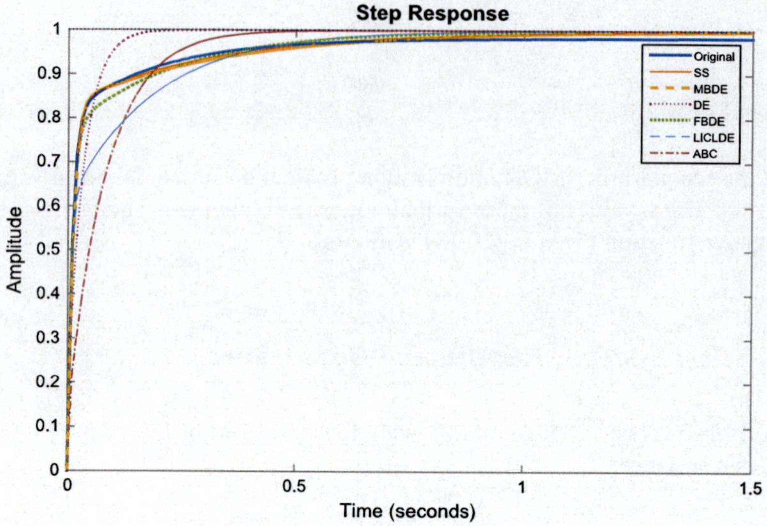

(a)

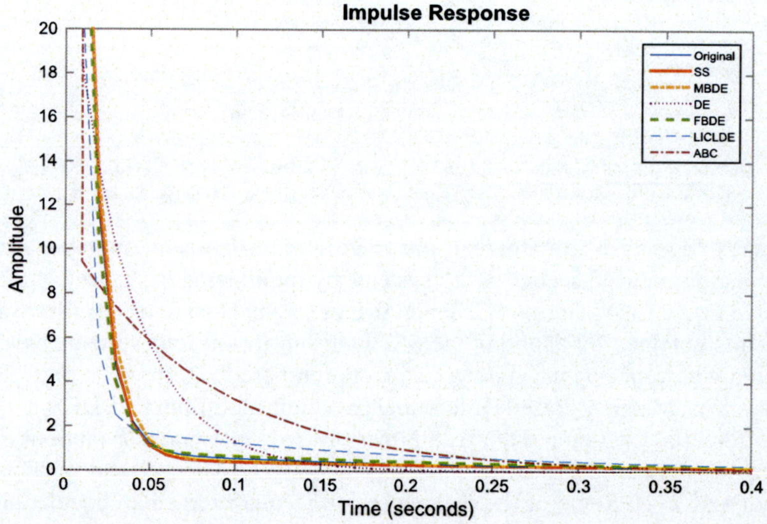

(b)

Fig. 19.5 Comparison of step and impulse responses of test system $G_1(s)$ **a** Step response of original test system $G_1(s)$ and its reduced systems obtained by different methods, **b** Impulse response of original test system $G_1(s)$ and its reduced systems obtained by different methods

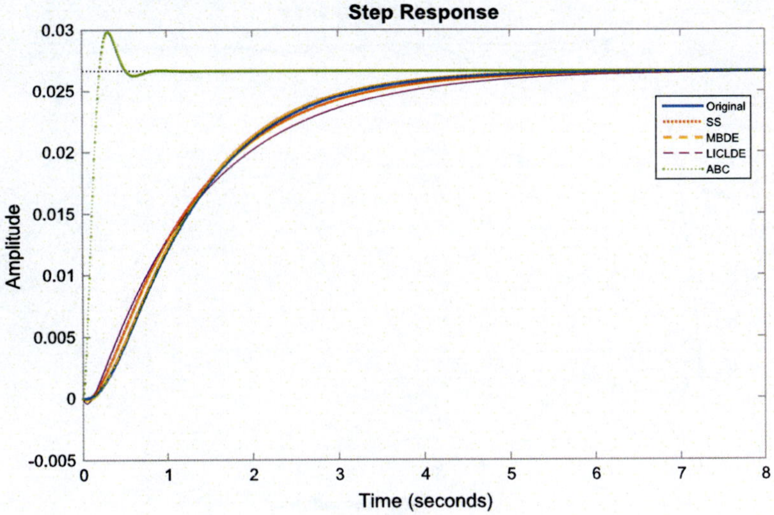

(a)

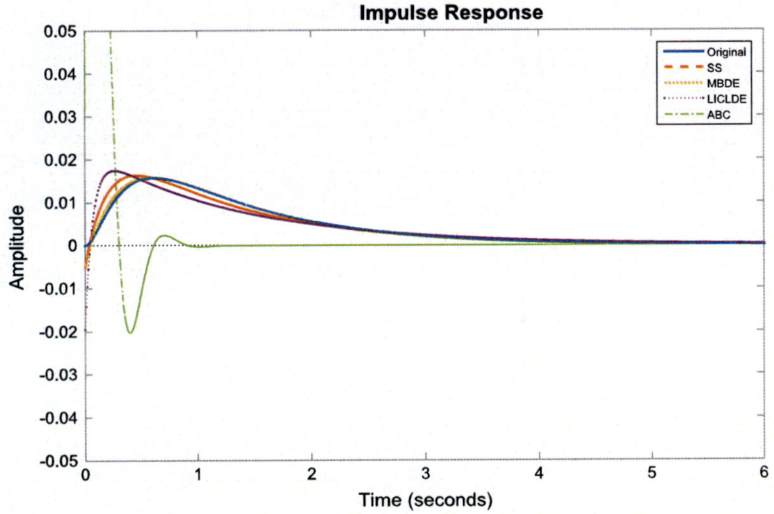

(b)

Fig. 19.6 Comparison of step and impulse responses of test system $G_2(s)$ **a** Step response of original test system $G_2(s)$ and its reduced systems obtained by different methods, **b** Impulse response of original test system $G_2(s)$ and its reduced systems obtained by different methods

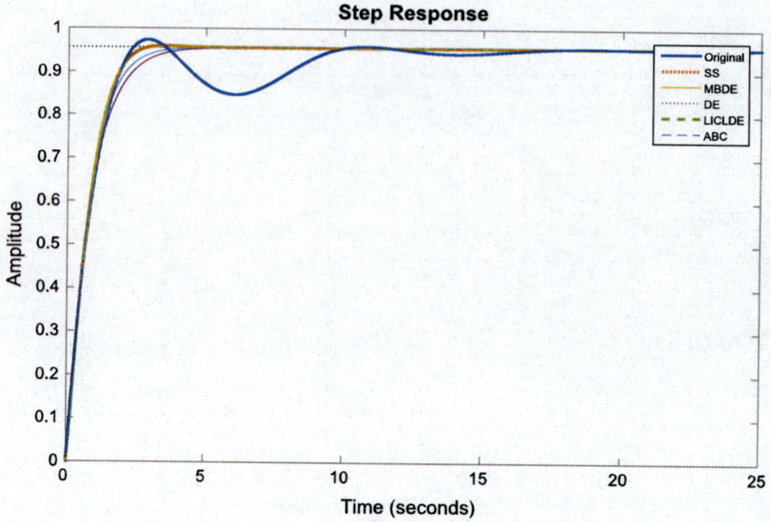

(a)

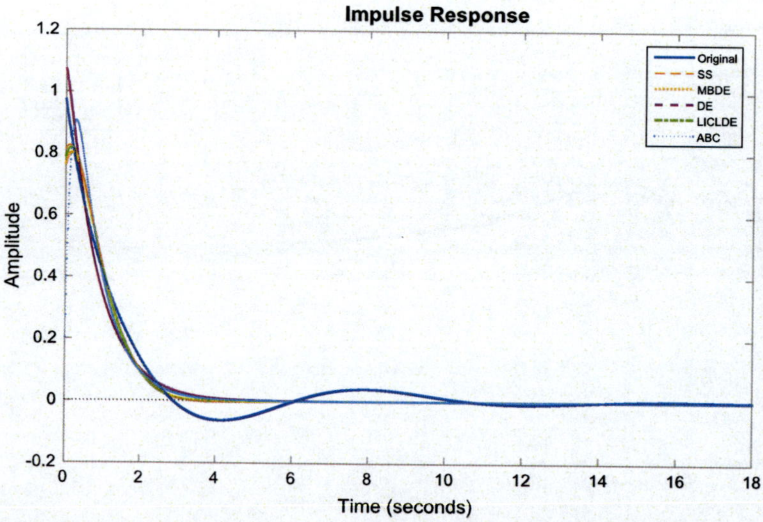

(b)

Fig. 19.7 Comparison of step and impulse responses of test system $G_3(s)$ **a** Step response of original test system $G_3(s)$ and its reduced systems obtained by different methods, **b** Impulse response of original test system $G_3(s)$ and its reduced systems obtained by different methods

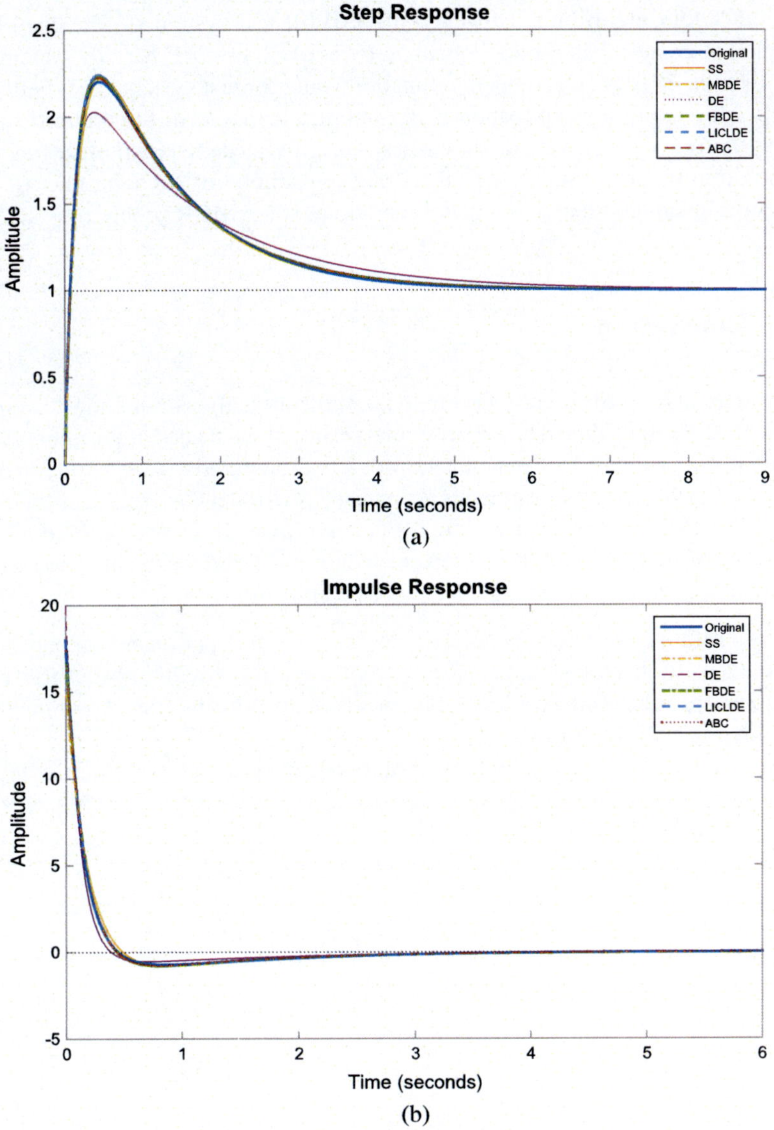

Fig. 19.8 Comparison of step and impulse responses of test system $G_4(s)$ **a** Step response of original test system $G_4(s)$ and its reduced systems obtained by different methods, **b** Impulse response of original test system $G_4(s)$ and its reduced systems obtained by different methods

using other four algorithms. However, the ISE for test system G_3 obtained by SS algorithm is greater than the ISE obtained by MBDE, but IRE for this system is far better than the MBDE. So, objective function value for test system G_3 is better than MBDE. Moreover, the step response and impulse response of the reduced second-order system are closely lying on the curves of original systems. On these test systems, SS algorithm outperforms all the other four algorithms for MOR problems. Thus, SS algorithm can be treated as a better method to solve MOR problems.

19.4 Conclusion

This chapter discussed recently developed optimization algorithm known as Spherical Search (SS) algorithm. The experimental results show that SS algorithm provides superior result on most of the benchmark problems compared to state-of-the-art meta-heuristics. Further, this algorithm has been tested on Model Order Reduction (MOR) problem which is a representative of real-life complex optimization problems. On the basis of the results obtained from all experiments and comparisons of characteristics with other algorithms, following conclusions can be drawn:

1. SS algorithm has single parameter, step-size control parameter, which is self-adaptive, and this algorithm is easy to implement for solving most of the unconstrained optimization problems. No problem-specific tuning of parameters is required in SS algorithm.
2. The quality and accuracy of obtained solutions, the rate of convergence, efficiency, and effectiveness of SS algorithm are better as compared to the state-of-the-art meta-heuristics.
3. It is expected that, due to its projection property, SS algorithm can avoid local minima.

Acknowledgements The authors would like to thank the referees for their constructive comments which significantly improved the presentation of the chapter.

References

1. Bansal, J.C., Sharma, H.: Cognitive learning in differential evolution and its application to model order reduction problem for single-input single-output systems. Memetic Comput. pp. 1–21 (2012)
2. Bansal, J.C., Sharma, H., Arya, K.: Model order reduction of single input single output systems using artificial bee colony optimization algorithm. In: Nature Inspired Cooperative Strategies for Optimization (NICSO 2011), pp. 85–100. Springer (2011)
3. Dorigo, M., Birattari, M.: Ant colony optimization. In: Encyclopedia of Machine Learning, pp. 36–39. Springer (2010)
4. Fogel, L.J., Owens, A.J., Walsh, M.J.: Artificial Intelligence Through Simulated Evolution. Wiley, New York (1966)

5. Jastrebski, G.A., Arnold, D.V.: Improving evolution strategies through active covariance matrix adaptation. In: IEEE Congress on Evolutionary Computation, CEC 2006, pp. 2814–2821. IEEE (2006)
6. Karaboga, D., Basturk, B.: A powerful and efficient algorithm for numerical function optimization: artificial bee colony (ABC) algorithm. J. Glob. Optim. **39**(3), 459–471 (2007)
7. Kennedy, J.: Particle swarm optimization. In: Encyclopedia of Machine Learning, pp. 760–766. Springer (2010)
8. Koza, J.R.: Genetic Programming: On the Programming of Computers by Means of Natural Selection, vol. 1. MIT press (1992)
9. Kumar, A., Misra, R.K., Singh, D., Mishra, S., Das, S.: The spherical search algorithm for bound-constrained global optimization problems. Appl. Soft Comput. **85**, 105734 (2019)
10. Liang, J., Qu, B., Suganthan, P.: Problem definitions and evaluation criteria for the CEC 2014 special session and competition on single objective real-parameter numerical optimization. Zhengzhou University, Zhengzhou China and Technical Report, Nanyang Technological University, Singapore, Computational Intelligence Laboratory (2013)
11. Mirjalili, S., Mirjalili, S.M., Lewis, A.: Grey wolf optimizer. Adv. Eng. Softw. **69**, 46–61 (2014)
12. Parouha, R.P., Das, K.N.: A memory based differential evolution algorithm for unconstrained optimization. Appl. Soft Comput. **38**, 501–517 (2016)
13. Rechenberg, I.: Evolution strategy: nature's way of optimization. In: Optimization: Methods and Applications, Possibilities and Limitations, pp. 106–126. Springer (1989)
14. Sharma, H., Bansal, J.C., Arya, K.: Fitness based differential evolution. Memetic Comput. **4**(4), 303–316 (2012)
15. Storn, R., Price, K.: Differential evolution-a simple and efficient heuristic for global optimization over continuous spaces. J. Glob. Optim. **11**(4), 341–359 (1997)
16. Tang, K.S., Man, K.F., Kwong, S., He, Q.: Genetic algorithms and their applications. Signal Process. Mag. IEEE **13**(6), 22–37 (1996)
17. Wolpert, D.H., Macready, W.G.: No free lunch theorems for optimization. IEEE Trans. Evol. Comput. **1**(1), 67–82 (1997)
18. Yang, X.S.: A new metaheuristic bat-inspired algorithm. In: Nature Inspired Cooperative Strategies for Optimization (NICSO 2010), pp. 65–74. Springer (2010)
19. Yao, X., Liu, Y., Lin, G.: Evolutionary programming made faster. IEEE Trans. Evol. Comput. **3**(2), 82–102 (1999)